Selected Works of A. N. Kolmogorov

Mathematics and Its Applications (*Soviet Series*)

Volume 26

Selected Works of A. N. Kolmogorov

Volume II

Probability Theory and Mathematical Statistics

edited by

A. N. Shiryayev

Translated from the Russian by G. Lindquist

SPRINGER-SCIENCE+BUSINESS MEDIA, B.V.

Library of Congress Cataloging-in-Publication Data

Kolmogorov, A. N. (Andreĭ Nikolaevich), 1903-
[Teoriía veroíatnosteĭ i matematicheskaía statistika. English]
Probability theory and mathematical statistics / edited by A.N.
Shiryayev ; translated from the Russian by G. Lindquist.
p. cm. -- (Selected works of A.N. Kolmogorov ; v. 2)
(Mathematics and its applications (Soviet Series) ; v. 26)
Translation of: Teoriía veroíatnosteĭ i matematicheskaía
statistika.
ISBN 978-94-010-5003-6 ISBN 978-94-011-2260-3 (eBook)
DOI 10.1007/978-94-011-2260-3

1. Probabilities. 2. Mathematical statistics. I. Shiríaev,
Al'bert Nikolaevich. II. Title. III. Series: Kolmogorov, A. N.
(Andreĭ Nikolaevich), 1903- Selections. English. 1991 ; v. 2.
IV. Series: Mathematics and its applications (Kluwer Academic
Publishers). Soviet series ; 26.
QA3.K792513 1991 v. 2
[QA273.18]
510 s--dc20
[519.2] 91-14856
CIP

ISBN 978-94-010-5003-6
ISBN 978-94-010-5003-6

Printed on acid-free paper

This is an annotated translation from the original work
Теория вероятностей и математическая статистика

Published by Nauka, Moscow, © 1986

SERIES EDITOR'S PREFACE

'Et moi, ..., si j'avait su comment en revenir, je n'y serais point allé.'

Jules Verne

The series is divergent; therefore we may be able to do something with it.

O. Heaviside

One service mathematics has rendered the human race. It has put common sense back where it belongs, on the topmost shelf next to the dusty canister labelled 'discarded non-sense'.

Eric T. Bell

Mathematics is a tool for thought. A highly necessary tool in a world where both feedback and non-linearities abound. Similarly, all kinds of parts of mathematics serve as tools for other parts and for other sciences.

Applying a simple rewriting rule to the quote on the right above one finds such statements as: 'One service topology has rendered mathematical physics ...'; 'One service logic has rendered computer science ...'; 'One service category theory has rendered mathematics ...'. All arguably true. And all statements obtainable this way form part of the raison d'être of this series.

This series, *Mathematics and Its Applications*, started in 1977. Now that over one hundred volumes have appeared it seems opportune to reexamine its scope. At the time I wrote

> "Growing specialization and diversification have brought a host of monographs and textbooks on increasingly specialized topics. However, the 'tree' of knowledge of mathematics and related fields does not grow only by putting forth new branches. It also happens, quite often in fact, that branches which were thought to be completely disparate are suddenly seen to be related. Further, the kind and level of sophistication of mathematics applied in various sciences has changed drastically in recent years: measure theory is used (non-trivially) in regional and theoretical economics; algebraic geometry interacts with physics; the Minkowsky lemma, coding theory and the structure of water meet one another in packing and covering theory; quantum fields, crystal defects and mathematical programming profit from homotopy theory; Lie algebras are relevant to filtering; and prediction and electrical engineering can use Stein spaces. And in addition to this there are such new emerging subdisciplines as 'experimental mathematics', 'CFD', 'completely integrable systems', 'chaos, synergetics and large-scale order', which are almost impossible to fit into the existing classification schemes. They draw upon widely different sections of mathematics."

By and large, all this still applies today. It is still true that at first sight mathematics seems rather fragmented and that to find, see, and exploit the deeper underlying interrelations more effort is needed and so are books that can help mathematicians and scientists do so. Accordingly MIA will continue to try to make such books available.

If anything, the description I gave in 1977 is now an understatement. To the examples of interaction areas one should add string theory where Riemann surfaces, algebraic geometry, modular functions, knots, quantum field theory, Kac-Moody algebras, monstrous moonshine (and more) all come together. And to the examples of things which can be usefully applied let me add the topic 'finite geometry'; a combination of words which sounds like it might not even exist, let alone be applicable. And yet it is being applied: to statistics via designs, to radar/sonar detection arrays (via finite projective planes), and to bus connections of VLSI chips (via difference sets). There seems to be no part of (so-called pure) mathematics that is not in immediate danger of being applied. And, accordingly, the applied mathematician needs to be aware of much more. Besides analysis and numerics, the traditional workhorses, he may need all kinds of combinatorics, algebra, probability, and so on.

In addition, the applied scientist needs to cope increasingly with the nonlinear world and the extra mathematical sophistication that this requires. For that is where the rewards are. Linear models are honest and a bit sad and depressing: proportional efforts and results. It is in the non-linear world that infinitesimal inputs may result in macroscopic outputs (or vice versa). To appreciate what I am hinting at: if electronics were linear we would have no fun with transistors and computers; we would have no TV; in fact you would not be reading these lines.

There is also no safety in ignoring such outlandish things as nonstandard analysis, superspace and anticommuting integration, *p*-adic and ultrametric space. All three have applications in both electrical engineering and physics. Once, complex numbers were equally outlandish, but they frequently proved the shortest path between 'real' results. Similarly, the first two topics named have already provided a number of 'wormhole' paths. There is no telling where all this is leading - fortunately.

Thus the original scope of the series, which for various (sound) reasons now comprises five subseries: white (Japan), yellow (China), red (USSR), blue (Eastern Europe), and green (everything else), still applies. It has been enlarged a bit to include books treating of the tools from one subdiscipline which are used in others. Thus the series still aims at books dealing with:

- a central concept which plays an important role in several different mathematical and/or scientific specialization areas;
- new applications of the results and ideas from one area of scientific endeavour into another;
- influences which the results, problems and concepts of one field of enquiry have, and have had, on the development of another.

The roots of much that is now possible using mathematics, the stock it grows on, much of that goes back to A.N. Kolmogorov, quite possibly the finest mathematician of this century. He solved outstanding problems in established fields, and created whole new ones; the word 'specialism' did not exist for him.

A main driving idea behind this series is the deep interconnectedness of all things mathematical (of which much remains to be discovered). Such interconnectedness can be found in specially written monographs, and in selected proceedings. It can also be found in the work of a single scientist, especially one like A.N. Kolmogorov in whose mind the dividing lines between specialisms did not even exist.

The present volume is the second of a three volume collection of selected scientific papers of A.N. Kolmogorov with added commentary by the author himself, and additional surveys by others on the many developments started by Kolmogorov. His papers are scattered far and wide over many different journals and they are in several languages; many have not been available in English before. If you can, as Abel recommended, read and study the masters themselves; this collection makes that possible in the case of one of <u>the</u> masters, A.N. Kolmogorov.

The shortest path between two truths in the real domain passes through the complex domain.

J. Hadamard

La physique ne nous donne pas seulement l'occasion de résoudre des problèmes ... elle nous fait pressentir la solution.

H. Poincaré

Never lend books, for no one ever returns them; the only books I have in my library are books that other folk have lent me.

Anatole France

The function of an expert is not to be more right than other people, but to be wrong for more sophisticated reasons.

David Butler

Bussum, December 1991

Michiel Hazewinkel

Contents

Series Editor's Preface v
From the publishers of the Russian edition xi
A few words about A.N. Kolmogorov xiii
1. On convergence of series whose terms are determined by random events 1
2. On the law of large numbers 11
3. On a limit formula of A. Khinchin 13
4. On sums of independent random variables 15
5. On the law of the iterated logarithm 32
6. On the law of large numbers 43
7. General measure theory and probability calculus 48
8. On the strong law of large numbers 60
9. On analytical methods in probability theory 62
10. The waiting problem 109
11. The method of the median in the theory of errors 115
12. A generalization of the Laplace-Lyapunov Theorem 118
13. On the general form of a homogeneous stochastic process 121
14. On computing the mean Brownian area 128
15. On the empirical determination of a distribution law 139
16. On the limit theorems of probability theory 147
17. On the theory of continuous random processes 156
18. On the problem of the suitability of forecasting formulas found by statistical methods 169
19. Random motions 176
20. Deviations from Hardy's formulas under partial isolation 179
21. On the theory of Markov chains 182
22. On the statistical theory of metal crystallization 188
23. Markov chains with a countable number of possible states 193
24. On the reversibility of the statistical laws of nature 209

25. Solution of a biological problem 216
26. On a new confirmation of Mendel's laws 222
27. Stationary sequences in Hilbert space 228
28. Interpolation and extrapolation of stationary random sequences . . 272
29. On the logarithmic normal distribution of particle sizes under grinding 281
30. Justification of the method of least squares 285
31. A formula of Gauss in the method of least squares 303
32. Branching random processes 309
33. Computation of final probabilities for branching random processes . 315
34. Statistical theory of oscillations with continuous spectrum 321
35. On sums of a random number of random terms 331
36. A local limit theorem for classical Markov chains 338
37. Solution of a probabilistic problem relating to the mechanism of bed formation 364
38. Unbiased estimators 369
39. On differentiability of transition probabilities of time-homogeneous Markov processes with a countable number of states 395
40. A generalization of Poisson's formula for a sample from a finite set . 404
41. Some recent work on limit theorems in probability theory 406
42. On A.V. Skorokhod's convergence 419
43. Two uniform limit theorems for sums of independent terms 429
44. Random functions and limit theorems 442
45. On the properties of P. Levy's concentration functions 459
46. Transition of branching processes to diffusion processes and related genetic problems 466
47. On the classes $\Phi^{(n)}$ of Fortet and Blanc-Lapierre 470
48. On conditions of strong mixing of a Gaussian stationary process . . 471
49. Random functions of several variables almost all realizations of which are periodic 478
50. An estimate of the parameters of a complex stationary Gaussian Markov process 479
51. On the approximation of distributions of sums of independent terms by infinitely divisible distributions 484
52. Estimators of spectral functions of random processes 505
53. On the logical foundations of probability theory 515

Comments 520
A.N. Kolmogorov. On the papers on probability theory and mathematical statistics 520
Analytical methods in probability theory (No. 9) (*A.D. Ventsel*) . . . 522
Markov processes with a countable number of states (No. 10) (*B.A. Sevast'yanov*) 528
Homogeneous random processes (No. 13) (*V.M. Zolotarev*) 528
Homogeneous Markov processes (No. 39) (*A.A. Yushkevich*) 530
Branching processes (Nos. 25, 32, 33, 46) (*B.A. Sevast'yanov*) 538
Stationary sequences (No. 27) (*Yu.A. Rozanov*) 539
Stationary processes (No. 48) (*V.A. Statulyavichus*) 542
Statistics of processes (No. 50) (*A.N. Shiryaev*) 544
Spectral theory of stationary processes (No. 34) (*A.M. Yaglom*) 545
Spectral representation of random processes (Nos. 47, 49) (*Yu.G. Balasanov and I.G. Zhurbenko*) 551
Brownian motion (Nos. 14, 19, 24) (*A.M. Yaglom*) 554
Markov chains with a countable number of states (No. 23) (*A.A. Yushkevich*) 559
Wald identities (No. 35) (*A.A. Novikov*) 567
S-Convergence (No. 42) (*A.V. Skorokhod*) 569
Uniform limit theorems (Nos. 43, 51) (*T.V. Arak*) 570
Concentration functions (No. 45) (*V.M. Kruglov*) 571
Empirical distributions (No. 15) (*E.V. Khmaladze*) 574
The method of least squares (Nos. 30, 31) (*M.B. Malyutov*) 583
Unbiased estimators (No. 38) (*Yu.K. Belyaev and Ya.P. Lumel'skii*) . . 585
Statistical prediction (No. 18) (*A.M. Yaglom*) 587
On inter-bed washout (No. 37) (*A.B. Vistelius*) 591

From the Publishers of the Russian Edition

In accordance with the decision of the Praesidium of the USSR Academy of Sciences, the first book of the selected works of Academician A.N. Kolmogorov "Mathematics and Mechanics" ("Nauka", Moscow) came out in 1985. As for the second book of Kolmogorov's works planned for publication, the editorial board decided to divide it into two parts: the first contains the papers on probability theory and mathematical statistics, and the second those on information theory and the theory of algorithms. So the articles marked with two asterisks in the list of references of the first book should be naturally divided into those given in the present second book and those prepared for publication in the third book.

The second and third books of the selected works of Kolmogorov were prepared for publication by Yu.V. Prokhorov and A.N. Shiryaev.

A few words about A.N. Kolmogorov*

The remarkably broad creative interests of A.N. Kolmogorov, the wide range and variety of fields of mathematics he worked in through different periods of his life — all this makes Andrei Nikolaevich distinguished among mathematicians in our country and all over the world. This diversity of interests makes him unique among the mathematicians of our time. In many fields of mathematics he obtained truly fundamental and principally important results. The problems were often most difficult to solve and required great creative endeavour. This is true for the results obtained by Andrei Nikolaevich in his young years on the theory of sets and functions, both the descriptive and the metrical theories; for example, the theory of operations on sets developed by him and his celebrated example of a divergent Fourier series.

This was followed by papers on general measure theory, both the abstract, that is, "the general theory proper", and the geometric theory. After that Kolmogorov started his fundamental work in various branches of probability theory, work that made him, beyond any doubt, the most outstanding among the researchers in this field all over the world.

Along with this, Kolmogorov wrote his first papers on mathematical logic and the foundations of mathematics. Later, these were supplemented by studies in information theory.

Andrei Nikolaevich made a very important contribution in topology. It suffices to say that simultaneously with the outstanding American topologist J.W. Alexander and quite independently of him, Kolmogorov came to the notion of cohomology and laid the foundation of the theory of cohomology operations, obtaining the results that transformed topology as a whole. The deep

---

* *Uspekhi Mat. Nauk* **38**:4 (1983), 7–8.

connections between topology and the theory of ordinary differential equations, celestial mechanics and, later, the general theory of dynamical systems are well known. These connections were established back in the first papers by H. Poincaré. The ideas of Kolmogorov in this whole vast area of mathematics, further developed by his numerous students, have largely determined the present state of this field. Finally, the studies by Kolmogorov in mechanics proper are worth mentioning, in particular, his famous work on the theory of turbulence, which reached out directly to the realm of experimental natural sciences. All this, which is by no means everything that could be said about Kolmogorov as a scientist, shows that he is one of the most outstanding representatives of modern mathematics in the broadest sense of the word, including applied mathematics. Andrei Nikolaevich is undoubtedly recognized as such by the international scientific community. In particular, this is reflected in the fact that he has been elected a member of many Academies and scientific societies, and many universities have made him an honorary member — no other Soviet scientist has ever received such recognition. Among these are the Paris Academy of Sciences, the London Royal Scientific Society, the Academy of Natural Sciences "Leopoldina", the Polish Academy of Sciences, the East German Academy, the Polish Mathematical Society, the London Mathematical Society, the USA National Academy of Sciences, the American Philosophical Society founded by B. Franklin, and the Universities of Paris, Berlin, and Warsaw.

A.N. Kolmogorov was born on 25th April, 1903 in Tambov, where his mother Mariya Yakovlevna Kolmogorova was staying on her way from the Crimea. Mariya Yakovlevna died giving birth to her son, and he was raised by her sister Vera Yakovlevna Kolmogorova, who really replaced his mother in his life. A.N. Kolmogorov loved her as his mother all her life, up to her death in 1950 in Komarovka at the age of 87. On his mother's side, A.N. Kolmogorov comes from nobility: his maternal grandfather, Yakov Stepanovich Kolmogorov, was marshal of the nobility in the province of Uglich. Andrei Nikolaevich's father was a son of a priest and worked as an agronomist with a higher special education, or, as they said in those years, as a "learned agronomist".

My friendship with A.N. Kolmogorov played a very special role in my life: in 1979 it was fifty years since we had become friends, and through more than half a century, this friendship has never been in question. We have never had a single misunderstanding of any kind on any matter of importance for our life and outlook. Even when our opinions on some question differed, we always respected each other's viewpoint and never lost our deep sympathy for each other.

As already mentioned, Kolmogorov has a lot of students in various fields of mathematics, and some of them have become famous in their areas. The older students of Kolmogorov are Sergei Mikhailovich Nikolskii (b. 1905) and the late Anatoli Ivanovich Mal'tsev (b. 1910), both Academicians. Next in age are Boris Vladimirovich Gnedenko (b. 1912), Academician of the Ukrainian Academy of Sciences and expert in probability theory of world-wide recognition, Acad. Mikhail Dmitrievich Millionshchikov (1913–1973) and Acad. Izrael' Moiseevich Gel'fand (b. 1913), elected as a foreign member of the USA National Academy of Sciences and Paris Academy of Sciences. Much younger, though also belonging to the older generation of Kolmogorov's students, are Acad. Aleksander Mikhailovich Obukhov (b. 1918) and Corresponding Member Andrei Sergeevich Monin (b. 1921).

They are followed by Vladimir Andreevich Uspenskii, Vladimir Mikhailovich Tikhomirov, Vladimir Mikhailovich Alekseev, Yakov Grigor'evich Sinai and Vladimir Igorevich Arnol'd, Corresponding Member since 1984.

The largest group of Kolmogorov's students works in probability theory and mathematical statistics. It includes Acad. Yurii Vasil'evich Prokhorov, Corresponding Member Login Nikolaevich Bol'shev, Acad. of the Uzbek Academy of Sciences Sagdi Khasanovich Sirazhdinov, Acad. of the Ukrainian Academy of Sciences and Corresponding Member of the USSR Academy of Sciences Boris Aleksandrovich Sevast'yanov, Juri Anatol'evich Rozanov, Al'bert Nikolaevich Shiryaev and Igor' Georgievich Zhurbenko. Of course, this list is by no means complete; from the very title of my note it is clear that this is neither a jubilee review of the life and activities of A.N. Kolmogorov, nor a traditional "jubilee article"; so it does not claim to be complete in any respect.

P.S. Aleksandrov

1. ON CONVERGENCE OF SERIES WHOSE TERMS ARE DETERMINED BY RANDOM EVENTS *

Jointly with A.Ya. Khinchin

Consider a series

$$y_1 + y_2 + \ldots + y_n + \ldots \qquad (1)$$

whose terms are *random variables*; denote the values taken by y_n (their number is finite or, possibly, countable) by $y_n^{(1)}, y_n^{(2)}, \ldots, y_n^{(i)}, \ldots$, and the corresponding probabilities by $p_n^{(1)}, p_n^{(2)}, \ldots, p_n^{(i)}, \ldots$, with $\sum_i p_n^{(i)} = 1$. Further, denote by

$$a_n = \sum_i y_n^{(i)} p_n^{(i)}$$

the expectation of y_n, and by

$$b_n = \sum_i \{y_n^{(i)} - a_n\}^2 \cdot p_n^{(i)}$$

the expectation of the square of the deviation $y_n - a_n$.

We first prove that, assuming the convergence of the series $\sum_n a_n$ and $\sum_n b_n$, the probability P of the convergence of the series (1) is 1. A particular case of this statement (namely when y_n takes only two values, $+c_n$ and $-c_n$, with the same probability $\frac{1}{2}$) was established by Rademacher in function-theoretic terms.[1] In §1 we prove a general theorem using a generalization of the method suggested by Rademacher. In §2 we prove it again by other methods that bring about the result more quickly. In §3 we establish that, for one important and broad class of cases, convergence of the series $\sum a_n, \sum b_n$ is not only a sufficient, but also a necessary condition for $P = 1$ to hold; this is the case when the values taken by y_n are uniformly bounded. Finally, in §4 we indicate a necessary and sufficient criterion for $P = 1$ to hold in general.

§1 is due to A.Ya. Khinchin, §§2, 3 and 4 to A.N. Kolmogorov.

§1

Let us reduce the problem to the function-theoretic form more suitable for us. For this, we construct a system of functions

$$\phi_1(x), \phi_2(x), \ldots, \phi_n(x), \ldots$$

---

* Über Konvergenz von Reihen, deren Glieder durch den Zufall bestimmt werden', *Mat. Sb.* **32** (1925), 668–677.

[1] *Math. Ann.* **87** (1922), 135.

by means of the following recurrence: *assume, without loss of generality*, that

$$p_n^{(i)} \geq p_n^{(i+1)}$$

for all n and i; divide the interval $0 \leq x \leq 1$ from left to right into parts of lengths

$$p_1^{(1)},\ p_1^{(2)}, \ldots,\ p_1^{(i)}, \ldots$$

and assume that on each of these subintervals $\phi_1(x)$ equals a constant, namely

$$\phi_1(x) = z_1^{(i)} = y_1^{(i)} - a_1$$

on the ith interval. Then we have

$$\int_0^1 \phi_1(x)dx = 0, \qquad \int_0^1 \{\phi_1(x)\}^2 dx = b_1.$$

If the function $\phi_{n-1}(x)$ is already determined, then the function $\phi_n(x)$ is determined as follows: each interval on which $\phi_{n-1}(x)$ is constant is divided into intervals whose lengths (from left to right) are proportional to the values

$$p_n^{(1)}, p_n^{(2)}, \ldots, p_n^{(i)}, \ldots$$

and on the ith subinterval we set

$$\phi_n(x) = z_n^{(i)} = y_n^{(i)} - a_n.$$

Then

$$\int_0^1 \phi_n(x)dx = 0, \qquad \int_0^1 \{\phi_n(x)\}^2 dx = b_n.$$

The values of the functions $\phi_n(x)$ at the ends of the subintervals are unimportant and can be taken arbitrarily.

Assuming that $\sum a_n$ converges, the probability that (1) converges is equal to that of the convergence of

$$\phi_1(x) + \phi_2(x) + \ldots + \phi_n(x) + \ldots, \tag{2}$$

which in our case is given by the (Lebesgue) measure of the set on which this series converges. Hence, our main result can be interpreted as follows: [2] convergence of $\sum b_n$ is a sufficient condition for the almost everywhere convergence of (2).

[2] Clearly, this statement is the same as that stated in the introduction. General considerations of this kind of relation can be found in Steinhaus's work (*Fund. Math.* **4** (1923), 286–310).

The proof given below generalizes Rademacher's proof for a special case. [3] Set

$$\pi_n(x) = \prod_{k=1}^{n} \{1 + \phi_k(x)\}, \quad f_n(x) = \int_0^x \pi_n(t)dt.$$

We have

$$|f_{n+1}(x) - f_n(x)| = \left|\int_0^x \{\pi_{n+1}(t) - \pi_n(t)\}dt\right| = \left|\int_0^x \pi_n(t)\phi_{n+1}(t)dt\right|.$$

Let $\delta(a,b)$ be an interval of constancy of the function $\phi_n(t)$, taken to be $a \leq x < b$. Clearly

$$\int_0^a \pi_n(t)\phi_{n+1}(t)dt = 0$$

and, consequently,

$$|f_{n+1}(x) - f_n(x)| = \left|\int_a^x \pi_n(t)\phi_{n+1}(t)dt\right| = |\pi_n(x)| \cdot \left|\int_a^x \phi_{n+1}(t)dt\right|, \tag{3}$$

since $\pi_n(t)$ is constant on (a,x). From the conditions

$$\sum_i z_n^{(i)} p_n^{(i)} = 0 \ (n = 1, 2, \ldots)$$

we obtain

$$|z_n^{(j)} p_n^{(j)}| = \left|\sum_{i \neq j} z_n^{(i)} p_n^{(i)}\right| \quad (n, j = 1, 2, \ldots)$$

and the Schwarz inequality implies

$$|z_n^{(j)} p_n^{(j)}| \leq \sqrt{\sum_{i \neq j} \{z_n^{(i)}\}^2 p_n^{(i)} \sum_{i \neq j} p_n^{(i)}} \leq \sqrt{b_n}\sqrt{1 - p_n^{(j)}},$$

$$|z_n^{(j)}| \leq \sqrt{b_n}\frac{\sqrt{1 - p_n^{(j)}}}{p_n^{(j)}} \quad (n, j = 1, 2, \ldots).$$

Hence, if on (a,b)

$$\phi_k(x) = z_k^{(j_k)} \quad (k = 1, 2, \ldots, n),$$

then

$$|\pi_n(x)| = \left|\prod_{k=1}^{n} \{1 + z_k^{(j_k)}\}\right| \leq \prod_{k=1}^{n} \left\{1 + \sqrt{b_k}\frac{\sqrt{1 - p_k^{(j_k)}}}{p_k^{(j_k)}}\right\}. \tag{4}$$

[3] See footnote 1.

On the other hand we have

$$\left|\int_a^x \phi_{n+1}(t)dt\right| \le \int_a^x |\phi_{n+1}(t)|dt \le \int_a^b |\phi_{n+1}(t)|dt =$$
$$= p_1^{(j_1)} p_2^{(j_2)} \dots p_n^{(j_n)} \int_0^1 |\phi_{n+1}(t)|dt \le \sqrt{b_{n+1}} \prod_{k=1}^n p_k^{(j_k)} \quad (5)$$

and

$$\left|\int_a^x \phi_{n+1}(t)dt\right| \le |x-a| \cdot |z_{n+1}^{(u)}| < \prod_{k=1}^n p_k^{(j_k)} \sqrt{b_{n+1}} \frac{\sqrt{1-p_{n+1}^{(u)}}}{p_{n+1}^{(u)}},$$

where $|z_{n+1}^{(u)}|$ is the maximal value of $|\phi_{n+1}(t)|$ for $a < t \le x$. If

$$\phi_{n+1}(x) = z_{n+1}^{(j_{n+1})}, \quad j_{n+1} \ge u,$$

then by our hypotheses the inequality $p_{n+1}^{(j_{n+1})} \le p_{n+1}^{(u)}$ holds; therefore *a fortiori* we have

$$\left|\int_a^x \phi_{n+1}(t)dt\right| < \prod_{k=1}^n p_k^{(j_k)} \sqrt{b_{n+1}} \frac{\sqrt{1-p_{n+1}^{(j_{n+1})}}}{p_{n+1}^{(j_{n+1})}}. \quad (6)$$

From (5) and (6) it follows that

$$\left|\int_a^x \phi_{n+1}(t)dt\right| \le 2\sqrt{b_{n+1}}\, p_1^{(j_1)} p_2^{(j_2)} \dots p_n^{(j_n)} \sqrt{1-p_{n+1}^{(j_{n+1})}}, \quad (7)$$

since this follows from (6) for $p_{n+1}^{(j_{n+1})} \ge \frac{1}{2}$ and from (5) for $p_{n+1}^{(j_{n+1})} < \frac{1}{2}$.

Now, (3), (4) and (7) imply

$$|\Delta_n(x)| = |f_{n+1}(x) - f_n(x)| \le$$
$$\le 2\sqrt{b_{n+1}}\sqrt{1-p_{n+1}^{(j_{n+1})}} \prod_{k=1}^n \left\{p_k^{(j_k)} + \sqrt{b_k}\sqrt{1-p_k^{(j_k)}}\right\}$$

or, substituting $1 - p_k^{(j_k)} = w_k^2$,

$$|\Delta_n(x)| \le 2\sqrt{b_{n+1}} w_{n+1} \prod_{k=1}^n \left\{1 - w_k^2 + w_k\sqrt{b_k}\right\} \le$$
$$\le (b_{n+1} + w_{n+1}^2) \prod_{k=1}^n \left\{1 - w_k^2 + \frac{w_k^2}{2} + \frac{b_k}{2}\right\} <$$
$$< b_{n+1} \prod_{k=1}^\infty \left(1 + \frac{b_k}{2}\right) + w_{n+1}^2 \prod_{k=1}^n \left(1 - \frac{w_k^2}{2}\right)(1 + b_k).$$

If we set

$$\prod_{k=1}^{\infty}(1+b_k)=B, \quad \tfrac{1}{2}w_k^2=v_k,$$

then

$$|\Delta_n(x)|<Bb_{n+1}+2Bv_{n+1}\prod_{k=1}^{n}(1-v_k). \tag{8}$$

Since for every positive integer N, putting $v_0=0$, we have

$$1-\sum_{n=0}^{N}v_{n+1}\prod_{k=1}^{n}(1-v_k)=\prod_{k=1}^{N+1}(1-v_k)>0,$$

the right-hand side of (8) contains the general term of a convergent series. Hence, the limit

$$\lim_{n\to\infty}f_n(x)=f(x)$$

exists.

In the special case considered by Rademacher, $f(x)$ turned out to be a monotone function, which is not necessarily so for the general case considered here.

However, monotonicity was used only to prove differentiability of $f(x)$, so it suffices to demonstrate that $f(x)$ is a function of bounded variation. This is easily proved as follows. In any case the total variation of $f(x)$ is not greater than the upper bound of the total variations of all the $f_n(x)$ and, consequently, it is not greater than

$$\limsup\int_0^1|\pi_n(t)|dt.$$

But

$$\int_0^1|\pi_n(t)|dt=\int_0^1\left|\prod_{k=1}^{n}\{1+\phi_k(t)\}\right|dt=\prod_{k=0}^{n}\int_0^1|1+\phi_k(t)|dt\le$$

$$\le\prod_{k=1}^{n}\sqrt{\int_0^1|1+\phi_k(t)|^2dt}=\sqrt{\prod_{k=1}^{n}(1+b_k)}<\sqrt{B},$$

where the admissibility of changing the order of taking the product and integrating follows directly from the particular structure of the functions $\phi_k(x)$. This gives the desired statement. Thus, the derivative $f'(x)$ exists almost everywhere. Let x be an arbitrary point of $(0,1)$, other than a subdivision point, at which $f'(x)$ exists, and let (a_n, b_n) denote the interval containing x on which $\phi_n(x)$ is constant; since $f(a_n) = f_n(a_n)$, $f(b_n) = f_n(b_n)$, we obtain

$$f'(x) = \lim_{n\to\infty} \frac{f(b_n) - f(a_n)}{b_n - a_n} = \lim_{n\to\infty} \frac{f_n(b_n) - f_n(a_n)}{b_n - a_n} = \lim_{n\to\infty} \pi_n(x),$$

since [4]

$$\frac{f_n(b_n) - f_n(a_n)}{b_n - a_n} = \pi_n(x)$$

because $\pi_n(x)$ is constant on (a_n, b_n).

Hence, $\lim_{n\to\infty} \pi_n(x)$ exists almost everywhere. On the other hand, the series

$$\sum_{k=1}^{\infty} \{\phi_k(x)\}^2$$

converges almost everywhere, since otherwise the series

$$\sum_{k=1}^{\infty} \int_0^1 \{\phi_n(x)\}^2 dx = \sum_{k=1}^{\infty} b_k$$

would diverge, which contradicts the assumption.

But the convergence almost everywhere of the sequence $\pi_n(x)$ and the series $\sum\{\phi(x)\}^2$ implies that $\sum \phi_n(x)$ converges almost everywhere, as required.

§2

We now prove this assertion in a different, much simpler way. We set

$$\sum_{k=1}^{n} \phi_k(x) = s_n(x).$$

If the series (2) diverges at every point of some set of positive measure, then there exists a set E of positive measure mE and a positive constant A such that

$$|s_p(x) - s_n(x)| > A$$

---

[4] If $\lim_{n\to\infty} (b_n - a_n) > 0$, then we have for sufficiently small $\epsilon > 0$

$$f'(x) = \frac{1}{2\epsilon}[f(x+\epsilon) - f(x-\epsilon)] =$$
$$\lim_{n\to\infty} \frac{1}{2\epsilon}[f_n(x+\epsilon) - f_n(x-\epsilon)] = \lim_{n\to\infty} \pi_n(x).$$

for each positive integer and each $x \in E$, for a suitable $p = p(x,n) > n$.

Since $s_p(x) - s_n(x)$ is constant on every interval on which $\phi_p(x)$ is constant, by construction of our functions, we can find a finite number of non-overlapping intervals δ such that:

1) $\sum \delta$ (the sum of the lengths of all δ's) is greater than $\frac{1}{2}mE$;
2) Each δ is an interval on which $\phi_p(x)$ $(p > n)$ is constant and

$$|s_p(x) - s_n(x)| > A \in \delta.$$

If now k is the largest of these numbers p (there is a finite number of them) and δ is an interval on which $s_p(x)$ $(p \leq k)$ is constant, then, by the construction of our functions,

$$\int_\delta \{s_k(x) - s_n(x)\}^2 dx = \int_\delta \{s_p(x) - s_n(x)\}^2 dx + \int_\delta \{s_k(x) - s_p(x)\}^2 dx \geq A^2\delta,$$

consequently,

$$\sum_{p=n+1}^{k} b_p = \int_0^1 \{s_k(x) - s_n(x)\}^2 dx \geq A^2 \sum \delta > \tfrac{1}{2}A^2 mE.$$

This implies the divergence of $\sum b_n$, contrary to the assumption.

§3

Thus, convergence of $\sum b_n$ is a sufficient condition for convergence of $\sum \phi_n(x)$ almost everywhere. In general this condition is not necessary, as can easily be demonstrated by an example.

Namely, if $\phi_n(x)$ is equal to 0, $+c_n$, $-c_n$ on sets of measure $1-\tau_n$, $\frac{1}{2}\tau_n$, $\frac{1}{2}\tau_n$ respectively and $\sum \tau_n$ converges, then $\phi_n(x)$ does not vanish almost everywhere only for a finite number of subscripts n, and therefore $\sum \phi_n(x)$ converges almost everywhere, whatever c_n. Since $b_n = c_n^2 \tau_n$, by suitably choosing the c_n, $\sum b_n$ can be made to diverge.

For the corresponding series (1) we obviously have $P = 1$, so the convergence of $\sum b_n$ is not necessary for $P = 1$. It is also easy to show that the convergence of $\sum a_n$ is not necessary for $P = 1$. To show this, put $y_n^{(1)} = c_n$, $y_n^{(2)} = 0$, $p_n^{(1)} = \tau_n$, $p_n^{(2)} = 1 - \tau_n$, with $\sum \tau_n$ convergent and $\sum c_n \tau_n$ divergent.

However, there is an important special case, when our condition is also necessary. In fact, if the absolute values of our functions $\phi_n(x)$ are uniformly bounded, then convergence almost everywhere of $\sum \phi_n(x)$ implies the convergence of $\sum b_n$. The situation with (1) is similar. If $P = 1$ and if the random variables y_n $(n = 1, 2, \ldots)$ are uniformly bounded, then both $\sum a_n$ and $\sum b_n$ must converge.

In order to prove this general result we first establish the following: if the $\phi_n(x)$ are uniformly bounded and if there exist a set E of positive measure and a constant K such that

$$|s_k(x_1) - s_k(x_2)| < K \tag{9}$$

for any k and all pairs of points (x_1, x_2) belonging to E, then $\sum b_n$ converges.

Denote by E_k the union of those intervals of constancy of $\phi_k(x)$ (hence of $s_k(x)$) that contain the points of E. Obviously, E_k contains E and E_{k+1}. It is also clear that (9) holds for any pair of points of E_k. Set

$$E_k - E_{k+1} = F_k,$$

then

$$\int_{E_{k+1}}\int_{E_{k+1}} \{s_{k+1}(x_1) - s_{k+1}(x_2)\}^2 dx_1 dx_2 = \int_{E_k}\int_{E_k} \{s_{k+1}(x_1)- -s_{k+1}(x_2)\}^2 dx_1 dx_2 - 2\int_{E_{k+1}}\int_{F_k} \{s_{k+1}(x_1) - s_{k+1}(x_2)\}^2 dx_1 dx_2 - -\int_{F_k}\int_{F_k} \{s_{k+1}(x_1) - s_{k+1}(x_2)\}^2 dx_1 dx_2.$$

But $\phi_{k+1}(x)$ is orthogonal to both $s_k(x)$ and 1 on E_k, hence

$$\begin{aligned}\int_{E_{k+1}}\int_{E_{k+1}} \{s_{k+1}(x_1)-s_{k+1}(x_2)\}^2 dx_1 dx_2 &= \int_{E_k}\int_{E_k} \{s_k(x_1)-s_k(x_2)\}^2 dx_1 dx_2 + \\ &+ \int_{E_k}\int_{E_k} \{\phi_{k+1}(x_1) - \phi_{k+1}(x_2)\}^2 dx_1 dx_2 - \\ &-2\int_{E_{k+1}}\int_{F_k} \{s_k(x_1) - s_k(x_2) + \phi_{k+1}(x_1) - \phi_{k+1}(x_2)\}^2 dx_1 dx_2 - \\ &- \int_{F_k}\int_{F_k} \{s_k(x_1) - s_k(x_2) + \phi_{k+1}(x_1) - \phi_{k+1}(x_2)\}^2 dx_1 dx_2 \geq \\ &\geq \int_{E_k}\int_{E_k} \{s_k(x_1) - s_k(x_2)\}^2 dx_1 dx_2 + 2b_{k+1}(mE_k)^2 - \\ &-(K+2M)^2\{2mE_{k+1} \cdot mF_k + (mF_k)^2\},\end{aligned}$$

where M is the upper bound of $|\phi_n(x)|$ and mE denotes the measure of E. In the inequality obtained, we set $k = 0, 1, \ldots, n-1$, add up and note that

$$\sum_{k=0}^{n-1}\{2mE_{k+1}\cdot mF_k + (mF_k)^2\} \leq 1.$$

As a result we have

$$\int_{E_n}\int_{E_n}\{s_n(x_1) - s_n(x_2)\}^2 dx_1 dx_2 \geq \sum_{k=0}^{n-1} 2b_{k+1}(mE_k)^2 - (K+2M)^2,$$

hence, since $|s_n(x_1) - s_n(x_2)| < K$, we obtain

$$\sum_{k=0}^{n-1} 2b_{k+1}(mE_k)^2 < 2(K+2M)^2$$

and *a fortiori*,

$$(mE)^2\sum_{k=0}^{n-1} b_{k+1} < (K+2M)^2,$$

which clearly proves the convergence of $\sum b_n$.

We now suppose that the random variables y_n are uniformly bounded and that $P = 1$, which implies convergence almost everywhere of $\sum(a_n + \phi_n(x))$. Since a_n is independent of x, for $\sum \phi_n(x)$ the assumptions of the auxiliary statement just proved hold.[5] Consequently $\sum b_n$ converges and by §1 or §2, $\sum \phi_n(x)$ converges almost everywhere. But this entails the convergence of $\sum a_n$, which proves all we need.

§4

Consider, together with (1), another similar series

$$u_1 + u_2 + \ldots + u_n + \ldots, \tag{10}$$

denote by Q the probability of its convergence, and by τ_n the probability of the event

$$y_n \neq u_n.$$

[5] The uniform boundedness of the y_n clearly implies that of the a_n and consequently that of the $\phi_n(x)$.

When $\sum \tau_n$ converges we call the series (1) and (10) equivalent.[6] Clearly, in this case $Q = P$.

Now we claim that a necessary and sufficient condition for $Q = 1$ to hold for (10) is the existence of an equivalent series (1) for which $\sum a_n$ and $\sum b_n$ converge. Clearly, only the necessity of the condition needs to be proved.

Thus we suppose that $Q = 1$ and define the series (1) as follows. Let $y_n = u_n$ if $|u_n| < 1$ and $y_n = 0$ otherwise. Hence, τ_n coincides with the probability of $|u_n(x)| \geq 1$. Since $Q = 1$, this immediately implies the convergence of $\sum \tau_n$, so that (1) and (10) are in fact equivalent. Thus, the constructed series (1) is already uniformly bounded, therefore according to §3 the corresponding series $\sum a_n$, $\sum b_n$ converge, and the required result is proved.

Note also that our proof not only justifies the existence of (1) but also gives an extremely simple rule for constructing this series.

Moscow, 3 December 1925

---

[6] This means that there is a certain relation between u_n and y_n.

2. ON THE LAW OF LARGE NUMBERS *

Let $E_1, E_2, \ldots, E_n, \ldots$ be a sequence of independent trials. The reasoning given below is true only for a finite number of distinct possible outcomes $E_k^{(1)}, E_k^{(2)}, \ldots$ of the trial E_k. However, the theorem to be proved holds also in the general case.

Let F_n be a variable which depends on the first n trials. If for any positive ϵ the probability of

$$|F_n - D_n| < \epsilon,$$

where D_n is the expectation of F_n, tends to 1 as n tends to infinity, then F_n is said to satisfy the *law of large numbers*, or that F_n is *stable*.

The simplest sets of conditions sufficient for the stability of F_n are expressed in terms of restrictions imposed on the variance of F_n.

It is natural to consider the upper bound of the difference

$$F_n\left(E_1^{(i_1)}, \ldots, E_{k-1}^{(i_{k-1})}, E_k^{(i_k)}, E_{k+1}^{(i_{k+1})}, \ldots, E_n^{(i_n)}\right) - \\ -F_n\left(E_1^{(i_1)}, \ldots, E_{k-1}^{(i_{k-1})}, E_k^{(j_k)}, E_{k+1}^{(i_{k+1})}, \ldots, E_n^{(i_n)}\right),$$

when $i_1, i_2, \ldots, i_n$ and j_k run independently through all possible values, as the *variance of F_n with respect to E_k*. Denote this variance by Ω_{nk}. With this definition we can formulate the following theorem.

Theorem. *If the sum*

$$\Omega_{n1}^2 + \Omega_{n2}^2 + \ldots + \Omega_{nn}^2$$

tends to zero as n increases, then F_n satisfies the law of large numbers.

Corollary. *The condition*

$$\Omega_{nk} = o(1/\sqrt{n}). \tag{1}$$

is also sufficient for F_n to be stable.

Proof of the theorem. Denote by B_n the expectation of the square of the deviation $F_n - D_n$. A well-known condition for the stability of F_n is

$$B_n \to 0. \tag{2}$$

* 'Sur la loi des grands nombres', *C. R. Acad. Sci. Paris* **185** (1927), 917–919. Presented by E. Borel.

To use this condition, a more effective expression for B_n is needed. For this, we denote by D_{nk} the expectation of F_n assuming that the results of the first k trials are known. Let $Z_{nk} = D_{nk} - D_{nk-1}$. Thus, Z_{nk} is the increment of the expectation of F_n when the result of the trial E_k is known. Clearly we have

$$D_n = D_{n0}, \quad F_n = D_{nn},$$

$$F_n - D_n = Z_{n1} + Z_{n2} + \ldots + Z_{nn}. \tag{3}$$

It can be proved that the expectation $Z_{ni}Z_{nk}$ $(i \neq k)$ is *zero.*

Let $k > i$. Assume that the result of $E_1, E_2, \ldots, E_{k-1}$ is known. Under this assumption $D_{nk} = D_{nk-1}$; therefore, $Z_{nk} = 0$ and since Z_{ni} is a constant, $Z_{ni}Z_{nk}$ is also zero. This is true for any results of the trials $E_1, E_2, \ldots, E_{k-1}$. Consequently, the expectation $Z_{ni}Z_{nk}$ itself is also zero.

Thus, denoting by β_{nk} the expectation Z_{nk}^2, we obtain

$$B_n = \beta_{n1} + \beta_{n2} + \ldots + \beta_{nn}. \tag{4}$$

It can be proved that

$$\beta_{nk} \leq \tfrac{1}{4}\Omega_{nk}^2, \quad B_n \leq \tfrac{1}{4}(\Omega_{n1}^2 + \Omega_{n2}^2 + \ldots + \Omega_{nn}^2). \tag{5}$$

Inequality (5) permits us to derive our theorem because of (2).

Remark. In the classical Chebyshev case the trial E_k consists in determining X_k, and F_k is equal to the arithmetic mean of $X_1, X_2, \ldots, X_n$. Denoting by d_k the expectation of X_k and by b_k the expectation of $(X_n - d_n)^2$, we obtain

$$\beta_{nk} = b_k/n, \quad \Omega_{nk} = (\max X_k - \min X_k)/n.$$

Condition (1) turns into the well-known condition

$$\max |X_n| = o(\sqrt{n}).$$

The above is a generalization of Chebyshev's case, where F_n is the sum of independent random variables, since it concerns sums of dependent variables. Very often the summands here are functions of certain other independent variables. We think that in this case the above method for arbitrary functions in independent variables would be very natural.

31 October 1927

3. ON A LIMIT FORMULA OF A. KHINCHIN *

Let

$$z_1, z_2, \ldots, z_n, \ldots \tag{1}$$

be a sequence of independent random variables. Without loss of generality assume that the expectation of z_n is zero. Denote by b_n the expectation of the square of z_n and by S_n and B_n the sums

$$S_n = \sum_{k=1}^{n} z_k, \quad B_n = \sum_{k=1}^{n} b_k.$$

A.Ya. Khinchin [1] suggested that under broad assumptions the probability that

$$\limsup \frac{S_n}{\sqrt{2B_n \log\log B_n}} = 1 \tag{2}$$

is 1. Khinchin himself proved this formula for some important special cases [2]. Our aim is to formulate sufficiently general conditions for this formula to hold. These conditions are as follows:

(I) $B_n \to \infty$,

(II) m_n-boundedness from above, $|z_n| \leq m_n = o\left(\sqrt{\dfrac{B_n}{\log\log B_n}}\right)$.

Clearly the first condition is necessary, except for the case $z_n = 0$. *If the z_n are uniformly bounded, then the second condition follows from the first; in this case this condition is necessary and sufficient.*

If we want to use only probabilistic relations that can be effectively observed, the meaning of formula (2) can be explained as follows:

1°. Whatever the positive numbers η and δ, there is an integer n such that the probability that all the inequalities

$$S_k < \sqrt{2B_k \log\log B_k}(1+\delta) \quad (k = n, n+1, \ldots, n+p)$$

hold is not less than $1 - \eta$, whatever p is.

* 'Sur une formule limite de M.A. Khintchine', *C. R. Acad. Sci. Paris* **186** (1928), 824–825. Presented by J. Hadamard.

2°. For any η, δ and m, there exists an integer p such that the probability that all the inequalities

$$S_k < \sqrt{2B_k \log\log B_k}(1-\delta) \quad (k = m, m+1, \ldots, m+p)$$

hold simultaneously does not exceed η.

A complete proof of statements 1° and 2° using only the conditions (I) and (II) will be published elsewhere.[1]

Applied to the problem of recurrent trials, our theorem yields: let p_n be the probability of an event ϵ at the nth trial and $\mu(n)$ the number of events ϵ during the first n trials. Then the probability that

$$\limsup \frac{\mu(n) - (p_1 + p_2 + \ldots + p_n)/n}{\sqrt{2\sum_{k=1}^{n} p_k(1-p_k) \log\log \sum_{k=1}^{n} p_k(1-p_k)}} = 1 \tag{3}$$

is 1, provided the series $\sum p_n(1-p_n)$ diverges.

References

1. A.Ya. Khinchin, *Basic laws of probability theory*, Moscow, 1927 (in Russian).
2. A.Ya. Khinchin, *Math. Ann.* **99** (1928), 152.

[1] See article No. 5 of this publication.

4. ON SUMS OF INDEPENDENT RANDOM VARIABLES *

§1. Notation

$P(A)$ denotes the probability of an event A;

$P_B(A)$ or $P(A|B)$ denotes the conditional probability of event A with respect to an event B;

$\mathbf{E}\xi$ denotes the expectation of a random variable ξ;

$\mathbf{E}_B\xi$ or $\mathbf{E}(\xi|B)$ denotes the conditional expectation of ξ with respect to an event B;

$\mathbf{E}(\xi : B) = P(B) \cdot \mathbf{E}(\xi|B)$ or, what is the same, is equal to the Lebesgue integral

$$\int_B \xi dP$$

over B;

$\mathbf{D}\xi = \mathbf{E}(\xi - \mathbf{E}\xi)^2$ denotes the variance of a random variable ξ.

If $\xi_1, \dots, \xi_n$ are random variables, we set

$$S_k = \sum_{i=1}^{k} \xi_i, \ S = S_n; \quad v = \max_{1 \leq k \leq n} |S_k|;$$

$$\zeta_k = \xi_k - \mathbf{E}\xi_k; \quad T_k = \sum_{i=1}^{k} \zeta_i, \ T = T_n;$$

$$u = \max_{1 \leq k \leq n} |T_k|, \quad M = \sup_k |\zeta_k|, \quad D = \mathbf{E}T^2 = \mathbf{D}S^2.$$

(We assume that $\mathbf{E}\xi_k$ and $\mathbf{E}\xi_k^2$ exist; $1 \leq k \leq n$.) Clearly $\mathbf{E}\zeta_k = 0$, $T_k = S_k - \mathbf{E}S_k$. If $\xi_1, \dots, \xi_k$ are independent, then

$$\mathbf{D}S_k = \sum_{i=1}^{k} \mathbf{D}\xi_i.$$

§2 Some theorems on finite sums

Assume that $\xi_1, \dots, \xi_n$ are mutually independent.

Theorem 1. *Let $R > 0$. Then*

$$\mathbf{P}\{\max_{1 \leq k \leq n} |T_k| \geq R\} \leq \mathbf{E}T^2/R^2.$$

---

* 'Über die Summen durch den Zufall bestimmter unabhängiger Grössen', *Math. Ann.* **99** (1928), 309–319; Bemerkungen ... — *Math. Ann.* **102** (1929), 484–488.

Theorem 2. *Let $\epsilon > 0$, m a non-negative integer, and*

$$R = \epsilon D + M.$$

Then

$$\mathbf{P}\Big\{\max_{1\le k\le n} |T_k| \ge mR\Big\} \le \epsilon^{-2m}.$$

Theorem 3. *Let $R > 0$. Then*

$$\mathbf{P}\Big\{\max_{1\le k\le n} |S_k| \le R\Big\} \le 4(M+R)^2/D^2.$$

Theorem 4.

$$\mathbf{P}\{T > 0\} \ge \frac{1}{16}\frac{D-M}{D+M}.$$

Theorem 5.

$$\mathbf{P}\{T > -3M\} \ge \tfrac{1}{48}.$$

Theorem 6.

$$\mathbf{P}\{|S| > R\} \ge \tfrac{1}{48}[1 - 4(4M+R)^2/D^2].$$

Proof of Theorem 1. Let

$$A_k = \{|T_i| < R,\ i < k,\ |T_k| \ge R\}$$

and $A = \sum_{k=1}^n A_k$. Then

$$\mathbf{P}\{\max_k |T_k| \ge R\} = P(A) = \sum_{k=1}^n P(A_k).$$

Since the variables ζ_i with $i > k$ do not occur in the definition of A_k, the hypothesis that $\xi_1, \ldots, \xi_n$ are independent implies

$$\mathbf{E}(T^2; A_k) = \mathbf{E}(T_k^2; A_k) + \sum_{i=k-1}^n \mathbf{E}(\zeta_i^2; A_k) \ge \mathbf{E}(T_k^2; A_k) \ge R^2 P(A_k),$$

hence $\mathbf{E}T^2 \ge R^2 P(A)$, which proves the desired inequality.

Remark. This inequality is an identity, for example, in the following case: $n = 1$ and

$$\mathbf{P}\{\zeta_1 = R\} = \frac{D^2}{2R^2}, \quad \mathbf{P}\{\zeta_1 = -R\} = \frac{D^2}{2R^2}, \quad \mathbf{P}\{\zeta_1 = 0\} = 1 - \frac{D^2}{R^2}.$$

Proof of Theorem 2. We set

$$A_{ik} = \{|T_j| < iR,\ j < k;\ |T_k| \geq iR\}$$

and

$$A_i = \sum_{k=1}^{n} A_{ik}.$$

On the set A_{ik} we have

$$|T_k| \leq iR + M.$$

Clearly

$$P(A_{i+1}|A_{ik}) = \mathbf{P}\{u \geq (i+1)R\} \leq \mathbf{P}\Big\{\max_{k+1 \leq p \leq n} \Big|\sum_{j=k+1}^{p} \zeta_j\Big| \geq \epsilon D\Big\}.$$

As in the proof of Theorem 1, we note that the variables ζ_j for $j > k$ do not occur in the definition of A_{ik}. Therefore,

$$P(A_{i+1}|A_{ik}) \leq D^2/\epsilon^2 D^2 = 1/\epsilon^2$$

and consequently

$$\begin{aligned} P(A_{i+1}|A_i) \leq 1/\epsilon^2, \quad \mathbf{P}\{u \geq mR\} = P(A_m) = \\ = P(A_{m-1}|A_m)P(A_{m-1}|A_{m-2}) \ldots P(A_1) \leq 1/\epsilon^{2m}, \end{aligned}$$

which proves Theorem 2.

Proof of Theorem 3. Let

$$\begin{aligned} B_k &= \{|S_i| \leq R,\ i \leq k\}, \\ B &= B_u = \{v \leq R\}, \\ C_k &= B_{k-1} \backslash B_k. \end{aligned}$$

Clearly

$$P(B) + \sum_{k=1}^{n} P(C_k) = 1. \tag{1}$$

Put

$$a_k = \mathbf{E}(T_k | B_k).$$

On the set B_k we have

$$|T_k - a_k| = |S_k - \mathbf{E}(S_k | B_k)| \leq 2R,$$
$$|a_k - a_{k-1}| = |\mathbf{E}(\zeta_k | B_k)| \leq M.$$

Now consider the expression

$$\begin{aligned}\mathbf{E}[(T_k - a_k)^2;\ B_{k-1}] &= \\ = \mathbf{E}[(T_k - a_k)^2;\ B_k] &+ \mathbf{E}[(T_{k-1} - a_{k-1} - (a_k - a_{k-1}) + \\ &+ \zeta_k)^2;\ C_k] \leq \mathbf{E}[(T_k - a_k)^2;\ B_k] + P(C_k) \cdot 4(R+M)^2. \quad (2)\end{aligned}$$

On the other hand

$$\begin{aligned}&\mathbf{E}[(T_k - a_k)^2;\ B_{k-1}] = \\ &= \mathbf{E}[((T_{k-1} - a_{k-1}) - (a_k - a_{k-1}) + \zeta_k)^2;\ B_{k-1}] = \\ &= \mathbf{E}[(T_{k-1} - a_{k-1})^2 + (a_k - a_{k-1})^2 + \zeta_k^2;\ B_{k-1}] \geq \\ &\geq \mathbf{E}[(T_{k-1} - a_{k-1})^2;\ B_{k-1}] + P(B)\mathbf{E}\zeta_k^2. \qquad (3)\end{aligned}$$

Comparing (2) and (3) we obtain

$$\begin{aligned}&\mathbf{E}[(T_{k-1} - a_{k-1})^2;\ B_{k-1}] + P(B)\mathbf{E}\zeta_k^2 \leq \\ &\leq \mathbf{E}[(T_k - a_k)^2;\ B_k] + P(C_k) \cdot 4(R+M)^2.\end{aligned}$$

Setting $k = 0, 1, \ldots, n$ in this inequality, summing the corresponding inequalities and taking into account (1), we obtain

$$\begin{aligned}P(B)D^2 &\leq \mathbf{E}[(T_n - a_n)^2;\ B_n] + \sum_{k=1}^{n} P(C_k) \cdot 4(R+M)^2 \leq \\ &\leq P(B) \cdot 4R^2 + \sum_{k=1}^{n} P(C_k) \cdot 4(R+M)^2 \leq 4(R+M)^2,\end{aligned}$$

which proves Theorem 3.

Remark. If we take z_k instead of ζ_k, the statement of the theorem will be valid with v replaced by u.

Proof of Theorem 4. We set

$$C_m = \{mR < T \leq (m+1)R\},$$

where $R = 8D + M$ and set

$$B_m = \sum_{k=m}^{\infty} C_k.$$

Clearly

$$B_0 = \{T > 0\}.$$

By Theorem 2,

$$P(B_m) \leq 1/8^{2m}$$

$$\mathbf{E}(T;\ B_0) = \sum_{m=0}^{\infty} \mathbf{E}(T;\ C_m) \leq \sum_{m=0}^{\infty} (m+1)RP(C_m) =$$
$$= R\sum_{m=0}^{\infty} P(B_m) \leq R\Big[P(B_0) + \sum_{m=1}^{\infty} \frac{1}{8^{2m}}\Big] \leq$$
$$\leq R\Big[P(B_0) + \frac{1}{32}\Big].$$

On the other hand, it is easy to prove that

$$\mathbf{E}(T;\ B_0) \geq \tfrac{1}{2}D.$$

Thus

$$\tfrac{1}{2}D \leq R\Big[P(B_0) + \frac{1}{32}\Big] = (8D+M)\Big[P(B_0) + \frac{1}{32}\Big];$$
$$P(B_0) \leq \frac{D}{2(8D+M)} - \frac{1}{32} \leq \frac{1}{16}\frac{D-M}{D+M},$$

as required.

Proof of Theorem 5. Let

$$D \geq \tfrac{3}{2}M.$$

Then by Theorem 4

$$\mathbf{P}\{T > 0\} \geq \frac{1}{16}\frac{D-M}{D+M} \geq \frac{1}{48}.$$

If

$$D \geq \tfrac{3}{2}M$$

then by Theorem 1,

$$\mathbf{P}\{|T| \leq 3M\} \leq D^2/9M^2 < 1/4.$$

In both cases these estimates prove the desired statement.

Proof of Theorem 6. We set

$$W = \max_k |S - \sum_{i=k}^{n} \zeta_i|.$$

Then from Theorem 3,

$$\mathbf{P}\{W > R + 3M\} \geq 1 - 4(4M + R)^2/D^2$$

and applying Theorem 5 to $\sum_{i=k}^{n} \zeta_i$ we find that

$$\mathbf{P}\{s > R | W > R + 3M\} \geq 1/48.$$

This immediately implies the desired statement.

§3. Convergence of series

Below necessary and sufficient conditions are considered for the convergence of series in independent random variables, previously established by other methods in our joint paper with A.Ya. Khinchin (see paper 1 of this volume).

Consider two sequences of random variables

$$\eta = (\eta_1, \eta_2, \ldots, \eta_n, \ldots), \quad \bar{\eta} = (\bar{\eta}_1, \bar{\eta}_2, \ldots, \bar{\eta}_n, \ldots).$$

The sequences η and $\bar{\eta}$ are called *equivalent* if [1]

$$\sum_{i=1}^{\infty} \mathbf{P}\{\eta_i \neq \bar{\eta}_i\} < \infty. \tag{4}$$

In what follows we assume that each of the sequences η and $\bar{\eta}$ consists of independent random variables.

[1] This definition is due to A.Ya. Khinchin.

Theorem 7. 1) *The probability P of convergence of*

$$\sum_{i=1}^{\infty} \eta_i \tag{5}$$

is 1 if there is a sequence $\bar{\eta} = (\bar{\eta}_n)$ equivalent to $\eta = (\eta_n)$ such that the series

$$\sum_{i=1}^{\infty} \mathbf{E}\bar{\eta}_i \tag{6}$$

and

$$\sum \mathbf{E}\bar{\zeta}_i^2 \tag{7}$$

(where $\bar{\zeta}_i = \bar{\eta}_i - \mathbf{E}\bar{\eta}_i$) converge.

2) *P is zero if there is no equivalent sequence for which the series (6) and (7) converge simultaneously.*

Proof. 1) According to Theorem 1, for every positive y,

$$\mathbf{P}\Big\{\max_{n \le p \le N}\Big|\sum_{k=n}^{p} \bar{\zeta}_k\Big| \ge y\Big\} \le \frac{1}{y^2}\sum_{k=n}^{N} \mathbf{E}\bar{\zeta}_k^2 \le \frac{1}{y^2}\sum_{k=n}^{\infty} \mathbf{E}\bar{\zeta}_k^2. \tag{8}$$

Since the series (7) converges, for every $\epsilon > 0$ we can find some (possibly large) m such that for $n \ge m$ the right-hand side of (8) is smaller than ϵ.

By definition,

$$P = \lim_{y \to 0}\lim_{n \to \infty}\lim_{N \to \infty} \mathbf{P}\Big\{\max_{n \le p \le N}\Big|\sum_{k=n}^{p} \eta_k\Big| < y\Big\}.$$

Therefore (8) implies that the probability of convergence of

$$\sum_{n=1}^{\infty} \bar{\zeta}_n \tag{9}$$

is 1. Since the series (6) converges, the same is true for the series

$$\sum_{n=1}^{\infty} \bar{\eta}_n. \tag{10}$$

Since in the final analysis the probabilities of convergence of equivalent series coincide, $P = 1$.

2) We set

$$\bar{\eta}_n = \begin{cases} \eta_n, & |\eta_n| \le 1, \\ 0, & |\eta_n| > 1. \end{cases} \tag{11}$$

First we assume that $\eta = (\eta_n)$ and $\bar{\eta} = (\bar{\eta}_n)$ are equivalent , that is,

$$\sum_{n=1}^{\infty} \mathbf{P}\{|\eta_n| > 1\} < \infty. \tag{12}$$

In this case one of the series (6) or (7) should diverge. If the series (7) diverges, then by Theorem 3 we see that for any y and each n,

$$\mathbf{P}\Big\{\max_{1\le p\le N}\Big|\sum_{k=n}^{\infty} \bar{\eta}_k\Big| \le y\Big\} \le 4(y+2)^2 \Big/ \sum_{k=n}^{N} \mathbf{E}\zeta_k^2 \to 0$$

as $N \to \infty$. This means that the probability of convergence of (10) is 0. By the equivalence of $\eta = (\eta_n)$ and $\bar{\eta} = (\bar{\eta}_n)$ this implies that $P = 0$.

If (7) converges, then the probability of convergence of (9) is 1, according to the first part of the theorem. Therefore, if (6) diverges, then $P = 0$.

Now consider the case when $\sum_{n=1}^{\infty} \mathbf{P}\{|\eta_n| > 1\} = \infty$. In this case we immediately have

$$\mathbf{P}\Big\{\max_{n\le k\le N} |y_k| \le 1\Big\} = \prod_{k=n}^{N} [1 - \mathbf{P}\{|\eta_k| > 1\}] \to 0$$

for every n as $N \to \infty$, hence $P = 0$.

Thus, our theorem is proved for all possible cases.

Remark. If it is known that there exists a certain sequence $\bar{\eta} = (\bar{\eta}_n)$ satisfying the conditions of the theorem, then the sequence determined by (11) also satisfies the conditions of this theorem.

§4. A generalized law of large numbers

Consider the system of random variables

$$\eta_{11}, \eta_{12}, \dots, \eta_{1m_1},$$

$$\eta_{21}, \eta_{22}, \dots, \eta_{2m_2},$$

$$\dots\dots\dots$$

$$\eta_{n1}, \eta_{n2}, \dots, \eta_{nm_n},$$

denoted by $\|\eta_{nk}\|$ for short. Suppose that in every row the variables are independent of one another, whereas in different rows they may be dependent.

We say that the means

$$\sigma_n = (\eta_{n1} + \ldots + \eta_{nm_n})/m_n, \ n \geq 1,$$

are *stable* if there exists a sequence of numbers $d_1, d_2, \ldots$, such that for any $\epsilon > 0$,

$$\mathbf{P}\{|\sigma_n - d_n| > \epsilon\} \to 0, \ n \to \infty.$$

We will give a necessary and sufficient condition for the stability of the means. We say that two systems $\|\eta_{nk}\|$ and $\|\bar{\eta}_{nk}\|$ are equivalent if

$$m_n = \overline{m}_n$$

and

$$\mathbf{P}\{\sigma_n \neq \bar{\sigma}_n\} \to 0, \ n \to \infty.$$

Clearly for equivalent systems the means are simultaneously either stable or unstable.

Theorem 8. *A necessary and sufficient condition for the stability of the means* σ_n, $n \geq 1$, *is the existence of some system* $\|\bar{\eta}_{nk}\|$ *equivalent to* $\|\eta_{nk}\|$, *for which*

$$\frac{1}{m_k^2} \sum_{k=1}^{m_n} \mathbf{E}\bar{\zeta}_{nk}^2 \to 0, \ n \to \infty. \tag{13}$$

Proof of the sufficiency. Theorem 1 implies that for any $\epsilon > 0$,

$$\mathbf{P}\{|\bar{\sigma}_n - \mathbf{E}\bar{\sigma}_n| \geq \epsilon\} = \mathbf{P}\left\{\left|\sum_{k=1}^{m_n} \bar{\zeta}_{nk}\right| \geq m_n \epsilon\right\} \leq \frac{1}{m_n^2 \epsilon^2} \sum_{k=1}^{m_n} \mathbf{E}\zeta_{nk}^2.$$

Therefore, (13) implies that

$$\mathbf{P}\{|\bar{\sigma}_n - \mathbf{E}\bar{\sigma}_n| \geq \epsilon\} \to 0,$$

and, from the equivalence of $\|\eta_{nk}\|$ and $\|\bar{\eta}_{nk}\|$ we obtain

$$\mathbf{P}\{|\bar{\sigma}_n - \mathbf{E}\bar{\sigma}_n| \geq \epsilon\} \to 0, \tag{14}$$

which proves the sufficiency.

Proof of the necessity. Suppose that the means σ_n, $n \geq 1$ are stable. For any random variable η_{nk} there exists a constant f_{nk} such that

$$\mathbf{P}\{\eta_{nk} > f_{nk}\} \leq \tfrac{1}{2}, \quad \mathbf{P}\{\eta_{nk} < f_{nk}\} \leq \tfrac{1}{2}. \tag{15}$$

It is easy to prove that for any positive $\epsilon > 0$,

$$\sum_{k=1}^{m_n} \mathbf{P}\{|\eta_{nk} - f_{nk}| \geq m_n \epsilon\} \to 0, \tag{16}$$

as $n \to \infty$.

Now set

$$\bar{\eta}_{nk} = \begin{cases} \eta_{nk}, & \text{if } |\eta_{nk} - f_{nk}| \leq m_n, \\ f_{nk}, & \text{if } |\eta_{nk} - f_{nk}| > m_n; \end{cases} \tag{17}$$

$$\bar{\eta}_{nk}(\epsilon) = \begin{cases} \eta_{nk}, & \text{if } |\eta_{nk} - f_{nk}| \leq m_n \epsilon, \\ f_{nk}, & \text{if } |\eta_{nk} - f_{nk}| > m_n \epsilon. \end{cases}$$

By (16) the systems $||\eta_{nk}||$, $||\bar{\eta}_{nk}||$ and $||\bar{\eta}_{nk}(\epsilon)||$ are equivalent, and the means $\bar{\sigma}_n$ and $\bar{\sigma}_n(\epsilon)$ are stable. Note that

$$|\bar{\zeta}_{nk}(\epsilon)| \leq 2\epsilon m_n.$$

Therefore by Theorem 6,

$$\mathbf{P}\{|\bar{\sigma}_n(\epsilon) - d_n| > \epsilon\} = \mathbf{P}\left\{\left|\sum_{k=1}^{m_n} \bar{\eta}_{nk}(\epsilon) - m_n d_n\right| > \epsilon m_n\right\} \geq$$
$$\geq \frac{1}{48}\left[1 - \frac{324 m_n^2 \epsilon^2}{\sum_{k=1}^{m_n} \mathbf{E}(\bar{\zeta}_{nk}^2(\epsilon))}\right].$$

For any $\epsilon > 0$ the left-hand side of this inequality converges to zero. Hence,

$$\limsup_n \frac{1}{m_n^2} \sum_{k=1}^{m_n} \mathbf{E}(\bar{\zeta}_{nk}^2(\epsilon)) \leq 324\epsilon^2. \tag{18}$$

For $\epsilon \leq 1$,

$$|\mathbf{E}\bar{\zeta}_{nk}^2 - \mathbf{E}\bar{\zeta}_{nk}^2(\epsilon)| \leq 8 m_n^2 \mathbf{P}\{|\eta_{nk} - f_{nk}| > \epsilon m_n\}$$

and, by (16),

$$\frac{1}{m_n^2} \sum_{k=1}^{m_n} |\mathbf{E}\bar{\zeta}_{nk}^2 - \mathbf{E}\bar{\zeta}_{nk}^2(\epsilon)| \to 0.$$

Therefore (13) follows from (18).

Remark. If there exists a system $\|\bar{\eta}_{nk}\|$ satisfying the condition of the theorem, then the system determined by (17) and (15) is such a system.

We say that the means σ_n, $n \geq 1$, have the property of *normal stability* if for any $\epsilon > 0$,

$$\mathbf{P}\{|\sigma_n - \mathbf{E}\sigma_n| \geq \epsilon\} \to 0, \ n \to \infty.$$

Theorem 9. *A necessary and sufficient condition for normal stability of the means σ_n, $n \geq 1$, is the existence of an equivalent system $\|\bar{\eta}_{nk}\|$ for which* (13) *holds and*

$$\mathbf{E}\sigma_n - \mathbf{E}\bar{\sigma}_n \to 0, \ n \to \infty.$$

The proof of the theorem follows directly from (14) which was established for any stable system.

Remark. For the case of normal stability, instead of $\|\bar{\eta}_{nk}\|$ determined by (17), (15), one can take a system

$$\bar{\eta}_{nk} = \begin{cases} \eta_{nk}, & \text{if } |\zeta_{nk}| \leq m_n, \\ \mathbf{E}\eta_{nk}, & \text{if } |\zeta_{nk}| > m_n. \end{cases}$$

§5. The law of large numbers

The condition of stability of the means

$$\sigma_n = (\eta_1 + \ldots + \eta_n)/n,$$

where $\eta_1, \eta_2, \ldots,$ are independent random variables, is a special case of Theorems 8 and 9 if we set $m_n = n$, $\eta_{nk} = \eta_k$.

Normal stability of such means is usually called the *law of large numbers.* In this case we have

Theorem 10. *In order that for any $\epsilon > 0$,*

$$\mathbf{P}\{|\sigma_n - \mathbf{E}\sigma_n| > \epsilon\} \to 0, \ n \to \infty, \tag{19}$$

it is necessary and sufficient that there exist a system $\|\bar{\eta}_{nk}\|$ such that

0. $\overline{m}_n = n.$

1. $\sum_{k=1}^{n} \mathbf{P}\{\bar{\eta}_{nk} \neq \eta_{nk}\} \to 0.$

2. $\frac{1}{n} \sum_{k=1}^{n} [\mathbf{E}\bar{\eta}_{nk} - \mathbf{E}\eta_{nk}] \to 0.$

3. $\frac{1}{n^2} \sum_{k=1}^{n} \mathbf{E}\bar{\zeta}_{nk}^2 \to 0.$

Remark. If there is a system $\|\bar{\eta}_{nk}\|$ satisfying the above conditions, then as such a system we can take the system

$$\bar{\eta}_{nk} = \begin{cases} \eta_k, & \text{if } |\zeta_k| \leq n, \\ \mathbf{E}\eta_k, & \text{if } |\zeta_k| > n. \end{cases}$$

Theorem 11. *In order that*

$$\mathbf{P}\{|\sigma_n - \mathbf{E}\sigma_n| > \epsilon\} \to 0, \ n \to \infty,$$

it is necessary and sufficient that

1. $\sum_{k=1}^{n} \mathbf{P}\{|\zeta_k| > n\} \to 0.$

2. $\frac{1}{n} \sum_{k=1}^{n} \mathbf{E}(\zeta_k; \ |\zeta_k| \leq n) \to 0.$

3. $\frac{1}{n^2} \sum_{k=1}^{n} \mathbf{E}(\epsilon_k^2; \ |\zeta_k| \leq n) \to 0.$

Moscow, 24 December 1927

REMARKS

The purpose of these remarks is, on the one hand, to refine and improve certain results (§§1, 2) and on the other hand, to study an important special case of the law of large numbers (§3).

§1

V.V. Nemytskii pointed out to me that the inequality

$$\mathbf{E}(T; B_0) \geq \tfrac{1}{2}D,$$

used in the proof of Theorem 4 is false. Indeed, it is easy to show that

$$\mathbf{E}(T; B_0) = \tfrac{1}{2}\mathbf{E}|T| \leq \tfrac{1}{2}\sqrt{\mathbf{E}T^2} = \tfrac{1}{2}D,$$

that is, the reverse inequality holds.

In this connection Theorems 4, 5 and 6 of this paper must be replaced by the following.

Theorem 4*. *Let $M \leq D$. Then*

$$\mathbf{P}\left\{|S| \geq \frac{D}{2}\right\} \geq \frac{1}{1600}.$$

Theorem 5*.

$$\mathbf{P}\{|S| \geq R\} \geq \frac{1}{1600}\left[1 - \frac{M^2 + 4R^2}{D^2}\right].$$

Theorems 7–11 remain unchanged.

Theorem 6 used only for proving Theorem 8 should be replaced by Theorem 6* (below).

Proof of Theorem 4*. Let

$$E = \{|S| \geq D/2\}$$

and let $\overline{E}$ be the complementary event. If $|MS| \geq 2D$, then

$$D^2 \geq \mathbf{E}(T^2; \overline{E}) \geq \mathbf{E}((2D - D/2)^2;\ \overline{E}) \geq \tfrac{9}{4}P(\overline{E}) \cdot D^2, \quad P(\overline{E}) \leq 1/2,$$

as required.

If $|MS| < 2D$, then we introduce the event

$$E_m = \{3mD \leq |T| < 3(m+1)D,\ |S| \geq D/2\}.$$

By Theorem 2 we have

$$P(E_m) \leq \frac{1}{2^{2m}}, \quad \mathbf{E}(S^2; E_m) \leq \frac{(m+1)^2}{2^{2m}} \cdot 25D^2.$$

On the set

$$E' = \sum_{m=0}^{5} E_m$$

we have

$$|T| \leq 18D, \ |S| \leq 20D.$$

Hence

$$\mathbf{E}(S^2; E') \leq 400D^2 P(E).$$

Since

$$\mathbf{E}(S^2; \overline{E}) \leq \tfrac{1}{4}D^2,$$

it follows that

$$\begin{aligned} D^2 \leq \mathbf{E}S^2 \leq \mathbf{E}\Big(S^2; \overline{E} + E' + \sum_{m=6}^{\infty} E_m\Big) \leq \\ \leq \tfrac{1}{4}D^2 + 400D^2P(E) + \sum_{m=6}^{\infty} \frac{(m+1)^2}{2^{2m}} 25D^2 \leq \\ \leq \tfrac{3}{4}D^2 + 400D^2P(E), \end{aligned}$$

whence

$$P(E) \geq \frac{1}{1600},$$

which proves Theorem 4*.

Proof of Theorem 5*. If $M > D$ or $R > \frac{1}{2}D$, then the required inequality holds, since then its right-hand side is negative.

If, however, $M \leq D$ and $R \leq \frac{1}{2}D$, then the required inequality follows from Theorem 4*.

§2

In proving Theorem 8 I used the formula

$$\sum_{k=6}^{m_n} \mathbf{P}\{|\eta_{nk} - f_{nk}| \geq m_n \cdot n\} \to 0. \tag{16}$$

For a rigorous proof of this formula we need the following

Theorem 6*. *Let $E_1, \ldots, E_n$ be a sequence of independent events and let U be an event such that*

$$P(U|E_k) \geq u, \quad P(E_1 + \ldots + E_n) \geq u.$$

Then

$$P(U) \geq \tfrac{1}{9}u^2.$$

Proof. If there exists k such that

$$P(E_k) \geq \tfrac{1}{3}u,$$

then

$$P(U) \geq P(E_k) \cdot P(U|E_k) \geq \tfrac{1}{3}u^2,$$

as required.

Now for all k let

$$P(E_k) < \tfrac{1}{3}u.$$

Let us show that in this case there exists a k such that

$$\tfrac{1}{3}u \leq P(E_1 + \ldots + E_k) \leq \tfrac{2}{3}u.$$

For this we put

$$F_i = E_1 + \ldots + E_{i-1}$$

and let $\overline{F}_i$ be the complementary event. Since F_i and E_i are independent, for $i \leq k$ we have

$$P(F_i|E_i) = P(F_i) \leq P(F_k) \leq \tfrac{2}{3}u,$$

$$P(U\overline{F}_i|E_i) \geq \tfrac{1}{3}u,$$

$$P(U\overline{F}_iE_i) \geq \tfrac{1}{3}uP(E_i),$$

$$P(U) \geq \sum_{i=1}^{k} P(U\overline{F}_iE_i) \geq \tfrac{1}{3}u\sum_{i=1}^{k} P(E_i) \geq \tfrac{1}{3}uP(F_k) \geq \tfrac{1}{9}u^2,$$

as required.

Proof of Formula (16). We set

$$U = \{|\sigma_n - d_n| \geq \epsilon/2\},$$

$$E_k = \{|\eta_{nk} - f_{nk}| \geq \epsilon m_n\},$$

$$F_k = \left\{\left|\left(\sum_{i\neq k} \eta_{ni} + f_{nk}\right) / m_n - d_n\right| \geq \frac{\epsilon}{2}\right\}.$$

Clearly

$$P(U|F_k) \geq \tfrac{1}{2}.$$

For sufficiently large n, in view of the stability of the means we have

$$P(U) \leq \tfrac{1}{4},$$

hence

$$P(F_k) \leq \tfrac{1}{2}.$$

If the event E_k takes place and F_k does not, then U takes place. Therefore

$$P(F_k|E_k) + P(U|E_k) \geq 1.$$

But the events E_k and F_k are independent, so for sufficiently large n,

$$P(F_k|E_k) = P(F_k) \leq \tfrac{1}{2},$$

$$P(U|E_k) \geq \tfrac{1}{2}.$$

By Theorem 6*

$$P(U) \geq \frac{1}{36} P^2(E_1 + \ldots + E_{m_n})$$

and, since $P(U) \to 0$ as $n \to \infty$,

$$P(E_1 + \ldots + E_{m_n}) \to 0,$$

$$\sum_{k=1}^{m_n} P(E_k) \to 0,$$

as desired.

§3

Here we study a special case of the law of large numbers, namely the case when the independent random variables $\eta_1, \eta_2, \ldots$ have the same distribution $\mathbf{P}\{\eta_1 < x\} = F(x)$. Then we have

Theorem 12. *The means $\sigma_n = (\eta_1 + \ldots + \eta_n)/n$, $n \geq 1$, are stable if and only if*

$$n\mathbf{P}\{|\eta_1| > n\} \to 0, \ n \to \infty.$$

Proof. Set

$$\bar{\eta}_{nk} = \begin{cases} \eta_n, & \text{if } |\eta_n| \leq K, \\ 0, & \text{if } |\eta_n| > K. \end{cases}$$

It is easy to see that if $n\mathbf{P}\{|\eta_1| > n\} \to 0$ as $n \to \infty$, then the systems $(\bar{\eta}_{nk})$ and (η_n) are equivalent and

$$\frac{1}{n^2}\sum_{k=1}^{n}\mathbf{E}\bar{\zeta}_{nk}^2 \leq \frac{1}{n^2}\sum_{k=1}^{n}\mathbf{E}\bar{\eta}_{nk}^2 \to 0, \ n \to \infty.$$

Hence by Theorem 8, the means σ_n, $n \geq 1$, are stable. Conversely, if the means σ_n, $n \geq 1$, are stable, then as has been shown, formula (16) holds. In the present case this formula takes the form

$$m\mathbf{P}\{|\eta_1 - f| \geq \epsilon n\} \to 0,$$

where f is a constant. This formula immediately implies the required condition.

If the random variables η_n have finite expectation,

$$\int_{-\infty}^{\infty} |x| dF(x) < \infty,$$

then the condition of Theorem 12 holds, since then

$$n\mathbf{P}\{|\eta_1| > n\} \leq \Big(\int_{-\infty}^{-n} + \int_{n}^{+\infty}\Big)|x| dF \to 0.$$

It can be shown that in this case we have normal stability. Thus the following theorem holds:

Theorem 13. *The stability of the means $\sigma_n = (\eta_1 + \ldots + \eta_n)/n$, $n \geq 1$, of a sequence of independent identically distributed random variables with singular distribution is normal if*[2] *and only if*

$$\mathbf{E}|\eta_1| < \infty.$$

The latter statement was earlier proved by A.Ya. Khinchin (*C.R. Acad. Sci. Paris* **188** (1929), 477).

8 February 1929

[2] The 'only if' part is trivial and follows from the fact that the definition of normal stability makes sense only if the expectation exists.

5. ON THE LAW OF THE ITERATED LOGARITHM *

We consider a sequence of independent random variables

$$z_1, z_2, \dots, z_n, \dots, \tag{1}$$

with zero expectations $\mathbf{E}z_n$. Further let

$$b_n = \mathbf{E}z_n^2, \quad B_n = \sum_{k=1}^{n} b_k, \quad S_n = \sum_{k=1}^{n} z_k.$$

Following A.Ya. Khinchin we say [1] that the sequence (1) obeys the *law of iterated logarithm* if the probability that

$$\limsup \frac{S_n}{\sqrt{2B_n \ln\ln B_n}} = 1 \tag{2}$$

is 1. In one important special case this law was established by Khinchin himself; [2] we will prove that it is applicable under the following conditions:

I. $B_n \to \infty$,

II. $|z_n| \leq m_n = o\left(\sqrt{\dfrac{B_n}{\ln\ln B_n}}\right)$.

If the case $z_n \equiv 0$ is excluded, the necessity of the first condition can be easily derived. *If the absolute values of all the z_n are uniformly bounded, then the second condition follows from the first; in this case the first condition is necessary and sufficient.*

Without using the probabilities of relations that cannot be observed directly, for example, the event (2), the law of the iterated logarithm can be formulated as follows:

1°. *For arbitrary positive numbers η and δ there exists a positive integer n such that the probability of at least one of the inequalities*

$$S_k > \sqrt{2B_k \ln\ln B_k}(1+\delta) \quad (k = n, n+1, \dots, n+p)$$

for any p remains smaller than η.

---

* 'Über das Gesetz des iterierten Logarithmus', *Math. Ann.* **101** (1929), 126–135.

[1] See his book *Basic laws of probability theory*, Moscow, 1927 (in Russian).

[2] *Math. Ann.* **99** (1928), 152.

2°. *For arbitrary η, δ and m there exists a non-negative integer p such that the probability that all the inequalities*

$$S_k < \sqrt{2B_k \ln\ln B_k}(1-\delta) \ (k = m, m+1, \dots, m+p)$$

hold simultaneously is smaller than η.

§1. Lemmas

Let

$$M_n = \max_{1\le k\le n} m_k, \quad U_n = \max_{1\le k\le n} S_k,$$

$$W_n(x) = \mathbf{P}\{S_n > x\}, \quad \overline{W}_n(x) = \mathbf{P}\{U_n > x\}.$$

In the following lemmas x denotes a *positive* number. Since in this section n is taken to be fixed, we shall omit it in our notation.

Lemma I. *If $xM \le B$, then*

$$W(x) < e^{-(x^2/2B)(1-\theta)}, \quad \theta = xM/2B.$$

Lemma II. *If $xM \ge B$, then*[3]

$$W(x) < e^{-x/4M}.$$

Proof of Lemmas I *and* II. Let $a > 0, aM \le 1$. Then

$$\begin{aligned} \mathbf{E}e^{az_k} &= \mathbf{E}1 + \mathbf{E}(az_k) + \mathbf{E}\Big(\frac{a^2z_k^2}{2}\Big) + \mathbf{E}\Big(\frac{a^3z_k^3}{6}\Big) + \dots \le \\ &\le 1 + \frac{a^2b_k}{2}\Big(1 + \frac{aM}{3} + \frac{a^2M^2}{12} + \dots\Big) \le \\ &\le 1 + \frac{a^2b_k}{2}\Big(1 + \frac{aM}{2}\Big) < \exp\Big[\frac{a^2b_k}{2}\Big(1+\frac{aM}{2}\Big)\Big], \end{aligned} \tag{3}$$

$$\mathbf{E}e^{aS} = \prod_{k=1}^{n} \mathbf{E}e^{az_k} < \exp\Big[\frac{a^2B}{2}\Big(1+\frac{aM}{2}\Big)\Big], \tag{4}$$

[3] See: S. Bernshtein, 'On a variation of Chebyshev's inequality and the error of Laplace's formula', *Uchen. Zap. Nauch.-Issl. Kafedr. Ukrain. Otd. Mat.* **1** (1924), 38–48 (in Russian). (See also: *Collected works*, Vol. 4, pp. 71–79.)

$$W(x)e^{ax} < \mathbf{E}e^{aS}. \tag{5}$$

Formulas (4) and (5) imply that

$$W(x) < \exp\Big[-ax + \frac{a^2B}{2}\Big(1 + \frac{aM}{2}\Big)\Big].$$

For $xM \le B$ we set $a = x/B$ in the following formula and obtain Lemma I; for $xM > B$ we set $a = 1/M$ and obtain

$$W(x) < \exp\Big[-\frac{x}{M} + \frac{B}{2M^2}(1 + \tfrac{1}{2})\Big] \le \exp\Big[-\frac{x}{4M}\Big].$$

In both cases the condition $aM \le 1$ is satisfied.

Lemma I immediately implies

Lemma III. *If $xM \le B$, then*

$$W(x) < e^{-x^2/4B}.$$

Lemma IV. *If*

$$xM/B = \omega < 1/256, \tag{6}$$

$$x^2/B = \lambda > 512, \tag{7}$$

then

$$W(x) > e^{-(x^2/2B)(1+\epsilon)},$$

where ϵ denotes the maximum of $32\sqrt{(\ln\lambda)/\lambda}$ and $64\sqrt{\omega}$.

Proof. Let $\delta = \frac{1}{8}\epsilon$, then

$$\delta^2 = \max\Big(64\,\omega, 16\frac{\ln\lambda}{\lambda}\Big), \tag{8}$$

and also, because of (6) and (7),

$$\delta^2 < \tfrac{1}{4}, \tag{9}$$

$$\delta < \tfrac{1}{2}, \tag{10}$$

$$\delta > 2\delta^2. \tag{11}$$

We further set

$$a = x/(B(1-\delta)).$$

Clearly,

$$x = aB(1-\delta), \tag{12}$$

$$x/B < a < 2x/B, \tag{13}$$

$$aM < 2\omega < 1/128, \tag{14}$$

$$a^2B > \lambda > 512. \tag{15}$$

Since for every positive u we have

$$1 + u > e^{u(1-u)},$$

it follows from (3) that

$$\mathbf{E}e^{az_k} \geq 1 + \frac{a^2b_k}{2}\Big(1 - \frac{aM}{3} - \frac{a^2M}{12} - \dots\Big) > 1 + \frac{a^2b_k}{2}\Big(1 - \frac{aM}{2}\Big) >$$
$$> \exp\Big[\frac{a^2b_k}{2}\Big(1 - \frac{aM}{2} - \frac{a^2b_k}{2}\Big)\Big] \geq \exp\Big[-\frac{a^2b_k}{2}(1 - aM)\Big]. \tag{16}$$

The condition $aM \leq 1$ assumed here is satisfied in our case in view of (4). Further, by (4) we have

$$\mathbf{E}e^{aS} > e^{(a^2B/2)(1-aM)}. \tag{17}$$

Because of (14) and (8),

$$aM < \delta^2/4,$$

so that we finally obtain

$$\mathbf{E}e^{aS} > \exp\Big[\frac{a^2B}{2}\Big(1 - \frac{\delta^2}{4}\Big)\Big]. \tag{18}$$

On the other hand we have

$$\mathbf{E}e^{aS} = -\int_{-\infty}^{\infty} e^{ay}dW(y) = a\int_{-\infty}^{\infty} e^{ay}W(y)dy =$$
$$= a\Big(\int_{-\infty}^{0} + \int_{0}^{aB(1-\delta)} + \int_{aB(1-\delta)}^{aB(1+\delta)} + \int_{aB(1+\delta)}^{8aB} + \int_{8aB}^{\infty}\Big) = \tag{19}$$
$$= a(J_1 + J_2 + J_3 + J_4 + J_5).$$

Since $W(y) \leq 1$, it follows that

$$aJ_1 \leq a \int_{-\infty}^{0} e^{ay}\,dy \leq 1. \tag{20}$$

By Lemma II and (14), for all $y \geq B/M$ we have

$$W(y) < e^{-y/4M} < e^{-2ay}.$$

By Lemma III we also have for $8aB \leq y \leq B/M$

$$W(y) < e^{-y^2/4B} \leq e^{-2ay}.$$

Therefore we obtain

$$aJ_5 < a \int_{8aB}^{\infty} e^{-ay}\,dy < 1. \tag{21}$$

Since by (18), (9) and (15),

$$\mathbf{E}e^{aS} > 8,$$

equations (20) and (21) imply

$$aJ_1 + aJ_5 < \tfrac{1}{4}\mathbf{E}e^{aS}. \tag{22}$$

To estimate aJ_2 and aJ_4 we apply Lemma I. Because of (6) and (8), for $y \leq 8aB$ we have the inequalities

$$\theta < \omega < \frac{1}{8}\delta^2,$$

$$W(y) < \exp\Big[-\frac{y^2}{2B}\Big(1 - \frac{1}{8}\delta^2\Big)\Big].$$

Therefore we obtain

$$a(J_2 + J_4) < a\Big(\int_0^{aB(1-\delta)} + \int_{aB(1+\delta)}^{8aB}\Big) \exp\Big[ay - \frac{y^2}{2B}\Big(1 - \frac{1}{8}\delta^2\Big)\Big]\,dy.$$

In both integration intervals the quantity

$$u(y) = ay - \frac{y^2}{2B}\Big(1 - \frac{1}{8}\delta^2\Big)$$

does not exceed

$$u(aB(1+\delta)) = a^2B(1+\delta) - \frac{a^2B}{2}(1+\delta)^2\Big(1 - \frac{1}{8}\delta^2\Big) =$$

$$= \frac{a^2B}{2}\Big[1 - \delta^2 + \frac{1}{8}\delta^2(1+\delta)^2\Big] < \frac{a^2B}{2}\Big(1 - \frac{1}{2}\delta^2\Big).$$

Hence we have

$$a(J_2+J_4) < a\int_0^{8aB} \exp\Big[\frac{a^2B}{2}\Big(1-\tfrac{1}{2}\delta^2\Big)\Big]dy = 8a^2B\exp\Big[\frac{a^2B}{2}\Big(1-\tfrac{1}{2}\delta^2\Big)\Big].$$

Since by (7), (8) and (15),

$$\begin{aligned} \ln(32\lambda) &< 2\ln\lambda \le \frac{1}{8}\lambda\delta^2, \\ \ln(32a^2B) &< (a^2B/8)\delta^2, \end{aligned} \tag{23}$$

we finally obtain

$$a(J_2+J_4) < \tfrac{1}{4}\exp\Big[\frac{a^2B}{2}\Big(1-\frac{\delta^2}{4}\Big)\Big] < \tfrac{1}{4}\mathbf{E}e^{aS}. \tag{24}$$

From (19), (22) and (24) it follows that

$$aJ_3 > \tfrac{1}{2}\mathbf{E}e^{aS} > \tfrac{1}{2}\exp\Big[\frac{a^2B}{2}(1-\delta)\Big]. \tag{25}$$

On the other hand, since $W(y)$ is monotone decreasing, (12) implies

$$aJ_3 = a\int_{aB(1-\delta)}^{aB(1+\delta)} e^{ay}W(y)dy < 2a^2Be^{a^2B(1+\delta)}W(x). \tag{26}$$

From (25) and (26) we conclude that

$$W(x) > \frac{1}{4a^2B}\exp\Big[-\frac{a^2B}{2}(1+3\delta)\Big].$$

Since, along with (23) and for similar reasons, the formula

$$\ln(4a^2B) < (a^2B/2)\delta$$

holds, finally we have

$$W(x) > \exp\Big[-\frac{a^2B}{2}(1+4\delta)\Big] > \exp\Big[-\frac{x^2}{2B(1-\delta)^2}(1+4\delta)\Big] >$$

$$> \exp\Big[-\frac{x^2}{2B}(1+8\delta)\Big] = \exp\Big[-\frac{x^2}{2B}(1+\epsilon)\Big].$$

Lemma V.

$$\overline{W}(x) \le 2W(x-\sqrt{2B}).$$

Proof. Let E denote the event

$$U \geq x$$

and E_k the event

$$S_i < U, \quad i < k$$

$$S_k = U \geq x.$$

Clearly, [4]

$$E = E_1 + E_2 + \ldots + E_n, \quad \overline{W}(x) = \mathbf{P}\{E\}.$$

Since the z_i with $i > k$ do not occur in the definition of the event E_k, we have

$$\mathbf{E}_{E_k}(\sigma_k^2) = \mathbf{E}\Big(\Big(\sum_{i=k+1}^{n} z_i\Big)^2 \,\Big|\, E_k\Big) = \sum_{i=k+1}^{n} \mathbf{E}z_i^2 < B,$$

$$\mathbf{P}_{E_k}\{|\sigma_k| \geq \sqrt{2B}\} \leq \tfrac{1}{2}.$$

Since $S = U + \sigma_k$, the following inequalities also hold:

$$\mathbf{P}_{E_k}\{S \geq U - \sqrt{2B}\} \geq \tfrac{1}{2},$$

$$W(x - \sqrt{2B}) = \mathbf{P}\{S \geq x - \sqrt{2B}\} \geq$$

$$\geq \overline{W}(x)\mathbf{P}_E\{S \geq U - \sqrt{2B}\} \geq \tfrac{1}{2}\overline{W}(x).$$

This proves our assertion.

§2. Proof of the first part (1°) of the main theorem

It suffices to consider the case

$$\delta < \tfrac{1}{2}. \tag{27}$$

Clearly, by Conditions I and II there exists an n_0 such that

$$B_{n_0} > e, \tag{28}$$

$$\sqrt{\ln\ln B_{n_0}} > 4/\delta \tag{29}$$

and for any $n \geq n_0$

$$M_n^2/B_n < \delta/16, \tag{30}$$

[4] The + sign is used instead of ∪ in order to stress the fact that the events in question are pairwise incompatible.

$$M_n\sqrt{\frac{\ln\ln B_n}{B_n}} < \frac{\delta}{4}. \tag{31}$$

After $n_0, n_1, \ldots, n_{k-1}$ have been defined, we choose n_k so that

$$B_{n_k}/B_{n_{k-1}} \le 1 + \delta/4, \tag{32}$$

$$B_{n_k+1}/B_{n_{k-1}} > 1 + \delta/4. \tag{33}$$

By (30) and (32) we have

$$\frac{B_{n_k}}{B_{n_{k-1}}} \ge \frac{B_{n_k+1} - M_{n_k+1}}{B_{n_{k-1}}} > 1 + \frac{\delta}{8}$$

and hence, by (28),

$$B_{n_k} > (1 + \delta/8)^k. \tag{34}$$

Suppose further that

$$\chi(t) = \sqrt{2t\ln\ln t}.$$

By (32) we have

$$\chi(B_{n_k})/\chi(B_{n_{k-1}}) < 1 + \delta/4. \tag{35}$$

The truth of at least one of the inequalities

$$S_n > \chi(B_n)(1+\delta) \quad (n_{k-1} \le n \le n_k)$$

implies

$$U_{n_k} > \chi(B_{n_{k-1}})(1+\delta). \tag{36}$$

Denote by V_k the probability of (36). Then clearly the convergence of the series

$$\sum_{k=1}^{\infty} V_k \tag{37}$$

suffices for Statement 1° to hold.

By (35) we have

$$V_k \le \mathbf{P}\{U_{n_k} > \chi(B_{n_k}(1+\delta/2)\}, \tag{38}$$

while according to Lemma V, the latter probability does not exceed

$$2W_{n_k}[\chi(B_{n_k})(1+\delta/2) - \sqrt{2B_{n_k}}].$$

Note that (29) implies

$$\sqrt{2B_{n_k}} < (\delta/4)\chi(B_{n_k}),$$

so that we have

$$V_k < 2W_{n_k}[\chi(B_{n_k})(1+\delta/4)]$$

and by Lemma I,

$$V_k < 2\exp\Big[-\frac{\chi^2(B_{n_k})}{2B_{n_k}}\Big(1+\frac{\delta}{4}\Big)^2(1-\theta)\Big] = 2\exp\Big[-\ln\ln B_{n_k}\Big(1+\frac{\delta}{4}\Big)^2(1-\theta)\Big],$$

where, by (31),

$$\theta = \frac{M_{n_k}}{2}\sqrt{\frac{\ln\ln B_{nk}}{B_{nk}}} < \frac{\delta}{8}.$$

This implies that

$$V_k < 2\exp\Big[-\ln\ln B_{n_k}\Big(1+\frac{\delta}{4}\Big)\Big] < 2(\ln B_{n_k})^{-(1+\delta/4)} <$$

$$< 2\Big[k\ln\Big(1+\frac{\delta}{8}\Big)\Big]^{-(1+\delta/4)} < Ck^{-(1+\delta/4)},$$

which proves the convergence of the series (37).

§3. Proof of the second part 2° of the main theorem

As before, suppose that (27) holds. Conditions I and II and the first part of the main theorem just proved imply that there exists an n_0 such that, first (28) holds, secondly, for every $n \geq n_0$,

$$M_n/B_n < \tfrac{1}{2} \tag{39}$$

and thirdly, the probability that all inequalities

$$|S_n| < 2\chi(B_n), \ (n = n_0, n_0+1, \dots, n_0+p) \tag{40}$$

simultaneously hold is greater than $1-\eta/2$ for any p.

Choose n_k so that

$$\frac{B_{n_k}}{B_{n_{k-1}}} > \frac{16}{\delta} \geq \frac{B_{n_k-1}}{B_{n_{k-1}}}. \tag{41}$$

Then because of (27) and (39), we have

$$B_{n_k}/B_{n_{k-1}} < 32/\delta, \quad B_{n_k} < B_{n_0}(32/\delta)^k. \tag{42}$$

Set

$$\sigma_k = S_{n_k} - S_{n_{k-1}}, \quad \beta_k = \mathbf{E}\sigma_k^2 = B_{n_k} - B_{n_{k-1}}.$$

Then (41) implies that

$$\beta_k > B_{n_k}(1-\delta/4), \quad \chi(\beta_k) > \chi(B_{n_k})(1-\delta/4), \tag{43}$$

$$\chi(\beta_k)(1-\delta/4) > \chi(B_{n_k})(1-\delta/2). \tag{44}$$

On the other hand, (41) also implies

$$\sqrt{B_{n_{k-1}}} < \frac{\delta}{4}\sqrt{B_{n_k}}, \quad \chi(B_{n_{k-1}}) < \frac{\delta}{4}\chi(B_{n_k}),$$

$$\chi(B_{n_k})(1-\delta/2) - 2\chi(B_{n_{k-1}}) > \chi(B_{n_k})(1-\delta). \tag{45}$$

Since $S_{n_k} = \sigma_k + S_{n_{k-1}}$, it follows that if (40) holds, then the inequality

$$\sigma_k > \chi(\beta_k)(1-\delta/4) \tag{46}$$

together with (44) and (45), implies

$$S_{n_k} > \chi(B_{n_k})(1-\delta). \tag{47}$$

Now we prove that for sufficiently large p the probability that at least one of the inequalities (46) holds for $k = 1, 2, \ldots, p$ is greater than $1 - \eta/2$. This clearly implies that the corresponding probability for the inequalities (47) is greater than $1 - \eta$, which proves the second part of the main theorem.

Since the σ_k are mutually independent, for our purpose it suffices to prove the divergence of the series

$$\sum_{k=1}^{\infty} V_k \tag{48}$$

of probabilities of the inequalities (46).

To estimate the probabilities V_k we make use of Lemma IV, setting

$$B = \beta_k, \quad M = M_{n_k},$$
$$x = \chi(\beta_k)(1-\delta/4).$$

For sufficiently large k, conditions (6) and (7) hold, since

$$\beta_k > \tfrac{1}{2} B_{n_k}$$

as a consequence of (43); therefore

$$\lim \frac{\chi(\beta_k) M_{n_k}}{\beta_k} \leq \lim \left(2M_{n_k} \sqrt{\frac{\ln \ln B_{n_k}}{B_{n_k}}} \right) = 0,$$

$$\lim \frac{\chi^2(\beta_k)}{\beta_k} \geq \lim \left(\tfrac{1}{2} \ln \ln B_{n_k} \right) = \infty.$$

From the latter formulas we can see that ϵ tends to zero as $k \to \infty$. Thus for sufficiently large k we have

$$(1+\epsilon)(1+\delta/4) < 1$$

and, because of (42),

$$V_k > \exp\left[-\frac{\chi^2(\beta_k)}{2\beta_k} \left(1 - \frac{\delta}{4} \right) \right] = \exp\left[-\ln \ln \beta_k \left(1 - \frac{\delta}{4} \right) \right] =$$

$$= (\ln \beta_k)^{-(1-\delta/4)} > (\ln B_{n_k})^{-(1-\delta/4)} > Ck^{-(1-\delta/4)},$$

which proves the divergence of (48).

Moscow, 24 November 1927

6. ON THE LAW OF LARGE NUMBERS *

By the law of large numbers in mathematical probability theory we usually mean either Poisson's theorem or Chebyshev's theorem, with their generalizations to the case of dependent trials. However, this law can be given a broader interpretation, more in agreement with its natural philosophical treatment, if it is used whenever it can be asserted with probability close to one that a certain variable differs only very slightly from a certain *a priori* given constant.[1] In a previous paper[2] I gave a theorem which, as I think, corresponds most closely to this intuitive concept. In this paper I would like to come back to this subject, formulate a somewhat more general statement, and also clarify some questions arising in this connection.

§1. For an accurate formulation of the problem, consider the sequence of real numbers[3]

$$X_1, X_2, \dots, X_n, \dots,$$

where X_n depends on the result of successive trials[4]

$$\mathcal{E}_1^{(n)}, \mathcal{E}_2^{(n)}, \dots, \mathcal{E}_n^{(n)}.$$

Let us agree to say that X_n is *stable*[5] or that X_n obeys the *law of large numbers*, if there is a sequence of constants

$$d_1, d_2, \dots, d_n, \dots$$

such that for every positive η,

$$\mathbf{P}\{|X_n - d_n| \geq \eta\} \to 0. \tag{1}$$

---

* 'Sur la loi des grands nombres', *Atti. Accad. Naz. Lincei. Rend.* **9** (1929), 470–474. Presented by G. Castelnuovo on 3rd March 1929.

[1] See: S.N. Bernshtein, *Probability theory*, p.142, where a more general definition is given.

[2] *C. R. Acad. Sci. Paris* **185** (1927), 917.

[3] Similar results can be obtained by considering vectors. V. Glivenko has informed me that similar arguments hold even for vectors in Hilbert space of functions.

[4] We assume throughout that these trials are independent.

[5] Concerning the definition of stability, see my paper: 'Über die Summen durch den Zufall bestimmter unabhängiger Grössen', *Math. Ann.* **99** (1928), 309 (Paper 4 in this volume.)

In most of the important cases d_n can be taken to be the expectation $\mathbf{E}(X_n)$. If it turns out that

$$P_n(\eta) = \mathbf{P}\{|X_n - \mathbf{E}(X_n)| \geq \eta\} \to 0 \tag{2}$$

for every positive η, then we say that X_n has ***normal stability***. It can be proved that if $X_n - \mathbf{E}(X_n)$ is uniformly bounded, then (2) follows from (1), that is, in this case the stability can only be normal.

According to Chebyshev we have

$$P_n(\eta) \leq B_n^2/\eta^2, \tag{3}$$

where

$$B_n^2 = \mathbf{E}[X_n - \mathbf{E}(X_n)]^2.$$

Thus the condition

$$B_n \to 0 \quad \text{(Markov's condition)} \tag{4}$$

is a sufficient condition for normal stability of X_n. If the absolute values of $X_n - \mathbf{E}(X_n)$ are uniformly bounded by a constant M, then, as can easily be shown,

$$P_n(\eta) = (B_n^2 - \eta^2)/(M^2 - \eta^2). \tag{5}$$

Hence, in this case Markov's condition is not only sufficient, but also necessary for the stability of X_n which, as has already been mentioned, can only be normal in the situation considered.

§2. In formulating Markov's condition we touched upon normal stability only. However, this condition is interesting since in most cases it is easier to calculate B_n than $P_n(\eta)$.

In the general case, we denote by $\mathbf{E}_k(X_n)$ the expectation of X_n and assume that the results of the first k trials $\mathcal{E}_1^{(n)}, \mathcal{E}_2^{(n)}, \ldots, \mathcal{E}_k^{(n)}$ are known. Then we have the following fundamental formula

$$B_n^2 = \sum_{k=1}^{n} \mathbf{E}[\mathbf{E}_k(X_n) - \mathbf{E}_{k-1}(X_n)]. \tag{6}$$

Indeed, it is clear that

$$\mathbf{E}_0(X_n) = \mathbf{E}(X_n), \quad \mathbf{E}_n(X_n) = X_n,$$

therefore, setting

$$Z_{nk} = \mathbf{E}_k(X_n) - \mathbf{E}_{k-1}(X_n),$$

we have

$$X_n - \mathbf{E}(X_n) = Z_{n1} + Z_{n2} + \ldots + Z_{nn}.$$

More generally, denoting by $\mathbf{E}_k(Y)$ the expectation of Y and assuming that the results of the trials $\mathcal{E}_1^{(n)}, \mathcal{E}_2^{(n)}, \ldots, \mathcal{E}_k^{(n)}$ are known, we have

$$\mathbf{E}_{k-1}(Z_{nk}) = \mathbf{E}_{k-1}[\mathbf{E}_k(X_n) - \mathbf{E}_{k-1}(X_n)] = \mathbf{E}_{k-1}(X_n) - \mathbf{E}_{k-1}(X_n) = 0.$$

Since Z_{ni} is constant for $i < k$, if the results of the first $k-1$ trials are fixed, we also have

$$\mathbf{E}_{k-1}(Z_{ni}Z_{nk}) = 0$$

and finally,

$$\mathbf{E}(Z_{ni}Z_{nk}) = \mathbf{E}[\mathbf{E}_{k-1}(Z_{ni}Z_{nk})] = 0.$$

This latter equality immediately implies [6]

$$B_n^2 = \beta_{n1}^2 + \beta_{n2}^2 + \ldots + \beta_{nn}^2, \quad \beta_{nk} = \sqrt{\mathbf{E}(Z_{nk}^2)}, \tag{7}$$

which differs from (6) only in notation.

Notice the meaning of β_{nk}: it is the mean squared deviation of the expectation of X_n when the result of the trial $\mathcal{E}_k^{(n)}$ is known. Therefore, β_{nk} is a natural measure of the dependence of X_n on the result of the trial [7] $\mathcal{E}_k^{(n)}$. This agrees with the concept of the law of large numbers formulated in the beginning of the paper. However, now we measure the dependence of X_n on each trial $\mathcal{E}_k^{(n)}$ using the moments β_{nk}. We have proved that the law of large numbers can be efficiently applied to X_n if the sum of the squares of β_{nk} is infinitesimally small. This is so if, in particular

$$\beta_{nk} = o(1/\sqrt{n}). \tag{8}$$

§3. Now we consider the case when the trials $\mathcal{E}_k^{(n)}$ determining the value of X_n are independent. Denote by $\mathbf{E}^k(Y)$ the expectation of Y when the results of

---

[6] In my article (see footnote 5) this formula is proved only for the case of independent trials.

[7] It is assumed that the trials $\mathcal{E}_k^{(n)}$ are sequentially carried out in accordance with the order of their indices k.

all trials $\mathcal{E}_i^{(n)}$ except $\mathcal{E}_k^{(n)}$ are known. In the case under consideration, that is, when the trials are independent, we have

$$\beta_{nk}^2 \leq \alpha_{nk} = \mathbf{E}[X_n - \mathbf{E}^k(X_n)], \tag{9}$$

$$B_n^2 \leq A_n^2 = \alpha_{n1}^2 + \alpha_{n2}^2 + \ldots + \alpha_{nn}^2. \tag{10}$$

Thus, for normal stability of X_n the condition [8]

$$A_n \to 0 \tag{11}$$

is sufficient.

Formula (9) can be proved as follows:

$$\mathbf{E}_k[\mathbf{E}^k(X_n)] = \mathbf{E}_{k-1}(X_n),$$

$$\mathbf{E}_k[X_n - \mathbf{E}^k(X_n)]^2 \geq \{\mathbf{E}_k(X_n) - \mathbf{E}_k[\mathbf{E}^k(X_n)]\}^2 = [\mathbf{E}_k(X_n) - \mathbf{E}_{k-1}(X_n)]^2,$$

$$\alpha_{nk}^2 = \mathbf{E}[X_n - \mathbf{E}^k(X_n)]^2 \geq \mathbf{E}[\mathbf{E}_k(X_n) - \mathbf{E}_{k-1}(X_n)]^2 = \beta_{nk}^2.$$

In the general case of dependent trials the first of these formulas does not hold.

Let $M^{(k)}(X_n)$ be the least upper bound of all possible values of X_n when the results of all trials $\mathcal{E}_i^{(n)}$ except $\mathcal{E}_k^{(n)}$ are known, and let $m^{(k)}(X_n)$ denote the corresponding greatest lower bound. We set

$$\Omega_{nk} = \sup |M^{(k)}(X_n) - m^{(k)}(X_n)|.$$

Thus, Ω_{nk} is the maximal deviation of X_n if only the result of the trial $\mathcal{E}_k^{(n)}$ is unknown. We see that

$$\alpha_{nk} \leq \tfrac{1}{2}\Omega_{nk}, \tag{12}$$

$$A_n^2 \leq \tfrac{1}{4}(\Omega_{n1} + \Omega_{n2} + \ldots + \Omega_{nn}). \tag{13}$$

This implies the condition for stability of X_n in the case of independent trials, given in a previous paper of mine:

$$\Omega_{n1}^2 + \Omega_{n2}^2 + \ldots + \Omega_{nn}^2 \to 0. \tag{14}$$

---

[8] This condition is especially interesting, since the definition of the moments α_{nk} does not involve the order of the trials $\mathcal{E}_k^{(n)}$.

Finally we remark that in the case considered by Chebyshev,

$$X_n = (x_1 + x_2 + \ldots + x_n)/n,$$

where x_k depends on the trial $\mathcal{E}_k^{(n)}$ and these trials are independent, we have

$$\beta_{nk}^2 = \alpha_{nk}^2 = \frac{1}{n^2}\mathbf{E}[X_n - \mathbf{E}(X_n)]^2 \frac{d^2(x_n)}{n^2},$$

which leads to the classical condition for stability of X_n:

$$\frac{d^2(x_1) + d^2(x_2) + \ldots + d^2(x_n)}{n^2} \to 0.$$

7. GENERAL MEASURE THEORY AND PROBABILITY CALCULUS *

Probability theory has become a topic of interest in modern mathematics not only because of its growing significance in natural sciences, but also because of the gradually emerging deep connections of this theory with many problems in various fields of pure mathematics. It seems that the formulas of probability calculus express one of the fundamental groups of general mathematical laws.

However, it would have been imprudent to assert that these facts indicate the dependence of notions used in pure mathematics on the notion of randomness dealt with in probability theory. For example, the fact that the distribution of the digits of decimal expansions of irrationals can be studied using formulas of probability calculus should not be interpreted as indicating dependence of these expansions on chance.

On the contrary, I believe that these facts indicate the possibility of constructing a very *general and purely mathematical theory* whose formulas can be applied both to probability calculus and to many other fields of pure and applied mathematics. To outline the contents of this theory, it suffices to single out from probability theory those elements that bring out its intrinsic logical structure, but have nothing to do with the specific meaning of the theory.

This way we come, first of all, to *general measure theory*. The general notion of measure of a set includes the notion of probability as a special case. A set of arbitrary elements considered from the viewpoint of the measure defined on its subsets will be called *purely metric space*, though perhaps it is an abuse of the term "space". In particular, in probability calculus we speak of the space of elementary events of a given problem and of the probabilities of various sets of these events.

Function theory in purely metric spaces is a further development of general measure theory. This theory studies the properties of functions that only depend on the measures of the sets on which the functions take a certain range of values. For instance, such are the properties of orthogonality of two functions or completeness of a system of orthogonal functions. If we consider variables depending on events (random variables) as functions defined on the space of elementary events, then all the relevant statements of probability theory appear to be particular cases of statements of such a general theory.

---

* *Trudy Kommunist. Akad. Razd. Mat.* 1 (1929), 8–21 (in Russian).

The strength of methods in probability theory as applied to problems in pure mathematics is largely based on the notion of *independence* of random variables. This notion has not yet achieved a purely mathematical formulation, though it is not so difficult to give one. We then define "independence" of a system of functions. This independence turns out to be a purely metric property.

Properties of sets that at first glance are quite different when considered merely from the viewpoint of the definition of measure established for their subsets, often turn out to be identical.

For example, it can be proved by elementary means that there exists a one-to-one map of a cube of arbitrary dimension onto an interval so that all the (L)-measurable sets of the cube are mapped into sets whose linear measures equal the measure of the initial sets. We say therefore that the cube of arbitrary dimension is *metrically equivalent* to an interval. Apart from the notion of metric equivalence, we define the more general notion of isometry of two spaces.

It is difficult to foresee whether we will need an early substantial development of the outlines of the theory or whether it will suffice merely to indicate that such is possible in principle. The latter is necessary anyway in order to give a correct outlook on the connections between various fields of mathematics. Besides, this gives us a method for transferring reasonings from one field to another field in which the statements of the general theory could be used.

I. ABSTRACT DEFINITION OF MEASURE

Consider a set A with elements a. We say that A is endowed with a measure M if a certain measure $M(E)$ is assigned to some of its subsets E. The set A together with the measure M is a *metric space*. We now give the axioms satisfied by all the measures we shall be considering.

First, we assume that the measure of a set is a real number, positive or zero.

Axiom I.

$$M(E) \geq 0.$$

Secondly, we assume that if two sets do not intersect, then the measure of their sum is equal to the sum of their measures.

Axiom II. *If*

$$E_1 \times E_2 = 0,$$

then

$$M(E_1 + E_2) = M(E_1) + M(E_2).$$

Here we assume that the existence of measures for two of the sets involved implies the existence of a measure for the third set.

Based on this axiom the general formula

$$M(E_1 + E_2) = M(E_1) + M(E_2 - E_1) \tag{1}$$

can be proved. One should not assume, however, that the existence of measures of two intersecting sets implies the existence of a measure for their sum or difference: there are certain important measures without this property.

Since the product of two sets can be defined as

$$E_1 \cdot E_2 = E_1 - (E_1 - E_2), \tag{2}$$

(1) can be used for deriving all the various relations among sums, differences and products of a finite number of sets.

For convenience we include the empty set in the domain of sets to be considered, and denote it by 0. If for a certain measure at least one set has a measure, then Axiom II implies that

$$M(0) = 0. \tag{3}$$

Finally, the third axiom introduces a quite arbitrary restriction on the measures considered: the measure of the whole space is 1.

Axiom III.

$$M(A) = 1.$$

This greatly simplifies the discussion and the general case can be easily derived from this particular case. In many applications, in particular, in probability calculus, this restriction is due to the essential nature of the subject.

Let us now consider some examples of measures satisfying these axioms.

1) The Lebesgue measure of point sets in an n-dimensional cube with side equal to 1.

2) The density of linear sets at 0 can be considered as their measure.

3) If a set A consists of positive integers, then the set density can be considered as its measure, that is, the limit of the ratio of the number of elements not greater than n, to n, as $n \to \infty$.

4) If A is a set of elementary events of a problem in probability calculus, then the measure of a set of elementary events is the probability that at least one of them occurs.

II. CLOSEDNESS OF A MEASURE

§1. A given measure M determines the system (E) of sets E having a measure. Suppose we know about this system that it includes the empty set 0 and the whole space A, and also that if

$$E_1 + E_2 = E_3, \quad E_1 \cdot E_2 = 0,$$

and two of these sets belong to (E), then so does the third one. Clearly, the complement of a set belonging to (E) also belongs to (E) since

$$E + \overline{E} = A, \quad E \cdot \overline{E} = 0.$$

§2. A measure is said to be complete, if it consists of (that is, is defined on) all the subsets of the space A.

A complete measure in which all one-element sets have zero measure is not known. To prove the existence of such a measure without using the axiom of choice seems to be a problem of great difficulty.

§3. We say that a measure M contains a measure M' given on the same set A if (E) contains (E') and both measures coincide on the sets belonging to (E').

One should not assume that any measure is contained in some complete one; counterexamples are elementary. Still, for none of the measures usually employed in mathematics has it been proved that it is impossible to consider it as contained in a complete measure. On the contrary, using the axiom of choice, S. Banach proved that linear Lebesgue measure is contained in a certain complete measure.[1] In view of the metric equivalence of spaces of arbitrary dimension the same is true for n-dimensional Lebesgue measure. But already

[1] *Fund. Math.* 4.

in the three-dimensional case, for a complete measure containing the Lebesgue measure the principle of equality of the measure of congruent sets cannot hold: this is proved by Hausdorff's example on the decomposition of a sphere into three sets, each of these sets being congruent to the sum of the two others to within a countable set.

Even if the main set A is countable, the construction of complete measures is quite difficult. For example, the generalized density problem for sequences of real numbers, that is, to find a complete measure containing the measure 3) of Part I, has difficulties similar to those encountered in constructing[2] a point set that is not (L)-measurable.

§4. Accordingly, when posing problems in probability calculus, it should be required to indicate for which events the probabilities are assumed to exist. It is clear, for example, that for geometrical probabilities it is not wise to assume the existence of the hitting probability of a point for each point set in space.

§5. In most cases, however, the system (E) can be assumed to possess certain closedness properties. We say that a system (E) is *finitely closed* if it contains all sums of pairs of sets belonging to this system. Clearly, since

$$E_1 - E_2 = \overline{\overline{E_1} + E_2}, \tag{4}$$

a closed system contains all the differences, and because of (2), it also contains the products of sets belonging to this system.

§6. For each system (E) there is a certain minimal system $F(E)$ that is closed and contains (E). If one can assign to all sets of the system $F(E)$ a measure coinciding with the measure given for the sets of (E) and satisfying our three axioms, then we say that the initial measure is *closable*.

It is not known whether every measure is closable. If closure is possible, then it is not necessarily closable in only one way; examples of the latter are elementary.

It would seem that it is very difficult to find a measure that closes the measure given above under 3).

It is also doubtful if a measure connected with some problem in probability calculus need be closed.

---

[2] My results in *C. R. Acad. Sci. Paris* (1925) point out the similarity of these two problems.

§7. A system (E) is called *countably closed* if it contains all possible countable sums of its elements. A countably closed system contains countable products of its elements, since

$$\prod E_n = \overline{\sum \overline{E}_n} \quad (n = 1, 2, \ldots). \tag{5}$$

The study of countable closedness is closely connected with normality of measures, and in fact makes sense only for normal measures.

III. NORMALITY OF A MEASURE

§8. A measure is *normal* if for sets having a measure the condition

$$E = \sum E_n, \quad E_n \cdot E_m = 0, \ n \neq m \quad (n = 1, 2, \ldots)$$

implies

$$M(E) = \sum M(E_n) \quad (n = 1, 2, \ldots).$$

Of the measures mentioned above, those in 2) and 3) are not normal. Measures corresponding to problems in probability calculus also need not be normal.

§9. For a finitely closed normal measure it is easy to prove the formulas

$$M\left(\sum E_n\right) = \sum M\left(E_n - \sum_{k=1}^{n-1} E_k\right), \tag{6}$$

$$M\left(\prod E_n\right) = \lim M\left(\prod_{k=1}^{n} E_k\right), \quad n = 1, 2, \ldots, \tag{7}$$

which are true under the single condition that the sets E_n and either their sum or product have a measure.

§10. Moreover, for finitely closed normal measures the following *Proposition on countable coverings* holds: if

$$E \subset \sum E_n \quad (n = 1, 2, \ldots),$$

then

$$M(E) \leq \sum M(E_n) \quad (n = 1, 2, \ldots).$$

Indeed, set

$$E'_n = E\Big(E_n - \sum_{k=1}^{n-1} E_k\Big) \quad (n = 1, 2, \ldots).$$

Then clearly

$$E'_n \cdot E'_m = 0, \quad n \neq m;$$

$$E = \sum E'_n,$$

$$\sum M(E_n) \geq \sum M(E'_n) = M(E) \quad (n = 1, 2, \ldots).$$

§11. A set E is called *measurable* with respect to a measure M if for any ϵ there exist two sequences of sets from (E),

$$E_1, E_2, \ldots, E_n, \ldots \quad E'_1, E'_2, \ldots, E'_n, \ldots$$

such that

$$E \subset \sum E_n, \quad \overline{E} \subset \sum E'_n,$$

$$\sum M(E_n) + M(E'_n) \leq 1 + \epsilon \quad (n = 1, 2, \ldots).$$

By the *generalized measure* of a measurable set E we mean the infimum of the sums

$$\sum M(E_n) \ (n = 1, 2, \ldots)$$

over the coverings of E consisting of sets from (E).

§12. Thus we obtain a certain new measure $L(M)$ which measures any set that is measurable with respect to M. *If M is finitely closed and normal, then $L(M)$ satisfies Axioms* I–III, *is normal and countably closed.* The methods for proving this are the same as those used for studying Lebesgue measure. Thus, for any normal finitely closed measure there exists a countably closed measure containing it.

§13. The measure $L(M)$ is not the minimal countably closed measure containing M. It is not too difficult to determine the minimal such measure $B(M)$ contained in $L(M)$.

However, the measure $L(M)$ has another remarkable property: *a set that is measurable with respect to M has the same measure with respect to any normal measure containing M as with respect to $L(M)$.*

§14. In particular, in probability calculus, if a system of events with certain probabilities is finitely closed and normal, then without contradiction we can uniquely define the probability of the countable product or the countable sum of events with probabilities. It would seem that only after this has been done can an argument involving probabilities of events such as convergence of series of random variables, etc., be considered justified.

IV. METRIC EQUIVALENCE AND ISOMETRY OF SPACES

§15. Two metric spaces are called *metrically equivalent* if they can be put into a one-to-one correspondence so that sets with a measure correspond to sets with a measure, and measures of corresponding sets are equal.

§16. The cube of arbitrary dimension with side equal to 1 and with Lebesgue measure on its subsets is metrically equivalent to an interval of length 1, also with Lebesgue measure. The proof of this is elementary.

§17. Two sets E_1 and E_2 in a metric space *have the same metric type* if

$$M[(E_1 - E_2) + (E_2 - E_1)] = 0.$$

In many cases when we can neglect sets of measure zero only the metric types, not sets themselves, need be involved in the argument.[3]

If a measure is finitely closed, then the type of the sum, product or difference of two sets with a measure depends only on their types and consequently we can speak of the sums, products and differences of the types themselves.

If a measure is countably closed and normal, then so are the countable sums and products.

§18. Two finitely closed spaces are called *isometric* if a one-to-one correspondence can be established between the metric types of their subsets with measures in such a way that the sum of two types corresponds to the sum of the corresponding types, and the measures of corresponding types are equal.

---

[3] Of course, everything that follows can be presented without introducing the notion of metric type as based on the "subdivision" axiom. Instead the notion of metric "equality" of sets can be used.

Obviously, the products and differences of two types also correspond to the products and differences of the corresponding types. If both measures are normal and countably closed, the same is also true for countable sums and products. Note, however, that a countably closed normal measure can be isometric to a non-normal measure.

§19. A measure has a ***countable basis*** if the system (E) has a countable subsystem (I) (a basis) such that any set of (E) is measurable for the basis (I), that is, this set and its complement each have a covering by sets of (I) in which the sum of the measures of the covering sets is greater than their measure by an arbitrary small value. Of course, this definition is interesting only when the space is normal.

The Lebesgue measure on an interval has a countable basis consisting of segments with rational ends points. Normal spaces with the cardinality of the continuum and with a countable basis are not known (except those determined using the axiom of choice). Such a space of cardinality 2^c can be easily constructed.

§20. *A measure M that is normal, finitely closed and with a countable basis is isometric to the measure M^1, which is the Lebesgue measure of the interval* $(0,1)$.

§21. In conclusion, since many properties of purely metric spaces depend only on relations between the metric types of their subsets, which can be identical for spaces that are most different, even for spaces of different cardinalities, it would be worth trying to set forth the theory of such spaces considered as systems of metric types that can be added, multiplied, etc., without assuming the existence "elements" of the space.

V. PURELY METRIC FUNCTION THEORY

§22. We will consider only functions that take real values at all points of a purely metric space. Such a function is called ***measurable*** if the set of points at which it takes values belonging to an interval is always measurable.

§23. Clearly, the set on which a function takes a certain value a is also measurable, since

$$E[b < f < a] + E[f = a] + E[a < f < c] = E[b < f < c].$$

If the measure in the space considered is countably closed, then the set on which a measurable function takes a (B)-measurable set of values has a measure, since it can be obtained by countable additions and multiplications of sets that obviously have a measure. But one cannot claim that the same is true for an (L)-measurable set of values of a function, even if the measure considered is of the type $L(M)$.

§24. A family of functions $[f]$ is called *equimeasurable* if, for an arbitrary choice of intervals corresponding to the functions, the set on which every function takes values in its corresponding interval is measurable.

For a finitely closed measure a finite family of measurable functions is measurable. For a countably closed measure the same is true for a countable family of functions.

For a countably closed measure the set on which the values of a sequence of functions determine the points of a (B)-measurable set in a countable dimensional space has a measure. This applies, for instance, to the set of points of convergence of a series of functions.

All this can be directly applied to random variables in probability theory.

§25. The *integral* of a function f over a set E can be defined in a purely metric theory only with respect to the Lebesgue measure as

$$\lim_{n\to\infty} \sum_{m=-\infty}^{\infty} \frac{m}{n} M\left[\left(\frac{m}{n} < f \leq \frac{m+1}{n}\right) E\right],$$

provided that the series converges absolutely and the limit exists. When considering problems connected with the notion of integral we assume the space to be normal.

§26. It can be proved in the usual way that for a bounded measurable function the integral exists.

§27. For an integral defined in such a way the following relations hold:

1. $\displaystyle\int_E (f_1 + f_2) = \int_E f_1 + \int_E f_2.$
2. $\displaystyle\int_E cf = c\int_E f.$
3. $\displaystyle\int_{E_1+E_2} f = \int_{E_1} f + \int_{E_2} f, \quad \text{if } E_1 \cdot E_2 = 0.$

4. $\lim \int_E f_n = \int_E f,$ if $f_1 < f_2 < \ldots < f_n \ldots \to f.$

§28. The notation of *expectation* which is close to that of the Lebesgue integral, has long been used in probability theory. If probability is considered as a measure on the set of elementary events, then the expectation of a random variable Z is

$$D(Z) = \int_A Z,$$

and its expectation under a hypothesis E is

$$D_E(Z) = \frac{1}{M(E)} \int_E Z.$$

By analogy with the notion of relative probability we introduce the notation

$$M_{E_1}(E_2) = M(E_1 E_2)/M(E_1);$$

then

$$D_E(Z) = \lim \sum_{m=-\infty}^{\infty} \frac{m}{n} M_E \Big[\frac{m}{n} < Z \leq \frac{m+1}{n} \Big].$$

§29. Based on this definition of integral the theory of *orthogonal functions* in purely metric spaces can be developed. For example, for a system of bounded orthogonal functions to have at most countable cardinality it is sufficient that the measure considered have a countable basis.

It would be interesting to see to what extent purely metric function theory can be presented in terms invariant for isometric spaces. In this case the very definition of a function should be different.

VI. INDEPENDENCE

§30. Two partitions of a space A into non-intersecting parts

$$A = \sum F', \quad A = \sum F''$$

are *independent* if for any measurable sets E' and E'' composed from the elements of certain sets F' and F'' respectively,

$$M(E' \cdot E'') = M(E') \cdot M(E'').$$

§**31.** The (finite or infinite) *product* of partitions $[F]$ is the partition of the space into the products of elements of the given partitions, with one element taken out of every partition.

A finite or infinite number of partitions $[F]$ are *mutually independent* if any two products of these partitions without common elements are independent. Mutual independence of partitions does not imply independence of these partitions.

§**32.** These general definitions allow a number of simplifications in particular cases. For instance, if a measure is finitely closed and the partitions (F') and (F'') consist of a finite number of elements with a measure, then they are independent if

$$M(F' \cdot F'') = M(F') \cdot M(F'')$$

for each pair F' and F''.

If the measure is normal, the same is true for partitions into a countable set of parts.

§**33.** Every function defines a partition of the space into the sets at which it takes a certain value. Functions are called mutually *independent* if their corresponding partitions are independent.

§**34.** Independence of partitions of a set (E) and independence of functions on a set are defined in exactly the same way.

For functions $f_1, f_2, \ldots, f_n$, independent on a set E, the following relation holds:

$$DE(f_1 f_2 \ldots f_n) = DE(f_1) DE(f_2) \ldots DE(f_n).$$

8. ON THE STRONG LAW OF LARGE NUMBERS *

Let

$$x_1, x_2, \ldots, x_n, \ldots \tag{1}$$

be a sequence of independent random variables with zero expectations $\mathbf{E}(x_n)$. Following Cantelli and Khinchin we say that (1) satisfies the *strong law of large numbers* (SLLN) if the probability of convergence to zero of the means

$$\sigma_n = \frac{s_n}{n} = \frac{x_1 + x_2 + \ldots + x_n}{n}$$

is equal to 1.

We will now prove that the SLLN holds if the second moments $\mathbf{E}(x_n^2) = b_n$ exist and the *series*

$$\sum_{n=1}^{\infty} \frac{b_n}{n^2} \tag{2}$$

converges. This condition cannot be replaced by any weaker one: if for some sequence of constants b_n (2) diverges, then we can construct a sequence (1) of independent random variables satisfying $\mathbf{E}(x_n) = 0$, $\mathbf{E}(x_n^2) = b_n$ but not satisfying the SLLN.

To prove this, we can use the lemma expressed by the formula

$$\mathbf{P}\Big\{\max_{2 \leq k \leq n} |s_k| \geq R\Big\} \leq \frac{1}{R^2} \sum_{k=1}^{k=n} b_k, \tag{3}$$

which was proved elsewhere (for $R \geq 0$).[1]

---

* 'Sur la loi forte des grands nombres', *C. R. Acad. Sci. Paris* **191** (1930), 910–912.

[1] 'Über die Summen durch den Zufall bestimmer unabhängiger Grössen', *Math. Ann.* **99** (1928), 310, Satz 1 (Paper 4 in this volume).

Clearly, according to (3),

$$P_m = \mathbf{P}\Big\{\max_{2^m \le n < 2^{m+1}} |\sigma_n| \ge \epsilon\Big\} \le$$

$$\le \mathbf{P}\Big\{\max_{1 \le n < 2^{m+1}} |s_n| \ge 2^m\epsilon\Big\} \le \Big(\frac{1}{2^m\epsilon}\Big)^2 \sum_{n=1}^{n<2^{m+1}} b_n,$$

$$= \mathbf{P}\{\limsup |\sigma_n| > \epsilon\} \le \sum_{m=0}^{\infty} P_m \le \frac{1}{\epsilon^2}\sum_{m=0}^{\infty}\Big(\frac{1}{2^m}\Big)^2 \sum_{n=1}^{n<2^{m+1}} b_n \le \tag{4}$$

$$\le \frac{1}{\epsilon^2}\sum_{i=0}^{\infty}\sum_{m=i}^{\infty}\Big(\frac{1}{2^m}\Big)^2 \sum_{n=2^i}^{n<2^{i+1}} b_n \le$$

$$\le \frac{1}{\epsilon^2}\sum_{i=0}^{\infty}\Big(\frac{1}{2^{i-1}}\Big)^2 \sum_{n=2^i}^{n<2^{i+1}} b_n \le \frac{8}{\epsilon^2}\sum_{n=1}^{\infty}\frac{b_n}{n^2}.$$

But P does not change if a finite number of first terms in (1) are replaced by 0. If (2) converges, then the last term in (4) can be made arbitrarily small. Hence $P = 0$. Since this is true for any $\epsilon > 0$, in the case under consideration, the SLLN holds.

To prove the second part of our statement, assume that (2) diverges.

For $b_n/n^2 \le 1$ let $x_n = n$, $x_n = -n$, $x_n = 0$ with probabilities $b_n/2n^2$, $b_n/2n^2$, $1 - b_n/n^2$ respectively, while for $b_n/n^2 \ge 1$ we set $x_n = \sqrt{b_n}$ and $x_n = -\sqrt{b_n}$ with probabilities $\frac{1}{2}$ and $\frac{1}{2}$. It can easily be seen that $\mathbf{E}(x_n) = 0$, $\mathbf{E}(x_n^2) = b_n$ and the SLLN does not hold.

9. ON ANALYTICAL METHODS IN PROBABILITY THEORY*

The object of investigation

A physical process (a change of a certain physical system) is called *stochastically determined* if, knowing a state X_0 of the system at a certain moment of time t_0 we also know the probability distribution for all the states X of this system at the moments $t > t_0$.

I systematically consider the simplest cases of stochastically determined processes, and primarily, processes continuous in time (this is what makes the method essentially new: so far, a stochastic process has usually been considered to be a discrete sequence of separate "events").

If the set $\mathfrak{A}$ of different possible states of the system is finite, then a stochastic process can be characterized using ordinary linear differential equations (Chapter II). If a state of the system depends on one or several continuous parameters, then the corresponding analytic apparatus reduces to partial differential equations of parabolic type (Chapter IV) and we obtain various distribution functions, the normal Laplace distribution being the simplest.

INTRODUCTION

1. In order to subject social or natural phenomena to mathematical treatment, these phenomena should first be schematized. The fact is that mathematical analysis can only be applied to studying changes of a certain system if every possible state of this system can be completely determined using known mathematical techniques, for example, by the values of a certain number of parameters. This mathematically defined system is not a reality itself, but a scheme that can be used to describe reality.

Classical mechanics makes use only of the schemes for which the state y of a system at time t is uniquely determined by its state x at any preceding time t_0. Mathematically this can be expressed by the formula

$$y = f(x, t_0, t).$$

If such a unique function f exists, as is always assumed in classical mechanics, then we say that our scheme is a *scheme of a purely deterministic*

---

* Über die analytischen Methoden in der Wahrscheinlichkeitsrechnung, *Math. Ann.* **104** (1931), 415–458.

process. These purely deterministic processes also include processes when the state y is not completely determined by giving a state x at a single moment of time t, but also essentially depends on the pattern of variation of this state x prior to t. However, usually it is preferred to avoid such a dependence on the preceding behaviour of the system, and to do this the notion of the state of the system at time t is generalized by introducing new parameters.[1]

Outside the realm of classical mechanics, along with the schemes of purely deterministic processes, one often considers schemes in which the state x of the system at a certain time t_0 only determines a certain probability of a possible event y to occur at a certain subsequent moment $t > t_0$. If for any given t_0, $t > t_0$, and x there exists a certain probability distribution for the states y, we say that our scheme is a *scheme of a stochastically determined process*. In the general case this distribution function can be represented in the form

$$P(t_0, x, t, \mathfrak{E})$$

where $\mathfrak{E}$ denotes a certain set of states y, and P is the probability of the fact that at time t one of the states y belonging to this set will be realized. Here we face a complication: in general, this probability cannot be determined for all sets $\mathfrak{E}$. A rigorous definition of a stochastically determined process which enables one to avoid this complication is given in §1.

As in the case of a purely deterministic process, we could also have considered here schemes in which the probability P essentially depends not only on the state x but also on the past behaviour of the system. Still, this influence of the past behaviour of the process can be bypassed using the same method as in the scheme of a purely deterministic process.

Note also that the possibility of applying a scheme of either a purely deterministic or a stochastically determined process to the study of some real processes is in no way linked with the question whether this process is deterministic or random.

2. In probability theory one usually considers only schemes according to which any changes of the states of a system are only possible at certain moments

[1] A well-known example of this method is to introduce, in addition to positions of points, the components of their velocities when describing a state of a certain mechanical system.

$t_1, t_2, \ldots, t_n, \ldots$ which form a discrete series. As far as I know, Bachelier[2] was the first to make a systematic study of schemes in which the probability $P(t_0, x, t, \mathfrak{E})$ varies continuously with time t. We will return to the cases studied by Bachelier in §16 and in the Conclusion. Here we note only that Bachelier's constructions are by no means mathematically rigorous.

Starting from Chapter II of this paper we mainly consider above-mentioned schemes that are continuous with respect to time. From the mathematical point of view these schemes have an important advantage: they allow one to introduce differential equations for P with respect to time and lead to simple analytic expressions which in the usual theory can be derived only as asymptotic formulas. As for the applications, first the new schemes can be directly applied to real processes, and secondly, from the solutions of differential equations for processes continuous with respect to time new asymptotic formulas for continuous schemes can be derived, as will be shown later in §12.

3. We do not start with the complete system of axioms of probability theory. Let us indicate, however, all the prerequisites we will use in our further discussion. We do not make any special assumptions about the set $\mathfrak{A}$ of possible states x. Mathematically, $\mathfrak{A}$ can be considered as an arbitrary set consisting of arbitrary elements. All assumptions concerning the system $\mathfrak{F}$ of sets and the function $P(t_0, x, t, \mathfrak{E})$ are given in §1. In what follows the theory is developed as a purely mathematical one.

CONTENTS

Chapter I. Generalities:

§1. General scheme of a stochastically determined process.

§2. The operator $F_1(x, \mathfrak{E}) * F_2(x, \mathfrak{E})$.

§3. Classification of particular cases.

§4. The ergodic principle.

Chapter II. Finite state systems:

§5. Preliminary remarks.

§6. Differential equations of a continuous stochastic process.

---

[2] I. 'Théorie de la spéculation', *Ann. École Norm. Supér.* **17** (1900), 21; II. 'Les probabilités à plusieurs variables', *Ann. École Norm. Supér.* **27** (1910), 339; III. *Calcul des probabilités*, Paris, 1912.

§7. Examples.

Chapter III. Countable state systems:

§8. Preliminary remarks. Discontinuous schemes.

§9. Differential equations of a process continuous in time.

§10. Uniqueness of solutions and their calculation for a process homogenous in time.

Chapter IV. Continuous state systems, the case of one parameter:

§11. Preliminary remarks.

§12. Lindeberg's method. Passage from discontinuous to continuous schemes.

§13. The first differential equation for processes continuous in time.

§14. The second differential equation.

§15. Statement of the uniqueness and existence problem of solutions of the second differential equation.

§16. Bachelier's case.

§17. A method for transforming distribution functions.

§18. Stationary distribution functions.

§19. Other possibilities.

Conclusion.

CHAPTER I. GENERALITIES

§1. General scheme of a stochastically determined process

Let $\mathfrak{S}$ be a system that can be in states $x, y, z, \ldots$, and $\mathfrak{F}$ a system of sets $\mathfrak{E}$ formed from the elements $x, y, z, \ldots$. A process of variation of the system $\mathfrak{S}$ is *stochastically determined with respect to* $\mathfrak{F}$ if for any choice of state x, set $\mathfrak{E}$ and moments t_1 and t_2 ($t_1 < t_2$) the probability $P(t_1, x, t_2, \mathfrak{E})$ of the fact that, if x takes place at t_1, then one of the states of $\mathfrak{E}$ takes place at t_2 exists. If $P(t_1, x, t_2, \mathfrak{E})$ is defined only for $t_2 > t_1 \geq t_0$, then we say that the process of variation is stochastically determined for $t \geq t_0$.

Regarding the system $\mathfrak{F}$, we assume that it is first additive (that is, it contains all the differences, as well as finite or countable sums of its elements), and secondly contains the empty set, the set $\mathfrak{A}$ of all possible states $x, y, z, \ldots$ and all the one-element sets. If the set $\mathfrak{A}$ is finite or countable, then clearly $\mathfrak{F}$ consists of all the subsets of $\mathfrak{A}$. In the most important case when $\mathfrak{A}$ is uncountable, the assumption that $\mathfrak{F}$ contains all the subsets of $\mathfrak{A}$ does not hold for any of the schemes known at present.

Of course we assume that

$$P(t_1, x, t_2, \mathfrak{A}) = 1 \tag{1}$$

and for the empty set $\mathfrak{N}$,

$$P(t_1, x, t_2, \mathfrak{N}) = 0.$$

We further assume that $P(t_1, x, t_2, \mathfrak{E})$ is additive as a function of $\mathfrak{E}$, that is, for any decomposition of $\mathfrak{E}$ into a finite or countable number of non-intersecting summands $\mathfrak{E}_n$ the following identity holds

$$\sum_n P(t_1, x, t_2, \mathfrak{E}_n) = P(t_1, x, t_2, \mathfrak{E}). \tag{2}$$

To formulate further assumptions on $P(t_1, x, t_2, \mathfrak{E})$ we need the notion of measurability of a function $f(x)$ with respect to the system $\mathfrak{F}$ and the definition of abstract Stieltjes integral. We given them here in a form suitable for our needs.[3]

A function $f(x)$ is called *measurable with respect to the system* $\mathfrak{F}$ if for any choice of real numbers a and b the set $\mathfrak{E}_\Delta(a < f(x) < b)$ of all x for which $f(x)$ satisfies the inequality in parentheses, belongs to $\mathfrak{F}$. It can easily be shown that if the system $\mathfrak{F}$ is additive and $f(x)$ is measurable with respect to $\mathfrak{F}$, then the set $\mathfrak{E}$ of all x for which $f(x)$ belongs to a given Borel-measurable set is contained in $\mathfrak{F}$.

Now let $f(x)$ be measurable with respect to $\mathfrak{F}$ and bounded, and let $\phi(\mathfrak{E})$ denote a non-negative additive function defined on $\mathfrak{F}$; then, as is known, the sum

$$\sum_m \frac{m}{n} \phi\Big(\frac{m}{n} \le f(x) < \frac{m+1}{n}\Big)$$

tends to a well defined limit as $n \to \infty$. This limit will be called the *integral*

$$\int_{\mathfrak{A}_x} f(x)\phi(d\mathfrak{A}).$$

[3] Concerning these notions, as well as additive sets of systems, etc., see, for example, M. Fréchet, 'Sur l'intégrale d'une fonctionnelle étendu à un ensemble abstrait', *Bull. Soc. Math. France* **43** (1915), 248.

This notation differs from the usual one only in the specification of the variable of integration and the place of the differential inside the parentheses.

In what follows we assume that $P(t_1, x, t_2, \mathfrak{E})$, as a function of the state x, is measurable with respect to the system $\mathfrak{F}$. Finally, $P(t_1, x, t_2, \mathfrak{E})$ must satisfy the fundamental equation

$$P(t_1, x, t_2, \mathfrak{E}) = \int_{\mathfrak{A}_y} P(t_2, y, t_3, \mathfrak{E}) P(t_1, x, t_2, d\mathfrak{A}) \tag{3}$$

for arbitrary t_1, t_2, t_3, $(t_1 < t_2 < t_3)$. If $\mathfrak{A}$ is a finite or countable set of elements $x_1, x_2, \ldots, x_n, \ldots$, then

$$\int_{\mathfrak{A}_y} P(t_2, y, t_3, \mathfrak{E}) P(t_1, x, t_2, d\mathfrak{A}) = \sum_n P(t_2, x_n, t_3, \mathfrak{E}) P(t_1, x, t_2, x_n)$$

and on the right-hand side we have the expression for the total probability $P(t_1, x, t_3, \mathfrak{E})$; therefore (3) is satisfied in this case. In case $\mathfrak{A}$ is uncountable we take (3) as a new axiom.

The above requirements completely define a stochastically determined process: the elements $x, y, z, \ldots$ of an arbitrary set $\mathfrak{A}$ can be considered as characteristics of a state of a certain system, and an arbitrary function $P(t_1, x, t_2, \mathfrak{E})$ satisfying the above requirements as the corresponding probability distribution.

A non-negative function $F(\mathfrak{E})$ defined on $\mathfrak{F}$, additive and such that

$$F(\mathfrak{A}) = 1, \tag{4}$$

will be called a *normal distribution function*. All the requirements imposed on $P(t_1, x, t_2, \mathfrak{E})$ can now be formulated in the following way: $P(t_1, x, t_2, \mathfrak{E})$, as a function of $\mathfrak{E}$, is a normal distribution function; as a function of x it is measurable with respect to the system $\mathfrak{F}$ and satisfies the integral equation (3).

Suppose now that at $t = t_0$ we have a normal distribution function $Q(t_0, \mathfrak{E})$ which gives the probability of the fact that the system $\mathfrak{S}$ at t_0 is in one of the states belonging to $\mathfrak{E}$. The distribution function $Q(t, \mathfrak{E})$ for $t > t_0$ is determined by means of the second fundamental equation

$$Q(t, \mathfrak{E}) = \int_{\mathfrak{A}_x} P(t_0, x, t, \mathfrak{E}) Q(t_0, d\mathfrak{A}). \tag{5}$$

We clearly have

$$Q(t, \mathfrak{A}) = \int_{\mathfrak{A}} Q(t_0, d\mathfrak{A}) = Q(t_0, \mathfrak{A}) = 1, \tag{6}$$

$$\int_{\mathfrak{A}_x} P(t_1, x, t_2, \mathfrak{E}) Q(t_1, d\mathfrak{A}) =$$

$$= \int_{\mathfrak{A}_x} P(t_1, x, t_2, \mathfrak{E}) \int_{\mathfrak{A}'_x} P(t_0, y, t_1, d\mathfrak{A}) Q(t_0, d\mathfrak{A}') =$$

$$= \int_{\mathfrak{A}'_y} \int_{\mathfrak{A}_x} P(t_1, x, t_2, \mathfrak{E}) P(t_0, y, t_1, d\mathfrak{A})) Q(t_0, d\mathfrak{A}') =$$

$$= \int_{\mathfrak{A}'_y} P(t_0, y, t_2, \mathfrak{E}) Q(t_0, d\mathfrak{A}') = Q(t_2, \mathfrak{E}). \tag{7}$$

Formula (5) is considered as the definition of $Q(t, \mathfrak{E})$, not as a new requirement imposed on $\mathfrak{S}$. Note, however, that (5) implies (3) as a particular case.

§2. The operator $F_1(x, \mathfrak{E}) * F_2(x, \mathfrak{E})$

Let $F_1(x, \mathfrak{E})$ and $F_2(x, \mathfrak{E})$ be two normal distribution functions which, considered as functions of x, are measurable with respect to $\mathfrak{F}$. Set

$$F(x, \mathfrak{E}) = F_1(x, \mathfrak{E}) * F_2(x, \mathfrak{E}) = F_1 * F_2(x, \mathfrak{E}) = \int_{\mathfrak{A}_y} F_2(y, \mathfrak{E}) F_1(x, d\mathfrak{A}); \tag{8}$$

It is easy to see that $F(x, \mathfrak{E})$ satisfies the same conditions of measurability and additivity as $F_1(x, \mathfrak{E})$ and $F_2(x, \mathfrak{E})$ and (4) also holds:

$$F(x, \mathfrak{A}) = \int_{\mathfrak{A}'_y} F_2(y, \mathfrak{A}) F_1(x, d\mathfrak{A}') = \int_{\mathfrak{A}'_y} F_1(x, d\mathfrak{A}') = 1;$$

consequently, $F(x, \mathfrak{E})$ is also a normal distribution function.

Further, the operator $F_1 * F_2$ is associative,

$$F_1 * (F_2 * F_3) = (F_1 * F_2) * F_3, \tag{9}$$

which can easily be seen by the following simple calculation:

$$F_1 * (F_2 * F_3)(x, \mathfrak{E}) = \int_{\mathfrak{A}_y} \int_{\mathfrak{A}'_z} F_3(z, \mathfrak{E}) F_2(y, d\mathfrak{A}') F_1(x, d\mathfrak{A}) =$$

$$= \int_{\mathfrak{A}'_z} F_3(z, \mathfrak{E}) \int_{\mathfrak{A}_y} F_2(y, d\mathfrak{A}') F_1(x, d\mathfrak{A}) = (F_1 * F_2) * F_3(x, \mathfrak{E}).$$

By contrast, $F_1 * F_2$ is not, in general, commutative.

Now we define the unit function $\mu(x, \mathfrak{E})$, which for any normal distribution function $F(x, \mathfrak{E})$ satisfies

$$\mu * F(x, \mathfrak{E}) = F * \mu(x, \mathfrak{E}) = F(x, \mathfrak{E}). \tag{10}$$

To this end it suffices to set $\mu(x, \mathfrak{E}) = 1$ when x belongs to $\mathfrak{E}$ and $\mu(x, \mathfrak{E}) = 0$ otherwise. We then have

$$\mu * F(x, \mathfrak{E}) = \int_{\mathfrak{A}_y} F(y, \mathfrak{E})\mu(x, d\mathfrak{A}) = F(x, \mathfrak{E}),$$
$$F * \mu(x, \mathfrak{E}) = \int_{\mathfrak{A}_y} \mu(y, \mathfrak{E})F(x, d\mathfrak{A}) = \int_{\mathfrak{E}} F(x, d\mathfrak{E}) = F(x, \mathfrak{E}).$$

The probability $P(t_1, x, t_2, \mathfrak{E})$ has been defined so far only for $t_2 > t_1$; now set for any t

$$P(t, x, t, \mathfrak{E}) = \mu(x, \mathfrak{E}). \tag{11}$$

In view of (10), this new definition does not contradict the fundamental equation (3), since (3) can be written as

$$P(t_1, x, t_2, \mathfrak{E}) * P(t_2, x, t_3, \mathfrak{E}) = P(t_1, x, t_3, \mathfrak{E}). \tag{12}$$

§3. Classification of particular cases

If the changes in the state of the system $\mathfrak{S}$ take place only at certain moments which form a discrete series

$$t_0 < t_1 < t_2 < \ldots < t_n < \ldots \to \infty,$$

then obviously

$$P(t', x, t'', \mathfrak{E}) = P(t_m, x, t_n, \mathfrak{E}) \tag{13}$$

for all moments t' and t'' such that

$$t_m \le t' < t_{m+1}, \quad t_n \le t'' < t_{n+1}.$$

Introducing the notation

$$P(t_m, x, t_n, \mathfrak{E}) = P_{mn}(x, \mathfrak{E}), \tag{14}$$
$$P_{n-1,n}(x, \mathfrak{E}) = P_n(x, \mathfrak{E}), \tag{15}$$

we have

$$P_{mn}(x, \mathfrak{E}) = P_{m+1} * P_{m+2} * \ldots * P_n(x, \mathfrak{E}). \tag{16}$$

Hence in this case the process of change of $\mathfrak{S}$ is totally determined by the elementary distribution functions $P_n(x, \mathfrak{E})$.

Now let $P_1(x, \mathfrak{E})$, $P_2(x, \mathfrak{E})$, ..., $P_n(x, \mathfrak{E})$ be arbitrary normal distribution functions which are assumed to be measurable as functions of x; further, let $t_0 < t_1 < \ldots < t_n < \ldots$ be a certain sequence of moments of time. Defining $P_{mn}(x, \mathfrak{E})$ and $P(t', x, t'', \mathfrak{E})$ by (16), (14) and (13), we also obtain normal distribution functions which satisfy the equations

$$P_{mn}(x, \mathfrak{E}) * P_{np}(x, \mathfrak{E}) = P_{mp}(x, \mathfrak{E}) \quad (m < n < p), \tag{17}$$

and hence the equation

$$P(t', x, t'', \mathfrak{E}) * P(t'', x, t''', \mathfrak{E}) = P(t', x, t''', \mathfrak{E}) \quad (t' < t'' < t''').$$

But this latter equation is none other than the fundamental equation (12) or (13). Thus we see that every sequence of arbitrary normal distribution functions $P_n(x, \mathfrak{E})$, measurable as functions of x, characterizes a certain stochastically determined process.

The schemes with discrete time defined above are those usually considered in probability theory. If all the distribution functions $P_n(x, \mathfrak{E})$ coincide,

$$P_n(x, \mathfrak{E}) = P(x, \mathfrak{E}), \tag{18}$$

we have a homogeneous scheme with discrete time; in this case (16) yields

$$P_{n,n+p}(x, \mathfrak{E}) = \underbrace{P(x, \mathfrak{E}) * P(x, \mathfrak{E}) * \ldots * P(x, \mathfrak{E})}_{p \text{ times}} = [P(x, \mathfrak{E})]_*^p = P^p(x, \mathfrak{E}). \tag{19}$$

As far back as 1900 Bachelier considered stochastic processes continuous in time.[4] There are good grounds for giving schemes with continuous time a central place in probability theory. It seems that most important here are schemes homogeneous in time, in which $P(t, x, t + \tau, \mathfrak{E})$ depends only on the difference $t_2 - t_1$:

$$P(t, x, t + \tau, \mathfrak{E}) * P(\tau_2, x, \mathfrak{E}) = P(\tau, x, \mathfrak{E}). \tag{20}$$

[4] See the first of the papers cited in footnote 2.

The fundamental equation in this case takes the form

$$P(\tau_1, x, \mathfrak{E}) * P(\tau_2, x, \mathfrak{E}) = P(\tau_1 + \tau_2, x, \mathfrak{E}). \tag{21}$$

Another series of particular cases is obtained under special requirements on the set $\mathfrak{A}$ of elementary states x. Here one should distinguish the cases of finite or countable sets $\mathfrak{A}$; in the continuous case the classification is performed with respect to the number of parameters determining the state of the system. The subsequent subdivision of the material in this paper is based on distinguishing such particular cases.

§4. The ergodic principle

Without special assumptions on the set $\mathfrak{A}$ of all possible states x, we can only prove several general theorems, namely those dealing with the ergodic principle. We say that a stochastic process obeys the *ergodic principle* if for any $t^{(0)}, x, y$ and $\mathfrak{E}$

$$\lim_{t\to\infty} [P(t^{(0)}, x, t, \mathfrak{E}) - P(t^{(0)}, y, t, \mathfrak{E})] = 0. \tag{22a}$$

For a scheme with discrete time (22a) is clearly equivalent to the following:

$$\lim_{n\to\infty} [P_{mn}(x, \mathfrak{E}) - P_{mn}(y, \mathfrak{E})] = 0; \tag{22b}$$

and in the latter case the following theorem holds:

Theorem 1. *If for any x, y, and $\mathfrak{E}$*

$$P_n(x, \mathfrak{E}) \geq \lambda_n P_n(y, \mathfrak{E}), \ \ \lambda_n \geq 0, \tag{23}$$

and the series

$$\sum_{n=1}^{\infty} \lambda_n \tag{24}$$

diverges, then the ergodic principle (22b) *holds and the limit in* (22b) *is uniform with respect to x, y and $\mathfrak{E}$.*

Proof. Let

$$\sup_x P_{kn}(x, \mathfrak{E}) = M_{kn}(\mathfrak{E}), \quad \inf_x P_{kn}(x, \mathfrak{E}) = m_{kn}(\mathfrak{E}).$$

For $i < k$ we clearly have

$$P_{in}(x, \mathfrak{E}) = \int_{\mathfrak{A}_y} P_{kn}(y, \mathfrak{E}) P_{ik}(x, d\mathfrak{A}) \leq M_{kn}(\mathfrak{E}) \int_{\mathfrak{A}_y} P_{ik}(x, d\mathfrak{A}) = M_{kn}(\mathfrak{E}) \tag{25}$$

and, similarly,

$$P_{in}(x, \mathfrak{E}) \geq m_{kn}(\mathfrak{E}). \tag{26}$$

By (23), for any x and y we have

$$\begin{gathered}
P_k(x, \mathfrak{E}) - \lambda_k P_k(y, \mathfrak{E}) \geq 0, \\
P_{k-1,n}(x, \mathfrak{E}) = \int_{\mathfrak{A}_z} P_{kn}(z, \mathfrak{E}) P_k(x, d\mathfrak{A}) = \\
= \int_{\mathfrak{A}_z} P_{kn}(z, \mathfrak{E})[P_k(x, d\mathfrak{A}) - \lambda_k P_k(y, d\mathfrak{A})] + \\
+ \lambda_k \int_{\mathfrak{A}_z} P_{kn}(z, \mathfrak{E}) P_k(y, d\mathfrak{A}) \geq \\
\geq m_{kn}(\mathfrak{E}) \int_{\mathfrak{A}_z} [P_k(x, d\mathfrak{A}) - \lambda_k P_k(y, d\mathfrak{A})] + \lambda_k P_{k-1,n}(y, \mathfrak{E}) = \\
= m_{kn}(\mathfrak{E})(1 - \lambda_k) + \lambda_k P_{k-1,n}(y, \mathfrak{E}), \\
P_{k-1,n}(y, \mathfrak{E}) - P_{k-1,n}(x, \mathfrak{E}) \leq (1 - \lambda_k)[P_{k-1,n}(y, \mathfrak{E}) - m_{kn}(\mathfrak{E})];
\end{gathered}$$

hence by (25),

$$P_{k-1,n}(y, \mathfrak{E}) - P_{k-1,n}(x, \mathfrak{E}) \leq (1 - \lambda_k)[M_{kn}(\mathfrak{E}) - m_{kn}(\mathfrak{E})]. \tag{27}$$

Since (27) holds for any x and y, we also have

$$M_{k-1,n}(\mathfrak{E}) - m_{k-1,n}(\mathfrak{E}) \leq (1 - \lambda_k)[M_{kn}(\mathfrak{E}) - m_{kn}(\mathfrak{E})]. \tag{28}$$

Setting $k = m+1, m+2, \ldots, n$ successively in (28) and multiplying all the resulting equalities we find

$$M_{mn}(\mathfrak{E}) - m_{mn}(\mathfrak{E}) \leq \prod_{k=m+1}^{n} (1 - \lambda_k). \tag{29}$$

The right-hand side of (29) tends to zero as $n \to \infty$; this proves the theorem.

For a homogeneous scheme with discontinuous time the following holds:

Theorem 2. *If for any x, y and $\mathfrak{E}$*

$$P(x, \mathfrak{E}) \geq \lambda P(y, \mathfrak{E}) \quad (\lambda > 0), \tag{30}$$

then $P^n(x, \mathfrak{E})$ converges uniformly to a certain distribution function $Q(\mathfrak{E})$.

Proof. We have

$$M_{n,n+p}(\mathfrak{E}) = \sup P^p(x, \mathfrak{E}) = M_p(\mathfrak{E}),$$

$$m_{n,n+p}(\mathfrak{E}) = \inf P^p(x, \mathfrak{E}) = m_p(\mathfrak{E}),$$

$$\lambda_n = \lambda,$$

and, by (29),

$$M_p(\mathfrak{E}) - m_p(\mathfrak{E}) \leq (1-\lambda)^p. \tag{31}$$

But (25) and (26) imply that for $q > p$,

$$p^q(x, \mathfrak{E}) = P_{0q}(x, \mathfrak{E}) \leq M_{q-p,q}(\mathfrak{E}) = M_p(\mathfrak{E}), \tag{32}$$

$$P^q(x, \mathfrak{E}) \geq m_p(\mathfrak{E}); \tag{33}$$

therefore,

$$M_p(\mathfrak{E}) \geq M_q(\mathfrak{E}) \geq m_q(\mathfrak{E}) \geq m_p(\mathfrak{E}). \tag{34}$$

Our theorem now follows immediately from (31) and (34).

Important particular cases of Theorem 2 were proved by Gostinskii and Hadamard.[5] As has been shown by Hadamard, in these particular cases $Q(\mathfrak{E})$ satisfies the integral equation

$$Q(\mathfrak{E}) = \int_{\mathfrak{A}_x} P(x, \mathfrak{E}) Q(d\mathfrak{A}). \tag{35}$$

For the most general stochastically determined scheme one has:

Theorem 3. *If for some sequence*

$$t_0 < t_1 < \ldots < t_n < \ldots \to \infty$$

and any x, y and $\mathfrak{E}$,

$$P(t_{n-1}, x, t_n, \mathfrak{E}) \geq \lambda_n P(t_{n-1}, y, t_n, \mathfrak{E}), \quad \lambda_n \geq 0, \tag{36}$$

and if the series $\sum_{n=1}^{\infty} \lambda_n$ diverges, then the ergodic principle (22a) *holds and the convergence in* (22a) *is uniform with respect to $x, y, \mathfrak{E}$.*

Proof. For a given $t^{(0)}$ let

$$\sup_x P(t^{(0)}, x, t, \mathfrak{E}) = M(t, \mathfrak{E}),$$

$$\inf_x P(t^{(0)}, x, t, \mathfrak{E}) = m(t, \mathfrak{E}).$$

---

[5] *C. R. Acad. Sci. Paris* **186** (1928), 59; **189**; 275.

If

$$t^{(0)} \leq t_m \leq t_n \leq t \leq t_{n+1},$$

then, in the same way as in the proof of Theorem 1, we obtain the analogous formula to (29),

$$M(t, \mathfrak{E}) - m(t, \mathfrak{E}) \leq \prod_{k=m+1}^{n} (1 - \lambda_k).$$

Since $n \to \infty$ together with t, it follows that $M(t, \mathfrak{E}) - m(t, \mathfrak{E}) \to 0$ as $t \to \infty$, which proves the theorem.

Finally in case of a scheme homogeneous in time one has the following theorem, analogous to Theorem 2:

Theorem 4. *If there exists σ such that for any $x, y, \mathfrak{E}$,*

$$P(\sigma, x, \mathfrak{E}) \geq \lambda P(\sigma, y, \mathfrak{E}) \quad (\lambda > 0), \tag{37}$$

then $P(\tau, x, \mathfrak{E})$ converges uniformly to a certain distribution function $Q(\mathfrak{E})$ as $\tau \to \infty$.

CHAPTER II. FINITE STATE SYSTEMS

§5. Preliminary remarks

Let us now assume that $\mathfrak{A}$ is formed from a finite number of elements

$$x_1, x_2, \ldots, x_n.$$

In this case set

$$P(t_1, x_i, t_2, x_j) = P_{ij}(t_1, t_2). \tag{38}$$

Since for any set $\mathfrak{E}$ we obviously have

$$P(t_1, x_i, t_2, \mathfrak{E}) = \sum_{x_k \subset \mathfrak{E}} P_{ik}(t_1, t_2), \tag{39}$$

we can confine ourselves to the probabilities $P_{ij}(t_1, t_2)$. The fundamental equation (3) now takes the form

$$\sum_j P_{ij}(t_1, t_2) P_{jk}(t_2, t_3) = P_{ik}(t_1, t_3), \tag{40}$$

whereas (1) can be written as

$$\sum_j P_{ij}(t_1, t_2) = 1. \tag{41}$$

Any non-negative functions $P_{ij}(t_1, t_2)$ satisfying the conditions (40) and (41) determine some stochastically determined process of variation of the systems $\mathfrak{S}$.

In this case the operator is defined as follows:

$$F_{ik} = F_{ik}^{(1)} * F_{ik}^{(2)} = \sum_j F_{ij}^{(1)} F_{jk}^{(2)}, \tag{42}$$

hence the fundamental equation (40) reduces to

$$P_{ik}(t_1, t_2) * P_{ik}(t_2, t_3) = P_{ik}(t_1, t_3). \tag{43}$$

For a scheme with discontinuous time we set

$$P_{pq}(x_i, x_j) = P_{ij}^{(pq)}, \quad P_p(x_i, x_j) = P_{ij}^{(p)}.$$

Then the probabilities $P_{ij}^{(p)}$ satisfy

$$\sum_j P_{ij}^{(p)} = 1, \tag{44}$$

and, conversely, arbitrary non-negative values $P_{ij}^{(p)}$ satisfying (44) can be considered as the corresponding values of the probabilities of a certain stochastically determined process.

The probabilities $P_{ij}^{(pq)}$ can be calculated by the formula

$$P_{ij}^{(pq)} = P_{ij}^{(p+1)} * P_{ij}^{(p+2)} * \ldots * P_{ij}^{(q)}. \tag{45}$$

For a homogeneous scheme with discontinuous time we have

$$P_{ij}^{(p)} = P_{ij}, \quad P_{ij}^{(pq)} = [P_{ij}]_*^{q-p} = P_{ij}^{q-p}.$$

If all the P_{ij} are positive, then obviously the conditions of Theorem 2 (§4) hold, hence P_{ij}^q tend to a certain limit Q_j as $q \to \infty$. The integral equation (35) transforms in our case into the system of equations

$$Q_i = \sum_j Q_j P_{ji} \quad (i = 1, \ldots, n). \tag{46}$$

These results were obtained by Gostinskii and Hadamard.[6]

§6. Differential equations of a continuous stochastic process

By (11) we have

$$P_{ii}(t,t) = 1, \quad P_{ij}(t,t) = 0, \; i \neq j. \tag{47}$$

If the variations of our system $\mathfrak{S}$ are possible at any time t, then it is natural to suppose that

$$\lim_{\Delta \to \infty} P_{ii}(t, t+\Delta) = 1, \quad \lim_{\Delta \to 0} P_{ij}(t, t+\Delta) = 0, \; i \neq j, \tag{47a}$$

that is, for small time intervals the probability of a change in the state of the system is small. This assumption is contained in the hypothesis of the continuity of the functions $P_{ij}(t_1, t_2)$ with respect to t_1 and t_2.

Now assume that the functions $P_{ij}(s,t)$ are continuous and differentiable with respect to t and s for $t \neq s$. We do not require differentiability of these functions at $t = s$. It would be imprudent to assume *a priori* the existence of a derivative at these special points.[7]

For $t > s$ we have

$$\begin{aligned} \frac{\partial P_{ik}(s,t)}{\partial t} &= \lim_{\Delta \to 0} \frac{P_{ik}(s,t+\Delta) - P_{ik}(s,t)}{\Delta} = \\ &= \lim_{\Delta \to 0} \frac{1}{\Delta}\Big[\sum_j P_{ij}(s,t)P_{jk}(t,t+\Delta) - P_{ik}(s,t)\Big] = \\ &= \lim_{\Delta \to 0}\Big[\sum_{j \neq k} P_{ij}(s,t)\frac{P_{ij}(t,t+\Delta)}{\Delta} + P_{ik}(s,t)\frac{P_{kk}(t,t+\Delta)-1}{\Delta}\Big]. \end{aligned} \tag{48}$$

If the determinant

$$\Xi = |P_{ij}(s,t)|$$

is non-zero, then the equations

$$\sum_{j \neq k} P_{ij}(s,t)\frac{P_{jk}(t,t+\Delta)}{\Delta} + P_{ik}(s,t)\frac{P_{kk}(t,t+\Delta)-1}{\Delta} = \alpha_{ik} \quad (i = 1, \dots, n)$$

can be solved:

$$\frac{P_{kk}(t,t+\Delta)-1}{\Delta} = \frac{\Lambda_{kk}}{\Xi}, \quad \frac{P_{jk}(t,t+\Delta)}{\Delta} = \frac{\Lambda_{jk}}{\Xi}, \; j \neq k. \tag{49}$$

---

[6] See Footnote 5.

[7] Compare with the functions $F(s,x,t,y)$ considered in Chapter 4, which necessarily have points of discontinuity at $t = s$.

Since by (48) α_{ik} tend to the limit values $\partial P_{ik}(s,t)/\partial t$ as $\Delta \to 0$, the values (49) tend to well defined limits[8]

$$\lim \frac{P_{kk}(t,t+\Delta)-1}{\Delta} = A_{kk}(t), \tag{50a}$$

$$\lim \frac{P_{jk}(t,t+\Delta)}{\Delta} = A_{jk}(t), \quad j \neq k. \tag{50b}$$

In fact it is evident from the relation

$$\lim_{s\to t} \Xi = 1, \tag{51}$$

which holds by (47) and the continuity of Ξ, that Ξ may be non-zero under a proper choice of $s < t$.

From (48) and (50) we immediately obtain the *first system of differential equations* for the function $P_{ik}(s,t)$:

$$\frac{\partial P_{ik}(s,t)}{\partial t} = \sum_j A_{jk}(t) P_{ij}(s,t) = P_{ik}(s,t) * A_{ik}(t). \tag{52}$$

In this case, by (47) and (50),

$$A_{jk}(t) = \left[\frac{\partial P_{jk}(t,u)}{\partial u}\right]_{u=t}, \tag{53}$$

$$A_{jk} \geq 0,\ j \neq k, \quad A_{kk} \leq 0, \tag{54}$$

and, by (41) and (50),

$$\sum_k A_{jk} = 0. \tag{55}$$

The equations (52) were established only for $s < t$; however, (47) and (53) show that these equations are valid also for $t = s$.

For $s < t$ we have

$$\frac{\partial P_{ik}(s,t)}{\partial s} = \lim_{\Delta\to 0} \frac{P_{ik}(s+\Delta,t) - P_{ik}(s,t)}{\Delta} =$$

$$= \lim_{\Delta\to 0} \frac{1}{\Delta}\Big[P_{ik}(s+\Delta,t) - \sum_j P_{ij}(s,s+\Delta)P_{jk}(s+\Delta,t)\Big] =$$

$$= -\lim_{\Delta\to 0}\Big[\frac{P_{ii}(s,s+\Delta)-1}{\Delta} P_{ik}(s+\Delta,t) +$$

$$+ \sum_{j\neq i} \frac{P_{ij}(s,s+\Delta)}{\Delta} P_{jk}(s+\Delta,t)\Big] \tag{56}$$

[8] We could equally well have taken the opposite approach: to assume *a priori* that the conditions (47a) and (50) hold and to derive from this the continuity and differentiability of the function $P_{ij}(s,t)$ with respect to t.

and, by (50), we obtain the *second system of differential equations*

$$\frac{\partial P_{ik}(s,t)}{\partial s} = -\sum_j A_{ij}(s)P_{jk}(s,t) = -A_{ik}(s) * P_{ik}(s,t). \tag{57}$$

If the functions $A_{ij}(s)$ are continuous, then clearly the equations (57) are also true for $s = t$.

Now assume that at t_0 we know the distribution function

$$Q(t_0, x_k) = Q_k(t_0), \quad \sum_k Q_k(t_0) = 1,$$

of the probabilities that the system $\mathfrak{S}$ at t_0 is in the state x_k. Then equation (5) takes the form

$$Q_k(t) = \sum Q_i(t_0)P_{ik}(t_0,t).$$

By (52) the functions $Q_k(t)$ satisfy

$$\frac{dQ_k(t)}{dt} = \sum_j A_{jk}(t)Q_j(t) \quad (k = 1, \ldots, n). \tag{58}$$

If the functions $A_{ik}(t)$ are continuous, then the functions $P_{ik}(s,t)$ form the unique system of solutions of (52) satisfying the initial conditions (47); consequently, the considered stochastic process is totally determined by all the $A_{ik}(t)$. The real meaning of the functions $A_{ik}(t)$ can be illustrated in the following way: for $i \neq k$ $A_{ik}(t)dt$ is the probability of passing from the state x_i to the state x_k during the time from t to $t + dt$, whereas

$$A_{kk}(t) = -\sum_{j \neq k} A_{kj}(t).$$

It can also be shown that if we have any continuous functions $A_{ik}(t)$ satisfying the conditions (54) and (55), then the solutions $P_{ik}(s,t)$ of the differential equations (52) under the initial conditions (47) are non-negative and satisfy the conditions (40) and (41); in other words, they determine a stochastic process.

Indeed, by (52) and (55) we have

$$\frac{\partial}{\partial t}\sum_k P_{ik}(s,t) = \sum \Big[\sum_k A_{jk}(t)\Big] P_{ij}(s,t) = 0, \tag{59}$$

and, by (47),

$$\sum_k P_{ik}(t,t) = 1.$$

Thus (59) implies (41).

For $t_1 < t_2$ we now assume that

$$P'_{ik}(t_1,t) = P_{ik}(t_1,t), \quad \text{if } t_1 \le t \le t_2, \tag{60}$$

$$P'_{ik}(t_1,t) = \sum_j P_{ij}(t_1,t_2)P_{jk}(t_2,t), \quad \text{if } t_2 < t. \tag{61}$$

The functions $P'_{ik}(t_1,t)$ are continuous and satisfy the differential equations (52); consequently, (60) holds for any t and not merely for $t \le t_2$; but then (61) with $t = t_3$ coincides with (40),

It remains to show that the solutions $P_{ik}(t_1,t)$ are non-negative. For this we assume that for fixed s,

$$\psi(t) = \min P_{ik}(s,t).$$

Choosing appropriate i and k we clearly have

$$D^+\psi(t) = \frac{\partial P_{ik}(s,t)}{\partial t}, \quad P_{ik}(s,t) = \psi(t),$$

and if $\psi(t) \le 0$, then by (54),

$$A_{kk}(t)P_{ik}(s,t) \ge 0,$$

$$A_{jk}(t)P_{ij}(s,t) \ge A_{jk}(t)\psi(t), \ j \ne k,$$

$$D^+\psi(t) = \frac{\partial P_{ik}(s,t)}{\partial t} = \sum_j A_{jk}(t)P_{ij}(s,t) \ge$$

$$\ge \sum_{j\ne k} A_{jk}(t)\psi(t) = R(t)\psi(t).$$

Since $\psi(s) = 0$, $\psi(t)$ is clearly greater than any negative solution of the equation

$$dy/dt = R(t)y,$$

and therefore it cannot be negative itself.

§7. Examples

In schemes homogeneous in time the coefficients $A_{ik}(t)$ appear to be independent of the time t; in this case the process is completely determined by the n^2 constants A_{ik}. Equations (52) now take the form

$$\frac{dP_{ik}(t)}{dt} = \sum_j A_{ik}P_{ij}(t); \tag{62}$$

and solving these equations is not difficult. If all the A_{ik} are non-zero, then the conditions of Theorem 4 (§4) hold and consequently, $P_{ik}(t)$ tends to a limit Q_k as $t \to \infty$. The quantities Q_k satisfy the equations

$$\sum_k Q_k = 1, \quad \sum_j A_{jk} Q_j = 0 \ (k = 1, \dots, n).$$

For example, let

$$n = 2, \quad A_{12} = A_{21} = A, \quad A_{11} = A_{22} = -A,$$

that is, the probabilities of transition from the state x_1 to the state x_2 and the reverse transition from x_2 to x_1 are the same. The differential equations (62) in our case give

$$P_{12}(t) = P_{21}(t) = \tfrac{1}{2}(1 - e^{-2At}),$$

$$P_{11}(t) = P_{22}(t) = \tfrac{1}{2}(1 + e^{-2At}).$$

We see that $P_{ik}(t)$ tends to the limit $Q_k = \frac{1}{2}$ as $t \to \infty$.

The following example shows that approaching the limit can be accompanied by oscillations damping with time:

$$n = 3, \quad A_{12} = A_{23} = A_{31} = A,$$

$$A_{21} = A_{32} = A_{13} = 0, \quad A_{11} = A_{22} = A_{33} = -A;$$

$$P_{11}(t) = P_{22}(t) = P_{33}(t) = \frac{2}{3} e^{-3/2At} \cos \alpha t + \frac{1}{3},$$

$$P_{12}(t) = P_{23}(t) = P_{31}(t) = e^{-3/2At} \Big(\frac{1}{\sqrt{3}} \sin \alpha t - \frac{1}{3} \cos \alpha t \Big) + \frac{1}{3},$$

$$P_{21}(t) = P_{32}(t) = P_{13}(t) = -e^{-3/2At} \Big(\frac{1}{\sqrt{3}} \sin \alpha t + \frac{1}{3} \cos \alpha t \Big) + \frac{1}{3},$$

$$\alpha = \frac{\sqrt{3}}{2} A.$$

Similar damping oscillations for schemes with discontinuous time were found by Romanovskii.

Chapter III. COUNTABLE STATE SYSTEMS

§8. Preliminary remarks. Discontinuous schemes

If $\mathfrak{A}$ consists of a countable set of elements

$$x_1, x_2, \dots, x_n, \dots,$$

all the notations and results of §5 of Chapter II remain valid. The convergence of the series

$$\sum_k P_{ik}(t_1,t_2) = 1, \quad \sum_k F_{ik} = 1$$

is assumed, and from this we derive convergence of the series (40), (42), (46); by contrast, we do not require that the series

$$\sum_i P_{ik}(t_1,t_2)$$

should converge.

We now make a few remarks on schemes with discontinuous time, in particular homogeneous ones. The conditions of our theorems concerning ergodic principles for schemes with a countable set of states fail in most cases, but nevertheless the principle itself often appears to be satisfied.

Consider, for example, a game studied recently by S.N. Bernshtein: in any separate trial a gambler wins only one rouble with probability A and loses it with probability B ($B > A$, $A + B \leq 1$), the latter, however, provided only that his cash is non-zero; otherwise he does not lose anything.

If we denote by x_n the state in which the cash of our gambler is $n-1$ roubles, then the conditions of the game can be written as follows:

$$P_{n,n+1} = A, \quad P_{n+1,n} = B \quad (n = 1,2,3,\ldots),$$
$$P_{11} = 1 - A, \quad P_{nn} = 1 - A - B \quad (n = 2,3,4,\ldots),$$
$$P_{ij} = 0 \quad \text{otherwise.}$$

It can easily be proved that

$$\lim_{p\to\infty} P_{ij}^p = \left(1 - \frac{A}{B}\right)\left(\frac{A}{B}\right)^{j-1} = Q_j, \quad \sum_j Q_j = 1,$$

which implies the ergodic principle in this case.

Note that the fact that the limits

$$\lim_{p\to\infty} P_{ij}^p = \Lambda_j$$

exist, implies the ergodic principle only if

$$\sum_j \Lambda_j = \Lambda = 1.$$

It can be shown that always $\Lambda \leq 1$ and that for $\Lambda < 1$ the ergodic principle fails.

If all the Λ_j exist and are zero, then there arises the question of the asymptotic expression for P_{ij}^p as $p \to \infty$. If such an expression exists independently of i:

$$P_{ij}^p = \lambda_j^p + o(\lambda_j^p),$$

then we say that the *local ergodic principle* holds. This principle seems to be of great significance in the case of a countable set of possible states.

Now let all possible states x be enumerated by the integers $(-\infty < n < +\infty)$. All the notation and formulas of §5 are then true, but now the sums run over all the integers. We consider the case

$$P_{ij} = P_{j-i}^p$$

in more detail. Clearly in this case we have

$$P_{ij}^p = P_{j-i}^p, \quad P_k^{p+1} = \sum_i P_i^p P_{k-i}^n, \quad P_k^{m+n} = \sum_i P_i^m P_{k-i}^n.$$

If the series

$$a = \sum_k k P_k, \quad b^2 = \sum_k k^2 P_k$$

are absolutely convergent, then there arises the question on the conditions of applicability of the generalized Laplace formula

$$P_k^p = \frac{1}{b\sqrt{2\pi p}} \exp\left[-\frac{(k-pa)^2}{2pb^2}\right] + o\left(\frac{1}{\sqrt{p}}\right). \tag{63}$$

All we know is that it holds in the Bernoulli case, when

$$P_0 = 1 - A, \quad P_1 = A, \tag{64}$$

and the other P_k vanish. Lyapunov's theorem is of no help for our problem, as is clear from the following example:

$$P_{+1} = P_{-1} = \tfrac{1}{2}, \quad P_k = 0, \ k \neq \pm 1,$$

where (63) is inapplicable. In order for (63) to hold, it is necessary[9] that for any integer m there exists k such that

$$k \not\equiv 0 \pmod{m}, \quad P_k \neq 0.$$

[9] More details on this question can be found in R. von Mises *Wahrscheinlichkeitsrechnung*, Berlin, 1931, especially the chapter on "local" limit theorems. (Remark by Russian editor.)

Note also that only for $a = 0$ does formula (63) actually give an asymptotic expression for P_k^p for a given k. In this case it follows from (63) that for a given

$$P_k^p = \frac{1}{b\sqrt{2\pi p}} + o\left(\frac{1}{\sqrt{p}}\right) \tag{65}$$

and given i and j,

$$P_{ij}^p = \frac{1}{b\sqrt{2\pi p}} + o\left(\frac{1}{\sqrt{p}}\right). \tag{66}$$

By (66) we obtain the ergodic principle in the case considered.

We obtain special approximation formulas for P_{ij}^p when the probabilities P_{ii}, that is, the probabilities of the facts that the state of the system does not vary at any particular moment, are very close to one. For example, in the Bernoulli case, for small A the approximate Poisson formula can be used:

$$P_k^p \sim \frac{A^k p^k}{k!} e^{-Ap}. \tag{67}$$

A general method for deriving such formulas can be obtained by using differential equations for processes continuous in time, as is shown in §10 for formula (67).

§9. Differential equations of a process continuous in time

As in §6, we assume that the functions $P_{ij}(s,t)$ are continuous and have derivatives with respect to t and s for $t \neq s$. In the case of a countable set of possible states formulas (48) and (56) still remain valid; but to prove the possibility of changing the order of the sum and the limit in these formulas and thus arrive at the differential equations (52) and (57), we have to introduce new restrictions, namely:

A) the existence of limit values in (50);

B) uniform convergence in (50b) with respect to j for a given k;

C) uniform convergence of the series

$$\sum_{k \neq j} \frac{P_{jk}(t, t+\Delta)}{\Delta} = \frac{1 - P_{jj}(t, t+\Delta)}{\Delta} \tag{68}$$

with respect to Δ (the fact that this series converges follows immediately from (41)).

In §6, for a finite number of states we deduced condition A) from the differentiability of $P_{ij}(s,t)$ for $t \neq s$; by contrast, in the case of a countable set of states this condition does not seem to follow from this property of P_{ij}. With regard to condition B), note that uniform convergence in (50b) with respect to k for a given j follows from the obvious inequality

$$P_{jk}(t,t+\Delta) \le 1 - P_{jj}(t,t+\Delta).$$

Note, further, that we do not require uniform convergence in (50b) for any j and k, nor do we require uniform convergence in (50a) with respect to k; these requirements would have been inconvenient for applications.

Since the factors $P_{ij}(s,t)$ in (48) form an absolutely convergent series, we can, in view of conditions A) and B), change the order of the signs lim and $\sum$ in this formula and obtain (52). Then the variables $A_{jk}(t)$ clearly satisfy the formulas of the last condition; moreover, since the factors $P_{jk}(s+\Delta,t)$ are uniformly bounded, we can change the order of the sum and limit signs in (56), which suffices for deducing (57).

§10. Uniqueness of solutions and their calculation for a process homogeneous in time

In the present case, (52) takes the form

$$\frac{dP_{ik}(t)}{dt} = \sum_j A_{jk}P_{ij}(t) = P_{ik}(t) * A_{ik}, \tag{69}$$

with constants A_{jk}. We will prove that if the series

$$\begin{gathered}\sum_j |A_{jk}| = B_k^{(1)},\\ \sum_j B_j^{(1)}|A_{jk}| = B_k^{(2)},\\ \dots\dots\dots \\ \sum_j B_j^{(n)}|A_{jk}| = B_k^{(n+1)},\\ \dots\dots\dots\end{gathered} \tag{70}$$

$$\sum_n \frac{B_k^{(n)}}{n!}x^n, \quad k=1,2,\dots, \quad |x| \le \theta\ (>0), \tag{71}$$

converge and the initial conditions

$$P_{ii}(0) = 1, \quad P_{ij}(0) = 0,\ i \ne j, \tag{72}$$

hold, then the equations (69) have the unique system of solutions $P_{ik}(t)$ satisfying the conditions of our problem.

Indeed, since always $P_{ij}(t) \leq 1$, (69) and (70) imply

$$|dP_{ik}(t)/dt| \leq B_k^{(1)};$$

therefore (69) can be differentiated term by term

$$\frac{d^2 P_{ik}(t)}{dt^2} = \sum A_{jk} \frac{dP_{ij}(t)}{dt} = \frac{d}{dt} P_{ik}(t) * A_{ik}.$$

In a similar way the general relations are obtained:

$$\left|\frac{d^n}{dt^n} P_{ik}(t)\right| \leq B_k^{(n)}, \tag{73}$$

$$\frac{d^{n+1}}{dt^{n+1}} P_{ik}(t) = \frac{d^n}{dt^n} P_{ik}(t) * A_{ik}. \tag{74}$$

From (73) and the assumption of convergence of the series (71) it follows that the functions P_{ik} are analytic. Further, by (69) and (74) we find that

$$\frac{d^n}{dt^n} P_{ik}(t) = P_{ik}(t) * [A_{ik}]_*^n; \tag{75}$$

in particular, for $t = 0$ we have, by (72),

$$\frac{d^n}{dt^n} P_{ik}(0) = [A_{ik}]_*^n, \tag{76}$$

which implies that the analytic functions $P_{ik}(t)$ are uniquely determined by the constants A_{ik}. Formulas (76) and (75) serve also for calculating the solutions of the system (69) using Taylor series.

For example, if

$$A_{i,i+1} = A, \quad A_{ii} = -A,$$

$$A_{ij} = 0 \text{ otherwise},$$

then we easily obtain

$$P_{mn}(t) = \frac{(At)^{n-m}}{(n-m)!} e^{-At}, \quad n \geq m,$$

$$P_{mn}(t) = 0, \quad m > n,$$

that is, the formula of the Poisson distribution: for $k = n - m$, $p = t$ the resulting formula coincides with (67).

If the ergodic principle holds and $P_{ik}(t) \to Q_k$ as $t \to \infty$, then obviously the constants Q_k satisfy the equations

$$\sum_k Q_k = 1, \quad \sum_i A_{ik} Q_i = 0 \quad (k = 1, 2, \ldots). \tag{77}$$

If, for example

$$\begin{aligned} A_{i,i+1} &= A, \quad A_{i+1,i} = B, \quad B > A, \\ A_{11} &= -A, \quad A_{ii} = -(A+B), \quad i > 1, \\ A_{ij} &= 0 \quad \text{otherwise}, \end{aligned}$$

then we easily obtain from (77)

$$Q_n = (1 - A/B)(A/B)^{n-1}.$$

As a second example, we set

$$\begin{aligned} A_{i,i+1} &= A, \quad A_{i+1,i} = iB, \\ A_{ii} &= -A - (i-1)B, \\ A_{ij} &= 0 \quad \text{otherwise}, \end{aligned}$$

so that from (77) we have

$$Q_{n+1} = \frac{1}{n!}\left(\frac{A}{B}\right)^n e^{-A/B},$$

which again is Poisson's formula.

CHAPTER IV. CONTINUOUS STATE SYSTEMS, THE CASE OF ONE PARAMETER

§11. Preliminary remarks

Suppose now that the state of the system considered is determined by the values of a certain real parameter x; in this case we denote by x both the state of the system and the value of the parameter corresponding to this state. If $\mathfrak{E}_y$ is the set of all states x for which $x \leq y$, then we set

$$P(t_1, x, t_2, \mathfrak{E}_y) = F(t_1, x, t_2, y).$$

As a function of y, $F(t_1, x, t_2, y)$ is monotone and right continuous and satisfies the boundary conditions

$$F(t_1, x, t_2, -\infty) = 0, \quad F(t_1, x, t_2, +\infty) = 1. \tag{78}$$

For the function $F(t_1, x, t_2, y)$ the fundamental equation (3) transforms into:

$$F(t_1, x, t_3, z) = \int_{-\infty}^{\infty} F(t_2, y, t_3, z) dF(t_1, x, t_2, y). \tag{79}$$

Thus we have to use integral distribution functions of random variables and ordinary Stieltjes integrals.

Integral (79) exists according to Lebesgue[10] if $F(t_2, y, t_3, z)$ is Borel-measurable with respect to y. In what follows we assume that the system $\mathfrak{F}$ (see §1) coincides with the system of all Borel sets, which implies Borel-measurability of $F(t_1, x, t_2, y)$ as a function of x. In this case, as is known, the additive set function $P(t_1, x, t_2, \mathfrak{E})$, for all Borel sets $\mathfrak{E}$, is uniquely determined by the corresponding function $F(t_1, x, t_2, y)$.

A function $F(y)$, monotone and right continuous, such that

$$F(-\infty) = 0, \quad F(+\infty) = 1$$

is called a *normal distribution function*. If $F_1(x, y)$ and $F_2(x, y)$, as functions of x, are Borel measurable and, as functions of y, are normal distribution functions, then the same is true for the function

$$F(x, y) = F_1(x, y) \oplus F_2(x, y) = \int_{-\infty}^{\infty} F_2(z, y) dF_1(x, z). \tag{80}$$

This operator $\oplus$, like $*$, obeys the associative law; using this law, the fundamental equation (79) can be expressed as

$$F(t_1, x, t_3, y) = F(t_1, x, t_2, y) \oplus F(t_2, x, t_3, y). \tag{81}$$

with $F_1(x, y) = V_1(y - x)$, $\quad F_2(x, y) = V_2(y - x)$. Then, as can easily be shown,

$$F_1(x, y) \oplus F_2(x, y) = V(y - x) = V_1(y - x) \odot V_2(y - x), \tag{82}$$

[10] H. Lebesgue, *Leçons sur l'intégration et la recherche des fonctions primitives*, Gauthier-Villars, Paris, 1928.

where

$$V(x) = V_1(x) \odot V_2(x) = \int_{-\infty}^{\infty} V_2(x-z)dV_1(z). \tag{83}$$

The associative law also holds for the operator $\odot$ while for normal distribution functions the commutative law holds as well; if $V_1(x)$ and $V_2(x)$ are considered as distribution functions of two independent random variables X_1 and X_2, then $V_1(x) \odot V_2(x)$, as is known, is the distribution function of the sum[11] $X = X_1 + X_2$.

If $F(t_1, x, t_2, y)$ is absolutely continuous as a function of y, then we have

$$F(t_1, x, t_2, y) = \int_{-\infty}^{y} f(t_1, x, t_2, y)dy. \tag{84}$$

In this case the non-negative function $f(t_1, x, t_2, y)$ is Borel measurable with respect to x and y and satisfies

$$\int_{-\infty}^{\infty} f(t_1, x, t_2, y) = 1, \tag{85}$$

$$f(t_1, x, t_3, z) = \int_{-\infty}^{\infty} f(t_1, x, t_2, y)f(t_2, y, t_3, z)dy. \tag{86}$$

Conversely, if (85), (86) hold for $f(t_1, x, t_2, y)$, then the function $F(t_1, x, t_2, y)$ defined by (84) satisfies (78) and (79): hence, such a function determines the scheme of a stochastic process. This function $f(t_1, x, t_2, y)$ will be called the *differential distribution function* for the random variable y.

Note also that the following mixed formulas hold:

$$F(t_1, x, t_3, z) = \int_{-\infty}^{\infty} F(t_2, y, t_3, z)f(t_1, x, t_2, y)dy, \tag{87}$$

$$f(t_1, x, t_3, z) = \int_{-\infty}^{\infty} f(t_2, y, t_3, z)dF(t_1, x, t_2, y). \tag{88}$$

When the scheme is discontinuous in time, the functions

$$F_{mn}(x, y) = F(t_m, x, t_n, y), \quad F_n(x, y) = F_{n-1,n}(x, y),$$

are considered; they satisfy the equations

$$F_{m,n+1}(x, y) = F_{mn}(x, y) \oplus F_{n+1}(x, y), \tag{89}$$

$$F_{kn}(x, y) = F_{km}(x, y) \oplus F_{mn}(x, y) \quad (k < m < n). \tag{90}$$

[11] See P. Lévy, *Calcul des probabilités*, Paris, 1927, p.187.

If

$$F_{mn}(x,y) = \int_{-\infty}^{y} f_{mn}(x,y)dy, \quad f_n(x,y) = f_{n-1,n}(x,y),$$

then, in addition, we have

$$f_{m,n+1}(x,z) = \int_{-\infty}^{\infty} f_{mn}(x,y) f_{n+1}(y,z)dy, \tag{91}$$

$$f_{kn}(x,z) = \int_{-\infty}^{\infty} f_{km}(x,y) f_{mn}(y,z)dy \quad (k < m < n). \tag{92}$$

§12. Lindeberg's method. Passage from discontinuous to continuous schemes

As we noted in §3, probability theory usually deals only with schemes that are discontinuous in time. For these schemes, the main problem is to find approximate expressions for the distributions $F_{mn}(x,y)$ for large $n-m$, or what is essentially the same, to construct asymptotic formulas for $F_{mn}(x,y)$ as $n \to \infty$. The Laplace-Lyapunov theorem is the most important result achieved in this direction. Now we will consider in more detail the proof of this theorem given by Lindeberg [12] with the purpose of outlining his main idea in as general a form as possible and thus obtaining a general method for constructing asymptotic expressions for $F_{mn}(x,y)$.

Let

$$F_n(x,y) = V_n(y-x),$$

$$a_n(x) = \int_{-\infty}^{\infty} (y-x)dF_n(x,y) = \int_{-\infty}^{\infty} y dV_n(y) = 0,$$

$$b_n^2(x) = \int_{-\infty}^{\infty} (y-x)^2 dF_n(x,y) = \int_{-\infty}^{\infty} y^2 dV_n(y) = b_n^2,$$

$$B_{mn}^2 = b_{m+1}^2 + b_{m+2}^2 + \ldots + b_n^2.$$

The Laplace-Lyapunov theorem states that under certain additional assumptions, for constant m and as $n \to \infty$ we have

$$F_{mn}(x,y) = \Phi\Big(\frac{y-x}{B_{mn}}\Big) + o(1)$$

[12] *Math. Z.* **15** (1922), 211.

uniformly with respect to x and y, where

$$\Phi(z) = \frac{1}{\sqrt{2\pi}} \int_{-\infty}^{z} e^{-z^2/2} dz.$$

Along with the stochastic process with discontinuous time determined by the functions $F_n(x, y)$, we will consider another one, with continuous time; we suppose that it is characterized by the function

$$\overline{F}(t', x, t'', y) = \Phi\Big(\frac{y - x}{\sqrt{t'' - t'}}\Big).$$

Further, let

$$t_0 = 0, \quad t_n = B_{0n}^2,$$
$$\overline{F}_{mn}(x, y) = \overline{F}(t_m, x, t_n, y), \quad \overline{F}_n(x, y) = \overline{F}_{n-1,n}(x, y).$$

Clearly we have

$$\overline{F}_n(x, y) = \Phi\Big(\frac{y - x}{b_n}\Big),$$
$$\bar{a}_n(x) = \int_{-\infty}^{\infty} (y - x) d\overline{F}_n(x, y) = 0,$$
$$\bar{b}_n^2(x) = \int_{-\infty}^{\infty} (y - x)^2 d\overline{F}_n(x, y) = b_n^2.$$

The first and second moments $\bar{a}_n(x)$ and $\bar{b}_n^2(x)$ of the distribution $\overline{F}_n(x, y)$ coincide with the corresponding moments $a_n(x)$ and $b_n^2(x)$ of the distribution $F_n(x, y)$. From this Lindeberg deduced that

$$F_{mn}(x, y) - \overline{F}_{mn}(x, y) \to 0 \quad \text{as } n \to \infty,$$

after which the Laplace-Lyapunov theorem follows directly from the obvious identity

$$\overline{F}_{mn}(x, y) = \Phi\Big(\frac{y - x}{B_{mn}}\Big).$$

In the general case of arbitrary functions $F_n(x, y)$ we can only apply Lindeberg's method if we know a function $\overline{F}(t', x, t'', y)$ characterizing a continuous stochastic process, and which, for a certain sequence of instants of time

$$t_0 < t_1 < t_2 < \ldots < t_n < \ldots$$

gives the moments $\bar{a}_n(x)$, $\bar{b}_n^2(x)$ which coincide with $a_n(x)$, $b_n^2(x)$, or are close to them. A general method for constructing such functions $\overline{F}$ is obtained by using differential equations of continuous processes, considered in the sections below. To pass from $\overline{F}$ to F we can use the following:

Transition Theorem. *Let the functions $F_n(x,y)$ and $\overline{F}_n(x,y)$ determine two stochastic processes with discontinuous time. If*

$$\int_{-\infty}^{\infty}(y-x)dF_n(x,y)=a_n(x) \quad \int_{-\infty}^{\infty}(y-x)d\overline{F}_n(x,y)=\bar{a}_n(x), \tag{93}$$

$$\int_{-\infty}^{\infty}(y-x)^2dF_n(x,y)=b_n^2(x), \quad \int_{-\infty}^{\infty}(y-x)^2d\overline{F}_n(x,y)=\bar{b}_n^2(x), \tag{94}$$

$$\int_{-\infty}^{\infty}|y-x|^3dF_n(x,y)=c_n(x), \quad \int_{-\infty}^{\infty}|y-x|^3d\overline{F}_n(x,y)=\bar{c}_n(x), \tag{95}$$

$$|a_n(x)-\bar{a}_n(x)|\le p_n, \quad |b_n(x)-\bar{b}_n^2(x)|\le q_n, \quad c_n(x)\le r_n, \quad \bar{c}_n(x)\le \bar{r}_n, \tag{96}$$

and if there exists a function $R(x)$ such that

$$\begin{aligned} R(x)&=0, \quad \textit{for } x\le 0,\\ 0\le R(x)&\le 1, \quad \textit{for } 0<x<l,\\ R(x)&=1, \quad \textit{for } l\le x, \end{aligned} \tag{97}$$

and for

$$U_{kn}(x,z)=\int_{-\infty}^{\infty}R(z-y)d\overline{F}_{kn}(x,y) \tag{98}$$

the inequalities

$$\begin{aligned} &\left|\frac{\partial}{\partial x}U_{kn}(x,z)\right|\le K_n^{(1)}, \quad \left|\frac{\partial^2}{\partial x^2}U_{kn}(x,z)\right|\le K_n^{(2)},\\ &\left|\frac{\partial^3}{\partial x^3}U_{kn}(x,z)\right|\le K_n^{(3)}, \quad (k=0,1,\dots,n), \end{aligned} \tag{99}$$

hold, then the relation

$$\overline{F}_{0n}(x,y-l)-\epsilon_n\le F_{0n}(x,y)\le \overline{F}_{0n}(x,y+l)+\epsilon_n, \tag{100}$$

holds, where

$$\epsilon_n=K_n^{(1)}\sum_{k=1}^{n}p_k+\frac{1}{2}K_n^{(2)}\sum_{k=1}^{n}q_k+\frac{1}{6}K_n^{(3)}\sum_{k=1}^{n}(r_k+\bar{r}_k).$$

In applying this theorem to the case when the moments $a(x), b(x), c(x)$ are unbounded as x increases, it is often possible to eliminate this unboundedness by introducing a new properly chosen variable $x' = \phi(x)$.

Proof of the Transition Theorem. By (98) we have

$$U_{k-1,n}(x,y) = \overline{F}_{k-1,n}(x,y) \oplus R(y-x) =$$
$$= \overline{F}_k(x,y) \oplus \overline{F}_{k+1}(x,y) \oplus \ldots \oplus \overline{F}_n(x,y) \oplus R(y-x) =$$
$$= \overline{F}_k(x,y) \oplus U_{kn}(x,y) \tag{101}$$

and by (93)–(95), (99),

$$U_{k-1,n}(x,y) = \int_{-\infty}^{\infty} U_{kn}(z,y)d\overline{F}_k(x,z) =$$
$$= \int_{-\infty}^{\infty} \Big[U_{kn}(x,y) + \frac{\partial}{\partial x}U_{kn}(x,y)\frac{z-x}{1} +$$
$$+ \frac{\partial^2}{\partial x^2}U_{kn}(x,y)\frac{(z-x)^2}{2} + \frac{\partial^3}{\partial x^3}U_{kn}(\xi,y)\frac{(z-x)^3}{6}\Big]d\overline{F}_k(x,z) =$$
$$= U_{kn}(x,y) + \frac{\partial}{\partial x}U_{kn}(x,y)\bar{a}_k(x) +$$
$$+ \frac{\partial^2}{\partial x^2}U_{kn}(x,y)\frac{\bar{b}_k^2(x)}{2} + \bar{\theta}K_n^{(3)}\frac{\bar{c}_k(x)}{6}, \quad |\bar{\theta}| \leq 1. \tag{102}$$

Setting

$$V_{k-1,n}(x,y) = F_k(x,y) \oplus U_{kn}(x,y), \tag{103}$$

we obtain a formula similar to (102),

$$V_{k-1,n}(x,y) = U_{kn}(x,y) + \frac{\partial}{\partial x}U_{kn}(x,y)a_k(x) +$$
$$+ \frac{\partial^2}{\partial x^2}U_{kn}(x,y)\frac{b_k^2(x)}{2} + \theta K_n^{(3)}\frac{c_k(x)}{6}, \quad |\theta| \leq 1. \tag{104}$$

From (102) and (104) and using (96) and (99) it follows that

$$|U_{k-1,n}(x,y) - V_{k-1,n}(x,y)| \leq K_n^{(1)}p_k + \frac{1}{2}K_n^{(2)}q_k + \frac{1}{6}K_n^{(3)}(r_k + \bar{r}_k). \tag{105}$$

Now let

$$W_{kn}(x,y) = F_{0k}(x,y) \oplus U_{kn}(x,y) =$$
$$= F_1(x,y) \oplus F_2(x,y) \oplus \ldots \oplus F_k(x,y) \oplus U_{kn}(x,y) =$$
$$= F_{0,k-1}(x,y) \oplus V_{k-1,n}(x,y). \tag{106}$$

Then by (105) we have

$$|W_{kn}(x,y) - W_{k-1,n}(x,y)| =$$
$$= |F_{0,k-1}(x,y) \oplus V_{k-1,n}(x,y) - F_{0,k-1}(x,y) \oplus U_{k-1,n}(x,y)| \leq$$
$$\leq \int_{-\infty}^{\infty} |V_{k-1,n}(z,y) - U_{k-1,n}(z,y)| dF_{0,k-1}(x,z) \leq$$
$$\leq \sup |V_{k-1,n}(z,y) - U_{k-1,n}(z,y)| \leq K_n^{(1)} p_k + \frac{1}{2} K_n^{(2)} q_k + \frac{1}{6} K_n^{(3)} (r_k + \bar{r}_k). \tag{107}$$

$$|W_{nn}(x,y) - W_{0n}(x,y)| \leq$$
$$\leq K_n^{(1)} \sum_{k=1}^{n} p_k + \frac{1}{2} K_n^{(2)} \sum_{k=1}^{n} q_k + \frac{1}{6} K_n^{(3)} \sum_{k=1}^{n} (r_k + \bar{r}_k) = \epsilon_n.$$

But

$$W_{nn}(x,y) = F_{0n}(x,y) \oplus R(y-x) = \int_{-\infty}^{\infty} R(y-z) dF_{0n}(x,z)$$

and

$$W_{0n}(x,y) = \overline{F}_{0n}(x,y) \oplus R(y-x) = \int_{-\infty}^{\infty} R(y-z) d\overline{F}_{0n}(x,z).$$

Taking into account (97) we obtain

$$\begin{aligned} W_{nn}(x,y) &\leq \int_{-\infty}^{y} dF_{0n}(x,z) = F_{0n}(x,y), \\ W_{nn}(x,y+l) &\geq \int_{-\infty}^{y} dF_{0n}(x,z) = F_{0n}(x,y), \\ W_{0n}(x,y) &\geq \int_{-\infty}^{y-l} d\overline{F}_{0n}(x,z) = \overline{F}_{0n}(x,y-l), \\ W_{0n}(x,y+l) &\leq \int_{-\infty}^{y+l} d\overline{F}_{0n}(x,z) = \overline{F}_{0n}(x,y+l), \end{aligned} \tag{108}$$

Formula (100) now follows immediately from (107) and (108). The details of the proof can be found in Lindeberg's paper referred to above.

§13. The first differential equation for processes continuous in time

If the state of our system $\mathfrak{S}$ can be changed at any moment t, then it is natural to assume that significant changes of the parameter x during small time intervals will occur very seldom, or, more exactly, that for any positive ϵ,

$$P(t, x, t+\Delta, \ |y-x| > \epsilon) \to 0 \quad \text{as } \Delta \to 0. \tag{109}$$

In most cases we may assume that the stricter condition

$$m^{(p)}(t, x, \Delta) = \int_{-\infty}^{+\infty} |y-x|^p dF(t, x, t+\Delta, y) \to 0 \quad \text{as } \Delta \to 0 \tag{110}$$

holds, at least for the first three moments $m^{(1)}, m^{(2)}$ and $m^{(3)}$. A general study of the possibilities that arise under these assumptions is of great interest; some remarks to this end will be given below in §19.

In the following sections we also assume that the following important condition holds:

$$\frac{m^{(3)}(t, x, \Delta)}{m^{(2)}(t, x, \Delta)} \to 0 \quad \text{as } \Delta \to 0. \tag{111}$$

This condition will certainly hold if in the definition of $m^{(3)}(t, x, \Delta)$ via (110) only infinitesimally small differences $y - x$ play a significant role for infinitesimally small Δ or, more precisely, if

$$\frac{\int_{x-\epsilon}^{x+\epsilon} |y-x|^3 dF(t, x, t+\Delta, y)}{\int_{-\infty}^{+\infty} |y-x|^3 dF(t, x, t+\Delta, y)} \to 1 \quad \text{as } \Delta \to 0. \tag{112}$$

Strictly speaking, only in this case is our process continuous in time. Formula (111) also implies that

$$\frac{m^{(2)}(t, x, \Delta)}{m^{(1)}(t, x, \Delta)} \to 0 \quad \text{as } \Delta \to 0.$$

Finally, we will also assume that for $s \neq t$ all the partial derivatives of the function $F(s, x, t, y)$ up to the fourth order exist, and that these derivatives for constant t, y are uniformly bounded with respect to s and x for $t - s > k > 0$. From (78) and (110) we conclude that for $s = t$ the function $F(s, x, t, y)$ is, by contrast, discontinuous. The function

$$f(s, x, t, y) = \frac{\partial}{\partial y} F(s, x, t, y), \tag{113}$$

clearly satisfies (84)–(86) and, at given t, y, has for $t - s > k > 0$ derivatives up to the third order that are uniformly bounded with respect to s and t. All further calculations are made for this differential distribution function $f(s, x, t, y)$.

Set

$$a(t,x,\Delta) = \int_{-\infty}^{\infty} (y-x) f(t,x,t+\Delta,y)dy, \tag{114}$$

$$b^2(t,x,\Delta) = \int_{-\infty}^{\infty} (y-x)^2 f(t,x,t+\Delta,y)dy = m^{(2)}(t,x,\Delta), \tag{115}$$

$$c(t,x,\Delta) = \int_{-\infty}^{\infty} |y-x|^3 f(t,x,t+\Delta,y)dy = m^{(3)}(t,x,\Delta). \tag{116}$$

By (85) and (86) we have

$$\begin{aligned} f(s,x,t,y) &= \int_{-\infty}^{\infty} f(s,x,s+\Delta,z) f(s,\Delta,z,t,y)dz = \\ &= \int_{-\infty}^{\infty} f(s,x,s+\Delta,z)\Big[f(s+\Delta,x,t,y) + \\ &\quad + \frac{\partial}{\partial x} f(s+\Delta,x,t,y)(z-x) + \\ &\quad + \frac{\partial^2}{\partial x^2} f(s+\Delta,x,t,y)\frac{(z-x)^2}{2} + \\ &\quad + \frac{\partial^3}{\partial \xi^3} f(s+\Delta,\xi,t,y)\frac{(z-x)^3}{6}\Big]dz = \\ &= f(s+\Delta,x,t,y) + \frac{\partial}{\partial x} f(s+\Delta,x,t,y)a(s,x,\Delta) + \\ &\quad + \frac{\partial^2}{\partial x^2} f(s+\Delta,x,t,y)\frac{b^2(s,x,\Delta)}{2} + \theta\frac{c(s,x,\Delta)}{6}, \quad |\theta| < C, \end{aligned} \tag{117}$$

where for $s + \Delta < \tau < t$, C can be chosen independently of Δ. From (117) we immediately obtain

$$\begin{aligned} &\frac{f(s+\Delta,x,t,y) - f(s,x,t,y)}{\Delta} = \\ &= -\frac{\partial}{\partial x} f(s+\Delta,x,t,y)\frac{a(s,x,\Delta)}{\Delta} - \\ &\quad -\frac{\partial^2}{\partial x^2} f(s+\Delta,x,t,y)\frac{b^2(s,x,\Delta)}{2\Delta} - \theta\frac{c(s,x,\Delta)}{6\Delta}. \end{aligned} \tag{118}$$

First we prove that if for given x and s the determinant

$$D(s,x,t',y',t'',y'') = \begin{vmatrix} \frac{\partial}{\partial x} f(s,x,t',y') & \frac{\partial}{\partial x} f(s,x,t'',y'') \\ \frac{\partial^2}{\partial x^2} f(s,x,t',y') & \frac{\partial^2}{\partial x^2} f(s,x,t'',y'') \end{vmatrix} \tag{119}$$

does not vanish identically for any t', y', t'', y'', then the ratios

$$a(s,x,\Delta)/\Delta \text{ and } b^2(s,x,\Delta)/2\Delta$$

tend to well defined limits $A(s,x)$ and $B^2(s,x)$ as $\Delta \to 0$.

Thus, let t', y', t'', y'' be chosen so that (119) is non-zero; in this case, for any sufficiently small Δ we also have

$$D(s+\Delta, x, t', y', t'', y'') \neq 0,$$

so that the equations

$$\begin{aligned} &\lambda(\Delta)\frac{\partial}{\partial x} f(s+\Delta,x,t',y') + \mu(\Delta)\frac{\partial}{\partial x} f(s+\Delta,x,t'',y'') = 0, \\ &\lambda(\Delta)\frac{\partial^2}{\partial x^2} f(s+\Delta,x,t',y') + \mu(\Delta)\frac{\partial^2}{\partial x^2} f(s+\Delta,x,t'',y'') = 1 \end{aligned} \tag{120}$$

have a unique solution. In this case $\lambda(\Delta)$ and $\mu(\Delta)$ tend to $\lambda(0)$ and $\mu(0)$ as $\Delta \to 0$. Further, by (118) we obtain

$$\begin{aligned} \lambda(\Delta)&\frac{f(s+\Delta,x,t',y') - f(s,x,t',y')}{\Delta} + \\ &+\mu(\Delta)\frac{f(s+\Delta,x,t'',y'') - f(s,x,t'',y'')}{\Delta} = \\ &= -\frac{b^2(s,x,\Delta)}{2\Delta} - (\theta' + \theta'')\frac{c(s,x,\Delta)}{6\Delta}. \end{aligned} \tag{121}$$

The left-hand side of formula (121) tends to

$$\Omega = \lambda(0)\frac{\partial}{\partial s} f(s,x,t',y') + \mu(0)\frac{\partial}{\partial s} f(s,x,t'',y'')$$

as $\Delta \to 0$, whereas on the right-hand side the second term is infinitesimally small as compared with the first one, by (111); therefore this term tends to the limit

$$B^2(s,x) = \lim_{\Delta\to 0} \frac{b^2(s,x,\Delta)}{2\Delta} = -\Omega. \tag{122}$$

It follows immediately from (122) and (111) that

$$c(s,x,\Delta)/\Delta \to 0 \quad \text{as } \Delta \to 0. \tag{123}$$

By (122) and (123), formula (118) for $\Delta = 0$ becomes:

$$\lim_{\Delta\to 0}\left[\frac{\partial}{\partial x} f(s,x,t,y)\frac{a(s,x,\Delta)}{\Delta}\right] = -\frac{\partial}{\partial s} f(s,x,t,y) - \\ -\frac{\partial^2}{\partial x^2} f(s,x,t,y) B^2(s,x).$$

Since $\partial f(s,x,t,y)/\partial x$ does not vanish identically for any t and y, the following limit also exists:

$$A(s,x) = \lim_{\Delta\to 0}\frac{a(s,x,\Delta)}{\Delta} = \\ = \frac{-\partial f(s,x,t,y)/\partial s - B^2(s,x)\partial^2 f(s,x,t,y)/\partial x^2}{\partial f(s,x,t,y)/\partial x}. \tag{124}$$

Passing to the limit in (118), (122), (123) and (124) we obtain the *first fundamental differential equation*

$$\frac{\partial}{\partial s} f(s,x,t,y) = -A(s,x)\frac{\partial}{\partial x} f(s,x,t,y) - B^2(s,x)\frac{\partial^2}{\partial x^2} f(s,x,t,y). \tag{125}$$

When the determinant $D(s,x,t',y',t'',y'')$ vanishes for any t',y',t'',y'', then the limits $A(s,x)$ and $B^2(s,x)$ do not in general exist, as is clear from the following example:

$$f(s,x,t,y) = \frac{3y^2}{2\sqrt{\pi(t-s)}} e^{-(y^3-x^3)^2/4(t-s)}. \tag{126}$$

Here, for $x = 0$ we have

$$b^2(s,x,\Delta)/2\Delta \to +\infty \quad \text{as } \Delta \to 0.$$

It can be shown, however, that these singular points (s,x) form a nowhere dense set in the (s,x)-plane.

The practical significance of these very important quantities $A(s,x)$ and $B(s,x)$ is as follows: $A(s,x)$ is the mean rate of variation of the parameter x over an infinitesimally small time interval and $B(s,x)$ is the differential variance of the process. The variance of the difference $y - x$ for the time interval Δ is

$$b(s,x,\Delta) = B(s,x)\sqrt{2\Delta} + o(\sqrt{\Delta}) = O(\sqrt{\Delta}); \tag{127}$$

and the expectation of this difference is

$$a(s,x,\Delta) = A(s,x)\Delta + o(\Delta) = O(\Delta). \tag{128}$$

It is worth mentioning that the expectation $m^{(1)}(t,x,\Delta)$ of $|y-x|$, like the variance $b(s,x,\Delta)$, is a quantity of order $\sqrt{\Delta}$.

As will be shown in the following sections, the functions $A(s,x)$ and $B(s,x)$ in some cases uniquely determine our stochastic scheme.

§14. The second differential equation

In this section we retain all the requirements imposed on $f(s,x,t,y)$ in §13 and, in addition, assume that $f(s,x,t,y)$ has continuous derivatives up to the fourth order. Then, from (120) it clearly follows that if the determinant (119) is non-zero, then $\lambda(0)$ and $\mu(0)$ have continuous derivatives with respect to s and x up to the second order; by (122) and (124) the same is also true for $B^2(s,x)$ and $A(s,x)$.

Now, assume that for a certain t we are given an interval $a \le y \le b$ such that at each point of this interval the determinant $D(t,y,u',z',u'',z'')$ does not vanish identically for any u', z', u'', z''. Next, let $R(y)$ be a function that is non-zero only on the interval $a < y < b$, is non-negative and has bounded derivatives up to the third order. In this case we have

$$\int_a^b \frac{\partial}{\partial t} f(s,x,t,y)R(y)dy = \frac{\partial}{\partial t}\int_a^b f(s,x,t,y)R(y)dy =$$

$$= \lim_{\Delta\to 0}\frac{1}{\Delta}\int_{-\infty}^{\infty}[f(s,x,t+\Delta,y) - f(s,x,t,y)]R(y)dy =$$

$$= \lim_{\Delta\to 0}\frac{1}{\Delta}\Big\{\int_{-\infty}^{\infty} R(y)\int_{-\infty}^{\infty} f(s,x,t,z)f(t,z,t+\Delta,y)dz\,dy -$$

$$- \int_{-\infty}^{\infty} f(s,x,t,y)R(y)dy\Big\} = \lim_{\Delta\to 0}\frac{1}{\Delta}\times$$

$$\times\Big\{\int_{-\infty}^{\infty} f(s,x,t,z)\int_{-\infty}^{\infty} f(t,z,t+\Delta,y)\Big[R(z) + R'(z)(y-z) +$$

$$+ R''(z)\frac{(y-z)^2}{2} + R'''(\xi)\frac{(y-z)^3}{6}\Big]dy\,dz -$$

$$- \int_{-\infty}^{\infty} f(s,x,t,z)R(z)dz\Big\} = \lim_{\Delta\to 0}\frac{1}{\Delta}\int_{-\infty}^{\infty} f(s,,x,t,z)\times$$

$$\times\left[R'(z)a(t,z,\Delta)+R''(z)\frac{b^2(t,z,\Delta)}{2}+\theta\frac{c(t,z,\Delta)}{6}\right]dz=$$

$$=\int_{-\infty}^{\infty}f(s,x,t,z)[R'(z)A(t,z)+R''(z)B^2(t,z)]dz=$$

$$=\int_a^b f(s,x,t,y)[R'(y)A(t,y)+R''(y)B^2(t,y)]dy, \tag{129}$$

$$|\theta|\le\sup|R'''(\xi)|.$$

The passage to the limit with respect to Δ in deriving these formulas is justified by the fact that $a(t,z,\Delta)/\Delta$, $b^2(t,z,\Delta)/2\Delta$ and $c(t,z,\Delta)/\Delta$ tend uniformly to $A(t,z)$, $B^2(t,z)$ and 0 respectively, and the integral of the factor $f(s,x,t,z)$ with respect to z is finite.

Integrating by parts we obtain

$$\int_a^b f(s,x,t,y)R'(y)A(t,y)dy=-\int_a^b\frac{\partial}{\partial y}[f(s,x,t,y)A(t,y)]R(y)dy. \tag{130}$$

In exactly the same way, integrating by parts twice, we obtain

$$\int_a^b f(s,x,t,y)R''(y)B^2(t,y)dy=\int_a^b\frac{\partial^2}{\partial y^2}[f(s,x,t,y)B^2(t,y)]R(y)dy, \tag{131}$$

since $R(a)=R(b)=R'(a)=R'(b)=0$. Formulas (129)–(131) immediately imply that

$$\int_a^b\frac{\partial}{\partial t}f(s,x,t,y)R(y)dy=\int_a^b\Big\{-\frac{\partial}{\partial y}[A(t,y)f(s,x,t,y)]+$$
$$+\frac{\partial^2}{\partial y^2}[B^2(t,y)f(s,x,t,y)]\Big\}R(y)dy. \tag{132}$$

However, since the function $R(y)$ can be chosen arbitrarily only if the above conditions are fulfilled, we easily see that for points (t,y) at which the determinant $D(t,y,u',z',u'',z'')$ does not vanish identically the *second fundamental differential equation* also holds:

$$\frac{\partial}{\partial t}f(s,x,t,y)=-\frac{\partial}{\partial y}[A(t,y)f(s,x,t,y)]+\frac{\partial^2}{\partial y^2}[B^2(t,y)f(s,x,t,y)]. \tag{133}$$

This second equation could also have been obtained without using the first one, using the methods described in §13 directly; then, however, new and more stringent restrictions (that we omit here) would have to be imposed on $f(s,x,t,y)$. In that case we start from the formula, similar to (118),

$$\frac{1}{\Delta}[f(s,x,t,y)-f(s,x,t-\Delta,y)]=$$

$$=f(s,x,t-\Delta,y)\frac{1}{\Delta}\Big[\int_{-\infty}^{\infty}f(t-\Delta,z,t,y)dz-1\Big]+$$

$$+\frac{\partial}{\partial y}f(s,x,t-\Delta,y)\frac{1}{\Delta}\int_{-\infty}^{\infty}f(t-\Delta,z,t,y)(z-y)dz+$$

$$\frac{\partial^2}{\partial y^2}f(s,x,t-\Delta,y)\frac{1}{2\Delta}\int_{-\infty}^{\infty}f(t-\Delta,z,t,y)(z-y)^2dz+$$

$$+\frac{1}{6\Delta}\int_{-\infty}^{\infty}f(t-\Delta,z,t,y)|z-y|^3dz. \tag{134}$$

Then we prove that

$$\lim_{\Delta\to 0}\frac{1}{\Delta}\int_{-\infty}^{\infty}f(t-\Delta,z,t,y)|z-y|^3dz=0$$

and that the limits

$$\lim_{\Delta\to 0}\frac{1}{2\Delta}\int_{-\infty}^{\infty}f(t-\Delta,z,t,y)|z-y|^2dz=\overline{B}^2(t,y), \tag{135}$$

$$\lim_{\Delta\to 0}\frac{1}{\Delta}\int_{-\infty}^{\infty}f(t-\Delta,z,t,y)(z-y)dz=\overline{A}(t,y), \tag{136}$$

$$\lim_{\Delta\to 0}\frac{1}{\Delta}\Big(\int_{-\infty}^{\infty}f(t-\Delta,z,t,y)dz-1\Big)=\overline{N}(t,y), \tag{137}$$

exist. Thus we would have obtained our second equation in the following form

$$\frac{\partial}{\partial t}f(s,x,t,y)=\overline{N}(t,y)f(s,x,t,y)+$$

$$+\overline{A}(t,y)\frac{\partial}{\partial y}f(s,x,t,y)+\overline{B}^2(t,y)\frac{\partial^2}{\partial y^2}f(s,x,t,y). \tag{138}$$

To show the identity of this equation with the one derived before, we would have to prove that

$$\overline{B}^2(t,y)=B^2(t,y), \tag{139}$$

$$\overline{A}(t,y)=-A(t,y)+\frac{\partial}{\partial y}B^2(t,y), \tag{140}$$

$$\overline{N}(t,y)=\frac{\partial}{\partial y}A(t,y)+\frac{\partial^2}{\partial y^2}B^2(t,y). \tag{141}$$

§15. Statement of the uniqueness and existence problem of solutions of the second differential equation

In order to define the function $f(s,x,t,y)$ uniquely by the differential equations (125) and (133) we have to set up some initial conditions. For the second equation (133) the following approach can be adopted: according to (85), the function $f(s,x,t,y)$ satisfies the condition

$$\int_{-\infty}^{\infty} f(s,x,t,y)dy = 1 \tag{142}$$

for every $t > s$ and, in view of (110), we also have

$$\int_{-\infty}^{\infty} (y-x)^2 f(s,x,t,y)dy \to 0, \quad \text{as } t \to s. \tag{143}$$

The main question regarding the uniqueness of solutions is as follows: under what conditions can we assert that for given s and x there exists a unique non-negative function $f(s,x,t,y)$ of the variables t, y defined for all y and $t > s$ and satisfying (133) and the conditions (142), (143) ? In certain important particular cases such conditions may be described: these are, for example, all the cases considered in the following two sections.

Now given the functions $A(t,y)$ and $B^2(t,y)$, the question is whether there exists a non-negative function $f(s,x,t,y)$ such that, on the one hand, it satisfies (85) and (86) (as was indicated in §11, these requirements are needed for $f(s,x,t,y)$ to determine a stochastic system), and on the other hand, after passing to the limit via formulas (122) and (124), it gives these functions $A(t,y)$ and $B^2(t,y)$.

To solve this problem, we can, for example, first determine some non-negative solution of our second differential equation (133) satisfying the conditions (142), (143) and then check if it is indeed a solution to our problem. In doing this, the following two general questions arise:

1) Under what conditions does there exist such a solution of equation (133)?

2) Under what conditions does this solution really satisfy (85) and (86)?

There are good grounds for assuming that these conditions are of a sufficiently general character.

§16. Bachelier's case

We now assume that $f(s,x,t,y)$ is a function of the difference $y-x$, depending arbitrarily on s and t, that is, that our process is homogeneous with respect to the parameter

$$f(s,x,t,y) = v(s,t,y-x). \tag{144}$$

In this case, clearly $A(s,t)$ and $B(s,t)$ depend only on s, so that the differential equations (125) and (133) are now expressed as

$$\frac{\partial f}{\partial s} = -A(s)\frac{\partial f}{\partial x} - B^2(s)\frac{\partial^2 f}{\partial x^2}, \tag{145}$$

$$\frac{\partial f}{\partial t} = -A(t)\frac{\partial f}{\partial y} + B^2(t)\frac{\partial^2 f}{\partial y^2}. \tag{146}$$

For the function $v(s,t,z)$, we obtain from (145) and (146):

$$\frac{\partial v}{\partial s} = A(s)\frac{\partial v}{\partial z} - B^2(s)\frac{\partial^2 v}{\partial z^2}, \tag{147}$$

$$\frac{\partial v}{\partial t} = -A(t)\frac{\partial v}{\partial z} + B^2(t)\frac{\partial^2 v}{\partial z^2}. \tag{148}$$

Equation (148) was found by Bachelier,[13] but strictly speaking, was not proved.

If we have $A(t) = 0$ and $B(t) = 1$ identically, then (133) (respectively (146)) turns into the heat equation

$$\partial f/\partial t = \partial^2 f/\partial y^2, \tag{149}$$

for which the only non-negative solution satisfying (142), (143) is given, as is well known, by Laplace's formula

$$f(s,x,t,y) = \frac{1}{\sqrt{\pi(t-s)}}e^{-(y-x)^2/4(t-s)}. \tag{150}$$

In general we assume that

$$x' = x - \int_a^s A(u)du, \quad y' = y - \int_a^t A(u)du,$$

$$s' = \int_a^s B^2(u)du, \quad t' = \int_a^t B^2(u)du.$$

[13] See papers Nos. 1 and 3 in footnote 2.

Then (146) turns into

$$\partial f/\partial t' = \partial^2 f/\partial y'^2,$$

and the conditions (142), (143) retain the same form in the new variables s', x', t', y' as in the variables s, x, t, y. Hence in the general case, the function

$$f(s,x,t,y) = \frac{1}{\sqrt{\pi(t'-s')}} e^{-(y'-x')^2/4(t'-s')} = \frac{1}{\sqrt{\pi\beta}} e^{-(y-\alpha)^2/4\beta}$$

$$\left(\beta = \int_s^t B^2(u)du, \quad \alpha = x + \int_s^t A(u)du\right) \tag{151}$$

is the only solution of equation (146) satisfying our conditions.

§17. A method of transforming distribution functions

Let

$$s' = \phi(s), \quad t' = \phi(t), \quad x' = \psi(s,x), \quad y' = \psi(t,y)$$

$$f(s,x,t,y) = (\partial\psi(t,y)/\partial y) f'(s',x',t',y'), \tag{152}$$

and assume that $\phi(t)$ is a continuous, nowhere decreasing function, whereas $\psi(t,y)$ is arbitrary with respect to t and has a continuous positive derivative with respect to y. If $f(s,x,t,y)$ satisfies (85) and (86), then the same is also true, as can easily be demonstrated, for the function f' with respect to the new variables s', x', t', y'; in other words, our transformation gives a new function $f'(s',x',t',y')$ which, like $f(s,x,t,y)$, determines a stochastic scheme.

If $\phi(t)$ and $\psi(t,y)$ have appropriate derivatives, then under transition to the new variables (125) and (133) turn into

$$\frac{\partial f'}{\partial s'} = -A'(s',x')\frac{\partial f'}{\partial x'} - B'^2(s',x')\frac{\partial^2 f'}{\partial x'^2}, \tag{153}$$

$$\frac{\partial f'}{\partial t'} = -\frac{\partial}{\partial y'}[A'f'] + \frac{\partial^2}{\partial y'^2}[B'^2 f'], \tag{154}$$

where we have set

$$A'(t',y') = \frac{(\partial^2\psi(t,y)/\partial y^2)B^2(t,y) + (\partial\psi(t,y)/\partial y)A(t,y) + \partial\psi(t,y)/\partial t}{\partial\phi(t)/\partial t},$$

$$B'^2(t',y') = \frac{(\partial\psi(t,y)/\partial y)^2 B^2(t,y)}{\partial\phi(t)/\partial t}. \tag{155}$$

With the help of the above transformation the solutions of (133) can be obtained for many new types of coefficients $A(t,y)$ and $B^2(t,y)$. For example, let

$$A(t,y) = a(t)y + b(t), \quad B^2(t,y) = c(t); \tag{156}$$

we set

$$\begin{aligned} \phi(t) &= \int c(t)e^{-2\int a(t)dt}dt, \\ \psi(t,y) &= ye^{-\int a(t)dt} - \int b(t)e^{-\int a(t)dt}dt \end{aligned} \tag{157}$$

and obtain in the new variables s', x', t', y', f' the simplest heat equation:

$$\partial f'/\partial t' = \partial^2 f'/\partial y'^2. \tag{158}$$

In this case the initial conditions (142) and (143) remain valid for $f'(s',x',t',y')$ as well; therefore the formula

$$f' = \frac{1}{\sqrt{\pi(t'-s')}}e^{-(y'-x')^2/4(t'-s')} \tag{159}$$

together with (157) and (152) gives the unique solution $f(s,x,t,y)$ of (133) with coefficients of the form (156) satisfying our conditions. It is easy to see that in this case the function $f(s,x,t,y)$ is of the form

$$\frac{1}{\sqrt{\pi\beta}}e^{-(y-\alpha)^2/4\beta}, \tag{160}$$

where α and β depend only on s, x and t, but not on y.

It is an important problem to find all possible types of coefficients $A(t,y)$ and $B^2(t,y)$ such that for any s, t, x we always obtain a function of the form (160), that is, the Laplace distribution function.

As a second example consider the case

$$A(t,y) = a(t)(y-c), \quad B^2(t,y) = b(t)(y-c)^2. \tag{161}$$

This time, setting

$$\phi(t) = \int b(t)dt, \quad \psi(t) = \ln(y-c) + \int [b(t) - a(t)]dt \tag{162}$$

we again obtain for $f'(s',x',t',y')$ equation (158) for which the solution (159) is already known. Note that here it suffices to consider only the values $x > c$,

$y > c$, since as x or y varies from c to $+\infty$ the variable x' (hence y') runs through all the values from $-\infty$ to $+\infty$. Certain complications arising in connection with this when transferring the conditions (142) and (143) to f' can easily be eliminated.

In particular, for

$$A(t,y) = 0, \quad B^2(t,y) = y^2 \tag{163}$$

we have the formula

$$f(s,x,t,y) = \frac{1}{y\sqrt{\pi(t-s)}} \exp\left\{-\frac{(\ln y + t - \ln x - s)^2}{4(t-s)}\right\}. \tag{164}$$

For applications the most important is the case when $A(t,y)$ and $B^2(t,y)$ depend only on y, but do not depend on the time t. The next step in this direction would be to solve our problem for coefficients of the form

$$A(y) = ay + b, \quad B^2(y) = cy^2 + dy + e. \tag{165}$$

§18. Stationary distribution functions

If at time t_0 the differential function of a probability distribution $g(t_0, y)$ is known, then, as for the general formula (5), the distribution function $g(t,y)$ is determined for any $t > t_0$ by the formula

$$g(t,y) = \int_{-\infty}^{\infty} g(t_0,x) f(t_0,x,t,y)\,dx. \tag{166}$$

Clearly, $g(t,y)$ satisfies the equation

$$\frac{\partial g}{\partial t} = -\frac{\partial}{\partial y}[A(t,y)g] + \frac{\partial^2}{\partial y^2}[B^2(t,y)g]. \tag{167}$$

We now assume that the coefficients $A(t,y)$ and $B^2(t,y)$ depend only on y (the process is homogeneous in time) and study the functions $g(t,y)$ which in this case do not change with time. It is clear that for such functions we have

$$-Ag + (B^2 g)' = C. \tag{168}$$

If we assume that g and g' tend to 0 so rapidly as $y \to \pm\infty$ that the entire left-hand side of (168) tends to 0, then clearly $C = 0$ and we have

$$g'/g = [A - (B^2)']/B^2. \tag{169}$$

Moreover the function $g(y)$ must also satisfy the condition

$$\int_{-\infty}^{\infty} g\,dy = 1. \tag{170}$$

In most cases it appears possible to prove that, if there exists a stationary solution $g(x)$, then $f(s,x,t,y)$ tends to $g(y)$ as $t \to \infty$ and for arbitrary constants s and x; thus, $g(y)$ appears to be not only a stationary, but also the limiting solution.

If the coefficients A and B^2 are of the form (165) , then (169) turns into the Pearson equation

$$\frac{g'}{g} = \frac{y-p}{q_0 + q_1 y + q_2 y^2}, \tag{171}$$

with

$$p = \frac{d-b}{a-2c}, \quad q_0 = \frac{e}{a-2c}, \quad q_1 = \frac{d}{a-2c}, \quad q_2 = \frac{e}{a-2c}. \tag{172}$$

Hence we can construct stochastic schemes for which any of the functions of the Pearson distribution is a stationary solution.

§19. Other possibilities

The theory presented in §§13–18 is essentially determined by the assumption (111). If we get rid of this assumption, then even when the condition (110) is retained, a number of new possibilities appear. For example, consider the scheme determined by the distribution function

$$F(s,x,t,y) = e^{-a(t-s)}\sigma(y-x) + (1 - e^{-a(t-s)}) \int_{-\infty}^{y} u(z)dz, \tag{173}$$

where $\sigma(z) = 0$ for $z < 0$ and $\sigma(z) = 1$ for $z \geq 0$, and $u(z)$ is a continuous non-negative function for which

$$\int_{-\infty}^{\infty} u(z)dz = 1$$

and the moments

$$\int_{-\infty}^{\infty} u(z)|z|^i dz \quad (i = 1,2,3)$$

are finite. It can easily be shown that the function $F(s,x,t,y)$ satisfies (78) and (79), as well as (110).

This scheme can be interpreted as follows: during an infinitely small time interval $(t, t+dt)$ the parameter y either remains constant with probability

$1 - adt$, or takes a value $y', z < y' < z + dz$ with probability $au(z)dt\,dz$. Thus a jump is possible in any time interval, and the distribution function of the values of the parameter after the jump does not depend on the values of this parameter prior to the jump.

This scheme could also be generalized in the following way: imagine that, during an infinitely small time interval $(t, t + dt)$ the parameter y retains its former value with probability $1 - a(t, y)dt$ and turns into y', $z < y' < z + dz$ with probability $u(t, y, z)dt\,dz$. Clearly we assume that

$$\int_{-\infty}^{\infty} u(t, y, z)dz = a(t, y). \tag{174}$$

In this case for $g(t, y)$ the integro-differential equation

$$\frac{\partial}{\partial t} g(t, y) = -a(t, y)g(t, y) + \int_{-\infty}^{\infty} g(t, z)u(t, z, y)dz \tag{175}$$

should hold.

If we wish to consider not only jumps but also continuous changes in y, then it is natural to expect that $g(t, y)$ satisfies

$$\frac{\partial}{\partial t} g(t, y) = -a(t, y)g(t, y) + \int_{-\infty}^{\infty} g(t, z)u(t, z, y)dz - \\ - \frac{\partial}{\partial y}[A(t, y)g(t, y)] + \frac{\partial^2}{\partial y^2}[B^2(t, y)g(t, y)], \tag{176}$$

provided (174) holds and the coefficients $A(t, y)$ and $B^2(t, y)$ are as indicated in §13.

CONCLUSION

If the state of the system under consideration is determined by n real parameters $x_1, x_2, \ldots, x_n$, then under certain conditions similar to those of §13 we have the following differential equations for the differential distribution function $f(s, x_1, \ldots, x_n, t, y_1, \ldots, y_n)$:

$$\frac{\partial f}{\partial s} = -\sum_{i=1}^{n} A_i(s, x_1, \ldots, x_n)\frac{\partial f}{\partial x_i} - \sum_{i=1}^{n}\sum_{j=1}^{n} B_{ij}(s, x_1, \ldots, x_n)\frac{\partial^2 f}{\partial x_i \partial x_j}, \tag{177}$$

$$\frac{\partial f}{\partial t} = -\sum_{i=1}^{n} \frac{\partial}{\partial y_i}[A_i(t, y_1, \ldots, y_n)f] + \\ + \sum_{i=1}^{n}\sum_{j=1}^{n} \frac{\partial^2}{\partial y_i \partial y_j}[B_{ij}(t, y_1, \ldots, y_n)f]. \tag{178}$$

For the case when $A_i(t, y_1, \ldots, y_n)$ and $B_{ij}(t, y_1, \ldots, y_n)$ depend only on t, these equations were discovered and solved by Bachelier.[14] In this case the solutions satisfying the conditions of our problem have the form

$$f = P \exp\Big\{-\frac{1}{Q} \sum p_{ij}(y_i - x_i - q_i)(y_j - x_j - q_j)\Big\}, \tag{179}$$

with P, Q, p_{ij} and q_i depending only on s and t.

It is also possible to consider mixed schemes, where the state of the system is determined by parameters some of which are discrete and others continuous.

Moscow, 26 July 1930

[14] See item II in footnote 2.

10. THE WAITING PROBLEM *

§1. The purpose of this paper is to show several applications of the general equations which I studied in my memoir submitted to "Mathematische Annalen" [1] (see No. 9 of the present publication). For this I give a new solution to the "Waiting problem" dealt with in an extensive memoir by Polyachek. [2]

The essence of the problem is as follows. (Naturally, the same mathematical problem can arise when studying other real phenomena, therefore such expressions as "telephone lines" or "conversation" are used here only for the sake of illustration.)

Assume that, in a telephone station, there are n lines over which telephone conversations can take place. At any moment there are m clients that either have a conversation or await their turn; the latter takes place only for $m > n$, and the length of the waiting line is $m - n$. Theoretically, m can take all non-negative integer values:

$$m = 0, 1, 2, \ldots .$$

We denote by

$$Q_0(t),\ Q_1(t),\ Q_2(t), \ldots, Q_m(t), \ldots$$

the probabilities of these values at time t. Clearly,

$$\sum_{m=0}^{\infty} Q_m(t) = 1. \tag{1}$$

If we know m at time t_0, then

$$Q_m(t_0) = 1, \quad Q_i(t_0) = 0,\ i \neq m. \tag{2}$$

The first problem to be solved is to calculate *the probabilities $Q_m(t)$ for $t > t_0$ if we know their initial values $Q_m(t_0)$, which, in particular, can take the form* (2). *The second problem is to determine the expectation and the distribution law for the waiting time for a client arriving at time t.*

---

* Sur le problème d'attente', *Mat. Sb.* **38**:1/2 (1931), 101–106.

[1] This article, however, does not suggest that the reader has read this memoir, apart from the proof of several auxiliary statements.

[2] *Math. Z.* **32** (1930), 64–100.

§2. We proceed from the following two assumptions on the character of the random events studied.

I. At any infinitely small time interval $(t, t+dt)$ a new client arrives at the station with probability $\alpha(t)dt$. More precisely, at any interval $(t, t+\Delta)$ the probability of a new client arriving at the station is

$$\alpha(t)\Delta + o(\Delta),$$

whereas the probability for more than one client to arrive is

$$o(\Delta);$$

here, as usual, $o(\Delta)$ is an infinitesimal with respect to Δ; we assume that the function $\alpha(t)$ is continuous. The probabilities relating to the interval $(t, t+\Delta)$ do not depend on the number of clients arriving earlier or on the times of their arrival.

II. The probability of finishing a conversation maintained by a client at a moment t during the time interval $(t, t+dt)$ equals βdt; it does not depend on the length of the conversation prior to the moment t. This gives us the exponential distribution for the length of a conversation: the probability that this length is between t and $t+dt$ is

$$p(t)dt = \beta e^{-\beta t}dt. \tag{3}$$

In this case the expected length of a conversation is

$$D = \int_0^\infty \beta t e^{-t\beta} dt = \frac{1}{\beta}. \tag{4}$$

Instead of distribution (3) Polyachek assumes that the length of a conversation is always equal to some constant D; this assumption is as arbitrary as ours.

§3. Denote by $Q_{mp}(t, t+\Delta)$ the conditional probability that at time $t+\Delta$ there are p clients at the station if at time t there were m clients. The probability that within the interval $(t, t+\Delta)$ more that one client will arrive, or more than one on-going conversation will be finished, is $o(\Delta)$. Therefore

$$Q_{mp}(t, t+\Delta) \leq \omega(\Delta) = o(\Delta), \quad |p-m| > 1, \tag{5}$$

where $\omega(\Delta)$ does not depend on m and p. For $|p-m| \leq 1$ we have

$$Q_{m,m+1}(t,t+\Delta) = \alpha(t)\Delta + o(\Delta), \tag{6}$$

$$Q_{m,m-1}(t,t+\Delta) = m\beta\Delta + o(\Delta), \quad \text{if } m \leq n, \tag{7}$$

$$Q_{m,m-1}(t,t+\Delta) = n\beta\Delta + o(\Delta), \quad \text{if } m \geq n, \tag{8}$$

therefore

$$Q_{mm}(t,t+\Delta) = 1-(\alpha(t)+m\beta)\Delta + o(\Delta), \quad \text{if } m \leq n, \tag{9}$$

$$Q_{mm}(t,t+\Delta) = 1-(\alpha(t)+n\beta)\Delta + o(\Delta), \quad \text{if } m \geq n. \tag{10}$$

By the formula for the total probability we have

$$Q_p(t+\Delta) = \sum_{m=0}^{\infty} Q_m(t)Q_{mp}(t+\Delta). \tag{11}$$

Substituting the value of $Q_{mp}(t,t+\Delta)$ from (5)–(10) into (11) we prove that the limit

$$Q'_p(t) = \lim_{\Delta\to 0} \frac{Q_p(t+\Delta)-Q_p(t)}{\Delta} \tag{12}$$

always exists and the following equations hold:

$$Q'_0(t) = \beta Q_1(t) - \alpha(t)Q_0(t), \tag{13}$$

$$Q'_i(t) = (i+1)\beta Q_{i+1}(t) - (\alpha(t)+i\beta)Q_i(t) + \alpha(t)Q_{i-1}(t), \; 1 \leq i < n, \tag{14}$$

$$Q'_i(t) = n\beta Q_{i+1}(t) - (\alpha(t)+n\beta)Q_i(t) + \alpha(t)Q_{i-1}(t), \quad n \leq i. \tag{15}$$

Thus we obtain for the functions $Q_i(t)$ a countable infinite system of differential equations. It suffices to find a solution $Q_i^{(m)}(t)$ with initial conditions of type (2). A general solution can be found with the help of the formulas for the total probability,

$$Q_i(t) = \sum_{m=0}^{\infty} Q_m(t_0)Q_i^{(m)}(t). \tag{16}$$

§4. Consider the simplest and most important case when the function $\alpha(t)$ is constant:

$$\alpha(t) = \alpha. \tag{17}$$

In this case the equations (13)–(15) have constant coefficients. In the paper mentioned above I proved the existence and uniqueness of the solution for these systems (under the assumption that the $Q_m(t)$ are non-negative and satisfy (1), which holds in our problem). Therefore the functions $Q_m(t)$ are uniquely determined by the equations (13)–(15) and their initial values $Q_m(t_0)$. It is not difficult to find approximate solutions for these equations; one method for obtaining such a solution is given in my paper mentioned above. But from the practical viewpoint it is preferable to go directly to the limit solution.

This solution only exists if

$$n\beta > \alpha. \tag{18}$$

If the reverse inequality

$$n\beta < \alpha \tag{19}$$

holds, then the waiting line in the station grows indefinitely. The case

$$n\beta = \alpha \tag{20}$$

is only of theoretical value, so we confine ourselves to the study of the case (18).

First we find the constants $Q_m(t) = Q_m$ satisfying the equations (13)–(15) and (1). In this case

$$\beta Q_1 - \alpha Q_0 = 0, \tag{21}$$

$$(i+1)\beta Q_{i+1} - (\alpha + \beta i)Q_i + \alpha Q_{i-1} = 0, \quad 1 \leq i < n, \tag{22}$$

$$n\beta Q_{i+1} - (\alpha + n\beta)Q_i + \alpha Q_{i-1} = 0, \quad n \leq i, \tag{23}$$

$$\sum_{m=0}^{\infty} Q_m = 1. \tag{1}$$

From (21)-(23) it is easy to obtain

$$Q_i = \omega_i Q_0,$$

$$\omega_i = \frac{1}{i!}\left(\frac{\alpha}{\beta}\right)^i, \quad 0 \leq i \leq n, \tag{24}$$

$$\omega_i = \frac{1}{n!n^{i-n}}\left(\frac{\alpha}{\beta}\right)^i, \quad n \leq i, \tag{25}$$

from which, by (1) we have

$$Q_i = \frac{\omega_i}{\Omega}, \tag{26}$$

$$\Omega = \sum_{i=0}^{\infty} \omega_i,$$

$$\Omega = \omega_0 + \omega_1 + \ldots + \omega_{n-1} + \frac{1}{n!}\left(\frac{\alpha}{\beta}\right)^n \frac{1}{1 - n^{-1}(\alpha/\beta)}. \tag{27}$$

Formulas (24)–(27) give the unique stationary solution of our problem.

It can be proved that for any other solution the functions $Q_m(t)$ tend to Q_m as $t \to \infty$, that is, the stationary solution is also the limit solution.

The following remark concerns formula (24)–(27). We say that the probability distribution $\{Q_m^{(1)}(t)\}$ exceeds the distribution $\{Q_m^{(2)}(t)\}$ if for any m,

$$R_m^{(1)}(t) \geq R_m^{(2)}(t), \tag{28}$$

where

$$R_m = \sum_{i=0}^{m} Q_i. \tag{29}$$

If the initial conditions $\{Q_m^{(1)}(t_0)\}$ exceed $\{Q_m^{(2)}(t_0)\}$, this relation persists at any moment $t > t_0$. Let $Q_m(t)$ be the solution of our problem with initial conditions

$$Q_0(t_0) = 1, \quad Q_m(t_0) = 0,\ m > 0. \tag{30}$$

These initial conditions exceed all other possible conditions. Therefore at any moment the solution $Q_m(t)$ exceeds any other solution, and in particular, the stationary solution.

Thus we see that if the coefficients α and β remain constant for a sufficiently long time, then the desired probabilities are given by the formulas (24)–(27). For example, if $n = 3$, $\alpha/\beta = 2$, then we have the table:

m	0	1	2	3	4	5	6
Q_m	1/9	2/9	2/9	4/27	8/81	16/243	32/729
R_m	1/9	1/3	5/9	19/27	65/81	211/243	665/729

In this case we see that if on the average two out of three lines are busy, then there is a free line with probability $R_2 = 5/9$, whereas the probability that $m > 6$, that is, the probability that more than three clients are awaiting their turn is equal to $1 - R_6 = 64/729 \sim 1/11$.

§5. If the limit solution is taken as the exact solution, then it is simple to calculate the expectation E of the waiting time. For this, the average length waiting time

$$\sigma = \sum_{m>n} (m-n) Q_m = \frac{1}{\Omega} \frac{1}{n!n} \left(\frac{\alpha}{\beta}\right)^{n+1} \left(1 - \frac{1}{n}\frac{\alpha}{\beta}\right)^{-2} \tag{31}$$

should be divided by $n\beta$:

$$E = \sigma / n\beta. \tag{32}$$

Thus, in our example, for $n = 3$, $\alpha/\beta = 2$ we find

$$\sigma = \frac{8}{9}, \quad E = \frac{8}{9.3\beta} = \frac{8}{27\beta} = \frac{8}{27} D.$$

This means that the average waiting time is 8/27 of the average length of a conversation.

§6. To be able to calculate the distribution law for the waiting time we have only to solve the following problem: under the assumption that at time t_0 there are m clients at the station and a new client A appears, we have to calculate the distribution law of the waiting time for this client. We are interested only in the case $m \geq n$, since otherwise there are free lines available. Set $m = n + p$ and denote by $P_q(t)$, $q = 0, 1, \ldots, p$, the probability that in the interval (t_0, t) precisely q conversations finish; denote by $P(t)$ the probability that this number exceeds p. Clearly, at time t_0,

$$P_0(t_0) = 1, \quad P_q(t_0) = 0, \ q > 0, \quad P(t_0) = 0. \tag{33}$$

The functions $P_0, P_1, \ldots, P_p, P$ satisfy the equations

$$\begin{aligned} &P_0'(t) = -n\beta P_0(t), \\ &P_q'(t) = n\beta(P_{q-1}(t) - P_q(t)), \quad q = 1, 2, \ldots, p, \\ &P'(t) = n\beta P_q(t). \end{aligned} \tag{34}$$

Thus we have found a finite system of linear equations, which can be easily solved. Hence we obtain the probability $P(t)$ that the waiting time for client A is at most $t - t_0$.

Paris, 24 November 1930

11. THE METHOD OF THE MEDIAN IN THE THEORY OF ERRORS *

Under the assumption that the error distribution is a normal law the method of the arithmetic mean, as is well known, is the best method for calculating the true value of an observable. The method of the median in this case is less effective, though not much less, as Haag has shown. However, if the hypothesis of normal distribution does not hold, then the problem arises of finding the best method for the given distribution law. In particular, in many cases when it is considered necessary to rule out "abnormal observations" it would be methodologically better to study a general distribution law and to find a more appropriate method for calculating the true value.

In this paper I show how, knowing the law of error distribution, one can determine the degree of accuracy of the method of the median and compare it with the accuracy of the method of the arithmetic mean. Which of the two methods is preferable depends on the nature of the distribution law adopted; however, by Theorem 2, *when the distribution law is unknown and can deviate markedly from the normal law, it is safer to use the method of the median.*

The method of study, in its essential features, is due to Haag [1] (who applied it, however, only to the normal distribution law).

Let the probability that the error of a single observation lies in $(x, x+dx)$ be $f(x)dx$. We shall assume that $f(x)$ is continuous and $f(m) \neq 0$ for the median m:

$$\int_{-\infty}^{m} f(x)dx = \frac{1}{2}.$$

Furthermore, let

$$a = \int_{-\infty}^{+\infty} x f(x)dx$$

be the expectation of the error. Denote by $x_1, x_2, \ldots, x_n$ the results of n successive observations, let a_n be their arithmetic mean and m_n their median. We restrict ourselves to the case of $n = 2k+1$, when $m_n = x_{k+1}$, assuming that the x_i are numbered in increasing order. The differences $(a_n - a)$ and $(m_n - m)$ are (as we shall see) of order $O(1/\sqrt{n})$, so that it is natural to set

$$\alpha_n = (a_n - a)\sqrt{n}, \quad \mu_n = (m_n - m)\sqrt{n}$$

and study the distribution laws for α_n and μ_n. As is known, for α_n we have a continuous distribution law: the probability that α_n lies in $(\alpha, \alpha + d\alpha)$ is

* *Mat. Sb.* **38**:3/4 (1931), 47–50.

equal to $\phi_n(\alpha)d\alpha$, where $\phi_n(\alpha)$ can be calculated by well-known formulas. If the variance

$$\sigma^2 = \int_{-\infty}^{+\infty} (x-a)^2 f(x)dx$$

exists, then $\phi_n(\alpha)$ converges to the normal distribution law

$$\phi(\alpha) = \frac{1}{\sqrt{2\pi}\sigma} e^{-\alpha^2/2\sigma^2}$$

with variance σ.

The probability that m_n lies in $(t, t+dt)$ is

$$u_n(t) = \frac{(2k+1)!}{(k!)^2}\left(\frac{1-\theta^2(t)}{4}\right)^k f(t)dt,$$

where [1]

$$\theta(t) = 2\int_m^1 f(s)ds.$$

The probability that μ_n lies in $(\mu, \mu + d\mu)$ is

$$\psi_n(\mu) = u_n(t)\frac{d\mu}{\sqrt{2k+1}},$$

$$\mu = (t-m)\sqrt{2k+1}, \quad t = m + \frac{\mu}{\sqrt{2k+1}}.$$

Theorem 1. *As n increases, $\psi_n(\mu)$ tends to the normal distribution law*

$$\psi(\mu) = \frac{1}{\sqrt{2\pi}\sigma_m} e^{-\mu^2/2\sigma_m^2}$$

with variance

$$\sigma_m = \tfrac{1}{2} f(m).$$

Indeed,

$$\psi_n(\mu) = \frac{\sqrt{2k+1}(2k)!}{(k!)^2 4k}\{1-\theta^2(t)\}^k f(t).$$

By Stirling's formula

$$\frac{\sqrt{2k+1}(2k)!}{(k!)^2 4^k} \to \sqrt{\frac{2}{\pi}}.$$

Further, it is clear that

$$f(t) \to f(m).$$

---

[1] See the derivation of the similar formula (4) in [1].

It only remains to consider the factor $\{1-\theta^2(t)\}^k$. First, we have

$$\theta(t) = m + \frac{\mu}{\sqrt{2k+1}} = \frac{\mu f(m)}{\sqrt{2k+1}}(1+\epsilon),$$

where $\epsilon \to 0$ as $n \to \infty$. Hence

$$\theta^2(t) = \frac{\mu^2 f(m)}{2k}(1+\epsilon'),$$

$$\{1-\theta^2(t)\}^k = e^{-\mu^2 f^2(m)/2}(1+\epsilon''),$$

where ϵ' and ϵ'' also tend to 0. Comparing the above expressions we obtain

$$\psi_n(\mu) \to \sqrt{\frac{2}{\pi}} f(m) e^{-\mu^2 f^2(m)/2},$$

which proves our theorem.

Thus we see that the variances of $m_n - m$ and $a_n - a$ are infinitesimally small as $n \to \infty$. Hence, m is the *systematic error* in the method of the median, while a is the systematic error in the method of the arithmetic mean. The accuracies of both methods are estimated by the relation

$$\lambda = \sigma_m/\sigma = \tfrac{1}{2} f(m)\sigma.$$

For a normal distribution $f(x)$ the ratio λ was calculated by Haag; it is

$$\lambda_g = \sqrt{\pi/2} \sim 5/4.$$

It is easy to construct examples of distribution laws for which λ takes any values between 0 and ∞. However, we have:

Theorem 2. *For distribution laws with one maximum the ratio λ can take any values in the interval*

$$0 < \lambda < \sqrt{3},$$

but it can not exceed $\sqrt{3}$.

The maximum value $\sqrt{3}$ is attained if

$$f(x) = a,\ x \le 1/2a, \quad f(x) = 0,\ x > 1/2a,$$

but this distribution law does not satisfy the continuity condition: $\lambda = 0$ if $\sigma = +\infty$.

References

1. J. Haag, *C. R. Acad. Sci. Paris* **179** (1924), 1388.
2. E. Borel, *Traité du calcul des probabilités et ses applications*, Paris, 1924.

12. A GENERALIZATION OF THE LAPLACE-LYAPUNOV THEOREM *

Let $x_1, x_2, \ldots$ be independent random variables with expectations

$$\mathbf{E}x_n = 0, \quad \mathbf{E}x_n^2 = 2b_n, \quad \mathbf{E}|x_n|^3 = d_n;$$

and suppose that the ratios d_n/b_n are smaller than a certain fixed constant

$$d_n/b_n \leq \mu. \tag{1}$$

Set

$$S_n = x_1 + \ldots + x_n, \quad t_n = b_1 + \ldots + b_n.$$

An important problem in probability theory is to study the dependence of the sums S_n as $n \to \infty$, on the properties of t_n as $\mu \to 0$. On the basis of the now classical law of Lyapunov [1] we have

$$\mathbf{P}\{a < S_n < b\} = \frac{1}{\sqrt{2\pi t_n}} \int_a^b e^{-s^2/4t_n} ds + \theta R(t_n, \mu), \quad |\theta| \leq 1, \tag{2}$$

and $R(t_n, \mu) \to 0$ uniformly with respect to μ if t_n is greater than some constant T.

Thus, for fixed n we have a formula for the asymptotic behaviour of S_n. Consider the following problem. Let $a(t)$ and $b(t)$ be functions of the parameter t. What is the probability that all the inequalities

$$a(t_k) < S_k < b(t_k), \quad k = 1, 2, \ldots, n \tag{3}$$

hold? Assume that $a(t)$ and $b(t)$ are continuously differentiable and that

$$a(t) < b(t), \quad a(0) < 0 < b(0).$$

Then we can obtain an asymptotic solution of our problem similar to (2).

All the sums $S_1, S_2, \ldots, S_n$ (up to S_n) can be divided into the following three classes K, K_1 and K_2:

K contains those sums for which all the inequalities (3) hold;

---

* 'Eine Verallgemeinerung des Laplace-Liapounoffschen Satzes', *Izv. Akad. Nauk SSSR*, OMEN (1931), 959–962. Presented by S.N. Bernshtein.

[1] A.M. Liapounoff, *Bull. Acad. Sci. St. Petersburg* **13** (1900), 359.

K_1 contains those sums for which there exists a k such that all inequalities

$$a(t_i) < S_i < b(t_i), \quad i = 1, 2, \dots, k-1.$$
$$S_k \le a(t_k)$$

hold;

K_2 contains those sums for which there exists k such that the inequalities

$$a(t_i) < S_i < b(t_i), \quad i = 1, 2, \dots, k-1,$$
$$b(t_k) \le S_k$$

hold. Denote the probabilities corresponding to these sets by $P_n, P_n^{(1)}, P_n^{(2)}$. Finally, denote by $P_n(x, y)$ the probability of the event

$$a(t_k) < S_k < b(t_k), \quad k = 1, 2, \dots, n-1,$$
$$x < S_n < y.$$

Clearly

$$P_n = P_n\{a(t_n), b(t_n)\}.$$

The inequalities

$$t > 0, \quad a(t) < s < b(t)$$

single out a region G in the (s,t)-plane. Denote by $g(s_0, t_0; s, t)$ the Green's function for the heat equation

$$\partial f/\partial t = \partial^2 f/\partial s^2 \tag{4}$$

in G and set

$$g(s,t) = g(0,0;\ s,t), \quad \partial g(s,t)/\partial s = u(s,t),$$
$$v_1(t) = -u[a(t), t], \quad v_2(t) = u[b(t), t].$$

Theorem. *The following asymptotic formulas hold*

$$P_n(x, y) = \int_x^y g(s, t_n)ds + \theta R(t_n, \mu), \tag{5}$$

$$P_n^{(1)} = \int_0^{t_n} v_1(t)dt + \theta_1 R_1(\mu), \tag{6}$$

$$P_n^{(2)} = \int_0^{t_n} v_2(t)dt + \theta_2 R_2(\mu), \tag{7}$$

$$|\theta| \le 1, \quad |\theta_1| \le 1, \quad |\theta_2| \le 1,$$

where R, R_1, R_2 *tend to zero uniformly with respect to* μ*; furthermore, for* $R(t,\mu)$ *this convergence is uniform with respect to* t *if* t *is greater than a certain constant.*

If, instead of (3) we have one-sided inequalities, for example,

$$a(t_k) < S_k, \quad k = 1, 2, \ldots, n, \tag{8}$$

then a similar result holds, which can be easily obtained from the above by passing to the limit.

Göttingen, 20 January 1931

13. ON THE GENERAL FORM OF A HOMOGENEOUS STOCHASTIC PROCESS*

(The problem of Bruno de Finetti)

Let $X(\lambda)$ be a real variable which varies randomly as a function of time λ, or more exactly, in such a way that the probability distribution

$$\Phi_\Delta(x) = \mathbf{P}\{X(\lambda_2) - X(\lambda_1) < x\} \; (\lambda_2 > \lambda_1)$$

depends only on $\Delta = \lambda_2 - \lambda_1$ and does not depend either on λ_1 or on $X(\lambda_1)$ or on the behaviour of $X(\lambda)$ for $\lambda < \lambda_1$. As usual $\mathbf{P}$ denotes the probability of the relation inside the braces. Then, as is well known, the distribution function $\Phi_\Delta(x)$ satisfies the functional equation [1]

$$\Phi_{\Delta_1+\Delta_2}(x) = \int_{-\infty}^{\infty} \Phi_{\Delta_1}(x-y)d\Phi_{\Delta_2}(y). \tag{1}$$

In the present paper, generalizing certain results of Bruno de Finetti, [2] we give a solution to equation (1) under the sole assumption that the first and second moments

$$m_\Delta = \int_{-\infty}^{\infty} x d\Phi_\Delta(x), \quad \sigma_\Delta^2 = \int_{-\infty}^{\infty} (x - m_\Delta)^2 d\Phi_\Delta(x)$$

are finite.

For non-specialists, we note that $\Phi_\Delta(x)$ is non-decreasing and left continuous and that $\Phi_\Delta(-\infty) = 0$, $\Phi_\Delta(\infty) = 1$.

Let

$$\psi_\Delta(t) = \int_{-\infty}^{\infty} e^{itx} d\Phi_\Delta(x)$$

---

* This article was presented on 6 March 1932 by the member of the Italian Academy of Sciences, G. Castelnuovo in the journal *Atti R. Accad. Naz. Lincei. Ser. Sesta. Rend.* and in the same year published in two parts in Italian. The first part was called 'Sulla forma generale di un processo stocastico omogeneo' **15**:10, 805–808). The translation of this part is the title of the present article. The second part was called 'Ancora sulla forma generale di un processo omogeneo' (**15**:12, 866–869).

[1] See, for example, my paper: 'Über die analytischen Methoden in der Wahrscheinlichkeitsrechnung', *Math. Ann.* **104** (1931), 415–458. (Paper 9 in this volume.)

[2] See: B. de Finetti, 'Le funzioni caratteristiche de legge istantanea', *Atti Accad. Naz. Lincei. Rend.* **12** (1930), 278–282.

be the characteristic function of $\Phi_\Delta(x)$. From the well known properties of characteristic functions we have

$$\psi_{1/n}(t) = [\psi_1(t)]^{1/n}, \quad \psi_{k/n}(t) = [\psi_1(t)]^{k/n}.$$

Since $\psi_\Delta(t)$ is continuous with respect to Δ,[3] we have

$$\psi_\Delta(t) = [\psi_1(t)]^\Delta. \tag{2}$$

Since the distribution $\Phi_\Delta(x)$ is uniquely determined by $\psi_\Delta(t)$, in order to describe it, it suffices to find the general form of $\psi_1(t)$.

According to de Finetti, we have[4]

$$\psi_1(t) = \lim e^{\frac{1}{\Delta}[\psi_\Delta(t)-1]}, \quad \log\psi_1(t) = \lim \frac{1}{\Delta}[\psi_\Delta(t) - 1], \ \Delta \to 0.$$

Moreover,

$$\frac{1}{\Delta}[\psi_\Delta(t) - 1] = \frac{1}{\Delta}\Big[\int_{-\infty}^{\infty} e^{itx} d\Phi_\Delta(x) - 1\Big] =$$
$$= \frac{1}{\Delta}\Big[it\int_{-\infty}^{\infty} x d\Phi_\Delta(x) + \int_{-\infty}^{\infty}(e^{itx} - 1 - itx)d\Phi_\Delta(x)\Big] =$$
$$= \frac{1}{\Delta}\Big[itm_\Delta + \int_{-\infty}^{\infty}(e^{itx} - 1 - itx)d\Phi_\Delta(x)\Big] =$$
$$= itm_1 + \frac{1}{\Delta}\int_{-\infty}^{\infty}(e^{itx} - 1 - itx)d\Phi_\Delta(x).$$

Now let

$$F_\Delta(x) = \frac{1}{\Delta}\int_{-\infty}^{x} y^2 d\Phi_\Delta(y). \tag{3}$$

The function $F_\Delta(x)$ is non-decreasing with respect to x; moreover, $F_\Delta(-\infty) = 0$ and

$$F_\Delta(\infty) = \frac{1}{\Delta}\int_{-\infty}^{\infty} x^2 d\Phi_\Delta(x) = \frac{1}{\Delta}(m_\Delta^2 + \sigma_\Delta^2) = \Delta m_1^2 + \sigma_1^2.$$

---

[3] If $\Delta \leq 1/n$, then $\sigma_\Delta^2 = \sigma_{1/n}^2 - \sigma_{1/n-\Delta}^2 \leq \sigma_{1/n}^2 = (1/n)\sigma_1^2$. Therefore $\sigma_\Delta^2 \to 0$ as $\Delta \to 0$ and $\psi_\Delta(t) \to 1$ accordingly. Taking into account the equality $\psi_{\Delta_1+\Delta_2}(t) = \psi_{\Delta_1}(t)\psi_{\Delta_2}(t)$ we conclude that $\psi_\Delta(t)$ is continuous with respect to Δ.

[4] $[\psi_\Delta(t) - 1]/\Delta$ is bounded for each fixed t, hence the equality $\psi_1(t) = 0$ is impossible, so that $\log\psi_1(t)$ is defined for any t.

For any interval (a, b) not containing 0, $a \neq 0$, $b \neq 0$, we have

$$\frac{1}{\Delta}\int_a^b (e^{itx} - 1 - itx)d\Phi_\Delta(x) = \int_a^b p(x,t)dF_\Delta(x),$$

where

$$p(x,t) = (e^{itx} - 1 - itx)/x^2 \quad \text{for } x \neq 0,$$
$$p(x,t) = -t^2/2 \quad \text{for } x = 0. \tag{4}$$

Since for any given t the function $p(x,t)$ is finite and continuous (including at $x = 0$), we have

$$\frac{1}{\Delta}\int_{-\infty}^{\infty} (e^{itx} - 1 - itx)d\Phi_\Delta(x) = \int_{-\infty}^{\infty} p(x,t)dF_\Delta(x).$$

We now choose a sequence of positive numbers $\Delta_1, \Delta_2, \dots, \Delta_n, \dots$ tending to zero. As is well known, given a sequence of functions

$$F_{\Delta_1}(x), F_{\Delta_2}(x), \dots, F_{\Delta_n}(x), \dots,$$

we can select a subsequence

$$F_{\Delta_{n_1}}(x),\ F_{\Delta_{n_2}}(x), \dots, F_{\Delta_{n_k}}(x), \dots$$

that converges to some function $F(x)$ at all points of continuity of the latter. Of course, $F(x)$ is non-decreasing and its extremal values are

$$F(-\infty) \geq \lim F_\Delta(-\infty) = 0,$$
$$F(\infty) \leq \lim F_\Delta(\infty) = \sigma_1^2 \quad (\text{as } \Delta \to 0). \tag{5}$$

Taking into account the fact that $p(x,t) \to 0$ for fixed t as $x \to \pm\infty$, we finally obtain

$$\int_{-\infty}^{\infty} p(x,t)dF_{\Delta_{n_k}}(x) \to \int_{-\infty}^{\infty} p(x,t)dF(x),$$

$$\log \psi_1(t) = \lim \frac{1}{\Delta}[\psi_\Delta(t) - 1] =$$

$$= \lim\Big[itm_1 + \int_{-\infty}^{\infty} (e^{itx} - 1 - itx)d\Phi_\Delta(x)\Big] =$$

$$= \lim\Big[itm_1 + \int_{-\infty}^{\infty} p(x,t)dF_\Delta(x)\Big] \quad (\text{as } \Delta \to 0);$$

$$\log \psi_1(t) = itm_1 + \int_{-\infty}^{\infty} p(x,t)dF(x). \tag{6}$$

Since the second moment σ_1^2 is finite, we have

$$\log \psi_1(t) = im_1 t - \frac{\sigma_1^2}{2} t^2 + o(t^2).$$

Moreover, (6) gives

$$\log \psi_1(t) = itm_1 + \int_{-\infty}^{\infty} \Big(-\frac{t^2}{2} + o(t^2)\Big) dF(x) =$$
$$= itm_1 - \frac{t^2}{2}[F(\infty) - F(-\infty)] + o(t^2),$$

implying that

$$F(\infty) - F(-\infty) = \sigma_1^2;$$

therefore, taking (5) into account,

$$F(-\infty) = 0, \quad F(\infty) = \sigma_1^2.$$

Let us assume that $F(x)$ is left continuous (though the values of $F(x)$ at points of discontinuity are of no importance). We have already mentioned that $F(x)$ is non-decreasing.

Formulas (6), (4) and (2) together with the well-known formula

$$\Phi_\Delta(x) - \Phi_\Delta(0) = \frac{1}{2\pi} \int_{-\infty}^{\infty} \frac{1 - e^{-itx}}{it} \psi_\Delta(t) dt \tag{7}$$

give the general solution of our problem.

The function $F(x)$ is completely determined by the distribution function $\Phi_\Delta(x)$. Indeed, the integral

$$\int_{-\infty}^{\infty} p(x,t) dF(x) = \log \psi_1(t) = itm_1$$

can be differentiated twice with respect to t and taking into account the equality

$$\frac{\partial^2}{\partial t^2} p(x,t) = -e^{itx},$$

we have

$$\int_{-\infty}^{\infty} e^{itx} dF(x) = -\frac{\partial^2}{\partial t^2} \log \psi_1(t) = \chi(t),$$
$$F(x) - F(0) = \frac{1}{2\pi} \int_{-\infty}^{\infty} \frac{1 - e^{-itx}}{it} \chi(t) dt.$$

From the latter equality we can determine $F(x)$ to within an additive constant which, in turn, is determined by the condition $F(-\infty) = 0$.

As already mentioned, from any sequence $F_{\Delta_n}(x)$ with $\Delta_n \to 0$ we can always select a subsequence $F_{\Delta_{n_k}}(x)$ converging to $F(x)$. Hence, $F_\Delta(x) \to F(x)$ as $\Delta \to 0$ and

$$F(x) = \lim F_\Delta(x) = \lim \frac{1}{\Delta} \int_{-\infty}^{x} y^2 d\Phi_\Delta(y). \tag{8}$$

However, (8) is only true at those points where $F(x)$ is continuous.

For $x < 0$, (3) implies that

$$\frac{1}{\Delta}\Phi_\Delta(x) = \int_{-\infty}^{x} y^{-2} dF_\Delta(y).$$

Further (8) implies that at each point of continuity $x < 0$ of $F(x)$ we have

$$P_1(x) = \lim \frac{1}{\Delta}\Phi_\Delta(x) = \int_{-\infty}^{x} y^{-2} dF(y). \tag{9}$$

Similarly, for $x > 0$ we have

$$\frac{1}{\Delta}[1 - \Phi_\Delta(x)] = \frac{1}{\Delta}[\Phi_\Delta(\infty) - \Phi_\Delta(x)] = \int_{x}^{\infty} y^{-2} dF_\Delta(y),$$

$$P_2(x) = \lim \frac{1}{\Delta}[1 - \Phi_\Delta(x)] = \int_{x}^{\infty} y^{-2} dF(y). \tag{10}$$

The meaning of $P_1(x)$ and $P_2(x)$ can be explained as follows. In general $X(\lambda)$ does not have to change continuously with time λ; this change can also have jumps and $P_2(x)d\lambda$ is the probability that during the time interval $d\lambda$ the function made a positive jump greater than x. Similarly, $P_1(x)d\lambda$ is the probability of a negative jump with absolute value greater than $|x|$.

It follows from (9) and (10) that

$$F(x) = \int_{-\infty}^{x} y^2 dP_1(y), \quad x < 0, \tag{11}$$

$$\sigma_1^2 - F(x) = F(\infty) - F(x) = \int_{x}^{\infty} y^2 dP_2(y), \quad x > 0. \tag{12}$$

Clearly, these formulas, as well as all the others in which $P_1(x)$ and $P_2(x)$ occur hold only when $F(x)$ is continuous.

Thus, the behaviour of $F(x)$ outside $x = 0$ depends only on the probability distribution of the jumps of $X(\lambda)$. At the same time, the discontinuity of $F(x)$ at the origin is associated with the continuous variation of $X(\lambda)$. To make this clear we set

$$\Omega(x) = F(x), \quad x \le 0,$$

$$\Omega(x) = F(x) - \sigma_0^2, \quad x > 0,$$

where σ_0^2 denotes the jump of $F(x)$ at $x = 0$. Clearly, $dF(x) = d\Omega(x)$ outside a neighbourhood of the origin. We can write (11) and (12) as follows

$$\Omega(x) = \int_{-\infty}^{x} y^2 dP_1(y), \quad x < 0, \tag{13}$$

$$\sigma_2^2 - \Omega(x) = \int_{x}^{\infty} y^2 dP_2(y), \quad x > 0, \tag{14}$$

where $\sigma_2^2 = \sigma_1^2 - \sigma_0^2$. Since $\Omega(x)$ must be continuous at the origin, $\sigma_2^2 = \Omega(\infty)$ and hence $\Omega(x)$ is completely determined by $P_1(x)$ and $P_2(x)$. We have

$$\int_{-\infty}^{\infty} p(x,t) dF(x) = \sigma_0^2 p(0,t) + \int_{-\infty}^{\infty} p(x,t) d\Omega(x).$$

Since $p(0,t) = -t^2/2$, we obtain the following final

Main formula

$$\log \psi_1(t) = i m_1 t - \frac{\sigma_0^2}{2} t^2 + \int_{-\infty}^{\infty} p(x,t) d\Omega(x). \tag{15}$$

If $\Omega(x) = 0$, that is, if $X(\lambda)$ can vary only continuously as a function of time, then $\sigma_0^2 = \sigma_1^2$. We have

$$\psi_\Delta(t) = \exp\Big\{ i m_1 t - \frac{\sigma_1^2}{2} t^2 \Big\},$$

$$\Phi_\Delta(x) = \frac{1}{\sigma_1 \sqrt{2\pi\Delta}} \int_{-\infty}^{x} \exp\Big\{ -\frac{y - m_1\Delta)^2}{2\Delta\sigma_1^2} \Big\} dy.$$

We show in conclusion that for any function $F(x)$ (non-decreasing, left continuous and with values $F(-\infty) = 0$, $F(\infty) = \sigma_1^2 < \infty$) (7), (6) and (2) give a solution to our problem.

In other words, we are going to prove that under the above conditions $\Phi_\Delta(x)$ given by (7) really is a distribution function for all $\Delta > 0$.

First we consider a step function $T(x)$ with jumps

$$\omega_k = T(x_k+0) - T(x_k),$$

only at a finite number of points $x_1, x_2, \ldots, x_n$. Denote by $\sigma_1^2 = \omega_1+\omega_2+\ldots+\omega_n$ the sum of these jumps (it is constant on every interval that does not contain any of the x_k). Suppose also that $x = 0$ is not one of the points x_k. Then [5]

$$\int_{-\infty}^{\infty} p(x,t)dT(x) = \sum_k p_k(e^{itx_k} - 1 - itx_k),$$

$$\log \bar{\psi}_1(t) = itP + \sum_k p_k[\chi_k(t) - 1],$$

$$P = m_1 - \sum_k p_k, \quad p_k = \frac{\omega_k}{x_k^2}, \quad \chi_k(t) = \exp(itx_k);$$

$$\bar{\psi}_\Delta(t) = [\bar{\psi}_1(t)]^\Delta = \exp\{itP\Delta + \sum \Delta p_k[\chi_k(t) - 1]\}.$$

Clearly $\chi_k(t) = \exp(itx_k)$ and $\exp(itP\Delta)$ are characteristic functions, which means that by (7) they generate distribution functions. The results of de Finetti imply that in this case $\exp\{\Delta p_k[\chi_k(t) - 1]\}$ and $\bar{\psi}_\Delta(t)$, which is a product of characteristic functions, are characteristic functions.

Under an appropriate choice of t_0 and for $\epsilon > 0$ one can always approximate any function $F(x)$ by a step function $T(x)$ so that the inequality

$$\left|\int_{-\infty}^{\infty} p(x,t)dF(x) - \int_{-\infty}^{\infty} p(x,t)dT(x)\right| < \epsilon \tag{16}$$

holds for all $|t| < t_0$. We now form a sequence $T_n(x)$ of functions of the form $T(x)$ such that the difference in (16) tends to 0 for all t, uniformly on every bounded interval. Then the corresponding functions $\bar{\psi}_\Delta^{(n)}(t)$ converge to $\psi_\Delta(t)$, which implies that $\psi_\Delta(t)$, being a limit of characteristic functions, is itself a characteristic function.

[5] When we replace $F(x)$ by $T(x)$ we write $\bar{\psi}_\Delta$ instead of ψ_Δ.

14. ON COMPUTING THE MEAN BROWNIAN AREA *

Jointly with M.A. Leontovich

In this paper we solve the following problem posed by S.I. Vavilov. Determine the expectation of the area covered during a given time by the projection onto a plane of a moving Brownian particle of finite size. The parts of the area covered by this projection several times must be counted only once. This problem will be reduced to the following more general problem: determine the probability that an infinitesimal Brownian particle moving in a region G and situated at $t = 0$ at a given point (x, y) hits the boundary R of this region at least once during the time t. In §1 we explain a method for solving this problem, and in §§2, 3 this method is applied to computing the mean Brownian area. §§1, 2 of this paper are written by A.N. Kolmogorov, §3 by M.A. Leontovich.

§1. First we consider the motion of an infinitesimal particle on a plane with Cartesian coordinates x and y. (The analogous three-dimensional problem can be studied in the same way.) $P(x, y; t)$ is the probability that the particle situated at time $t = 0$ at a point (x, y) of G hits the boundary R of this region at least once during time t. Further, let $P_L(x, y; t)$ be the probability that the particle situated at time $t = 0$ at the point (x, y) hits the boundary R at least once during time t in such a way that it first hits a part L of R. Then, clearly,

$$P_L(x, y; t) \leq P(x, y; t).$$

Now let $p(x, y; \xi, \eta; t)d\xi d\eta$ be the probability that the particle which was at the point (x, y) at the moment $t = 0$ is in the region $(\xi, \xi + d\xi; \eta, \eta + d\eta)$ at time t and does not hit the boundary during time t. Then, clearly,

$$\int_G p(x, y; \xi, \eta; t)d\xi d\eta + P(x, y; t) = 1. \tag{1}$$

Moreover, it can easily be seen that

$$P(x, y; t + \tau) = P(x, y; \tau) + \int_G p(x, y; \xi, \eta; \tau)P(\xi, \eta; t)d\xi d\eta, \tag{2}$$

$$P_L(x, y; t + \tau) = P_L(x, y; \tau) + \int_G p(x, y; \xi, \eta; \tau)P_L(\xi, \eta; t)d\xi d\eta, \tag{3}$$

---

* 'Zur Berechnung der mittleren Brounschen Fläche', *Phys. Z. Sow.* 4 (1933), 1–13.

We now make the following assumptions:

I. The functions $P(x,y;t)$ and $P_L(x,y;t)$ can be expanded into Taylor series in a neighbourhood of every interior point of G, so that

$$\begin{aligned} P(\xi,\eta;t) = P(x,y;t) &+ \frac{\partial P(x,y;t)}{\partial x}(\xi - x) + \frac{\partial P(x,y;t)}{\partial y}(\eta - y) + \\ &+\tfrac{1}{2}\Big\{\frac{\partial^2 P(x,y;t)}{\partial x^2}(\xi - x)^2 + 2\frac{\partial^2 P(x,y;t)}{\partial x \partial y}(\xi - x)(\eta - y) + \\ &+\frac{\partial^2 P(x,y;t)}{\partial y^2}(\eta - y)^2\Big\} + \theta\{(\xi - x)^2 + (\eta - y)^2\}^{3/2}, \end{aligned} \tag{4}$$

$$\begin{aligned} P_L(\xi,\eta;t) = P_L(x,y;t) &+ \frac{\partial P_L(x,y;t)}{\partial x}(\xi - x) + \frac{\partial P_L(x,y;t)}{\partial y}(\eta - y) + \\ &+\tfrac{1}{2}\Big\{\frac{\partial^2 P_L(x,y;t)}{\partial x^2}(\xi - x)^2 + 2\frac{\partial^2 P_L(x,y;t)}{\partial x \partial y}(\xi - x)(\eta - y) + \\ &+\frac{\partial^2 P_L(x,y;t)}{\partial y^2}(\eta - y)^2\Big\} + \theta\{(\xi - x)^2 + (\eta - y)^2\}^{3/2}, \end{aligned} \tag{5}$$

where $|\theta| \le M$, $|\theta'| \le M'$, M and M' being independent of ξ and η (but may depend on x, y and t);

II. At every interior point of G,

$$\lim_{\tau \to 0} \frac{P(x,y;\tau)}{\tau} = 0. \tag{6}$$

Since $P_L \le P$, this also implies that

$$\lim_{\tau \to 0} \frac{P_L(x,y;\tau)}{\tau} = 0. \tag{7}$$

III. The function $p(x,y;\xi,\eta;t)$ satisfies the following conditions:

$$\lim_{\tau \to 0}\Big\{\frac{1}{\tau}\int_G p(x,y;\xi,\eta;\tau)(\xi - x)d\xi d\eta\Big\} = A_1(x,y), \tag{8}$$

$$\lim_{\tau \to 0}\Big\{\frac{1}{\tau}\int_G p(x,y;\xi,\eta;\tau)(\eta - y)d\xi d\eta\Big\} = A_2(x,y), \tag{9}$$

$$\lim_{\tau \to 0}\Big\{\frac{1}{2\tau}\int_G p(x,y;\xi,\eta;\tau)(\xi - x)^2 d\xi d\eta\Big\} = B_{11}(x,y), \tag{10}$$

$$\lim_{\tau\to 0}\Big\{\frac{1}{2\tau}\int_G p(x,y;\xi,\eta;\tau)(\eta-y)^2 d\xi d\eta\Big\} = B_{22}(x,y), \tag{11}$$

$$\lim_{\tau\to 0}\Big\{\frac{1}{2\tau}\int_G p(x,y;\xi,\eta;\tau)(\xi-x)(\eta-y) d\xi d\eta\Big\} = B_{12}(x,y), \tag{12}$$

$$\lim_{\tau\to 0}\Big\{\frac{1}{\tau}\int_G p(x,y;\xi,\eta;\tau)[(\xi-x)^2+(\eta-y)^2]^{3/2} d\xi d\eta\Big\} = 0, \tag{13}$$

where $A_1, A_2, B_{11}, B_{12}, B_{22}$ are functions defined by these relations.

Under the assumptions I–III the functions $P(x,y;t)$ and $P_L(x,y;t)$ are solutions to the following differential equation:

$$\frac{\partial P}{\partial t} = B_{11}\frac{\partial^2 P}{\partial x^2} + 2B_{12}\frac{\partial^2 P}{\partial x\partial y} + B_{22}\frac{\partial^2 P}{\partial y^2} + A_1\frac{\partial P}{\partial x} + A_2\frac{\partial P}{\partial y}. \tag{14}$$

We prove this for $P(x,y;t)$, since for $P_L(x,y;t)$ the proof is quite similar. By (2) and (4) we have

$$\frac{P(x,y;t+\tau)}{\tau} - \frac{P(x,y;t)}{\tau} =$$

$$= \frac{P(x,y;\tau)}{\tau} + \frac{1}{\tau}\int_G p(x,y;\xi,\eta;\tau)P(\xi,\eta;t)d\xi d\eta - \frac{P(x,y;t)}{\tau} =$$

$$= \frac{P(x,y;\tau)}{\tau} + \frac{1}{\tau}\Big\{\int_G p(x,y;\xi,\eta;\tau)d\xi d\eta - 1\Big\}P(x,y;t) +$$

$$+\frac{1}{\tau}\int_G p(x,y;\xi,\eta;\tau)(\xi-x)d\xi d\eta\frac{\partial P(x,y;t)}{\partial x} +$$

$$+\frac{1}{\tau}\int_G p(x,y;\xi,\eta;\tau)(\eta-y)d\xi d\eta\frac{\partial P(x,y;t)}{\partial y} +$$

$$+\frac{1}{2\tau}\int_G p(x,y;\xi,\eta;\tau)(\xi-x)^2 d\xi d\eta\frac{\partial^2 P(x,y;t)}{\partial x^2} +$$

$$+\frac{1}{2\tau}\int_G p(x,y;\xi,\eta;\tau)(\xi-x)(\eta-y)d\xi d\eta\frac{\partial^2 P(x,y;t)}{\partial x\partial y} +$$

$$+\frac{1}{2\tau}\int_G p(x,y;\xi,\eta;\tau)(\eta-y)^2 d\xi d\eta\frac{\partial^2 P(x,y;t)}{\partial y^2} +$$

$$+\frac{\theta''}{\tau}\int_G p(x,y;\xi,\eta;\tau)\{(\xi-x)^2+(\eta-y)^2\}^{3/2}d\xi d\eta, \tag{15}$$

where $|\theta''| \le M''$. By (6) and (13) the first and the last terms on the right-hand side of (15) tend to 0 as $\tau \to 0$. But, since by (1)

$$1 - \int_G p(x,y;\xi,\eta;t)d\xi d\eta = P(x,y;t),$$

the second term on the right-hand side of (15) also tends to 0 as $\tau \to 0$. The coefficients of the derivatives of P with respect to x and y have the limits $A_1, A_2, B_{11}, 2B_{12}, B_{22}$; therefore the right-hand side of (15) tends to the right-hand side of (14) as $\tau \to 0$. The left-hand side of (15) also has the limit $\partial P/\partial t$. This proves (14).

Equation (14) satisfied by P is adjoint to the Fokker equation. This equation can also be derived under much weaker assumptions (cf. [1]–[4]).

We now assume that, in addition to I–III the following two conditions hold:

IV. For fixed t (where $t > 0$) the probability $P(x, y; t)$ tends to 1 as (x, y) approaches the boundary of G, and $P(x, y; t) \equiv 0$ at $t = 0$ for any interior point (x, y) of G;

V. For fixed t (where $t > 0$) the probability $P_L(x, y; t)$ tends to 1 as (x, y) tends to an interior point of the part L of the boundary and tends to 0 as (x, y) tends to a point of the boundary outside L, and $P_L(x, y; t) \equiv 0$ at $t = 0$ for every interior point (x, y) of G.

Taking into account that both P and P_L are bounded (namely, they are non-negative and not greater than 1), it can easily be shown that both P and P_L are *uniquely defined by* (14) *and the conditions* IV *and* V.

§2. Computation of the area covered by a Brownian particle of finite size. Assume that the projection of a Brownian particle can be represented by a disk of radius 1 (under an appropriate scaling). In what follows we only consider such projections, so for simplicity we call them particles. Note, however, that an entirely similar treatment can be given in the corresponding three-dimensional problem of computing the volume filled in by a Brownian particle.

The centre of the particle obeys the laws stated in §1. Symmetry considerations make it natural to assume that $A_1 = A_2 = B_{12} = 0$ and $B_{11} = B_{22} = D$. Then (14) takes the form

$$\frac{\partial P}{\partial t} = D\Big(\frac{\partial^2 P}{\partial x^2} + \frac{\partial^2 P}{\partial y^2}\Big).$$

By (10) the diffusion coefficient D is defined by the relation

$$D = \lim_{\tau \to 0} \frac{\mathbf{E}(\xi - x)^2}{2\tau} = \lim_{\tau \to 0} \frac{\mathbf{E}(\eta - y)^2}{2\tau},$$

where $\mathbf{E}$ denotes the expectation. It can easily be seen that D coincides with the analogous constant for a three-dimensional Brownian particle.

Our problem is *to determine the expectation of the area covered by successive positions of a particle at times from $t = 0$ to a given t.* As the origin we take the point that coincides with the particle's centre at $t = 0$. Denote by $W(x, y; t)$ the probability that during time t a given point (x, y) is at least once covered by this particle. When the distance from (x, y) to the origin is at most 1, then clearly, $W(x, y; t) = 1$. If this distance exceeds 1, then (x, y) is covered by the particle at least once if and only if the route of the particle's centre falls at least once inside the disk with radius 1 and centre at (x, y). By symmetry considerations the probability of such an event is also equal to the probability that the centre of the Brownian particle situated at (x, y) for $t = 0$ will during time t touch the boundary of the disk S of radius 1 with centre at the origin at least once. This probability can be computed using the considerations of §1.

Thus, inside S we have

$$W(x, y; t) = 1, \tag{16}$$

while outside this disk $W(x, y; t)$ must satisfy the differential equation

$$\frac{\partial W}{\partial t} = D\Big(\frac{\partial^2 W}{\partial x^2} + \frac{\partial^2 W}{\partial y^2}\Big). \tag{17}$$

Moreover, on the boundary of S the boundary condition

$$W(x, y; t) = 1 \tag{18}$$

must hold, as well as the initial condition

$$W(x, y; 0) = 0. \tag{19}$$

To determine the expectation of the Brownian area, we must reason as follows. Let $\mathfrak{F}$ be the region covered by the particle during time t and F the area of this region. Assume that $\delta(x, y; \mathfrak{F}) = 1$ for (x, y) in the interior of $\mathfrak{F}$, and $\delta(x, y; \mathfrak{F}) = 0$ otherwise. Clearly,

$$\mathbf{E}\{\delta(x, y; \mathfrak{F})\} = W(x, y; t),$$

and the Brownian area F can be defined by the relation

$$F = \iint \delta(x, y; \mathfrak{F}) dx dy.$$

This implies that

$$\mathbf{E}(F) = \iint \mathbf{E}\{\delta(x,y;\mathfrak{F})\}dxdy = \iint W(x,y;t)dxdy. \tag{20}$$

§3. To solve the desired problem with boundary conditions we pass to polar coordinates (r,ϕ). Clearly, W depends only on the position vector r, and not on the angle ϕ. We now choose the time unit in such a way that D takes the value 1. In this case our problem with boundary conditions for $W(r,t)$ takes the form

$$\frac{\partial W}{\partial t} = \frac{\partial^2 W}{\partial r^2} + \frac{1}{r}\frac{\partial W}{\partial r}, \tag{21}$$

$$W(1,t) = 1; \quad W(\infty,t) = 0; \tag{22}$$

$$W(r,0) = 0, \ r > 1. \tag{23}$$

To solve this problem we use the Laplace transformation. Multiply (21) by e^{-vt} and integrate over t from 0 to ∞. Integrating by parts and using the initial condition (23) for

$$U(r,v) = \int_0^\infty W(r,t)e^{-vt}dt \tag{24}$$

we obtain the following differential equation:

$$\frac{d^2U}{dr^2} + \frac{1}{r}\frac{dU}{dr} - vU = 0. \tag{25}$$

The boundary conditions (22) for W yield the following conditions for U:

$$U(1,v) = 1/v; \quad U(\infty,v) = 0. \tag{25'}$$

A solution of (25) under the conditions (25′) is of the form

$$U(r,v) = \frac{1}{v}\frac{K(r\sqrt{v})}{K(\sqrt{v})},$$

where

$$K(x) = \frac{i\pi}{2}H_0^{(1)}(ix),$$

and $H_0^{(1)}$ is the first Hankel function of order zero (see [5], §17.71).

By (24), in order to find $W(r,t)$ we have to solve the linear integral equation of the first kind

$$\int_0^\infty W(r,t)e^{-vt}\,dt = \frac{1}{v}\frac{K(r\sqrt{v})}{K(\sqrt{v})}. \tag{26}$$

The right-hand side of this equation for $\operatorname{Re} v > 0$ is an entire function of v. Indeed, by a known theorem (see, for example, [6], §15.7), $K(x)$ does not vanish for $|\arg x| \le \pi/2$. A solution of our integral equation when the right-hand side satisfies this condition is well-known (see, for example, [7], §6.7), and is of the following form:

$$W(r,t) = \frac{1}{2\pi i}\int_{\alpha-i\infty}^{\alpha+i\infty} \frac{e^{vt}}{v}\frac{K(r\sqrt{v})}{K(\sqrt{v})}\,dv, \tag{27}$$

where the integration is taken over a straight line parallel to the imaginary axis and $\alpha > 0$. This gives the solution of our problem with boundary conditions. Equation (27) holds for $r > 1$, whereas if $r \le 1$, then by (16) we have

$$W(r,t) = 1. \tag{27$'$}$$

For the expectation we have from (20)–(23) and (27′),

$$\mathbf{E}(F) = 2\pi\int_0^\infty W(r,t)r\,dr = \pi + 2\pi\int_1^\infty W(r,t)r\,dr =$$

$$= \pi + 2\pi\int_1^\infty\int_0^t \frac{\partial W}{\partial t} r\,dr\,dt = \pi + 2\pi\int_0^t dt\int_1^\infty r\,dr\Big(\frac{\partial^2 W}{\partial r^2} + \frac{1}{r}\frac{\partial W}{\partial r}\Big) = .$$

$$= \pi - 2\pi\int_0^t \frac{\partial W(1,t)}{\partial r}\,dt, \tag{28}$$

and from (27)

$$\mathbf{E}(F) = i\int_{\alpha-i\infty}^{\alpha+i\infty} \frac{e^{vt}}{v^{3/2}}\frac{K'(\sqrt{v})}{K(\sqrt{v})}\,dv + I_0, \tag{28$'$}$$

where

$$I_0 = \pi - i\int_{\alpha-i\infty}^{\alpha+i\infty} \frac{1}{v^{3/2}}\frac{K'(\sqrt{v})}{K(\sqrt{v})}\,dv.$$

Now we derive an asymptotic formula for $\mathbf{E}(F)$ for large t. It is known (see [5], §17.71) that

$$K(x) = -J_0(ix)\ln\frac{\gamma x}{2} + x^2P(x),$$

$$K'(x) = -\frac{J_0(ix)}{x} - i\ln\frac{\gamma x}{2}J_0'(ix) + x[2P(x) + xP'(x)],$$

where $\gamma = e^C$, C is Euler's constant (that is, $\gamma = 1.7810\ldots$) and $P(x)$ is an entire function.

Hence,

$$\frac{K'(x)}{K(x)} = \frac{1}{x\ln(\gamma x/2)} + xR(x);$$

$$R(x) = \frac{i\ln(\gamma x/2)J_0'(ix)/2 - 2P(x) - P'(x)x + P(x)/\ln(\gamma x/2)}{J_0(ix)\ln(\gamma x/2) - x^2P(x)}.$$

Since $K'(x)/K(x)$ does not have singularities for $|\arg x| < \pi/2$ (except $x = 0$) and tends to 0 as $|x| \to \infty$ (which follows from the well known asymptotic formulas $K(x) \sim (\pi/2)^{1/2}e^{-x}/x^{1/2}$ and $K'(x) \sim -(\pi/2)^{1/2}e^{-x}/x^{1/2}$), $R(x)$ does not have other singularities in the same region, except at $x = 2/\gamma$. As $x \to 0$

$$R(x) \to -\tfrac{1}{2} + \frac{A}{\ln(\gamma x/2)} + \frac{B}{\ln(\gamma x/2)},$$

hence $|R(x)|$ is bounded.

By (28′) we may set

$$\mathbf{E}(F) = I_0 + I_1 + I_2, \tag{29}$$

where

$$I_i = 2i\int_{\alpha-i\infty}^{\alpha+i\infty}\frac{e^{vt}dv}{v^2\ln(\gamma^2v/4)}, \quad I_2 = i\int_{\alpha-i\infty}^{\alpha+i\infty}\frac{e^{vt}}{v}R(\sqrt{v})dv.$$

To avoid the singularity of $R(\sqrt{v})$ at $v = 4/\gamma^2$ we choose as the path of integration a straight line that intersects the real axis between 0 and $4/\gamma^2$; in other words, we assume that $\alpha < 4/\gamma^2$. It is then easy to estimate I_2. Setting $vt = \xi$ we obtain

$$I_2 = i\int_{\alpha-i\infty}^{\alpha+i\infty}\frac{e^{vt}}{v}R(\sqrt{v})dv = i\int_{\alpha t-i\infty}^{\alpha t+i\infty}\frac{e^{\xi}}{\xi}R\Big(\sqrt{\frac{\xi}{t}}\Big)d\xi =$$

$$= \int_{\beta-i\infty}^{\beta+i\infty}\frac{e^{\xi}}{\xi}R\Big(\sqrt{\frac{\xi}{t}}\Big)d\xi,$$

where β can be considered as being independent of t. The expression

$$\int\frac{e^{\xi}}{\xi}R\Big(\sqrt{\frac{\xi}{t}}\Big)d\xi$$

is bounded for any t. This can easily be seen by applying the Second Mean Value theorem to its real and imaginary parts.

Thus we see that

$$I_2 = O(1). \tag{30}$$

The interval I_0 does not depend on t, so that

$$I_0 = O(1). \tag{31}$$

It merely remains to estimate the integral

$$I_1 = 2i \int_{\alpha - i\infty}^{\alpha + i\infty} \frac{e^{vt} dv}{v^2 \ln(\gamma^2 v/4)} = 2i\left(\frac{\gamma^2}{4}\right) \int_{\alpha_2 - i\infty}^{\alpha_1 + i\infty} \frac{e^{z\tau} dz}{z^2 \ln z},$$

where $z = v\gamma^2/4$, $\tau = 4t/\gamma^2$, $\alpha_1 = \gamma^2\alpha/4$. Transforming the integrand of I_1 using the identity

$$\frac{1}{z^2 \ln z} = \frac{e^{-2\ln z} - 1}{\ln z} + \frac{1}{\ln z} = -2\int_0^1 e^{-2a \ln z} da + \frac{1}{\ln z},$$

we obtain

$$I_1 = -i\gamma^2 \int_0^1 da \int_{\alpha_1 - i\infty}^{\alpha_1 + i\infty} e^{-2a\ln z + z\tau} dz + i\frac{\gamma^2}{2} \int_{\alpha_1 - i\infty}^{\alpha_1 + i\infty} \frac{e^{z\tau} dz}{\ln z}.$$

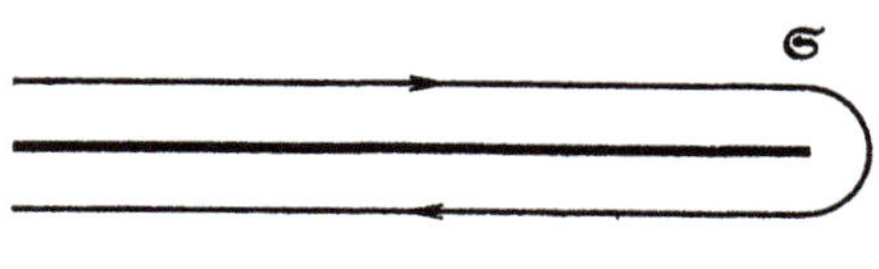

Fig. 1

In the term proportional to $\int \frac{e^{z\tau} dz}{\ln z}$ the integration can be carried out along the imaginary axis. It is then easy to show (for example, by applying the Second Mean Value theorem to the real and imaginary parts of the integral), that this term is of order $O(1)$. In the expression $\int e^{-2a\ln z + z\tau} dz$ the integration path can be deformed so that it coincides with the loop $\mathfrak{S}$ encircling the negative part of the real axis (see Fig. 1). This expression can then be transformed in the following way:

$$\int e^{-2a \ln z + z\tau} dz = \tau^{2a-1} \int \zeta^{-2a} e^{\zeta} d\zeta; \ \zeta = z\tau.$$

We use the relation (see [5], §12.22)

$$\int_{\mathfrak{S}} \zeta^{-2a} e^{\zeta} d\zeta = \frac{2\pi i}{\Gamma(2a)},$$

from which we obtain

$$I_1 = \frac{2\pi\gamma^2}{\tau} \int_0^1 \frac{e^{2a \ln \tau}}{\Gamma(2a)} da + O(1).$$

The integral on the right-hand side of the latter relation is given by

$$\int_0^1 \frac{e^{2a\ln\tau}}{\Gamma(2a)} da = \frac{\tau^2}{2} \int_0^2 \frac{e^{-y\ln\tau} dy}{\Gamma(2-y)} = \frac{\tau^2}{2} \Big\{ \int_0^1 \frac{e^{-y\ln\tau} dy}{\Gamma(2-y)} + $$
$$+ \frac{1}{\tau} \int_0^1 \frac{(1-y) e^{-y\ln\tau} dy}{\Gamma(2-y)} \Big\}.$$

Hence, by (29), (30) and (31) it follows that

$$\mathbf{E}(F) = 4\pi t \int_0^1 \frac{e^{-y\ln\tau} dy}{\Gamma(2-y)} + O(1).$$

For large $\ln\tau$ the following asymptotic relation holds:

$$\int_0^1 \frac{e^{-y\ln\tau} dy}{\Gamma(2-y)} \sim \frac{1}{\ln\tau} = \frac{1}{\ln(4t/\gamma^2)},$$

and therefore

$$\mathbf{E}(F) \sim \frac{4\pi t}{\ln(4t/\gamma^2)}. \qquad (32)$$

For $0 < y < 1$ we have

$$1 < \frac{1}{\Gamma(2-y)} < \frac{1}{0.88},$$

and therefore,

$$\frac{1}{\ln\tau} < \int_0^1 \frac{e^{-y\ln\tau} dy}{\Gamma(2-y)} < \frac{1}{0.88 \ln\tau}.$$

In order to obtain an asymptotic expansion for $\mathbf{E}(F)$ we can, by successive integration by parts, represent the latter integral as

$$\int_0^1 \frac{e^{-y\ln\tau} dy}{\Gamma(2-y)} = \frac{1}{\ln\tau \cdot \Gamma(2)} \Big\{ 1 + \frac{1}{\ln\tau} \frac{\Psi(2)}{\Gamma(2)} + \frac{1}{\ln^2\tau} \frac{\Psi^2(2) - \Psi'(2)\Gamma(2)}{\Gamma(2)} + \dots \Big\},$$

where $\Psi(x) = \Gamma'(x-1)/\Gamma(x-1)$. Substituting here the numerical values of Γ and Ψ we obtain

$$\mathbf{E}(F) = \frac{4\pi t}{\ln 1.26t} \Big\{ 1 + \frac{0.423}{\ln 1.26t} - \frac{0.467}{\ln^2 1.26t} + \dots \Big\} + O(1).$$

If times and lengths are measured in arbitrary units, then t must be replaced by Dt/σ^2, and F by F/σ^2, where σ is the particle's radius. Thus, finally we obtain

$$\mathbf{E}(F) = \frac{4\pi Dt}{\ln(1.26Dt/\sigma^2)}\left\{1 + \frac{0.423}{\ln(1.26Dt/\sigma^2)} - \frac{0.467}{\ln^2(1.26Dt/\sigma^2)} + \dots\right\} + O(1),$$

which is the asymptotic solution of our problem.

References

1. A.N. Kolmogoroff, 'Über die analytischen Methoden in der Wahrscheinlichkeitsrechnung', *Math. Ann.* **104** (1931), 415–458. (Paper 9 of this volume)
2. A.N. Kolmogoroff, 'Zur Theorie der stetigen zufälligen Prozesse', *Math. Ann.* **108** (1933), 149–160. (Paper 17 of this volume)
3. A.N. Kolmogoroff, 'Eine Verallgemeinerung des Laplace-Liapounoffschen Satzes', *Izv. Akad. Nauk SSSR* OMEN, (1931), 959–962. (Paper 12 of this volume)
4. A.N. Kolmogoroff, 'Über die Grenzwertsätze der Wahrscheinlichkeitsrechnung', *Izv. Akad. Nauk SSSR*, OMEN (1933), 366–372. (Paper 16 of this volume)
5. E.T. Whittaker and G.N. Watson, *A course of modern analysis*, Cambridge Univ. Press, 1927.
6. G.N. Watson, *A treatise on the theory of Bessel functions*, Cambridge Univ. Press, 1922.
7. R.H. Fowler, *Statistical mechanics*, Cambridge Univ. Press, 1929.

15. ON THE EMPIRICAL DETERMINATION OF A DISTRIBUTION LAW *

We consider the possibility of determining a distribution law based on a finite number of tests.

§1. Let $X_1, X_2, \ldots, X_n$ be the results of n mutually independent observations, ordered increasingly, that is, $X_1 \leq X_2 \leq \ldots \leq X_n$ and let

$$F(x) = \mathbf{P}\{X \leq x\}$$

be the distribution law corresponding to this sequence. The *empirical distribution law* is the function $F_n(x)$ defined by the relations

$$F_n(x) = 0, \quad x < X_1;$$

$$F_n(x) = k/n, \ X_k \leq x < X_{k+1}, \quad k = 1, 2, \ldots, n-1;$$

$$F_n(x) = 1, \quad X_n \leq x.$$

Hence, $nF_n(x)$ is the number of the X_k not exceeding x. A natural question is: does $F_n(x)$ approach $F(x)$ for large n ? A theorem related to this question was formulated by von Mises [1] and is called the ω^2-method. However, the fundamental statement that the probability of the inequality

$$\mathbf{D} = \sup |F_n(x) - F(x)| < \epsilon$$

tends to 1 as $n \to \infty$ for any ϵ has not, in my view, been explicitly formulated until now, though it could have been proved in various simple ways.

I shall obtain this statement as a direct consequence of the theorem proved below. I should add that the setting of the problem considered below is motivated by recent studies of Glivenko [2].

Theorem 1. *For any continuous distribution function $F(x)$, the probability $\Phi_n(\lambda)$ of the inequality*

$$\mathbf{D} = \sup |F_n(x) - F(x)| < \lambda/\sqrt{n}$$

* 'Sulla determinazione empirica di una legge di distribuzione', *Giorn. Ist. Ital. Attuar.* 4:1 (1933), 83–91.

tends to

$$\Phi(\lambda) = \sum_{-\infty}^{+\infty}(-1)^k e^{-2k^2\lambda^2} \tag{1}$$

uniformly in λ *as* $n \to \infty$.

Below the table of some values of $\Phi(\lambda)$ is given; these were calculated by N. Kozhevnikov.

λ	$\Phi(\lambda)$	λ	$\Phi(\lambda)$	λ	$\Phi(\lambda)$
0.0	0.0000	1.0	0.7300	2.0	0.99932
0.2	0.0000	1.2	0.8877	2.2	0.99986
0.4	0.0028	1.4	0.9603	2.4	0.999973
0.6	0.1357	1.6	0.9888	2.6	0.9999964
0.8	0.4558	1.8	0.9969	2.8	0.99999966

As is clear from the table, $D \leq 2.4/\sqrt{n}$ can be considered to be practically certain. Moreover, it turns out that

$$\Phi(0.83) \sim 0.5.$$

If λ is small, then (1) converges very slowly; in this case the following asymptotic formula may be used

$$\Phi(\lambda) \sim \frac{\sqrt{2\pi}}{\lambda} e^{-\pi^2/8\lambda^2}.$$

At $\lambda = 0.6$ this formula yields

$$\Phi(0.6) \sim 0.1327,$$

instead of the value

$$\Phi(0.6) \sim 0.1357$$

computed using the exact formula (1).

§2. Lemma. *The probability function* $\Phi_n(\lambda)$ *does not depend on the distribution function* $F(x)$ *if the latter is assumed to be continuous.*

Proof. Let X be a random variable with continuous distribution law $F(x)$; clearly, corresponding to the random variable $Y = F(X)$ there is the distribution law $F^{(0)}(x)$ for which

$$F^{(0)}(x) = 0, \quad x \leq 0;$$
$$F^{(0)}(x) = x, \quad 0 \leq x \leq 1;$$
$$F^{(0)}(x) = 1, \quad x \geq 1.$$

If $F_n(x)$ and $F_n^{(0)}(x)$ are the empirical distribution laws for X and Y for n observations, then

$$F_n(x) - F(x) = F_n^{(0)}[F(x)] - F^{(0)}[F(x)] = F_n^{(0)}(y) - F^{(0)}(y),$$
$$\sup|F_n(x) - F(x)| = \sup|F_n^{(0)}(y) - F^{(0)}(y)|.$$

This implies that the function $\Phi_n(\lambda)$ corresponding to an arbitrary continuous distribution function $F(x)$ is identical to the function corresponding to $F^{(0)}(x)$. Therefore, in proving our theorem, we can confine ourselves to the case $F(x) = F^{(0)}(x)$.

In what follows we will write $F_n(x)$ instead of $F_n^{(0)}(x)$ and confine ourselves to x for which $0 \leq x \leq 1$ and $F^{(0)}(x) = x$. Our problem then reduces to finding the probability $\Phi_n(\lambda)$ of the inequality

$$\sup|F_n(x) - x| < \lambda/\sqrt{n}, \quad 0 \leq x \leq 1. \tag{2}$$

Let λ be of the form

$$\lambda = \mu/\sqrt{n},$$

where μ is an integer. Substituting it into (2) we obtain

$$\Phi_n(\lambda) = \mathbf{P}\Big\{\sup|F_n(x) - x| \leq \frac{\mu}{n}\Big\}, \quad \lambda = \frac{\mu}{\sqrt{n}}. \tag{3}$$

The values $F_n(x)$ are multiples of $1/n$; for example, let $F_n(x) = i/n$ and $x = j/n + \epsilon$ $(0 \leq \epsilon < 1/n)$. Taking into account the fact that $F_n(x)$, being a distribution function, is monotone, we immediately obtain

$$F_n(x) - x = \frac{i-j}{n} - \epsilon,$$

$$F_n\Big(\frac{j}{n}\Big) - \frac{j}{n} \leq F_n(x) - (x - \epsilon) = \frac{i-j}{n},$$

$$F_n\Big(\frac{j+1}{n}\Big) - \frac{j+1}{n} \geq F_n(x) - \Big(x + \frac{1}{n} - \epsilon\Big) = \frac{i-j-1}{n}.$$

Therefore the inequality

$$|F_n(x) - x| = \left|\frac{i-j}{n} - \epsilon\right| \geq \frac{\mu}{n}$$

holds if and only if at least one of the following inequalities holds:

$$F_n\left(\frac{j}{n}\right) - \frac{j}{n} \leq \frac{i-j}{n} \leq -\frac{\mu}{n},$$

$$F_n\left(\frac{j+1}{n}\right) - \frac{j+1}{n} \geq \frac{i-j-1}{n} \geq \frac{\mu}{n}.$$

This implies that (3) may be replaced by the following:

$$\Phi_n(\lambda) = \mathbf{P}\left\{\max\left|F_n\left(\frac{i}{n}\right) - \frac{i}{n}\right| < \frac{\mu}{n}\right\}, \quad i = 0, 1, \ldots, n. \tag{4}$$

Now let P_{ik} be the probability of simultaneous fulfillment of the relations

$$\begin{aligned} &|F_n(j/n) - j/n| < \mu/n, \quad j = 0, 1, \ldots, k; \\ &|F_n(k/n) - k/n| = i/n. \end{aligned} \tag{5}$$

Note that

$$\Phi_n(\lambda) = P_{0n}. \tag{6}$$

For $k = 0$ we clearly have

$$P_{00} = 1, \quad P_{i0} = 0 \ (i \neq 0) \tag{7}$$

More generally,

$$P_{ik} = 0 \tag{8}$$

for $|i| \geq \mu$, since in this case the inequalities (5) are contradictory. Furthermore,

$$P_{ik+1} = \sum_j P_{jk} Q_{ji}^{(k)}, \quad |i| < \mu, \tag{9}$$

where $Q_{ji}^{(k)}$ denotes the probability of E_{ik+1} under the condition that E_{jk} holds, that is, the probability of

$$F_n\left(\frac{k+1}{n}\right) - F_n\left(\frac{k}{n}\right) = \frac{i-j+1}{n} \tag{10}$$

subject to the condition

$$F_n(k/n) = (k+j)/n. \tag{11}$$

Relation (11) means that, of the results of the n observations $X_1, X_2, \dots, X_n$ exactly $n-k-j$ belong to the interval $k/n < x \leq 1$; (10) can therefore be true only when $i-j+1$ of the results of these $n-k-j$ observations belong to the interval $k/n < x \leq (k+1)/n$.

Provided that X_m is uniformly distributed, we obtain the following expression for our desired probability:

$$Q_{ji}^{(k)} = \binom{n-k-j}{i-j+1}\left(1-\frac{1}{n-k}\right)^{n-k-i-1}\left(\frac{1}{n-k}\right)^{i-j+1}. \tag{12}$$

Formulas (7)–(9), (12) and (6) enable us to find the probability $\Phi_n(\lambda)$ for the case $\lambda = \mu/\sqrt{n}$.

These formulas can be replaced by other, more convenient ones. For this we set

$$R_{ik} = \frac{(n-k-i)!n^n}{(n-k)^{n-k-i}n!}e^{-k}P_{ik}. \tag{13}$$

Conditions (7) and (8) turn into the following:

$$R_{00} = 1, \quad R_{i0} = 0, \ i \neq 0; \tag{14}$$

$$R_{ik} = 0, \ |i| \geq \mu. \tag{15}$$

Simplifications reduce (9) to

$$R_{i,k+1} = \sum_j R_{jk}\frac{1}{(i-j+1)!}e^{-1}. \tag{16}$$

Finally, from (6) and (13) we obtain

$$\Phi_n(\lambda) = \frac{n!}{n^n}e^n R_{0n}. \tag{17}$$

Formulas (14)–(17) also enable us to find $\Phi_n(\lambda)$ for the case $\lambda = \mu/\sqrt{n}$.

§3. Now let $Y_1, Y_2, \dots, Y_n$ be a sequence of independent random variables with distribution law given by the formula

$$\mathbf{P}\left\{Y_k = \frac{i-1}{\mu}\right\} = \frac{1}{i!}e^{-i}, \ i = 1, 2, \dots. \tag{18}$$

Setting

$$S_k = Y_1 + Y_2 + \ldots + Y_k,$$

we easily see that the probability $\overline{R}_{ik}$ that the relations

$$|S_j| < \mu, \quad j = 0, 1, \ldots, k; \qquad S_k = i/\mu, \tag{19}$$

hold simultaneously satisfies the same conditions (14)–(16) that R_{ik} does, in other words, $\overline{R}_{ik} = R_{ik}$. This allows us to give an asymptotic expression for R_{ik} as $n \to \infty$. For this we state the following general theorem.

Theorem 2. *Let $Y_1, Y_2, \ldots, Y_n$ be a sequence of independent random variables whose values can only be multiples of a constant ϵ.*

Let

$$\mathbf{E}(Y_k) = 0, \quad \mathbf{E}(Y_k^2) = 2b_k, \quad \mathbf{E}(|Y_k^3|) = d_k,$$
$$S_k = Y_1 + Y_2 + \ldots + Y_k,$$
$$t_k = b_1 + b_2 + \ldots + b_k$$

and let $a(t), b(t)$ be continuously differentiable functions satisfying

$$a(t) < b(t), \quad a(0) < 0 < b(0).$$

Denoting by R_{in} the probability that the relations

$$a(t_k) < S_k < b(t_k), \ k = 1, 2, \ldots, n,$$
$$S_n = i\epsilon$$

hold simultaneously and by $u(\sigma, \tau, s, t)$ the Green's function of the heat equation

$$\partial f/\partial t = \partial^2 f/\partial s^2$$

in the domain G defined by the inequalities

$$a(t) < s < b(t),$$

we have

$$R_{in} = \epsilon \cdot \{u(0, 0, i\epsilon, t_n) + \Delta\},$$

where Δ tends uniformly to 0 as $\epsilon \to 0$ if the following conditions hold:

a) *$a(t)$ and $b(t)$ remain constant and the t_n are bounded:*

$$0 < T_1 < t_n < T_2;$$

b) *there exists a constant $C > 0$ such that*

$$d_k/b_k \leq C\epsilon;$$

c) *there exists a constant $K > 0$ such that for any k there exists an i_k for which*

$$\mathbf{P}\{Y_k = i_k\epsilon\} > K, \quad \mathbf{P}\{Y_k = (i_k+1)\epsilon\} > K;$$

d) *there exists a constant A such that*

$$a(t_n) + A < i\epsilon < b(t_n) - A.$$

Apart from these restrictions, the Y_k, as well as their number n and the integer i, can depend on ϵ in arbitrary fashion.

This theorem falls into the same scope of ideas as the one given in [3].

However, the assertion of the above theorem is stronger: the theorem in [3] allows us only to assert that

$$\sum_{i=p}^{i=q} R_{in} = \int_{p\epsilon}^{q\epsilon} u(0,0,z,t_n)dz + \Delta',$$

where Δ' tends to 0 as $\epsilon \to 0$ under the conditions a) and b). The condition c), which is essential in our new theorem, had already been used by von Mises in similar considerations.

In our case

$$\epsilon = 1/\mu,$$

$$\mathbf{E}(Y_k) = 0, \quad \mathbf{E}(Y_k^2) = 2b_k = 1/\mu^2, \quad \mathbf{E}(Y_k^3) = d_k = C/\mu^3,$$

$$d_k/b_k = C/\mu = C\epsilon, \ t_n = n/2\mu^2 = 1/\lambda^2, \ a(t) = -1, \ b(t) = +1,$$

$$R_{0n} = \epsilon\Big\{u\Big(0,0,0,\frac{1}{\lambda^2}\Big) + \Delta\Big\},$$

$$u(0,0,s,t) = \frac{1}{2\sqrt{\pi t}} \sum_{-\infty}^{+\infty} (-1)^k \exp\left[-\frac{(s-2k)^2}{4t}\right],$$

$$\Phi_n(\lambda) = P_{0n} = \frac{n!}{n^n} e^n R_{0n} =$$

$$= \{\sqrt{2\pi n} + \delta\} \frac{1}{\mu} \left\{ \frac{\lambda}{\sqrt{2\pi}} \sum_{-\infty}^{+\infty} (-1)^k e^{-2k^2\lambda^2} + \Delta \right\} =$$

$$= \sum_{-\infty}^{+\infty} (-1)^k e^{-2k^2\lambda^2} + R = \Phi(\lambda) + R.$$

In this formula the remainder R tends uniformly to 0 as $n \to \infty$ if λ is greater than some $\lambda_0 > 0$. For in this case $\epsilon = 1/\mu = 1/(\lambda\sqrt{n})$ tends to 0 as $n \to \infty$.

Thus, Theorem 1 is proved for values λ of the form $\mu/\sqrt{n}$, provided that $\lambda > \lambda_0$. Since the limit function $\Phi(\lambda)$ is continuous and its limit value is $\Phi(0) = 0$, it is easy to see that these restrictions are essential. We prove Theorem 2 elsewhere.

References

1. R. von Mises, *Wahrscheinlichkeitsrechnung und ihre Anwendung in der Statistik und theoretischen Physik*, Fr. Denticke, Leipzig, Vienna, 1931.
2. V. Glivenko, 'Sulla determinazione empirica della leggi di probabilità', *Giorn. Ist. Ital. Attuar.* **6**:1 (1933), 92–99.
3. A. Kolmogoroff, 'Eine Verallgemeinerung des Laplace-Liapounoffschen Satzes', *Izv. Akad. Nauk SSSR*, OMEN, 1931, 959–962 (Paper 12 in this volume).

16. ON THE LIMIT THEOREMS OF PROBABILITY THEORY *

As is well known, the fundamental ideas of P. Laplace [1] and P.L. Chebyshev [2] were subsequently developed in papers by A.A. Markov [3] and A.M. Lyapunov [4], culminating in a very general statement (the so-called Lyapunov theorem) on the limit of probability distributions for sums of a large number of small independent random variables. Further studies by Markov [5] and S.N. Bernshtein [6] have demonstrated that in many cases a similar statement also holds for sums of independent variables. These generalizations are of special importance for applied purposes, but in principle they do not go much further; in all cases studied by these authors only summands that are near to each other are strongly dependent, whereas if the sum is decomposed into sufficiently long partial sums, the later ones will be almost independent. Of much greater consequence is the two- (and multi-)dimensional generalization of Lyapunov's theorem to the case of sums of random vectors, which was first rigorously proved by Bernshtein [6].

What still remains after all these studies is to determine the limits of various kinds of probabilities related to the whole set of partial sums of a given sequence of random variables: these limits are unknown even in the simplest case of independent summands. In any case, the early results of Laplace and Poisson [7] on the probability of a gambler's ruination belong to this field; they were further elaborated by P. Lévy [8]. Recently I published a general statement of this kind [9]. In the meantime some partial two-dimensional problems of similar type were studied by R. Lüneburg [10]. Both these papers clarify the connection between these problems and differential heat equations [11].

Several months ago I.G. Petrovskii in Moscow found a general method of reducing problems in probability theory on sums of small random variables to corresponding differential equations in a very general situation. In what follows this method is applied to proving Lyapunov's theorem (§1) and my theorem mentioned above (§§2, 3). The applications to random walk problems can be found in a yet unpublished work by Petrovskii, where the setting given by Lüneburg is considerably generalized.

Apparently this method also can be used when the distribution law for each term x_{k+1} depends on the sums of all previous terms $s_k = x_1+x_2+\ldots+x_k$. This

* 'Über die Grenzwertsätze der Wahrscheinlichkeitsrechnung', *Izv. Akad. Nauk SSSR*, OMEN, 1933, 366–372. Presented by S.N. Bernshtein.

case is crucial for a rigorous mathematical justification of diffusion theory, in which the external forces as well as the diffusion coefficients are point functions. For this justification it was necessary to prove that the distribution laws of sums can be approximately represented by the corresponding solutions of Fokker-Planck differential equations [12], [13].

§1. Let

$$x_1, x_2, \ldots, x_n$$

be independent random variables with expectations

$$\mathbf{E}(x_k) = 0, \quad \mathbf{E}(x_k^2) = 2b_k, \quad \mathbf{E}(|x_k|^3) = d_k$$

and suppose that the quotients $d_k : b_k$ are uniformly bounded by a fixed constant

$$d_k / b_k \leq \mu. \tag{1}$$

By the well known inequality for moments, $b_k^3 \leq d_k^2$, (1) implies that

$$b_k \leq \mu^2. \tag{2}$$

We now set

$$x_1 + x_2 + \ldots + x_k = s_k, \; s_n = S,$$

$$b_1 + b_2 + \ldots + b_k = t_k, \; t_n = T.$$

Let T be fixed (and n variable); then Lyapunov's theorem asserts that

$$\mathbf{P}\{a < S < b\} = \frac{1}{2\sqrt{\pi T}} \int_a^b e^{-s^2/4T} ds + \theta R(\mu), \; |\theta| \leq 1, \tag{3}$$

where $\mathbf{P}$ denotes the probability of the inequality in parentheses and $R(\mu)$ tends to 0 together with μ.

To prove this, consider the probability

$$P_k(s) = \mathbf{P}\{a < S - s_k + s < b\}$$

so that $a < S < b$ under the assumption $s_k = s$. Clearly, the desired probability $\mathbf{P}\{a < S < b\}$ coincides with $P_0(0)$; $P_k(s)$ satisfies the equation [1]

$$P_k(s) = \int P_{k+1}(s + x) dT_{k+1}(x), \; k = 0, 1, 2, \ldots, n-1, \tag{4}$$

[1] Formula (4) is proved in the same way as that for the distribution function of the sum of two independent variables; see, for example, P. Lévy, *Calcul des probabilités*, Paris, 1927, p.187.

where T_{k+1} is the distribution function of x_{k+1}, and the integral is taken from $-\infty$ to $+\infty$. Moreover,

$$\begin{aligned} P_n(s) &= 1, \quad \text{if } a < s < b, \\ P_n(s) &= 0 \quad \text{if } s \le a \text{ or } s \ge b. \end{aligned} \tag{5}$$

The initial conditions (5) and equations (4) uniquely define $P_k(s)$ for all k $(0 \le k \le n)$.

We call $P_k^*(s)$ an upper function if the inequalities

$$P_k^*(s) \ge \int P_{k+1}^*(s+x)dT_{k+1}(x), \tag{6}$$

$$P_n^*(s) \ge P_n(s) \tag{7}$$

hold. It is not difficult to see that

$$P_k^*(s) \ge P_k(s) \tag{8}$$

holds also for $k < n$. Lower functions are defined similarly. We now wish to construct a function $u^*(s,t)$ which for any sufficiently small μ leads to the upper function

$$P_k^*(s) = u^*(s, t_k). \tag{9}$$

For this we first determine a four times continuously differentiable function $v(s)$ satisfying the following inequalities:

$$\begin{gathered} v(s) = 0, \quad s \le a - \epsilon, \quad 0 \le v(s) \le 1, \quad b \le s \le b + \epsilon, \\ 0 \le v(s) \le 1, \quad a - \epsilon \le s \le a, \quad v(s) = 0, \quad b + \epsilon \le s, \\ v(s) = 1, \quad a \le s \le b. \end{gathered} \tag{10}$$

Now let

$$\bar{u}(s,t) = \frac{1}{2\sqrt{\pi(T-t)}} \int v(s+x) e^{-x^2/4(T-t)} dx, \tag{11}$$

$$u^*(s,t) = \bar{u}(s,t) + \epsilon(T-t), \tag{12}$$

where $\epsilon > 0$ in both (10) and (12). The function $u(s,t)$ satisfies (for $t \le T$) the differential equation

$$\partial u/\partial t = -\partial^2 u/\partial s^2 \tag{13}$$

and its derivatives up to the fourth order with respect to s and the second order with respect to t are bounded. Let M be the least upper bound of these derivatives. If $P_k^*(s)$ is defined by (9), then we obtain

$$P_n^*(s) = u^*(s,T) = \bar{u}(s,T) = v(s) \geq P_n(s).$$

Thus, (7) is fulfilled. Therefore it suffices to prove (6) for small μ. For this we consider the difference

$$\Delta = P_k^*(s) - \int P_{k+1}^*(s+x)dT_{k+1}(x) = \epsilon(T-t_k) + \bar{u}(s,t_k) -$$

$$-\epsilon(T-t_{k+1}) - \int \bar{u}(s+x,t_{k+1})dT_{k+1}(x) = \epsilon(t_{k+1}-t_k) +$$

$$+\bar{u}(s,t_k) - \int\Big\{\bar{u}(s,t_{k+1}) + \frac{\partial}{\partial s}\bar{u}(s,t_{k+1})x + \frac{\partial^2}{\partial s^2}\bar{u}(s,t_{k+1})\frac{x^2}{2} +$$

$$+\theta M|x^3|\Big\}dT_{k+1}(x) = \epsilon b_{k+1} + \bar{u}(s,t_k) - \bar{u}(s,t_{k+1}) -$$

$$-\frac{\partial^2}{\partial s^2}\bar{u}(s,t_{k+1})b_{k+1} + \theta' M d_{k+1}, \ |\theta| \leq 1, \quad |\theta'| \leq 1. \tag{14}$$

But since

$$\bar{u}(s,t_k) - \bar{u}(s,t_{k+1}) = -\frac{\partial}{\partial t}\bar{u}(s,t_{k+1})(t_{k+1}-t_k) + \theta'' M(t_{k+1}-t_k)^2 =$$

$$= -\frac{\partial}{\partial t}\bar{u}(s,t_{k+1})b_{k+1} + \theta''' M b_{k+1}\mu^2, \quad |\theta''| \leq 1, \quad |\theta'''| \leq 1,$$

and in view of (1) and (13), (14) implies that

$$\Delta = \epsilon b_{k+1} + \Big\{-\frac{\partial}{\partial t}\bar{u}(s,t_{k+1}) = \frac{\partial^2}{\partial s^2}\bar{u}(s,t_{k+1})\Big\}b_{k+1} + \theta^{\text{iv}} M b_{k+1}\mu +$$

$$+\theta''' M b_{k+1}\mu^2 = b_{k+1}(\epsilon + \theta^{\text{iv}} M\mu + \theta''' M\mu^2), \quad |\theta^{\text{iv}}| \leq 1. \tag{15}$$

Formula (15) immediately implies that if μ is sufficiently small, then

$$\Delta \geq 0,$$

which proves (6). Therefore $P_k^*(s)$ is an upper function for a sufficiently small μ and

$$\mathbf{P}\{a < s_n < b\} = P_0(0) \leq P_0^*(0) = u^*(0,0) =$$

$$= \epsilon T + \frac{1}{2\sqrt{\pi T}}\int v(x)e^{-x^2/4T}dx = \frac{1}{2\sqrt{\pi T}}\int_a^b e^{-x^2/4T}dx + R, \tag{16}$$

with R arbitrarily small for an appropriate choice of ϵ. Consideration of lower functions gives the reverse inequality. Thus our main formula (3) is completely proved.

§**2.** Now let $a(t)$ and $b(t)$ be four times continuously differentiable functions of t satisfying the conditions

$$a(t) < b(t), \quad a(0) < 0 < b(0). \tag{17}$$

What is the probability P that

$$a(t_k) < s_k < b(t_k), \quad k = 1, 2, 3, \ldots, n \tag{18}$$

holds ? The inequalities

$$0 \leq t < T, \quad a(t) < s < b(t) \tag{19}$$

determine a certain region G in the (s,t)-plane. Let $u(s,t)$ be a bounded solution of the equation

$$\partial u/\partial t = -\partial^2 u/\partial s^2 \tag{13}$$

in G with boundary conditions

$$\begin{aligned} &u(s,T) = 1, \quad a(T) < s < b(T), \\ &u[a(t),t] = 0, \quad 0 \leq t < T, \\ &u[b(t),t] = 0, \quad 0 \leq t < T. \end{aligned} \tag{20}$$

Our purpose is to show that

$$P = u(0,0) + \theta R(\mu), \ |\theta| \leq 1, \tag{21}$$

where $R(\mu)$ tends to zero together with μ.

For this consider the more general case in which the inequalities

$$a(t_i) < s_i - s_k + s < b(t_i) \tag{22}$$

hold for all i, $k \leq i \leq n$. The desired probability is $P_0(0)$. If

$$a(t_k) < s < b(t_k), \ k < n, \tag{23}$$

then $P_k(s)$ satisfies the same equation

$$P_k(s) = \int P_{k+1}(s+x)dT_{k+1}(x), \tag{4}$$

as in §1. Conditions (5) are now replaced by the following:

$$\begin{aligned} P_n(s) &= 1 \quad \text{if } a(T) < s < b(T), \\ P_n(s) &= 0 \quad \text{if } s \le a(T_k) \text{ or } s \ge b(T_k). \end{aligned} \tag{24}$$

Equations (4) and (24) uniquely determine $P_k(s)$ for all k $(0 \le k \le n)$.

We now call $P_k^*(s)$ an upper function if (23) implies

$$P_k^*(s) \ge \int P_{k+1}^*(s+x)dT_{k+1}(x) \tag{25}$$

and

$$\begin{aligned} &P_n^*(s) \ge 1 \quad \text{for } a(T) < s < b(T), \\ &P_n^*(s) \ge 0 \quad \text{for } s \le a(T_k) \text{ and, correspondingly, } s \ge b(T_k). \end{aligned} \tag{26}$$

As in §1, we prove that for $k < n$ the inequality

$$P_k^*(s) \ge P_k(s)$$

holds.

To construct an upper function we need a function $u(s,t)$ which is defined in G as a solution of the differential equation (13) with the following boundary conditions:

$$\begin{aligned} &\bar{u}(s,T) = 1, \quad a(T) < s < b(T), \\ &\bar{u}\{a(t),t\} = v(t), \quad 0 \le t \le T, \\ &\bar{u}\{b(t),t\} = v(t), \quad 0 \le t \le T, \end{aligned} \tag{27}$$

where $v(t)$ is a four times continuously differentiable function such that

$$\begin{aligned} &v(t) = 0, \quad 0 \le t \le T - \epsilon, \\ &0 \le v(t) \le 1, \quad T - \epsilon \le t \le T, \\ &v(t) = 1, \quad v'(T) = v''(T) = 0. \end{aligned} \tag{28}$$

Since $\bar{u}(s,t)$ has continuous derivatives up to the fourth order (including on the boundary of G), this function can be extended beyond G so that the derivatives with respect to s up to the fourth order are bounded on the whole (s,t)-plane and the inequality

$$\bar{u}(s,t) > -\epsilon \tag{29}$$

holds everywhere. We now set

$$u^*(s,t) = \bar{u}(s,t) + \epsilon + \epsilon(T-t), \tag{30}$$

$$P_k^*(s) = u^*(s,t_k). \tag{9}$$

Apart from obvious changes, the proof of the fact that $P_k^*(s)$ is an upper function for any sufficiently small μ is the same as in §1. Thus for the indicated μ we obtain

$$P = P_0(0) \leq P_0^*(0) = u^*(0,0)$$

and $u^*(0,0) \to u(0,0)$ as $\epsilon \to 0$. To complete the proof of (21) we merely have to make similar estimates of the probability P from below, which is done using the same method with lower functions.

§3. Now let $P^{(1)}$ (respectively, $P^{(2)}$) denote the probability of the existence of k such that

$$\begin{gathered} a(t_i) < s_i < b(t_i), \quad i = 1,2,\ldots,k-1, \\ s_k \leq a(t_k), \end{gathered} \tag{31}$$

and

$$\begin{gathered} a(t_i) < s_i < b(t_i), \ i = 1,2,\ldots,k-1, \\ b(t_k) \leq s_k, \end{gathered} \tag{32}$$

respectively. Finally, let $P(x,y)$ be the probability that all the inequalities

$$\begin{gathered} a(t_k) < s_k < b(t_k), \ k = 1,2,3,\ldots,n-1, \\ x < s_n < y, \end{gathered} \tag{33}$$

hold with $a(T) \leq x < y \leq b(T)$.

Then the following limit relations hold:

$$P^{(1)} = u^{(1)}(0,0) + \theta^{(1)} R^{(1)}(\mu), \quad |\theta^{(1)}| \leq 1, \tag{34}$$

$$P^{(2)} = u^{(2)}(0,0) + \theta^{(2)} R^{(2)}(\mu), \quad |\theta^{(2)}| \leq 1, \tag{35}$$

$$P(x,y) = u_{x,y}(0,0) + \theta R(\mu), \qquad |\theta| \leq 1, \tag{36}$$

where $R^{(1)}(\mu), R^{(2)}(\mu)$ and $R(\mu)$ tend to 0 together with μ, and where

$$u^{(1)}(s,t),\ u^{(2)}(s,t) \text{ and } u_{x,y}(s,t)$$

are bounded solutions of (13) determined from the following boundary conditions:

$$\begin{aligned} &u^{(1)}(s,T) = 0, \quad a < s < b,\\ &u^{(1)}\{a(t),t\} = 1, \quad 0 \le t < T, \\ &u^{(1)}\{b(t),t\} = 0, \quad 0 \le t < T,\end{aligned} \tag{37}$$

$$\begin{aligned} &u^{(2)}(s,T) = 0, \quad a < s < b,\\ &u^{(2)}\{a,(t),t\} = 0, \quad 0 \le t < T, \\ &u^{(2)}\{b(t),t\} = 1, \quad 0 \le t < T,\end{aligned} \tag{38}$$

$$\begin{aligned} &u_{x,y}(s,T) = 0,\ a < s < x, \quad u_{x,y}\{a(t),t\} = 0,\ 0 \le t < T,\\ &u_{x,y}(s,T) = 1,\ x < s < y, \quad u_{x,y}\{b(t),t\} = 0,\ 0 \le t < T. \\ &u_{x,y}(s,T) = 0,\ y < s < b,\end{aligned} \tag{39}$$

The proof is the same as in §2. In [9] other expressions[2] are derived for the same values $u^{(1)}(0,0)$, $u^{(2)}(0,0)$ and $u_{x,y}(0,0)$.

References

1. P. Laplace, *Théorie analytique des probabilités*, 3rd ed. Paris, 1886.
2. P.L. Tchebycheff, *Sur deux théorèmes relatifs aux probabilités*, Oeuvres de Tchebycheff, St. Pétersburg, 1907, vol. 2, pp. 481–491.
3. A.A. Markov, 'The law of large numbers and the method of least squares', *Izv. Fiz.-Mat. Obshch. Kazan. Univ.* **8** (1898) (in Russian).

[2] In [9] and [14] multiple differentiability of $a(t)$ and $b(t)$ is not assumed. If $a(t)$ and $b(t)$ are continuously differentiable only once, then in the proof one must approximate them by four times continuously differentiable functions. The proof of the fact that the remainders in (3), (21), (34)–(36) tend to zero uniformly with respect to T when T is greater than some fixed constant is simple if the random variable is multiplied by $\sqrt{T_0/T}$ and in this way, the case of arbitrary T reduces to the case $T = T_0$. A simple argument then shows that in (34) and (35) the remainders tend to zero together with μ even without assuming that $T > T_0$.

4. A.M. Liapounoff, 'Sur une proposition de la théorie des probabilités', *Bull. Acad. Sci. St.-Pétersburg* **13** (1900), 359; 'Nouvelle forme du théorème sur la limite de probabilité', *Mém. Acad. Sci., St. Pétersburg* **12**:5 (1901).
5. A.A. Markov, 'Generalizing limit theorems in probability calculus to sums of variables related by a chain', *Zap. Akad. Nauk* **22** (1910) (in Russian).
6. S.N. Bernshtein, 'Generalizing a limit theorem in probability theory to sums of independent variables', *Uspekhi Mat. Nauk* **10** (1944), 65–114 (in Russian); *Math. Ann.* **97** (1926), 1–59 (in German).
7. S.-D. Poisson, *Recherches sur la probabilité*, Paris, 1837.
8. P. Lévy, 'Nuove formule relative al giuoco di testa e croce', *Giorn. Ist. Ital. Attuar.* **2** (1931), 3.
9. A.N. Kolmogoroff, 'Eine Verallgemeinerung des Laplace-Liapounoffschen Satzes', *Izv. Akad. Nauk SSSR*, OMEN (1931), 959–962 (Paper 12 of this volume).
10. R. Lüneburg, 'Das Problem der Irrfahrt', *Math. Ann.* **104** (1931), 700.
11. R. Courant, K. Friedrichs and H. Lewy, 'Über die partiellen Differenzengleichungen', *Math. Ann.* **100** (1928), 32–74.
12. A. Fokker, *Ann. Phys.* **43** (1914), 812.
13. M. Planck, *Sitzungsber. Preuss. Akad. Wiss.* (1917).
14. A.N. Kolmogoroff, *Math. Ann.* **104** (1931), 415; in particular, §§12–14 (Paper 9 of this volume).

17. ON THE THEORY OF CONTINUOUS RANDOM PROCESSES *

Let $\mathfrak{S}$ be a physical system with n degrees of freedom; this means that the admissible states x of $\mathfrak{S}$ form a Riemannian manifold $\mathfrak{R}$ of dimension n. The process of variation of $\mathfrak{S}$ is said to be *stochastically determined* if under an arbitrary choice of x, the region $\mathfrak{E}$ (in $\mathfrak{R}$) and times t' and t'' ($t' < t''$), the probability $P(t', x, t'', \mathfrak{E})$ that the system in state x at time t' will be in one of the states of $\mathfrak{E}$ at time t'' is defined. It is further assumed that the probability $P(t', x, t'', \mathfrak{E})$ can be given by the formula

$$P(t', x, t'', \mathfrak{E}) = \int_{\mathfrak{E}} f(t', x, t'', y) dV_y, \tag{1}$$

where dV_y denotes the volume element. Here $f(t', x, t'', y)$ has to satisfy the following fundamental equations:

$$\int_{\mathfrak{R}} f(t', x, t'', y) dV_y = 1, \tag{2}$$

$$f(t_1, x, t_3, y) = \int_{\mathfrak{R}} f(t_1, x, t_2, z) f(t_2, z, t_3, y) dV_z, \quad t_1 < t_2 < t_3. \tag{3}$$

The integral equation (3) was studied by Smolukhovskii and then by other authors.[1] In the paper 'Über die analytischen Methoden in der Wahrscheinlichkeitsrechnung'[2] I have proved that, under certain additional conditions, $f(t', x, t'', y)$ satisfies certain differential equations of parabolic type.[3] But in A.M. there was no answer to the question[4] as to what extent $f(t', x, t'', y)$ is uniquely determined by the coefficients $A(t, x)$ and $B(t, x)$. In this paper the theory is developed in the general case of a Riemannian manifold $\mathfrak{R}$ and the question of uniqueness is answered affirmatively for a closed manifold $\mathfrak{R}$.

§1. The first differential equation

Let $\mathfrak{R}$ be a Riemannian manifold of dimension n. Because of the assumptions made, $f(t', x, t'', y)$ is defined for $t' < t''$ and any pair of points x, y. Moreover,

* 'Zur Theorie der stetigen zufälligen Prozesse', *Math. Ann.* **108** (1933), 149–160.

[1] See bibliography in: B. Hostinsky, 'Méthodes générales du calcul des probabilités', *Mem. Sci. Math.* **52** (1931).

[2] *Math. Ann.* **104** (1931), 415–458. Referred to in the present paper as A.M. (see No. 9 of this book).

[3] These differential equations were introduced by Fokker and Planck independently of Smolukhovskii's integral equation. See: A. Fokker, *Ann. Phys.* **43** (1914), 812; M. Planck, *Sitzungsber. Preuss. Acad. Wiss.* (1917) 10 May.

[4] See A.M. §15.

we assume that $f(t', x, t'', y)$ has continuous derivatives up to the third order with respect to all the arguments (t', t'' and the coordinates $x_1, x_2, \ldots, x_n$, $y_1, y_2, \ldots, y_n$ of the points x and y) and satisfies the continuity condition

$$\frac{\int_{\mathfrak{R}} f(t, x, t+\Delta, z)\rho^3(x, z)dV_z}{\int_{\mathfrak{R}} f(t, x, t+\Delta, z)\rho^2(x, z)dV_z} \to 0 \quad \text{as } \Delta \to 0, \tag{4}$$

where $\rho(x, z)$ denotes the geodesic distance[5] between x and z.

We choose a coordinate system $z = (z_1, \ldots, z_n)$ in a neighbourhood $\mathfrak{A}$ of x. Then we set

$$\int_{\mathfrak{A}} f(s, x, s+\Delta, z)(z_i - x_i)dV_z = a_i(s, x, \Delta), \tag{5}$$

$$\int_{\mathfrak{A}} f(s, x, s+\Delta, z)(z_i - x_i)(z_j - x_j)dV_z = b_{ij}(s, x, \Delta), \tag{6}$$

$$\int_{\mathfrak{R}} f(s, x, s+\Delta, z)\rho^2(x, z)dV_z = \beta(s, x, \Delta), \tag{7}$$

$$\int_{\mathfrak{R}} f(s, x, s+\Delta, z)\rho^3(x, z)dV_z = \nu(s, x, \Delta). \tag{8}$$

Our purpose is to prove that the ratios

$$a_i(s, x, \Delta)/\Delta, \quad b_{ij}(s, x, \Delta)/\Delta$$

tend to limits $A_i(s, x)$ and $B_{ij}(s, x)$ as $\Delta \to 0$, independently of $\mathfrak{A}$. Below this is proved under the assumption that all $N = n + n(n+1)/2$ functions

$$\frac{\partial}{\partial x_i} f(s, x, t, y), \quad \frac{\partial^2}{\partial x_i \partial x_j} f(s, x, t, y)$$

of y and t (for fixed s and x) are linearly independent, that is, that $t_1, y_1, t_2, y_2, \ldots$ $\ldots, t_k, y_k, \ldots, t_N, y_N$ can be chosen so that the determinant

$$D^N(s, x) = \begin{vmatrix} \dfrac{\partial}{\partial x_i} f(s, x, t_k, y_k) \\ \dfrac{\partial^2}{\partial x_i \partial x_j} f(s, x, t_k, y_k) \end{vmatrix} \tag{9}$$

is non-zero.[6]

[5] See A.M., §13, formula (112).
[6] See A.M., §13, determinant (119).

In $\mathfrak{A}$ we have

$$\rho^2(x,z) = \sum g_{ij}(z_i - x_i)(z_j - x_j) + \Theta\rho^3(x,z), \quad |\Theta| \leq C,$$

while outside $\mathfrak{A}$ we clearly have

$$\rho^2(x,z) = \Theta'\rho^3(x,z), \quad |\Theta'| \leq C',$$

where C' and C are constants independent of z. Hence

$$\begin{aligned}\beta(s,x,\Delta) &= \int_{\mathfrak{R}} f(s,x,s+\Delta,z)\rho^2(x,z)dV_z = \\ &= \sum g_{ij}\int_{\mathfrak{A}} f(s,x,s+\Delta,z)(z_i-x_i)(z_j-x_j)dV_z + \\ &\quad + \int_{\mathfrak{A}} f(s,x,s+\Delta,z)\Theta\rho^3(x,z)dV_z + \\ &\quad + \int_{\mathfrak{R}-\mathfrak{A}} f(s,x,s+\Delta,z)\Theta'\rho^3(x,z)dV_z = \\ &\sum g_{ij}b_{ij}(s,x,\Delta) + \Theta''\nu(s,x,\Delta), \quad |\Theta''| \leq C''. \end{aligned} \tag{10}$$

But since, by the continuity condition (4),

$$\frac{\beta(s,x,\Delta)}{\nu(s,x,\Delta)} \to +\infty \quad \text{as } \Delta \to 0, \tag{11}$$

formula (10) implies that

$$\frac{\sum g_{ij}b_{ij}(s,x,\Delta)}{\nu(s,x,\Delta)} \to +\infty \quad \text{as } \Delta \to 0. \tag{12}$$

Now, for fixed x,y,s,τ,t, $s < \tau < t$, we consider only Δ so small that $s+\Delta < \tau$. Then $f(s+\Delta,z,t,y)$ and its derivatives with respect to z up to the fourth order are uniformly bounded and continuous in $\mathfrak{A}$ (we assume that $\mathfrak{A}$ is compact). Hence, for every point z in $\mathfrak{A}$ we have

$$\begin{aligned}f(s+\Delta,z,t,y) - f(s+\Delta,x,t,y) &= \sum(z_i-x_i)\frac{\partial}{\partial x_i}f(s+\Delta,x,t,y) + \\ +\tfrac{1}{2}\sum(z_i-x_i)(z_j-x_j)\frac{\partial^2}{\partial x_i\partial x_j}&f(s+\Delta,x,t,y) + \Theta\rho^3(x,z), \quad |\Theta| \leq C, \end{aligned} \tag{13}$$

where C does not depend on Δ or z. On the other hand, the fundamental equation (3) implies that

$$f(s,x,t,y) = \int_{\mathfrak{R}} f(s,x,s+\Delta,z) f(s+\Delta,z,t,y)\,dV_z =$$
$$= \int_{\mathfrak{R}} f(s,x,s+\Delta,z) f(s+\Delta,x,t,y)\,dV_z +$$
$$+ \int_{\mathfrak{A}} f(s,x,s+\Delta,z)\{f(s+\Delta,z,t,y) - f(s+\Delta,x,t,y)\}\,dV_z +$$
$$+ \int_{\mathfrak{R}-\mathfrak{A}} f(s,x,s+\Delta,z)\{f(s+\Delta,z,t,y) - f(s+\Delta,x,t,y)\}\,dV_z =$$
$$= I_1 + I_2 + I_3. \tag{14}$$

By (2),

$$I_1 = \int_{\mathfrak{R}} f(s,x,s+\Delta,z) f(s+\Delta,x,t,y)\,dV_z =$$
$$= f(s+\Delta,x,t,y) \int_{\mathfrak{R}} f(s,x,s+\Delta,z)\,dV_z = f(s+\Delta,x,t,y). \tag{15}$$

Then (13), (5) and (6) imply that

$$I_2 = \int_{\mathfrak{A}} f(s,x,s+\Delta,z)\{f(s+\Delta,z,t,y) - f(s+\Delta,x,t,y)\}\,dV_z =$$
$$= \int_{\mathfrak{A}} f(s,x,s+\Delta,z)\Big\{\sum (z_i - x_i)\frac{\partial}{\partial x_i} f(s+\Delta,x,t,y) +$$
$$+\tfrac{1}{2}\sum (z_i - x_i)(z_j - x_j)\frac{\partial^2}{\partial x_i \partial x_j} f(s+\Delta,x,t,y) +$$
$$+\Theta\rho^3(x,z)\Big\}\,dV_z = \sum a_i(s,x,\Delta)\frac{\partial}{\partial x_i} f(s+\Delta,x,t,y) +$$
$$+\tfrac{1}{2}\sum b_{ij}(s,x,\Delta)\frac{\partial^2}{\partial x_i \partial x_j} f(s+\Delta,x,t,y) +$$
$$+ \int_{\mathfrak{A}} f(s,x,s+\Delta,z)\Theta\rho^3(x,z)\,dV_z. \tag{16}$$

Finally, since throughout $\mathfrak{R}-\mathfrak{A}$ we have

$$\rho^3(x,z) > K > 0,$$

where K does not depend on z, in $\mathfrak{R}-\mathfrak{A}$ we can set

$$f(s+\Delta,z,t,y) - f(s+\Delta,x,t,y) = \Theta'\rho^3(x,z).$$

Then

$$I_3 = \int_{\mathfrak{R}-\mathfrak{A}} f(s,x,s+\Delta,z)\{f(s+\Delta,z,t,y) - f(s+\Delta,x,t,y)\}dV_z =$$
$$= \int_{\mathfrak{R}-\mathfrak{A}} f(s,x,s+\Delta,z)\Theta'\rho^3(x,z)dV_z, \quad |\Theta'| \le C' = \frac{1}{K}. \tag{17}$$

Substituting (15)–(17) into (14) we finally obtain

$$f(s,x,t,y) = f(s+\Delta,x,t,y) + \sum a_i(s,x,\Delta)\frac{\partial}{\partial x_i}f(s+\Delta,x,t,y)+$$
$$+\tfrac{1}{2}\sum b_{ij}(s,x,\Delta)\frac{\partial^2}{\partial x_i \partial x_j}f(s+\Delta,x,t,y)+$$
$$+\int_{\mathfrak{R}} f(s,x,s+\Delta,z)\Theta''\rho^3(x,z)dV_z, \quad |\Theta''| \le C''. \tag{18}$$

If we also take into account the obvious equality

$$\int_{\mathfrak{R}} f(s,x,s+\Delta,z)\Theta''\rho^3(x,z)dV_z = \Theta''' \int_{\mathfrak{R}} f(s,x,s+\Delta,z)\rho^3(x,z)dV_z =$$
$$= \Theta'''\nu(s,x,\Delta), \quad |\Theta'''| \le C''',$$

then (18) can be rewritten as follows:

$$\frac{f(s+\Delta,x,t,y) - f(s,x,t,y)}{\Delta} = -\sum \frac{a_i(s,x,\Delta)}{\Delta}\frac{\partial}{\partial x_i}f(s+\Delta,x,t,y)-$$
$$-\sum \frac{b_{ij}(s,x,\Delta)}{2\Delta}\frac{\partial^2}{\partial x_i \partial x_j} - f(s+\Delta,x,t,y) - \Theta'''\frac{\nu(s,x,\Delta)}{\Delta}. \tag{19}$$

The left-hand side in (19) tends to $\frac{\partial}{\partial s}f(s,x,t,y)$ as $\Delta \to 0$.

Suppose that the determinant $D^N(s,x)$ is non-zero for $t_1, y_1, t_2, y_2, \dots$ $\dots, t_N, y_N$. Then $D^N(s+\Delta,x) \neq 0$ for sufficiently small Δ. Hence, there exist $\lambda_k(\Delta)$, $k = 1,2,\dots,N$, such that

$$\begin{aligned} &\sum_k \lambda_k(\Delta)\frac{\partial}{\partial x_i}f(s+\Delta,x,t_k,y_k) = \alpha_i, \\ &\sum_k \lambda_k(\Delta)\frac{\partial^2}{\partial x_i \partial x_j}f(s+\Delta,x,t_k,y_k) = \alpha_{ij}. \end{aligned} \tag{20}$$

If we multiply (19) by $\lambda_k(\Delta)$ with $t = t_k$ and $y = y_k$ and sum all the N equalities thus obtained, then we have

$$\sum_k \lambda_k(\Delta)\frac{f(s+\Delta,x,t_k,y_k) - f(s,x,t_k,y_k)}{\Delta} =$$
$$= -\sum_i \alpha_i \frac{a_i(s,x,\Delta)}{\Delta} - \sum_{i,j} \alpha_{ij}\frac{b_{ij}(s,x,\Delta)}{2\Delta} - \sum_k \lambda_k(\Delta)\Theta_k'''\frac{\nu(s,x,\Delta)}{\Delta}. \tag{21}$$

If Δ tends to zero, then the $\lambda_k(\Delta)$, as solutions of (20), tend to the solution $\lambda_k(0)$ of the equations

$$\sum_k \lambda_k(0)\frac{\partial}{\partial x_i} f(s,x,t_k,y_k) = \alpha_i,$$
$$\sum_k \lambda_k(0)\frac{\partial^2}{\partial x_i \partial x_j} f(s,x,t_k,y_k) = \alpha_{ij}. \tag{22}$$

Hence, the left-hand side of (21) has a finite limit

$$\Lambda_0 = \sum_k \lambda_k(0)\frac{\partial}{\partial s} f(s,x,t_k,y_k) \tag{23}$$

as $\Delta \to 0$.

In particular, if we set $\alpha_i = 0$, $\alpha_{ij} - g_{ij}$, then

$$\frac{\sum g_{ij} b_{ij}(s,x,\Delta)}{2\Delta} + \sum \lambda_k(\Delta)\Theta_k''' \frac{\nu(s,x,\Delta)}{\Delta} \to \Lambda_0 \quad \text{as } \Delta \to 0. \tag{24}$$

By (12), the second term in (24) is infinitesimally small as compared with the first one (since the $\lambda_k(\Delta)$ are bounded). Hence we have

$$\sum g_{ij} b_{ij}(s,x,\Delta)/2\Delta \to \Lambda_0 \quad \text{as } \Delta \to 0. \tag{25}$$

But (25) and (12) imply

$$\nu(s,x,\Delta)/\Delta \to 0 \quad \text{as } \Delta \to 0. \tag{26}$$

If we now equate all but one of the coefficients α_i and α_{ij} in (21) to zero, then a similar passage to the limit using (26) shows that all the limits

$$A_i(s,x) = \lim \frac{a_i(s,x,\Delta)}{\Delta} \quad \text{as } \Delta \to 0, \tag{27}$$

$$B_{ij}(s,x) = \lim \frac{b_{ij}(s,x,\Delta)}{2\Delta} \quad \text{as } \Delta \to 0, \tag{28}$$

exist and do not depend on the choice[7] of $\mathfrak{A}$. Then (27), (28), (26) and (19) immediately imply the *first differential equation*

$$\frac{\partial}{\partial s} f(s,x,t,y) = -\sum A_i(s,x)\frac{\partial}{\partial x_i} f(s,x,t,y) - \\ - \sum B_{ij}(s,x)\frac{\partial^2}{\partial x_i \partial x_j} f(s,x,t,y). \tag{29}$$

---

[7] See A.M., §13, formulas (122)–(124).

Certainly the condition that $D_N(s,x)$ does not vanish identically can be replaced by the direct requirement that the limits (27) and (28) exist, since (28) implies the existence of a finite limit (25) and therefore of (26).

At certain exceptional points the limits (27) and (28) need not exist. This was illustrated in A.M.[8] by the following example: $\mathfrak{R}$ is the ordinary number axis and

$$f(s,x,t,y) = \frac{3y^2}{2\sqrt{\pi(t-s)}} \exp\left[-\frac{(y^3-x^3)^2}{4(t-s)}\right]; \tag{30}$$

for $x = 0$ we easily obtain

$$b(s,x,\Delta)/2\Delta \to +\infty \quad \text{as } \Delta \to 0.$$

Hence there is no finite limit $B(s,x)$.

§2. The second differential equation

Assume now that in a neighbourhood $\mathfrak{A}$ of the point y_0 for a given t the limits $A_i(t,y)$ and $B_{ij}(t,y)$ exist uniformly and that $\nu(t,y,\Delta)/\Delta$ tends uniformly to 0 in $\mathfrak{A}$. Suppose further that $R(y)$ is a non-negative function vanishing outside $\mathfrak{A}$ with bounded derivatives up to the third order. Then for $y \in \mathfrak{A}$, $z \in \mathfrak{A}$ we have

$$\begin{aligned} R(y) = R(z) + \sum (y_i - z_i)\frac{\partial}{\partial z_i}R(z) + \\ + \tfrac{1}{2}\sum (y_i - z_i)(y_j - z_j)\frac{\partial^2}{\partial z_i \partial z_j}R(z) + \\ + \Theta' \rho^3(y,z), \quad |\Theta'| \le C', \end{aligned} \tag{31}$$

whereas for $y \in \mathfrak{R} - \mathfrak{A}$ and $z \in \mathfrak{A}$,

$$R(y) = R(z) + \Theta'' \rho^3(y,z), \quad |\Theta''| \le C''. \tag{32}$$

Finally, for $y \in \mathfrak{R} - \mathfrak{A}$, $z \in \mathfrak{R} - \mathfrak{A}$

$$R(y) = 0. \tag{33}$$

[8] See A.M., §13, formula (126).

If in the corresponding regions $R(y)$ is replaced by (31)–(33), we obtain

$$\int_{\mathfrak{A}} R(y)\frac{\partial}{\partial t}f(s,x,t,y)dV_y =$$

$$= \frac{\partial}{\partial t}\int_{\mathfrak{A}} R(y)f(s,x,t,y)dV_y = \frac{\partial}{\partial t}\int_{\mathfrak{R}} R(y)f(s,x,t,y)dV_y =$$

$$= \lim \frac{1}{\Delta}\int_{\mathfrak{R}} R(y)[f(s,x,t+\Delta,y) - f(s,x,t,y)]dV_y =$$

$$= \lim \frac{1}{\Delta}\Big\{\int_{\mathfrak{R}} R(y)\int_{\mathfrak{R}} f(s,x,t,z)f(t,z,t+\Delta,y)dV_z dV_y -$$

$$- \int_{\mathfrak{R}} R(y)f(s,x,t,y)dV_y\Big\} =$$

$$= \lim \frac{1}{\Delta}\Big\{\int_{\mathfrak{R}} f(s,x,t,z)\int_{\mathfrak{R}} R(y)f(t,z,t+\Delta,y)dV_y dV_z -$$

$$- \int_{\mathfrak{R}} R(z)f(s,x,t,z)dV_z\Big\} =$$

$$= \lim \frac{1}{\Delta}\Big\{\int_{\mathfrak{A}} f(s,x,t,z)\int_{\mathfrak{R}} R(z)f(t,z,t+\Delta,y)dV_y dV_z +$$

$$+ \int_{\mathfrak{A}} f(s,x,t,z)\int_{\mathfrak{A}}\Big[\sum (y_i - z_i)\frac{\partial}{\partial z_i}R(z) +$$

$$+\tfrac{1}{2}\sum (y_i - z_i)(y_j - z_j)\frac{\partial^2}{\partial z_i \partial z_j}R(z)\Big] f(t,z,t+\Delta,y)dV_y dV_z +$$

$$+ \int_{\mathfrak{A}} f(s,x,t,z)\int_{\mathfrak{R}} \Theta''' \rho^3(y,z)f(t,z,t+\Delta,y)dV_y dV_z -$$

$$- \int_{\mathfrak{A}} R(z)f(s,x,t,z)dV_z\Big\} = \lim \frac{1}{\Delta}\Big\{\int_{\mathfrak{A}} f(s,x,t,z)R(z)dV_z +$$

$$+ \int_{\mathfrak{A}} f(s,x,t,z)\Big[\sum a_i(t,z,\Delta)\frac{\partial}{\partial z_i}R(z) +$$

$$+\tfrac{1}{2}\sum b_{ij}(t,z,\Delta)\frac{\partial^2}{\partial z_i \partial z_j}R(z)\Big] dV_z +$$

$$+\Theta \int_{\mathfrak{A}} f(s,x,t,z)\nu(t,z,\Delta)dV_z - \int_{\mathfrak{A}} f(s,x,t,z)R(z)dV_z\Big\} =$$

$$= \int_{\mathfrak{A}} f(s,x,t,z)\Big[\sum A_i(t,z)\frac{\partial}{\partial z_i}R(z)+$$

$$+\sum B_{ij}(t,z)\frac{\partial^2}{\partial z_i \partial z_j}R(z)\Big]dV_z.$$

Replacing z by y in the right-hand side of the equation we obtain

$$\int_{\mathfrak{N}} R(y)\frac{\partial}{\partial t}f(s,x,t,y)dV_y = \int_{\mathfrak{A}} f(s,x,t,y)\Big[\sum A_i(t,y)\frac{\partial}{\partial y_i}R(y)+$$

$$+\sum B_{ij}(t,y)\frac{\partial^2}{\partial y_i \partial y_j}R(y)\Big]dV_y. \qquad (34)$$

Now assume that $A_i(t,z)$ and $B_{ij}(t,z)$ are twice continuously differentiable in $\mathfrak{A}$. Then we set

$$Q(t,y) = |g_{ij}(t,y)|$$

and after integration by parts, we obtain

$$\int_{\mathfrak{A}} f(s,x,t,y)A_i(t,y)\frac{\partial}{\partial y_i}R(y)dV_y =$$

$$= \int_{\mathfrak{A}} f(s,x,t,y)A_i(t,y)Q(t,y)\frac{\partial}{\partial y_i}R(y)dy_1dy_2\ldots dy_n =$$

$$= -\int_{\mathfrak{A}} \frac{\partial}{\partial y_i}[f(s,x,t,y)A_i(t,y)Q(t,y)]R(y)dy_1dy_2\ldots dy_n. \qquad (35)$$

Double integration by parts (since all the derivatives vanish on the boundary of $\mathfrak{A}$) yields

$$\int_{\mathfrak{A}} f(s,x,t,y)B_{ij}(t,y)\frac{\partial^2}{\partial y_i \partial y_j}R(y)dV_y =$$

$$= \int_{\mathfrak{A}} \frac{\partial^2}{\partial y_i \partial y_j}[f(s,x,t,y)B_{ij}(t,y)Q(t,y)]R(y)dy_1dy_2\ldots dy_n. \qquad (36)$$

Formulas (34)–(36) immediately imply that

$$\int_{\mathfrak{A}} R(y)Q(t,y)\frac{\partial}{\partial t}f(s,x,t,y)dy_1dy_2\ldots dy_n =$$

$$= \int_{\mathfrak{A}} R(y)\Big\{-\sum\frac{\partial}{\partial y_i}[A_i(t,y)Q(t,y)f(s,x,t,y)]+$$

$$+\sum\frac{\partial^2}{\partial y_i \partial y_j}[B_{ij}(t,y)Q(t,y)f(s,x,t,y)]\Big\}dy_1dy_2\ldots dy_n.$$

Since $R(y)$ is arbitrary, apart from the above conditions, it is easy to conclude that at interior points of $\mathfrak{A}$ the *second differential equation*

$$Q(t,y)\frac{\partial}{\partial t}f(s,x,t,y) = -\sum \frac{\partial}{\partial y_i}[A_i(t,y)Q(t,y)f(s,x,t,y)]+$$

$$+\sum \frac{\partial^2}{\partial y_i \partial y_j}[B_{ij}(t,y)Q(t,y)f(s,x,t,y)] \tag{37}$$

also holds.

If at time t_0 the differential function of the probability distribution is given, that is, a non-negative function $g(t_0, y)$ of y satisfying the condition

$$\int_{\mathfrak{R}} g(t_0, y)dV_y = 1, \tag{38}$$

then for arbitrary $t > t_0$ the distribution function $g(t, y)$ is given by the formula

$$g(t,y) = \int_{\mathfrak{R}} g(t_0, x)f(t_0, x, t, y)dV_x. \tag{39}$$

The function $g(t, y)$ satisfies the equation [9]

$$Q\frac{\partial g}{\partial t} = -\sum \frac{\partial}{\partial y_i}(A_i Q g) + \sum \frac{\partial^2}{\partial y_i \partial y_j}(B_{ij} Q g). \tag{40}$$

§3. Uniqueness

Under a change of the coordinate system the coefficients $A_i(s, x)$ and $B_{ij}(s, x)$ are transformed in the following way:

$$A_i' = \sum \frac{\partial x_i'}{\partial x_k} A_k + \sum \frac{\partial^2 x_i'}{\partial x_k \partial x_l} B_{kl}, \tag{41}$$

$$B_{ij}' = \sum \frac{\partial x_i'}{\partial x_k}\frac{\partial x_j'}{\partial x_l} B_{kl}. \tag{42}$$

Here we always have

$$B_{ii} = \lim \frac{b_{ii}(s,x,\Delta)}{2\Delta} = \lim \frac{1}{2\Delta}\int_{\mathfrak{A}} f(s,x,s+\Delta,z)(z_i - x_i)^2 dV_z \geq 0. \tag{43}$$

Hence the *quadratic form*

$$\sum B_{ij}\xi_i\xi_j \tag{44}$$

9 See A.M., §18, formulas (169) and (170).

is non-negative. This is crucial in the proof of the following theorem.[10]

Uniqueness Theorem 1. *If* $\mathfrak{R}$ *is closed, then* (40) *has at most one solution* $g(t,y)$ *with given continuous initial condition* $g(t_0,y) = g(y)$.

Proof. Clearly it suffices to consider the initial condition $g(t_0,y) = 0$ and prove that $g(t,y) = 0$ also for $t > t_0$. We can transform (40) into the form

$$\frac{\partial g}{\partial t} = \sum B_{ij}\frac{\partial^2 g}{\partial y_i \partial y_j} + \sum S_i \frac{\partial g}{\partial y_i} + Tg. \tag{45}$$

Now set

$$v(t,y) = g(t,y)e^{-ct}.$$

The function $v(t,y)$ satisfies the equation

$$\frac{\partial v}{\partial t} = \sum B_{ij}\frac{\partial^2 v}{\partial y_i \partial y_j} + \sum S_i \frac{\partial v}{\partial y_i} + Tv - cv. \tag{46}$$

For fixed t_0 and t_1 the constant c can be chosen so large that

$$T(t,y) - c < 0$$

for all y and t, $t_0 \le t \le t_1$. Under these conditions $v(t,y)$ cannot have a positive maximum at any point (t,y), $t_0 < t < t_1$, since at such a maximum

$$\frac{\partial v}{\partial t} = 0, \quad \frac{\partial v}{\partial y_i} = 0, \quad \sum B_{ij}\frac{\partial^2 v}{\partial y_i \partial y_j} \le 0, \quad (T-c)v < 0,$$

which contradicts (46). Neither can there be a negative minimum of $v(t,y)$ within these limits. Since $v(t_0,y) = 0$ at $t = t_0$, we obtain for $t_0 < t < t_1$,

$$v(t,y) < \max v(t_1,y) = e^{-ct_1}\max g(t_1,y)$$

$$g(t,y) < e^{-c(t_1-t)}\max g(t_1,y).$$

Since c was arbitrarily large, it follows that

$$g(t,y) = 0.$$

---

[10] See: E. Rothe, 'Über die Wärmeleitungsgleichung', *Math. Ann.* **104** (1931), 353–354 (uniqueness proof).

Uniqueness Theorem 2. *Let $\mathfrak{R}$ be closed. Then there is at most one non-negative continuous solution $f(s,x,t,y)$ for* (2) *and* (3) *that satisfies* (29) *with given twice continuously differentiable coefficients $A_i(t,y)$ and $B_{ij}(t,y)$, and the continuity condition* (4).

The continuity condition (4) *can be replaced by the following, weaker one:*

$$\int_{\mathfrak{R}} f(s,x,t,y)\rho^2(x,y)dV_y \to 0 \quad \text{as } t \to s. \tag{47}$$

Proof. Assume that two different functions $f_1(s,x,t,y)$ and $f_2(s,x,t,y)$ satisfy all our conditions. Then we can choose s and a continuous function $g(x)$ such that

$$g_1(t,y) = \int_{\mathfrak{R}} g(x)f_1(s,x,t,y)dV_x,$$

$$g_2(t,y) = \int_{\mathfrak{R}} g(x)f_2(s,x,t,y)dV_x$$

are also different. By (2) and (47), $g_1(t,y)$ and $g_2(t,y)$ tend to $g(y)$ as $t \to s$. Since the functions $g_1(t,y)$ and $g_2(t,y)$ satisfy (40), this contradicts Uniqueness Theorem 1.

§4. An example

The following example, which is interesting also for applications, demonstrates that the quadratic form (44) need not be positive definite: let $\mathfrak{R}$ be the usual Euclidean plane and let

$$f(s,x_1,x_2,t,y_1,y_2) = \frac{2\sqrt{3}}{\pi(t-s)^2}\exp\Big\{-\frac{(y_1-x_1)^2}{4(t-s)} - \frac{3[y_2-x_2-(t-s)(y_1+x_2)/2]^2}{(t-s)^3}\Big\}. \tag{48}$$

A simple computation shows that

$$B_{11}=1, \quad B_{12}=0, \quad B_{22}=0, \quad A_1=0, \quad A_2(s,x)=x_1.$$

§5. The limit solution

Let $\mathfrak{R}$ be closed and $f(s,x,t,y)$ everywhere positive and dependent only on the difference $t-s$:

$$f(s,x,t,y) = \phi(t-s,x,y). \tag{49}$$

Then general ergodic theorems[11] imply the existence of the limit probability distribution. In other words, for any distribution $g(t,y)$ determined by (38) and (39) and any region $\mathfrak{E}$ the relation

$$\int_{\mathfrak{E}} g(t,y)dV_y \to P(\mathfrak{E}) \quad \text{as } t \to +\infty, \tag{50}$$

holds, where $P(\mathfrak{E})$ does not depend on $g(t_0,y)$. It can easily be proved that $g(t,y)$ is uniformly continuous for large t. From this we deduce that[12]

$$P(\mathfrak{E}) = -\int_{\mathfrak{E}} g(y)dV_y, \tag{51}$$

$$g(t,y) \to g(y) \quad \text{as } t \to +\infty. \tag{52}$$

Clearly, $g(y)$ and $P(\mathfrak{E})$ do not depend on $g(t_0,y)$.

Now, let $g(y)$ be the solution of the equations

$$-\sum \frac{\partial}{\partial y_i}[A_i(y)Q(y)g(y)] + \sum \frac{\partial^2}{\partial y_i \partial y_j}[B_{ij}(y)Q(y)g(y)] = 0, \tag{53}$$

$$\int_{\mathfrak{R}} g(y)dV_y = 1. \tag{53a}$$

Setting $g(t_0,y) = g(y)$ it can easily be seen that $g(t,y) = g(y)$ also for $t > t_0$ (see (40) and Uniqueness Theorem 1). From this we deduce that the *solution of* (53) *and* (53a) (*if it exists*) *is uniquely determined and coincides with the limit function* $g(y)$.

As a particular case, (52) implies

$$f(s,x,t,y) \to g(y) \quad \text{as } t \to +\infty. \tag{54}$$

Klyazma, near Moscow, 12 April 1932

---

[11] See A.M., §4, Theorem IV.

[12] See footnote 1.

18. ON THE PROBLEM OF THE SUITABILITY OF FORECASTING FORMULAS FOUND BY STATISTICAL METHODS *

§1. A number of authors (Exner, Baur, Vize) have made attempts, by means of statistical analysis of the data gathered during the last 30–50 years, to obtain formulas relating the deviation Δy of some meteorological factor y, to be forecast, from its mean $\bar{y}$ (averaged over many years) , with similar deviations $\Delta x_1, \Delta x_2, \ldots, \Delta x_k$ of various factors that can be determined beforehand. So far only linear formulas have been used for this purpose:

$$\Delta y = a_1 \Delta x_1 + a_2 \Delta x_2 + \ldots + a_k \Delta x_k. \tag{1}$$

By choosing appropriate factors $x_1, x_2, \ldots, x_k$, whose number is somewhere between three and seven, and coefficients $a_1, a_2, \ldots, a_k$ it can be shown that in many cases the correlation coefficient between the actually observed Δy and its value computed by formula (1), calculated from the data during the same 30–50 years that served for deriving the formula, reaches 0.60–0.75. However, when verification of the formula was based on the observations over years other than the years used in deriving the formula (Vize), the correlation coefficient between the computed and the observed Δy turned out to be 0.30–0.40. This correlation coefficient has no practical value, especially since forecasts of the degree of reliability can be obtained in a much simpler way.

The theory for obtaining regression equations of type (1) is based on the following assumptions. It is assumed that y, $x_1, x_2, \ldots, x_k$ are random variables with a certain distribution law $w(y, x_1, x_2, \ldots, x_k)$ which does not change over the years. It is assumed further, that the probabilities that $y, x_1, x_2, \ldots, x_k$ take certain values in a given year do not depend on the values taken by these variables in previous years. Without the first condition (of stability) the problem of formulating a regression equation has no meaning at all. If the first condition is fulfilled, then there exist determinate coefficients $a_1, a_2, \ldots, a_k$ that minimize the expectation

$$\mathsf{E}(u^2) = \mathsf{E}(\Delta y - a_1 \Delta x_1 - a_2 \Delta x_2 - \ldots - a_k \Delta x_k)^2.$$

The relation

$$\delta = \frac{\mathsf{E}(\Delta y^2) - \mathsf{E}(u^2)}{\mathsf{E}(\Delta y^2)}$$

* *Zh. Geofiz.* **3** (1933), 78–82.

is called the *true correlation coefficient*. The second condition (of independence) is essential for proving that the coefficients $a_1, a_2, \dots, a_k$ of the empirical regression equation (1) computed from the observations during a sufficiently large number of years n are close to the theoretical coefficients $a_1, a_2, \dots, a_k$, and also for proving that the empirical correlation coefficient R is close to the theoretical one, δ, also for sufficiently large n.

There can certainly be some doubts as to whether these two conditions can be applied to meteorological phenomena. Moreover, if we consider the existence of secular or periodic multi-year climate fluctuations an established fact, the first condition is clearly wrong. This fact is often used for explaining the discrepancy mentioned above: it is supposed that the regression equation (1) does indeed reflect the regularities that took place during the period under study with great accuracy, but these regularities themselves are subject to change as a result of long-term climatic fluctuations. I am going to prove that these excessively high empirical correlation coefficients are quite explicable also under the assumption of complete stability and independence of the studied factors from year to year, in other words, that the methods used by the above researchers inevitably lead to a certain "blow-up" of correlation coefficients. This is discussed in §2. In §3 we consider whether stability and independence in meteorological series are sufficient to make statistical determination of regression equations possible and useful; in §4 we give some ideas on the techniques for finding regression equations.

§2. The mathematical apparatus needed for solving our problem was developed quite recently by Fisher [1] in his study on the distribution law for the correlation coefficient under multiple correlation (see also the review by Rider [2]). It is assumed that the distribution law $w(y, x_1, x_2, \dots, x_k)$ is normal.

Suppose that we are given a regression equation for a certain variable y. Assume that y is related to $x_1, x_2, \dots, x_k$ in such a way that the true correlation coefficient is δ. Fisher's analysis makes it possible to compute the distribution law for the empirical correlation coefficient R from δ and the number of observations n (number of years during which observation took place). However, this distribution law by itself still gives no answer to our question. Indeed, $x_1, x_2, \dots, x_k$ can be chosen so that the correlation coefficient δ is as large as possible; then although at most 5–7 values are introduced into the regression equation, the stock of values from which these 5–7 values can be chosen is very large.

We assume therefore that there are i groups of values

$$x_1^{(1)}, x_2^{(1)}, \dots, x_k^{(1)},$$

$$x_1^{(2)}, x_2^{(2)}, \dots, x_k^{(2)},$$

$$\dots\dots$$

$$x_1^{(i)}, x_2^{(i)}, \dots, x_k^{(i)},$$

of k values each. Assume for simplicity that y is related to each of these groups with true correlation coefficient δ. For any λ Fisher's formulas allow us to compute the probability that in each individual case the empirical correlation coefficient R exceeds λ. Let this probability be p. It is natural to assume that with probability

$$P = 1 - (1-p)^i \tag{2}$$

the inequality $R > \lambda$ holds for at least one of the groups: this is so under the assumption of independence of the deviations of R from δ corresponding to different groups.[1] If, for example, $i = 14$ and $p = 1/20$, then

$$P = 1 - (1 - 1/20)^{14} \approx 1/2.$$

The values of λ corresponding to $p = 1/20$ for given δ, k and R can be computed using Fisher's tables. For this we set

$$m = n - k - 1, \quad \beta = \sqrt{m} \tanh^{-1} \delta,$$

then find B from β and m via Fisher's tables[2] and, finally, find λ by the formula

$$B = \sqrt{m} \tanh^{-1} \lambda.$$

For example, set $n = 42$ and $k = 5$. Then, after appropriate calculations, we find:

δ	0.20	0.30	0.40	0.50
λ	0.56	0.61	0.66	0.72

[1] This assumption is quite arbitrary, but in reality the number of groups from which to choose might be even greater than $i = 14$ taken for the calculation in the example; this, I hope, justifies the arbitrariness of our assumption.

[2] Tables on p.665 in [1].

The immediate significance of these results is as follows: by computing the five-term regression formula from 42 observations for $\delta = 0.20;\ 0.30;\ 0.40;\ 0.50$ in one twentieth of the cases we find that R exceeds the corresponding λ.

If, however, from the 14 regression formulas we choose the one with the largest R, then R exceeds the computed values of λ with probability greater than $\frac{1}{2}$.

In conclusion we discuss Fisher's general formulas. The probability that R^2 lies between v and $v + dv$ is, according to Fisher,

$$df = \frac{[\frac{1}{2}(m+k-1)]!}{[\frac{1}{2}(m-2)]![\frac{1}{2}(k-3)]!}\frac{(1-\delta^2)^{(m+k)/2}}{\pi}\times$$

$$\times v^{(k-2)/2}(1-v)^{(m-2)/2}\int_0^{\pi} d\phi \int_{-\infty}^{\infty} \frac{\sin^{k-2}\phi dz}{(\cosh z - \delta R\cos\phi)^{m+k}} dv =$$

$$= \frac{[\frac{1}{2}(m+k-2)]!}{[\frac{1}{2}(m-2)]![\frac{1}{2}(k-2)]!}(1-\delta^2)^{(m+k)/2}\times$$

$$\times F\left[\tfrac{1}{2}(m+k), \tfrac{1}{2}(m+k), \tfrac{1}{2}k, \delta^2 v\right] v^{(k-2)/2}(1-v)^{(m-2)/2}dv,$$

where $F(p,q,r,x)$ denotes the hypergeometric function. These relations can be simplified in special cases and replaced by relations asymptotic in m, but even after simplification they are still too complicated to be used directly.

§3. The established fact of "blowing up" of the correlation coefficient does not necessarily mean that these statistical regression formulas cannot be used. We must make the number k of values involved in the regression equation and the stock of values from which they are chosen as low as possible; in this case the risk of obtaining an artificially "blown-up" correlation coefficient can be significantly diminished.

More essential for evaluating future prospects for the statistical establishment of forecasting equations is to find out to what extent meteorological series satisfy the stability and independence conditions. The very existence of secular changes or periodic fluctuations does not in itself preclude the use of studies stemming from assumptions on stability and independence, if the role of these secular or periodic fluctuations in forming the deviations $\Delta y, \Delta x_1, \Delta x_2, \ldots, \Delta x_k$ is insignificant.

The simplest method for checking the stability of the series and the independence of its terms is the following: let $\Delta y^{(i)}$ be the value of the deviation Δy

for the ith year. We need to compute, over a long period of n years, the correlation coefficients between $\Delta y^{(i)}$ and $\Delta y^{(i+1)}$, between $\Delta y^{(i)}$ and $\Delta y^{(i+2)}$, etc. If these correlation coefficients deviate from zero within the bounds corresponding to theoretical computations made under the assumption of independence and stability, this confirms our hypothesis. Indeed, secular change, or periodic fluctuation with period of four years or more, inevitably lead to positive correlation between $\Delta y^{(i)}$ and $\Delta y^{(i+4)}$, whereas short fluctuations during two or three years lead to positive correlation between $\Delta y^{(i)}$ and $\Delta y^{(i+2)}$ or $\Delta y^{(i)}$ and $\Delta y^{(i+3)}$. In the case of stability and independence the expectation of the square of each of the above correlation coefficients is approximately $1/(n-3)$, for n not too small.

Such computations were made for the average monthly temperatures in Leningrad. For each month the correlation coefficients between $\Delta y^{(i)}$ and $\Delta y^{(i+1)}$, $\Delta y^{(i+2)}$ and $\Delta y^{(i+3)}$ were computed from the data over a hundred years. The mean square of these three correlation coefficients over twelve months turned out to be 0.065, 0.064 and 0.110, which is in good agreement with the theoretical values:

$$1/\sqrt{n-3} = 1/\sqrt{97} = 0.102.$$

Now consider two series $y^{(i)}$ and $x^{(i)}$. If y is related to x with positive correlation, then $\Delta y \Delta x$ has positive expectation. To study the stability of correlation between $y^{(i)}$ and $x^{(i)}$ we must form the correlation coefficients between $\Delta y^{(i)} - \Delta x^{(i)}$ and $\Delta x^{(i+1)} - \Delta y^{(i+1)}$, $\Delta x^{(i+2)} - \Delta y^{(i+2)}$, etc. If the double series is stable and the pairs $y^{(i)}, x^{(i)}$ relating to different years are independent, then the square of this correlation coefficient has expectation $1/(n-3)$, as in the first case. In this way the stability of the correlation between average temperatures of two successive months in Leningrad was studied (it is known to be a significant positive correlation): the mean square of the correlation coefficient between $\Delta y^{(i)} - \Delta x^{(i)}$ and $\Delta y^{(i+1)} - \Delta x^{(i+1)}$ over twelve pairs of adjacent months was 0.071, which is also in good agreement with the theoretical value 0.102. Similar treatment can be given to the stability of the variability of y by forming the correlation coefficients between $(\Delta y^{(i)})^2$ and $(\Delta y^{(i+1)})^2$, $(\Delta y^{(i+2)})^2$, etc.

Only such systematic studies of the stability of meteorological series can solve the problem, whereas separate remarks to the effect that a certain correlation coefficient appeared to be unstable can only distort our ideas on the real situation. If some separately chosen correlation coefficient between two meteorological factors has very different values for two adjacent periods of twenty-five years, this by no means proves that the general regularities of the climate that occurred during the first twenty-five years changed over the second twenty-five year period. Rather, this discrepancy can be considered to be of a purely random nature.

Clearly, a few figures are far from sufficient for the solution of our problem, but it is quite possible that further studies will reveal fairly good agreement with the stability and independence hypotheses. The point is that in order to prove secular climate changes or multi-year periodic fluctuations, highly averaged values are usually studied, based on observations made by individual stations, for which the absolute values of the deviations Δy are comparatively small and therefore the role of long-term fluctuations is more significant.

§4. We now assume that the requirements of stability and independence for our series are fulfilled. If the number of observation years is large enough, then under our assumption, even quite complex regression equations selected on the basis of these observations, would reflect with sufficient accuracy the regularities that are true for the entire series. However, for $n = 30$ or even 50 or even 100 the situation changes considerably: if too many values are introduced, from which we make our choice for entering in our equation, then the danger is that there randomly appear nice combinations whose use for forecasting would be quite unjustified.

On the contrary, if serious theoretical observations show that the value y to be forecast can to a considerable degree be determined, for example, by three values x_1, x_2, x_3, then the computation of the regression formula (1) which relates Δy to $\Delta x_1, \Delta x_2$ and Δx_3, based on 50 years' observations, should be considered reliable. Here we mean theoretical considerations dynamically justified and not "theories" of purely statistical origin. However, even purely statistical techniques might give a hint of where to look for non-random correlation relations. Such an attempt on a broad scale was made by Baur [3]. He computed correlation coefficients between average monthly temperatures in Iceland and atmospheric pressure in previous months at various points of the Earth. It appeared that for the stations on the Southern hemisphere the absolute values of the correlation coefficients do not exceed on the average (over different stations and twelve months) the values predicted theoretically under the assumption of independence of both events (temperature in Iceland and pressure at a given station in the Southern hemisphere). This makes these rarely occurring correlation coefficients considerably exceeding the general norm quite doubtful. In contrast to this, the pressure at the stations of the Northern hemisphere show correlation with subsequent temperatures in Iceland, which is systematically much greater than it would have been at random. Baur remarks, however, that perhaps the influence of pressure anomalies at distant points of the Earth tells something about temperatures in Iceland only over times longer than one month. Baur does not give sufficient proof of this hypothesis. The question can be solved only by systematically continuing Baur's studies on the average value of correlation coefficients, not by looking for new special cases of high correlation.

These studies should be guided by Fisher's result by virtue of which, for independent x and y the empirical correlation coefficient r has a distribution law which, after the transformation

$$z = \tan^{-1} r,$$

turns into the normal law for z with centre at the origin and variance

$$\sigma = 1/\sqrt{n-3},$$

where n is the number of observations.

References

1. R.A. Fisher, 'The general sampling distribution of the multiple correlation coefficient', *Proc. London Roy. Soc. A* **121** N787 (1928), 654–673.
2. P.R. Rider, 'A survey of the theory of small samples', *Ann. of Math.* **31**:4 (1930), 577–628.
3. F. Baur, 'Statistische Untersuchungen über Auswirkungen und Bedingungen der grossen Störungen der allgemeinen atmosphärischen Zirkulation', *Ann. Hydrogr. Maritimen Meteor.* **53**:1 (1925), 1–6; **53**:8, 241–258.

19. RANDOM MOTIONS *

On the theory of Brownian motion

In my two earlier papers [1], [2], I developed a general theory of continuous random processes. Under very general conditions, it was proved that if the state of a physical system at every given moment is fully determined by n parameters $x_1, x_2, \dots, x_n$, and if these n parameters continuously [1] change with time t, then the corresponding distribution functions satisfy the Fokker-Planck differential equation. In the general case of such random processes the increments Δx_i of the parameter x_i are of the same order as $(\Delta t)^{1/2}$. This implies that in the general case $\Delta x_i : \Delta t \to \infty$ as $\Delta t \to 0$, so that we cannot speak about a definite rate of variation of x_i. We will now show how to apply this general theory to random motions, for which we assume that not only the system's coordinates, but also their derivatives with respect to time vary continuously.

Let $q_1, q_2, \dots, q_n$ be the coordinates of a system with n degrees of freedom. Assume that if we know q and $\dot{q}$ and the time t, then we can determine the probability density

$$G(t, q_1, \dots, q_n, \dot{q}_1, \dots, \dot{q}_n, t', q'_1, \dots, q'_n, \dot{q}'_1, \dots, \dot{q}'_n)$$

of q' and $\dot{q}'$ for any $t' > t$. We assume, moreover, that G does not depend on the behaviour of the system before time t.

It is natural to assume that [2]

$$\mathsf{E}|\Delta q_i - \dot{q}_i \Delta t| = o(\Delta t), \tag{1}$$

$$\mathsf{E}(\Delta q_i)^2 = o(\Delta t), \tag{2}$$

where $\Delta t = t' - t$ and E is the expectation symbol. Relations (1) and (2) imply

$$\mathsf{E}(\Delta q_i) = \dot{q}_i \Delta t + o(\Delta t), \tag{3}$$

$$\mathsf{E}(\Delta q_i \Delta q_j) \le \sqrt{\mathsf{E}(\Delta q_1)^2 \mathsf{E}(\Delta q_j)^2} = o(\Delta t) \tag{4}$$

---

* 'Zufällige Bewegungen (Zur Theorie der Brownschen Bewegung)', *Ann. of Math.* **35** (1934), 116–117.

[1] On the precise meaning of the condition of continuity of random processes, see [1], §13.

[2] Since Δq_i should be of the same order as Δt.

In [2], under very general assumptions it was proved that in addition to (2)–(4) the following relations hold:

$$\mathsf{E}(\Delta \dot{q}_i) = f_i(t, q, \dot{q})\Delta t + o(\Delta t), \tag{5}$$

$$\mathsf{E}(\Delta \dot{q}_i)^2 = k_{ii}(t, q, \dot{q})\Delta t + o(\Delta t), \tag{6}$$

$$\mathsf{E}(\Delta \dot{q}_i \Delta \dot{q}_j) = k_{ij}(t, q, \dot{q})\Delta t + o(\Delta t), \tag{7}$$

where f and k are continuous functions. In many problems of physics the assumptions (5)–(7) also allow direct verification. In particular, (2) and (6) imply that

$$\mathsf{E}(\Delta q_i \Delta \dot{q}_j) \leq \sqrt{\mathsf{E}(\Delta q_i)^2 \mathsf{E}(\Delta \dot{q}_j)^2} = o(\Delta t). \tag{8}$$

Under the assumptions (2)–(8), which are quite natural from the point of view of physics, it is clear that G is a fundamental solution of the following differential equation of Fokker-Planck type[3]

$$\frac{\partial g}{\partial t'} = -\sum \dot{q}'_i \frac{\partial}{\partial q'_i} g - \sum \frac{\partial}{\partial \dot{q}'_i}\{f_i(t', q', \dot{q}')g\} + \\ + \sum\sum \frac{\partial^2}{\partial \dot{q}_i \partial \dot{q}'_j}\{k_{ij}(t', q', \dot{q}')g\}. \tag{9}$$

Thus, for $n = 1$ we obtain

$$\frac{\partial g}{\partial t'} = -\dot{q}' \frac{\partial}{\partial q'} g - \frac{\partial}{\partial \dot{q}'}\{f(t', q', \dot{q}')g\} + \frac{\partial^2}{\partial \dot{q}'^2}\{k(t', q', \dot{q}')g\}. \tag{10}$$

If f and k are constants, then the fundamental solution of (10) can be represented by

$$g = \frac{2\sqrt{3}}{\pi k^2 (t' - t)^2} \exp\Big\{-\frac{[\dot{q}' - \dot{q} - f(t' - t)]^2}{4k(t' - t)} - \\ - \frac{3\Big[q' - q - \frac{\dot{q}' + \dot{q}}{2}(t' - t)\Big]^2}{k(t' - t)^3}\Big\}. \tag{11}$$

[3] The proof can be found in [2].

Clearly $\Delta\dot{q}$ is of the order $(\Delta t)^{1/2}$, that is, it behaves like Δx in the case of a general continuous process. However, for Δq we obtain[4]

$$\Delta q = \dot{q}\Delta t + O(\Delta t)^{3/2}. \tag{12}$$

It can be proved that the latter relation also holds in the case of the general equation (9).

References

1. A.N. Kolmogoroff, 'Über die analytischen Methoden in der Wahrscheinlichkeitsrechnung', *Math. Ann.* **104**:3 (1931), 415–458 (No. 9 in this volume).
2. A.N. Kolmogoroff, 'Zur Theorie der stetigen zufälligen Prozesse', *Math. Ann.* **108**:1 (1933), 149–160 (No. 17 in this volume).

[4] This result means that for any constant K, however large, the probability of $|\Delta q - \dot{q}\Delta t| \geq K(\Delta t)$ is smaller than any fixed $\epsilon > 0$ uniformly in Δt.

20. DEVIATIONS FROM HARDY'S FORMULAS UNDER PARTIAL ISOLATION *

The profound mathematical studies by R. Fisher [1] and S. Wright [2] deal with the evolution of gene concentration in a population in which free crossing dominates. The purpose of this paper is to give a method for obtaining similar results for a population consisting of a large number of partial populations weakly connected with each other. Mathematical analysis was applied to the following scheme: a population with a constant number of individuals N consisting of s partial populations of n individuals each ($N = sn$) with free crossing in each partial population and in which in every generation on the average k "wandering" individuals are isolated from every population; regardless of their origin, wandering individuals randomly join any of the partial populations, where they take part in creating the next generation. This scheme was indicated as a possible one by N.P. Dubinin and D.D. Romashov. A number of other, not less interesting, schemes of restricted crossing do not yet succumb to mathematical treatment.

§1. The case of no selection. Denote by $\bar{p}$ the concentration of a certain gene in the large population and by p the concentration of the same gene in a partial population.

In this section we assume that the studied gene is not subject to selection. The formulas below are asymptotic, which are true as $n\bar{p}, n/k^2$ and s tend to infinity. In practice, they are applicable if $n\bar{p}$, n/k^2 and s are sufficiently large.

Denote by Δp the increment of the concentration p in a certain partial population over one generation. After Fisher and Wright, we assume that the expectations of Δp and (Δp^2) satisfy the formulas

$$A = \mathbf{E}(\Delta p) = \frac{k}{n}(\bar{p} - p), \quad B = \mathbf{E}(\Delta p)^2 = \frac{pq}{2n},$$

where $q = 1 - p$. Since s is large, the variation of the total concentration $\bar{p}$ will proceed much more slowly than those of the partial concentrations p. Therefore, $\bar{p}$ can temporarily be taken constant. The concentrations p in partial populations deviate from $\bar{p}$ in either direction. After sufficiently long time the fluctuations of p around $\bar{p}$ result in a certain stationary probability distribution for the concentrations p. This stationary distribution $u(p)$ satisfies the

* *Dokl. Akad. Nauk SSSR* **3** (1935), 129–132. Presented by S.N. Bernshtein.

differential equation

$$\frac{1}{2}\frac{\partial^2}{\partial p^2}(Bu) - \frac{\partial}{\partial p}(Au) = 0.$$

Solving this equation we obtain

$$u(p) = \frac{p^{4k\bar{p}-1}q^{4k\bar{q}-1}}{B(4k\bar{p}, 4k\bar{q})}. \tag{1}$$

Wright obtained (1) using a different method. It can be proved that (1) not only gives the probability distribution for the concentration p but, for a sufficiently large number s of partial populations, the actually observed distribution of partial populations over concentrations p will also be given by (1).

Under these assumptions Hardy's formula holds for every partial population, that is, the concentrations of individuals of the types AA, Aa and aa are equal to $q^2, 2pq$ and p^2. The concentrations of individuals of the type AA, Aa and aa in a large population can be computed by the formulas:

$$\overline{AA} = \int_0^1 q^2 u(p)dp,$$

$$\overline{Aa} = 2\int_0^1 pqu(p)dp,$$

$$\overline{aa} = \int_0^1 p^2 u(p)dp.$$

The computations give

$$\overline{AA} = \frac{4k}{4k+1}\bar{q}^2 + \frac{1}{4k+1}\bar{q},$$

$$\overline{Aa} = 2\frac{4k}{4k+1}\bar{p}\bar{q}, \tag{2}$$

$$\overline{aa} = \frac{4k}{4k+1}\bar{p}^2 + \frac{1}{4k+1}\bar{p}.$$

Formulas (2) give the solution to the problem raised in the title of the paper.

Due to the absence of selection, the expectation $\mathbf{E}(\Delta\bar{p})$ of the increment of the total gene concentration (in the large population) over one generation is 0, while the intensity of random concentration fluctuations in the large population is $\mathbf{E}(\Delta\bar{p})^2$. This latter value is computed by the formula

$$\mathbf{E}(\Delta\bar{p})^2 = \frac{1}{s}\int_0^1 \mathbf{E}(\Delta p)^2 u(p)dp = \frac{1}{2sn}\int_0^1 pqu(p)dp.$$

Computations give

$$\mathsf{E}(\Delta\bar{p})^2 = \frac{4k}{4k+1}\frac{\bar{p}\bar{q}}{2N}. \tag{3}$$

In the case of free crossing we have, according to Fisher and Wright,

$$\mathsf{E}(\Delta p)^2 = \bar{p}\bar{q}/2N.$$

§**2. The effect of selection**. Formula (1), Hardy's formula for partial populations and its substitute (2) for a large population remain valid in the presence of selection when the selection coefficient α is much less than $1/n$. We consider only this case: accordingly, the formulas below are merely asymptotic formulas, true as $n\alpha \to 0$. Of special interest is the case of a recessive gene. In this case the mean increment of gene concentration due to selection is $\alpha p^2 q$ for any partial population. Therefore, for the total concentration we have

$$\mathsf{E}(\Delta\bar{p}) = \alpha\int_0^1 p^2 q u(p)dp,$$

or

$$\mathsf{E}(\Delta\bar{p}) = \alpha\frac{4k}{(4k+2)(4k+1)}\bar{p}\bar{q}(4k\bar{p}+1). \tag{4}$$

Formula (4) can be used to confirm, for the above scheme, the general statement on the existence of an optimum of partial isolation for the selection of recessive genes, which was suggested and quantitatively justified by A.A. Malinovskii. It is easy to compute the k corresponding to the highest rate of selection. For small $\bar{p}$ this optimal k is

$$k_0 = \tfrac{1}{4}\sqrt{2} = 0.35.$$

Mathematical Institute of Moscow State University
18 June 1935

References

1. R.A. Fisher, *The genetic theory of natural selection*, Oxford, 1930.
2. S. Wright, *Genetics*, London, 1931, pp. 97–157.

21. ON THE THEORY OF MARKOV CHAINS*

The considerations given below, though simple, are, I believe, new and interesting for certain physical applications, especially in the analysis of reversibility of statistical laws of nature made by Schrödinger[1] for one particular case. In what follows it is a matter of indifference which of the two following assumptions is made: either the time variable t runs through all real values, or only through the integers. The classical understanding of Markov chains corresponds to the second possibility.

1. The notion of a Markov chain

Consider a physical system which at any given time t can be in one of the states of a finite set $E_1, E_2, \dots, E_N$. Assume that for any pair of states E_i and E_j and every pair of moments t and s, $t \leq s$, the conditional probability $P_{ij}(t,s)$ that the state E_j takes place at time s under the assumption that at t the system was in state E_i, is defined. A significant, but not always explicitly stated assumption is that the conditional probability $P_{ij}(t,s)$ is independent of the prehistory of the system before t. This assumption holds in deducing the fundamental equation of the theory of Markov chains,

$$P_{ik}(s,t) = \sum_j P_{ij}(s,u) P_{jk}(u,t), \quad s \leq u \leq t, \tag{1}$$

with the conditions

$$P_{ij}(t,s) \geq 0, \tag{2}$$

$$\sum_j P_{ij}(t,s) = 1, \tag{3}$$

$$P_{ij}(t,t) = \delta_{ij}, \tag{4}$$

where $\delta_{ij} = 0$ is the Kronecker symbol.

2. Absolute probabilities

So far we have only considered condition transition probabilities $P_{ij}(t,s)$. There arises the question whether it is possible, knowing the transition probabilities

* 'Zur Theorie der Markoffschen Ketten', *Math. Ann.* **112** (1936), 155–160.

1 *Berliner Berichte* (1931), 144.

$P_{ij}(t,s)$, to give the probabilities of various states at each instant t ? If we assume that the process of change for our system starts at a certain moment t_0, then the various states $E_1, E_2, \dots, E_N$ are associated at time t_0 with arbitrary probabilities $Q_1(t_0), Q_2(t_0), \dots, Q_N(t_0), Q_k(t_0) \geq 0$, $\sum_k Q_k(t_0) = 1$. For each moment $t > t_0$ we obtain the following expressions for the probabilities $Q_k(t)$ of the states E_k $(k = 1, 2, \dots, N)$:

$$Q_k(t) = \sum_i Q_i(t_0) P_{ik}(t_0, t).$$

Then for $t_0 \leq t \leq s$ we automatically also obtain the formula

$$Q_k(s) = \sum_i Q_i(t) P_{ik}(t, s). \tag{5}$$

The introduction of absolute probabilities is not so trivial if we do not assume a certain beginning of the process, but instead try to find the absolute probabilities $Q_k(t)$ at all times t, $-\infty < t < +\infty$. But even here the following assertion also holds:

For arbitrary transition probabilities $P_{ik}(t,s)$ given for all t and s $(t \leq s)$, it is possible to determine in at least one way for all t, $-\infty < t < +\infty$, the absolute probabilities $Q_k(t)$ corresponding to these transition probabilities $P_{ik}(t,s)$.

Purely analytically, the theorem means the following. For each choice of $P_{ik}(t,s)$ corresponding to (1)–(4), the infinite system of equations (5) has at least one solution satisfying the additional conditions

$$Q_k(t) \geq 0, \quad \sum_k Q_k(t) = 1. \tag{6}$$

To prove this theorem we first note that for each t_0, as we have already seen, there exists at least one system $Q_k^{(t_0)}(t)$ determined for all $t \geq t_0$ and satisfying (5) and (6) for $s \geq t \geq t_0$. Using a diagonal process, we can extract from the sequence

$$t_0 = -1, -2, -3, \dots$$

a subsequence

$$t_0 = \lambda_1, \lambda_2, \lambda_3, \dots$$

such that $\lambda_n \to -\infty$, so that for each k and each integer t the quantities $Q_k^{(\lambda_n)}(t)$ defined at each fixed t for all sufficiently large n tend to a certain limit $Q_k^*(t)$ as $n \to \infty$. From (6) it is easy to deduce that for all real non-integer t there exist limit values $Q_k^*(t)$. These limit values, as can be proved by passing to the limit, satisfy (5) and (6). This proves our theorem.

Of special interest is the case when the absolute probabilities are uniquely defined by the transition probabilities $P_{ik}(t,s)$. A necessary and sufficient condition for such uniqueness is the following:

For arbitrary fixed k and s, $P_{ik}(t,s)$ tends to a certain limit $Q_k^*(s)$, independent of i, as $t \to -\infty$. If this condition holds, then it is precisely these limits $Q_k^*(s)$ that form the desired unique system of absolute probabilities.

Let us first prove sufficiency. Let $Q_k(t)$ be certain absolute probabilities compatible with the transition probabilities $P_{ik}(t,s)$. Then

$$Q_k(s) = \sum_i Q_i(t) P_{ik}(t,s); \tag{5}$$

but since $P_{ik}(t,s) \to Q_k^*(s)$ as $t \to -\infty$, the right-hand side of (5) tends to

$$\sum_i Q_k^*(s) Q_i(t) = Q_k^*(s)$$

as $t \to -\infty$, which implies that

$$Q_k(s) = Q_k^*(s).$$

Now suppose that our condition fails. Then we can choose i_1 and i_2 and two sequences,

$$t_1' > t_2' > \ldots > t_n' > \ldots \to -\infty,$$

$$t_1'' > t_2'' > \ldots > t_n'' > \ldots \to -\infty,$$

such that $P_{i_1k}(t_n', s)$ and $P_{i_2k}(t_n'', s)$ tend to limits $Q_k'(s)$ and $Q_k''(s)$ respectively, as $n \to +\infty$, for arbitrary k and s; furthermore, $Q_k'(s)$ and $Q_k''(s)$ are not identically equal for all s. Then it is easy to prove that both $Q_k'(s)$ and $Q_k''(s)$ can be taken as absolute probabilities, implying the necessity of the condition.

3. Inverse probabilities

We now assume that there is a certain fixed system of absolute probabilities $Q_k(s)$ with all the $Q_k(s)$ positive. In this case, along with the transition probabilities $P_{ik}(t,s)$ we can also define the inverse conditional probabilities.

Namely, we denote by $\Pi_{ik}(t,s)$ the conditional probability of the state E_i at time t under the assumption that at a certain later time $s, s \geq t$, the state E_k is observed. Clearly,

$$\Pi_{ik}(t,s) = \frac{Q_i(t)}{Q_k(t)} P_{ik}(t,s). \tag{7}$$

It is easy to derive the following formulas analogous to (1)–(5).

$$\Pi_{ik}(s,t) = \sum_j \Pi_{ij}(s,u)\Pi_{jk}(u,t), \quad s \leq u \leq t, \tag{1*}$$

$$\Pi_{ik}(s,t) \geq 0, \tag{2*}$$

$$\sum_k \Pi_{ik}(s,t) = 1, \tag{3*}$$

$$\Pi_{ik}(t,t) = \delta_{ik}, \tag{4*}$$

$$Q_i(s) = \sum_k Q_k(t)\Pi_{ik}(s,t), \quad s \leq t. \tag{5*}$$

Note that if we want to bypass a certain new principle of "independence of the future", then (1*) should be derived from the above formulas, not directly, as for (1). This latter principle can then be deduced from (1*).

4. Inversion of the laws of nature

We now assume that the transition probabilities $P_{ik}(s,t)$ depend only on the difference $t-s$:

$$P_{ik}(s,t) = P_{ik}(t-s).$$

As is known, in this case there exists at least one system of absolute probabilities independent of time t:

$$Q_k(t) = Q_k.$$

For a fixed system of such probabilities Q_k the inverse probabilities $\Pi_{ik}(s,t)$ depend also only on $t-s$:

$$\Pi_{ik}(s,t) = \Pi_{ik}(t-s).$$

The question is: *under what conditions do we have the equality*

$$\Pi_{ik}(r) = P_{ki}(r)\ ? \tag{8}$$

Since

$$\Pi_{ik} = (Q_i/Q_r)P_{ik}(r),$$

it follows that (8) holds if and only if

$$Q_i\Pi_{ik}(r) = Q_kP_{ki}(r). \tag{9}$$

It should be noted, however, that necessary and sufficient conditions for (8) to hold can be expressed even without using Q_k, in terms of the transition probabilities $P_{ik}(r)$ only. Indeed, von Mises [2] proved that all states can be divided into classes so that the following conditions hold: 1) $P_{ik}(r)$ can be non-zero only when E_i and E_k belong to the same class; 2) for any pair E_i and E_k from the same class, the ratio $Q_i : Q_k$ is uniquely determined by the transition probabilities $P_{ik}(r)$. This implies that if E_i and E_k belong to the same class, then (9) is true regardless of the absolute probabilities Q_j; if, however, E_i and E_k are taken from different classes, then (9) trivially follows from the equations $P_{ik}(r) = P_{ki}(r) = 0$. We also see that (9) does not depend on the system Q_k.

A necessary and sufficient condition, independent of Q_k, for (8) to hold may be formulated as follows:

Relation (8) *holds if and only if for arbitrary* $r, q, k_1, k_2, \ldots, k_q$,

$$P_{k_1k_2}(r)P_{k_2k_3}(r)\ldots P_{k_{q-1}k_q}P_{k_qk_1} = P_{k_1k_q}(r)P_{k_qk_{q-1}}\ldots P_{k_3k_2}P_{k_2k_1}. \tag{10}$$

In the discrete case (when t runs only through the integers) it suffices to require (10) only for $r = 1$. The proof is elementary, and we leave it to the reader.

In particular, if the transition probabilities are symmetric:

$$P_{ik}(r) = P_{ki}(r), \tag{11}$$

then (10) holds. Hence, the symmetry condition (11) is sufficient for (8).

5. Conclusion

These almost trivial facts find many physical applications. Here we confine ourselves to an example which is different from Schrödinger's original example. Suppose that a circle is divided into a very large number M of identical intervals. A large number L of mobile particles move along this circle so that

[2] R. von Mises, *Wahrscheinlichkeitsrechnung*, Berlin, 1931, §16.

each particle, independently of others, passes at every step into the neighbouring interval, either to the right or the left, each possibility with probability $\frac{1}{2}$. There are $M^L = N$ various possible patterns of L particles on M intervals. The absolute probabilities Q_k corresponding to these N possibilities are all equal to each other and to $1/N$. The transition probabilities $P_{ik}(r)$ are clearly symmetric, and therefore, the invertibility equation (8) holds. If we consider the "macroscopic" distribution of the particles along the circle, then with probability very close to one, it will be uniform. If it is known (though *a priori* this is highly unlikely) that at a certain time t_0 a considerable deviation from this uniform distribution takes place, then with probability very close to 1 we can assert that this non-uniformity for $t > t_0$ will die out approximately in accordance with the diffusion differential equation. Formula (8) now implies the same probability of uniformity for $t < t_0$ with the same differential equation, but with the opposite sign of the time variable.

Mathematical Institute of Moscow State University
20 May 1935

22. ON THE STATISTICAL THEORY OF METAL CRYSTALLIZATION *

This paper gives a rigorous solution to the problem of the rate of a crystallization process under certain schematic, but still sufficiently general assumptions.

The study of the process of crystal growth after random formation of crystallization centres is of significant importance for metallurgy. In this connection it is fairly difficult to take account of collisions between the crystal grains appearing around separate crystallization centres. These collisions disturb the grain form by preventing the growth of crystals in certain directions. In papers by F. Göler and G. Sachs [1], G. Tammann [2], B.V. Stark, I.L. Mirkin and A.N. Romanovskii [3], and others only rough approximation formulas for the growth of crystal matter are given. In this paper I give, under rather wide assumptions, an exact formula for the probability $p(t)$ that a randomly taken point P of the volume filled with a crystallized substance is inside the crystal body within the crystallization period. With a sufficient approximation rate it can be considered that the amount of substance crystallized over time t is also $p(t)$. In conclusion I determine the number of crystallization centres formed throughout the entire crystallization process.

I extend my gratitude to I.L. Mirkin who interested me in this problem and kindly provided me with all the necessary material.

§1. Mathematical setting of the problem

Let V be a certain volume. Initially ($t = 0$) it is occupied by "the mother phase". After time t a certain part $V_1(t)$ of the volume V is occupied by crystallized substance. The volume $V_1(t)$ grows with time t as follows.

1. In the free part $V - V_1$ new crystallization centres appear. For any volume $V' < V - V_1$ the probability of forming *one* crystallization centre during the time between t and $t + \Delta t$ is

$$\alpha(t)V'\Delta t + o(\Delta t),$$

and that of *more than one* centre is $o(\Delta t)$, where $o(\Delta t)$ is an infinitesimal with respect to Δt. These probabilities do not depend on the distribution of the crystallization centres formed prior to time t provided that it can be guaranteed (see later) that at the time t there is no crystal bulk in V'.

* *Izv. Akad. Nauk SSSR Ser. Mat.* **3** (1937), 355–360. Presented by S.N. Bernshtein.

2. This crystal bulk grows around newly-formed crystallization centres and the whole crystal at a linear rate

$$c(t, n) = k(t)c(n),$$

depending on time t and direction n. We assume that the ends of the vectors of length $c(n)$ measured in the direction n from the origin form a convex surface.

Under these conditions an essential restriction is that although $c(t, n)$ may depend on n, this dependence should be the same at all points. In other words, the formulas given below hold either under the simplifying assumption of uniform growth in all directions, or for crystals of arbitrary shape similarly oriented in space.

§2. Determination of the probability $p(t)$

The quantity c is defined by the equality

$$c^3 = \frac{1}{4\pi} \int_S c^3(n)d\sigma,$$

where we integrate over the surface of the unit sphere S with centre at the origin. Clearly, at $t > t_0$ the volume of the crystal growing freely round a centre formed at the moment t_0 is given by

$$\frac{4\pi}{3} c^3 \Big(\int_{t_0}^t k(\tau)d\tau \Big)^3.$$

Now consider an arbitrary point P of V, lying at a distance greater than

$$\max c(n) \int_0^t k(\tau)d\tau$$

from the boundary of V.

In order for P to get into the crystallized bulk by the time t it is necessary and sufficient that a crystallization centre is formed at a certain time $t' < t$ at some point P' whose distance from P is less than

$$c(\bar{n}) \int_{t'}^t k(\tau)d\tau,$$

where $\bar{n}$ is the direction $\overline{P'P}$. For a fixed t' the volume occupied by the points P' satisfying our conditions is

$$V'(t') = \frac{4\pi}{3} c^3 \Big(\int_{t'}^t k(\tau)d\tau \Big)^3.$$

The probability of forming a crystallization centre during time $\Delta t'$ in $V'(t')$ is

$$\alpha(t')V'(t')\Delta t' + o(\Delta t'),$$

and the probability that this does not happen is

$$q(t) = \prod_{i=1}^{s}\{1 - \alpha(t_i)V'(t_i)\Delta t'\} + o(1).$$

Therefore the probability that the point P *does not* belong to the crystallized bulk at the moment t is

$$q(t) = \prod_{i=1}^{s}\{1 - \alpha(t_i)V'(t_i)\Delta t'\} + o(1), \tag{1}$$

where $t = s\Delta t'$, $t_i = i\Delta t'$, and $o(1)$ is infinitesimal if $\Delta t'$ is infinitesimal.

Taking the logarithm of (1) we obtain

$$\log q(t) = \sum_{i=1}^{s}\alpha(t_i)V'(t_i)\Delta t' + o(1) = -\int_0^t \alpha(t')V'(t')dt' =$$

$$= -\frac{4\pi}{3}c^3\int_0^t \alpha(t')\Big(\int_{t'}^t k(\tau)d\tau\Big)^3 dt'. \tag{2}$$

For the desired probability

$$p(t) = 1 - q(t)$$

for P to belong to the crystallized bulk we finally obtain

$$p(t) = 1 - \exp\{-\frac{4\pi}{3}c^3\Omega\}, \tag{3}$$

where

$$\Omega = \int_0^t \alpha(t')\Big(\int_{t'}^t k(\tau)d\tau\Big)^3 dt'. \tag{4}$$

§3. Conclusions

When V is large enough as compared with the sizes of the individual grains, we may set $V_1(t) = Vp(t)$ or, in view of (3),

$$V_1(t) = V\Big(1 - \exp\Big\{-\frac{4\pi}{3}c^3\Omega\Big\}\Big), \tag{5}$$

where Ω is defined by (4). The formula (5) for the volume$V_1(t)$ of the substance crystallized over the time t gives a solution of the first problem posed in the introduction. If $\alpha(t)$ and $c(t,n)$ do not depend on time, then we can set

$$\alpha(t) = \alpha, \quad k(t) = 1.$$

In this case

$$\Omega = \alpha t^4/4 \tag{4a}$$

and (5) gives

$$V_1(t) = V\Big(1 - \exp\Big\{-\frac{\pi}{3}c^3\alpha t^4\Big\}\Big). \tag{5a}$$

For a sufficiently large volume V the following formula holds for the number $N(t)$ of crystallization centres formed over the period t:

$$N(t) = V\int_0^t \alpha(\tau)q(\tau)d\tau. \tag{6}$$

For a constant $\alpha(t) = \alpha$ and $k = 1$ we obtain from (6)

$$N(t) = V\alpha\int_0^t \exp\Big\{-\frac{\pi}{3}c^3\alpha\tau^4\Big\}d\tau, \tag{7}$$

or

$$N(t) = V\sqrt[4]{\frac{3\alpha^3}{\pi c^3}}\int_0^x e^{-\xi^4}d\xi, \tag{6a}$$

where

$$x = \sqrt[4]{\pi c^3\alpha/3}t.$$

For $t = +\infty$ (6) and (6a) give the total number of crystallization centres throughout the whole process. In particular, for constant $\alpha(t) = \alpha$ and $k = 1$ we obtain

$$N(+\infty) = V\sqrt[4]{\frac{3\alpha^3}{\pi c^3}}\int_0^\infty e^{-\xi^4}d\xi \sim 0.9V\Big(\frac{\alpha}{c}\Big)^{3/4}. \tag{7a}$$

Note also the special case when all the crystallization centres are formed at the very beginning, with an average of β centres per unit volume. The corresponding formulas are derived from the general formulas by passage to the limit. Instead of (4) we obtain

$$\Omega = \beta\Big(\int_0^t k(\tau)d\tau\Big)^3, \tag{4b}$$

formula (5) still holds and (6) is replaced for any $t > 0$ by the trivial identity $N = V\beta$. If we assume, moreover, that $k = 1$ (that is, $c(t, n)$ is independent of t), then we obtain

$$\Omega = \beta t^3, \tag{4c}$$

$$V_1(t) = V\left(1 - \exp\left\{-\frac{4\pi}{3}c^3\beta t^3\right\}\right). \tag{5c}$$

Mathematical Institute of Moscow State University

20 April 1937

References

1. F. Göler and G. Sachs, *Z. Phys.* **77** (1932), 281.
2. G. Tammann, *Z. Andrg. Chemie* **214** (1933), 407.
3. B.V. Stark, I.L. Mirkin and A.N. Romanovskii, 'Metals and thermal treatment', *Mosk. Inst. Stal.* **7** (1935), 5–38 (in Russian).

23. MARKOV CHAINS WITH A COUNTABLE NUMBER OF POSSIBLE STATES *

In [1] I made some general assertions concerning the asymptotic behaviour of transition from one state to another in an unbounded number of steps for Markov chains with a countable set of possible states.

This paper gives a broader exposition of the same questions, complete with proofs of the main theorems announced earlier without detailed proofs, and some new facts. The way of dealing with these questions used in this paper may be of some interest also for Markov chains with a finite number of possible states. This is indicated in the note by W. Doeblin [2] which has since appeared. As I know from his letter, W. Doeblin independently proved some of the theorems on Markov chains with a countable number of states that I published.

§1. Notation

Denote by E_i the various possible states of the system under study, where i runs through all positive integers. We note that the following account is also applicable for the case when i takes a finite number of values; in this case a number of the statements can be simplified. The probabilities p_{ij} of transition from E_i to E_j in one step are, as usual, assumed to be subject to the conditions

$$p_{ij} \geq 0 \tag{1}$$

$$\sum_j p_{ij} = 1. \tag{2}$$

The probabilities $P_{ij}^{(n)}$ of transition from E_i to E_j in n steps are determined inductively from the equalities

$$P_{ij}^{0} = \delta_{ij}, \tag{3}$$

$$P_{ij}^{(n+1)} = \sum_k P_{ik}^{(n)} p_{kj}, \tag{4}$$

where

$$\delta_{ij} = \begin{cases} 1 & \text{for } i = j \\ 0 & \text{for } i \neq j \end{cases}$$

* *Bul. MGU Mat. Mekh.* 1:3 (1937), 1–16 (in Russian).

Further, let $K_{ij}^{(n)}$ be the probability of passing, in n steps, from E_i to E_j without visiting E_j in a fewer number of steps. Clearly,

$$K_{ij}^{(n)} = P_{ij}^{(n)} - K_{ij}^{(1)} P_{jj}^{(n-1)} - K_{ij}^{(2)} P_{jj}^{(n-2)} - \ldots - K_{ij}^{(n-1)} P_{jj}^{(1)}. \tag{5}$$

In particular, $K_{ii}^{(n)}$ is the probability of returning to E_i for the first time in n steps. We now set

$$\sum_{n=1}^{\infty} K_{ij}^{(n)} = L_{ij}. \tag{6}$$

Clearly, $L_{ij} \leq 1$. If $L_{ij} = 1$ then, starting from E_i the system will inevitably, sooner or later, visit the state E_j. The expectation of the number of steps needed in this case for the transition from E_i to E_j is

$$M_{ij} = \sum_{n=1}^{\infty} n K_{ij}^{(n)}. \tag{7}$$

In particular, M_{ii} is the expectation of the number of steps prior to the first return to E_i, provided we started from E_i. The expectation M_{ij} can be both finite and infinite.

§2. Inessential states, classes and subclasses of essential states

A state E_i is called *inessential* if there exist a j and an n such that $P_{ij}^{(n)} > 0$ and $P_{ji}^{(m)} = 0$ for all m, that is, there is a transition from E_i to E_j without return to E_i. All other states are called *essential.* Clearly, if two states E_i and E_j are essential and if there exists an n with $P_{ij}^{(n)} > 0$ then there also exists an m with $P_{ji}^{(m)} > 0$. If there are such n and m, then the essential states E_i and E_j are called *communicating.* If E_i communicates with E_j and E_j communicates with E_k, then E_i also communicates with E_k. Therefore all essential states fall into *classes* $S^{(\alpha)}$, such that states belonging to one class communicate and those belonging to different classes do not communicate. It is also clear that for an essential state E_i and an inessential state E_j, $P_{ij}^{(n)}$ is always 0. Thus, once having fallen into a state of $S^{(\alpha)}$, our system can never leave this class of states.

Consider now an essential state E_i. Let $\mathfrak{M}_i$ be the set of indices n for which $P_{ii}^{(n)} > 0$. Since E_i is essential, the set $\mathfrak{M}_i$ is non-empty. If n and m occur in $\mathfrak{M}_i$, then so does $n+m$. Let d_i be the greatest common divisor of all numbers in $\mathfrak{M}_i$. The set $\mathfrak{M}_i$ consists only of multiples of d_i. It can easily be shown that all sufficiently large multiples of d_i occur in $\mathfrak{M}_i$. The number d_i is called the *period* of the state E_i.

It can easily be shown that *all states belonging to the same class* $S^{(\alpha)}$ *have the same period* which we denote by $d(\alpha)$ and call the *period of* $S^{(\alpha)}$. Indeed, for two states E_i and E_j of the same class $S^{(\alpha)}$ let there exist n and m such that $P_{ij}^{(n)} > 0$ and $P_{ji}^{(m)} > 0$ (such n and m do exist, as indicated above). Then $P_{jj}^{(kd_j)} > 0$ for sufficiently large k. Hence, for sufficiently large k,

$$P_{ii}^{(kd_j+n+m)} \geq P_{ij}^{(n)} P_{jj}^{(kd_j)} P_{ji}^{(m)} > 0,$$

that is, all sufficiently large numbers of the form $kd_j + n + m$ occur in $\mathfrak{M}_i$, which is only possible if d_j is divisible by d_i. Conversely, d_i is also divisible by d_j and therefore $d_i = d_j$.

For two states E_i and E_j belonging to the same class $S^{(\alpha)}$, we simultaneously have $P_{ij}^{(n)} > 0$ and $P_{ij}^{(m)} > 0$ only if $n \equiv m (\operatorname{mod} d(\alpha))$. Therefore, having chosen a certain state E_{i_0} of class $S^{(\alpha)}$, we obtain for any state E_i of the same class a well-defined number $\beta(E_i) = 1, 2, \ldots, d(\alpha)$ such that $P_{i_0 i}^{(n)} > 0$ is possible only for $n \equiv \beta(E_i) (\operatorname{mod} d(\alpha))$. All the states E_j with given $\beta(E_j)$ belong to the subclass $S_\beta^{(\alpha)}$. Thus, $S^{(\alpha)}$ is divided into $d(\alpha)$ subclasses $S_\beta^{(\alpha)}$. With every step our system inevitably goes from the states of $S_\beta^{(\alpha)}$ into one of the states of $S_{\beta+1}^{(\alpha)}$ and in the case $\beta = d(\alpha)$, into one of the states of $S_1^{(\alpha)}$. Thus if E_i and E_j belong respectively to subclasses $S_\beta^{(\alpha)}$ and $S_\gamma^{(\alpha)}$, then $P_{ij}^{(n)} \neq 0$ only if $n \equiv \gamma - \beta (\operatorname{mod} d(\alpha))$. On the other hand, for sufficiently large n satisfying the latter congruence we actually have $P_{ij}^{(n)} > 0$.

§3. Recurrent and non-recurrent classes

Apart from the probability L_{ij} that, having started from the state E_i, we visit at least once, sooner or later, the state E_j, we introduce the probability Ω_{ij} that, starting from the state E_i we visit the state E_j an infinite number of times. Clearly,

$$\Omega_{ij} \leq L_{ij}. \tag{8}$$

On the other hand, we now prove the following lemma:

Lemma 1. $L_{ii} = 1$ *implies* $\Omega_{ii} = 1$.

Proof. Denote by $T_{(k)}$ the probability that, starting from a state E_i, we return to it no fewer than k times. Clearly, always

$$T_{(1)} = L_{ii}, \quad T_{(k+1)} = T_{(k)}L_{ii}, \quad \Omega_{ii} = \lim_{k\to\infty} T_{(k)}.$$

For $L_{ii} = 1$ these formulas imply that $\Omega_{ii} = 1$.

In this and the following sections we shall be concerned only with relations that hold within each class of essential elements; in other words, it will be assumed that all indices referring to states run only through the values corresponding to the states of one class. In the statements of the theorems this will be indicated by the words "within one class".

Theorem 1a. *Within one class either* $\Omega_{ij} < 1$ *or* $\Omega_{ij} = 1$.

Proof. Clearly it suffices to establish that for any i, j, k,

1) $\Omega_{ij} = 1$ implies $\Omega_{ik} = 1$;

2) $\Omega_{ji} = 1$ implies $\Omega_{ki} = 1$.

If these two points have been established, then for any i, j, i', j', $\Omega_{ij} = 1$ implies $\Omega_{ij'} = 1$ and $\Omega_{i'j'} = 1$.

Now let us prove 1). Assume that $\Omega_{ij} = 1$, that is, starting form E_i we return to E_j an infinite number of times with probability 1. Consider the interval between the sth and the $(s+1)$th revisits. Since all states under consideration belong to one class, the probability of the event $\mathfrak{Y}_s$ that in the interval between the sth and the $(s+1)$th visit to E_j the system will get to a fixed state E_k is positive. It can easily be seen that this probability does not depend on s, and that the events $\mathfrak{Y}_s$ for different s are independent. However, given an infinite sequence of possible independent events $\mathfrak{Y}_s$ with the same positive probability, an infinite number of these events will indeed take place with probability **1**, that is, $\Omega_{ik} = 1$ as required.

It now remains to prove 2). For this we use the formula

$$\Omega_{ji} = \sum_k P_{jk}^{(n)} \Omega_{ki}, \tag{9}$$

which holds for any i, j, k, n. Since $\sum_k P_{jk}^{(n)} = 1$, we have: $\Omega_{ji} = 1$ implies $\Omega_{ki} = 1$ for all k for which $P_{jk}^{(n)} > 0$. But for any j and k (within one class) such an n does indeed exist. This proves 2) and Theorem 1a.

If all the $\Omega_{ij} = 1$, then by (8) all the $L_{ij} = 1$. If, however, all the $\Omega_{ij} < 1$, then by Lemma 1, all the $L_{ii} < 1$. This implies:

Theorem 1b. *Within one class either all the $L_{ii} < 1$ or all the $L_{ij} = 1$.*

It should be noted that when all the $L_{ii} < 1$, there might still be some $L_{ij} = 1$ $(i \neq j)$.

If all the $L_{ij} = 1$ and all the $\Omega_{ij} = 1$, then the class is called *recurrent*. If, conversely, all the $L_{ii} < 1$ and all the $\Omega_{ij} < 1$, then the class is called *non-recurrent*. It is easy to see that if a state E_j is of non-recurrent class and E_i is arbitrary, then

$$\lim_{n \to +\infty} P_{ij}^{(n)} = 0.$$

§4. Positive and zero classes

The considerations of this section refer to phenomena observed within each of the recurrent classes. As a result of these considerations the recurrent classes will be divided into *positive* and *zero* classes. Note, however, that in the *following sections all non-recurrent classes will also be referred to as zero classes.*

The expectations M_{ij} (see (7)) are of primary importance in the study of recurrent classes. In this connection we will consider the means

$$\pi_{ij}^{(n)} = \frac{1}{n}(P_{ij}^{(1)} + P_{ij}^{(2)} + \ldots + P_{ij}^{(n)}). \tag{10}$$

Lemma 2a. *For any E_i from a recurrent class with finite M_{ii}*

$$\lim_{n \to \infty} \pi_{ii}^{(n)} = 1/M_{ii}\,;$$

while if $M_{ii} = +\infty$, then

$$\lim_{n \to \infty} \pi_{ii}^{(n)} = 0.$$

Proof. Starting from E_i we return to this state an infinite number of times with probability 1. Let the first return to E_i take place at the n_1th step, the second return at the n_2th step and the kth return at the n_kth step. The differences

$$x_1 = n_1,\ x_2 = n_2 - n_1,\ \ldots,\ x_k = n_k - n_{k-1},\ \ldots$$

form a sequence of random variables independent of each other with the same distribution law: $x_k = s$ with probability $K_{ii}^{(s)}$. Clearly, M_{ii} is none other than the expectation of each of the x_k.

Assume first that the expectation M_{ii} of the random variables x_k is finite. Then, according to Khinchin's theorem [3], the sequence $\{x_k\}$ satisfies the law of large numbers, that is, for any $\epsilon > 0$ there exists a k_0 such that for $k \geq k_0$ the probability of the inequality

$$\left|\frac{1}{k}\sum_{j=1}^{k} x_j - M_{ii}\right| = \left|\frac{n_k}{k} - M_{ii}\right| \geq \frac{\epsilon}{2}$$

is less than ϵ. Let $\epsilon < \frac{1}{2}$ and $n \geq n_0 = 2k_0 M_{ii}$. Set

$$k' = \frac{n}{M_{ii}}(1-\epsilon), \quad k'' = \frac{n}{M_{ii}}(1+\epsilon).$$

It can easily be seen that $k' \geq k_0$, $k'' \geq k_0$. Therefore, with probability greater than $1-\epsilon$ we can assert that

$$|n_{k'}/k' - M_{ii}| < \epsilon,$$

which implies (since $M_{ii} \geq 1$)

$$|n_{k'} - n(1-\epsilon)| < k'\epsilon \leq n\epsilon,$$

hence,

$$n_{k'} < n.$$

Similarly, with probability greater than $1-\epsilon$ we have

$$|n_{k''}/k'' - M_{ii}| < \epsilon/2, \quad |n_{k''} - n(1+\epsilon)| < k''\epsilon/2 \leq n\epsilon, \quad n_{k''} > n.$$

Thus if $n \geq n_0$, then with probability greater than $1-2\epsilon$ we have the inequality

$$n_{k'} < n < n_{k''},$$

that is, the number ψ_n of returns to the state E_i within the first n steps is between k' and k''. It can easily be seen that the expectation of the *frequency* ψ_n/n of returns to E_i within the first n steps is $\pi_{ii}^{(n)}$. Since for $n \geq n_0$ we have $k'/n < \psi_n/n < k''/n$ with probability greater than $1-2\epsilon$:

$$\frac{k'}{n} = \frac{1}{M_{ii}}(1-\epsilon) < \frac{\psi_n}{n} < \frac{1}{M_{ii}}(1+\epsilon) = \frac{k''}{n},$$

and since always $0 \le \psi_n/n \le 1$ and $M_{ii} \ge 1$, we finally obtain for $n \ge n_0$,

$$|\pi_{ii}^{(n)} - \frac{1}{M_{ii}}| \le \mathbf{E}|\frac{\psi_n}{n} - \frac{1}{M_{ii}}| \le \frac{\epsilon}{M_{ii}} + 2\epsilon,$$

which implies that

$$\lim_{n\to+\infty} \pi_{ii}^{(n)} = 1/M_{ii}.$$

Now consider the case $M_{ii} = +\infty$. Then, whatever $M < +\infty$ and $\epsilon > 0$, there exists a k_0 such that for $k \ge k_0$ the probability that the inequality

$$\frac{1}{k}\sum_{j=1}^{k} x_j = \frac{n_k}{k} \le M$$

holds is less than ϵ. To prove this it suffices to introduce a new sequence of random variables x'_k, independent of each other, smaller than the corresponding variables x_k ($x'_k \le x_k$ for every k) with equal expectations, for example, $2M$, and apply the law of large numbers in the above statement to this new sequence.

Setting $k = [n/M] + 1 > n/M$ we see that $k \ge k_0$ for $n \ge k_0 M$ and hence, with probability greater than $1 - \epsilon$,

$$\frac{n_k}{k} > M, \quad n_k > kM \ge n, \quad \psi_n \le k \le \frac{n}{M} + 1, \quad \frac{\psi_n}{n} \le \frac{1}{M} + \frac{1}{n}.$$

Hence, for $n \ge k_0 M$

$$\pi_{ii}^{(n)} = \mathbf{E}\Big(\frac{\psi_n}{M}\Big) \le \frac{1}{M} + \frac{1}{n} + \epsilon,$$

which brings us to the conclusion that in this case

$$\lim_{n\to+\infty} \pi_{ii}^{(n)} = 0.$$

Theorem 2. *Within one class either all the M_{ii} are infinite or all finite.*

Proof. For any two states E_i and E_j of the same class there exist k and m such that $P_{ij}^{(k)} > 0$, $P_{ji}^{(m)} > 0$. Furthermore it is clear that for any n,

$$P_{jj}^{(n+k+m)} \ge P_{ji}^{(m)} P_{ii}^{(n)} P_{ij}^{(k)}.$$

This inequality directly implies that

$$\lim_{n\to\infty} \pi_{jj}^{(n)} \ge P_{ji}^{(m)} P_{ij}^{(k)} \lim_{n\to\infty} \pi_{ii}^{(n)}.$$

Hence, the limits of $\pi_{ii}^{(n)}$ for all i (corresponding to a given class) are either zero or positive. Therefore, according to Lemma 2a, the M_{ii} are also either all infinite or all finite.

It only remains to prove that the finiteness of all the M_{ii} implies the finiteness of all the M_{ij}. We denote by $R_{ij}^{(n)}$ the probability that, starting from E_i we visit E_j $(j \neq i)$ in n steps without visiting E_i meanwhile. Then[1]

$$M_{ii} = \sum_{m=1}^{n} mK_{ii}^{(m)} + \sum_{j \neq i} R_{ij}^{(n)}(M_{ji} + n).$$

But within one class for any i and j we can find an n for which $R_{ij}^{(n)} > 0$. Hence $M_{ji} = +\infty$ would imply $M_{ii} = +\infty$, which proves our theorem.

The classes with all the M_{ij} finite are called *positive*, and with all the $M_{ii} = +\infty$ are called *zero*. Note that in zero classes some of the M_{ij} $(i \neq j)$ may be finite.

Theorem 3. *In a zero class $P_{ij}^{(n)} \to 0$, as $n \to +\infty$, for any E_i, E_j in the given class.*

To prove Theorem 3 we need the following:

Lemma 2b. *In a zero class, for any E_j the quantity*

$$\pi_{jj}^{(n,m)} = \frac{1}{m}(P_{jj}^{(n+1)} + P_{jj}^{(n+2)} + \ldots + P_{jj}^{(n+m)}) \tag{11}$$

tends to 0 uniformly with respect to n as $m \to +\infty$.

Proof. Denote by $H^{(s)}$ the probability that, with initial state E_j, starting from the $(n+1)$th step, the state E_j appears first at the $(n+s)$th step regardless

---

[1] *Proof.* For $r > n$,

$$K_{ii}^{(r)} = \sum_{j \neq i} R_{ij}^{(n)} K_{ji}^{(r-n)}.$$

Therefore

$$M_{ii} = \sum_{m=1}^{\infty} mK_{ii}^{(m)} = \sum_{m=1}^{n} mK_{ii}^{(m)} + \sum_{r>n} r \sum_{j \neq i} R_{ij}^{(n)} K_{ji}^{(r-n)} =$$

$$= \sum_{m=1}^{n} mK_{ii}^{(m)} + \sum_{j \neq i} R_{ij}^{(n)} \sum_{s=1}^{\infty} (n+s) K_{ji}^{(s)} = \sum_{m=1}^{n} mK_{ii}^{(m)} + \sum_{j \neq i} R_{ij}^{(n)}(M_{ji} + n).$$

of whether or not there was a return to E_j within the first n steps. Then

$$P_{jj}^{(n+k)} = \sum_{s=1}^{k} H^{(s)} P_{jj}^{(k-s)},$$

$$\pi_{jj}^{(n,m)} = \frac{1}{m}\sum_{k=1}^{m} P_{jj}^{(n+k)} = \frac{1}{m}\sum_{s=1}^{m} H^{(s)} \sum_{k=0}^{m-s} P_{jj}^{(k)} = \tag{12}$$

$$= \sum_{s=1}^{m} H^{(s)} \frac{m-s}{m}\pi_{jj}^{(m-s)} + \frac{1}{m}\sum_{s=1}^{m} H^{(s)}.$$

Choose an r_0 such that for $r \geq r_0$ we always have $\pi_{jj}^{(r)} < \epsilon$. (This is possible by Lemma 2a.) Choose $m_0 > r_0/\epsilon$. Also let $m \geq m_0$. Then for $m - s \leq r_0$ we have $(m-s)/m < \epsilon$, while for $m - s \geq r_0$ we have $\pi_{jj}^{(m-s)} < \epsilon$.

Since always $(m-s)/m \leq 1$, $\pi_{jj}^{(m-s)} \leq 1$, it follows that for all $m \geq m_0$, we have

$$\frac{m-s}{m}\pi_{jj}^{(m-s)} < \epsilon.$$

Since

$$\sum_{s=1}^{m} H^{(s)} \leq 1,$$

we see from (12) that $\pi_{jj}^{(n,m)} < \epsilon + 1/m$ as soon as $m \geq m_0$. Since $\epsilon > 0$ is arbitrary and does not depend on n, our lemma is proved.

Proof of Theorem 3. First note that to prove this theorem it suffices to prove it for $i = j$, that is, to prove that the probabilities $P_{jj}^{(n)}$ tend to 0 as $n \to +\infty$. To see this, we choose an m such that $P_{ji}^{(m)} > 0$. Then, clearly,

$$P_{jj}^{(n+m)} \geq P_{ji}^{(m)} P_{ij}^{(n)}.$$

Let n tend to $+\infty$ for m constant; since $P_{jj}^{(n+m)} \to 0$, it follows that $P_{ij}^{(n)} \to 0$ as well.

Now assume that

$$\limsup_{n\to+\infty} P_{jj}^{(n)} = \lambda > 0.$$

If this assumption leads to a contradiction, Theorem 3 has been proved.

Choose an integer a such that

$$K_{jj}^{(a)} = A > 0.$$

For any $\epsilon > 0$ there exists n_0 such that $n \geq n_0$ implies $P_{jj}^{(n)} \leq \lambda + \epsilon$. For some $\delta > 0$ choose $m_0 \geq a$ such that

$$\sum_{m>m_0} K_{jj}^{(m)} < \delta.$$

Then the assumptions

$$n \geq n_0 + m_0, \quad P_{jj}^{(n)} \geq \lambda - \eta \quad (\eta > 0)$$

imply that

$$P_{jj}^{(n-a)} \geq \lambda - (\eta + \epsilon + \delta)/A.$$

In fact,

$$\lambda - \mu \leq P_{jj}^{(n)} = K_{jj}^{(n)} P_{jj}^{(0)} + K_{jj}^{(n-1)} P_{jj}^{(1)} + \ldots + K_{jj}^{(1)} P_{jj}^{(n-1)} =$$

$$= K_{jj}^{(a)} P_{jj}^{(n-a)} + \sum_{a \neq m < m_0} K_{jj}^{(m)} P_{jj}^{(n-m)} + \sum_{m_0 \leq m \leq n} K_{jj}^{(m)} P_{jj}^{(n-m)} \leq$$

$$\leq A P_{jj}^{(n-a)} + (1 - A)(\lambda + \epsilon) + \delta.$$

But for $n \geq n_0 + m_0 + sa$ the inequality

$$P_{jj}^{(n)} \geq \lambda - \eta$$

implies that

$$P_{jj}^{(n-a)} \geq \lambda - (\eta + \epsilon + \delta)/A = \lambda - \eta_1,$$

$$P_{jj}^{(n-2a)} \geq \lambda - (\eta_1 + \epsilon + \delta)/A = \lambda - \eta_2,$$

$$\ldots\ldots\ldots\ldots\ldots\ldots\ldots\ldots\ldots$$

$$P_{jj}^{(n-sa)} \geq \lambda - (\eta_{s-1} + \epsilon + \delta)/A = \lambda - \eta_s,$$

where

$$\eta < \eta_1 < \eta_2 < \ldots < \eta_s.$$

Now, for any s we can choose $\eta > 0$, $\epsilon > 0$, $\delta > 0$ so that $\eta_s < \lambda/2$. By choosing n_0 and m_0 suitably we obtain for all $n \geq n_0 + m_0 + sa$ such that $P_{jj}^{(n)} > \lambda - \eta$ the inequalities

$$P_{jj}^{(n)} > \lambda/2, \quad P_{jj}^{(n-a)} > \lambda/2, \ldots, P_{jj}^{(n-sa)} > \lambda/2,$$

$$\pi_{jj}^{(n-sa,sa)} > (1/sa) \cdot s \cdot \lambda/2 = \lambda/2a.$$

Since $\lambda/2a$ is constant, s is arbitrarily large and n can be assumed arbitrarily large for fixed s, we arrive at a contradiction to Lemma 2b.

§5. Asymptotic relations within a positive class

Theorem 3 completes in a first approximation the study of zero classes. As for positive classes, we have the following general theorem.

Theorem 4a. *In a positive class $S^{(\alpha)}$, for any E_i from a subclass $S_\beta^{(\alpha)}$ and E_j from a subclass $S_\gamma^{(\alpha)}$, the probability $P_{ij}^{(n)}$ tends to a limit*

$$P_j = d(\alpha)/M_{jj},$$

which is independent of i, when $n \to \infty$ runs through the values $n \equiv \gamma - \beta \,(\operatorname{mod} d(\alpha))$.

Remark. If $n \not\equiv \gamma - \beta \,(\operatorname{mod} d(\alpha))$, then, as has already been noted, $P_{ij}^{(n)} = 0$.

To prove Theorem 4 we need several lemmas.

Lemma 3. *In a positive class for any i, j and $\epsilon > 0$ there exists m such that for any n the probability that, starting from E_i, we visit E_j at least once between the nth and the $(n+m)$th step is greater than $1 - \epsilon$.*

Proof. The probability that, starting from E_i, we do not visit E_j within the time interval indicated is given by

$$P = \sum_{p=n+m}^{\infty} K_{ij}^{(p)} + \sum_{k=1}^{n-1} P_{ij}^{(k)} \sum_{p=n+m-k}^{\infty} K_{jj}^{(p)} =$$

$$= \sum_{p=n+m}^{\infty} K_{ij}^{(p)} + \sum_{p=m+1}^{n+m-1} K_{jj}^{(p)} \sum_{k=n+m-p}^{n-1} P_{ij}^{(k)} + \sum_{p=n+m}^{\infty} K_{jj}^{(p)} \sum_{k=1}^{n-1} P_{ij}^{(k)} \leq$$

$$\leq \sum_{p=m}^{\infty} K_{ij}^{(p)} + \sum_{p=m+1}^{\infty} pK_{jj}^{(p)} = U^{(m)}.$$

But if

$$M_{jj} = \sum_{p=1}^{\infty} pK_{jj}^{(p)}$$

is finite, then $U^{(m)} \to 0$ as $m \to +\infty$. Since $U^{(m)}$ does not depend on n, our Lemma is proved.

Lemma 4a. *In a positive class consisting of one subclass,*

$$\liminf_{n\to+\infty} P_{ij}^{(n)} > 0.$$

Proof. If the class consists of one subclass, then there exists k_0 such that always $P_{jj}^{(k)} > 0$ for $k \geq k_0$. In accordance with Lemma 3, we choose m such that for given i and j the number ϵ from Lemma 3 can be taken equal to $\frac{1}{2}$. We set

$$\lambda = \inf\{P_{jj}^{(k_0)}, P_{jj}^{(k_0+1)}, \ldots, P_{jj}^{(k_0+m)}\}.$$

Clearly, $\lambda > 0$. Now let $n' > m + k_0$. Set $n' = n + m + k_0$. The probability that, starting from E_i, we visit E_j between the nth and the $(n+m)$th step is greater than $\frac{1}{2}$.

Suppose that the first arrival into E_j between the nth and the $(n+m)$th step takes place after the $(n+s)$th step $(s < m)$. Then the conditional probability of revisting E_j in $n' = n + m + k_0$ steps is $P_{jj}^{(k_0+m-s)} \geq \lambda$. This inequality holds for any s $(1 \leq s \leq m)$. Therefore, the total probability $P_{ij}^{(n')}$ that, starting from E_i, we visit E_j in $n' = n + m + k_0$ steps, satisfies the inequality

$$P_{ij}^{(n')} > \tfrac{1}{2}\lambda,$$

which, since n is arbitrary, proves our Lemma.

Lemma 4b. *In a positive class consisting of one subclass*

$$\lim_{n\to\infty} P_{ii}^{(n)} = 1/M_{ii}.$$

Proof. First we prove the existence of the limit of $P_{ii}^{(n)}$ as $n \to +\infty$. Let

$$\liminf_{n\to+\infty} P_{ii}^{(n)} = a \quad \limsup_{n\to+\infty} P_{ii}^{(n)} = b.$$

By Lemma 4a we have $b \geq a > 0$. Let $\epsilon > 0$. Choose m satisfying the conditions of Lemma 3 for $j = i$. Then choose k_0 such that for $k \geq k_0$ the inequalities

$$a - \epsilon < P_{ii}^{(k)} < b + \epsilon$$

hold. Let $n > m + k_0$ be such that $P_{ii}^{(n)} < a + \epsilon$, and $n' > n + k_0$ such that $P_{ii}^{(n')} > b - \epsilon$ (such n and n' always exist). Put $n' - n = k$. Then

$$P_{ii}^{(n')} = P_{ii}^{(k)} P_{ii}^{(n)} + A^{(1)} P_{ii}^{(n-1)} + A^{(2)} P_{ii}^{(n-2)} + \ldots + A^{(n)} P_{ii}^{(0)},$$

where $A^{(s)}$ is the probability that, starting from E_i, we visit after the kth step E_i for the first time only in $(k+s)$ steps. Clearly,

$$P_{ii}^{(k)} + \sum_{s=1}^{n} A^{(s)} \leq 1.$$

Moreover, by Lemma 3,

$$P_{ii}^{(k)} + \sum_{s=1}^{m} A^{(s)} > 1 - \epsilon.$$

These two inequalities imply that

$$\sum_{s=m+1}^{n} A^{(s)} < \epsilon.$$

Note also that for $s \leq m$ we have $n - s > k$, and hence $P_{ii}^{(n-s)} < b + \epsilon$. Therefore,

$$\begin{aligned} P_{ii}^{(n')} &= P_{ii}^{(k)} P_{ii}^{(n)} + \sum_{s=1}^{m} A^{(s)} P_{ii}^{(n-s)} + \sum_{s=m+1}^{n} A^{(s)} P_{ii}^{(n-s)} \leq \\ &\leq P_{ii}^{(k)}(a+\epsilon) + (1 - P_{ii}^{(k)})(b+\epsilon) + \epsilon = b + 2\epsilon - P_{ii}^{(k)}(b-a). \end{aligned}$$

Bearing in mind that $P_{ii}^{(k)} > a - \epsilon$, $P_{ii}^{(n')} > b - \epsilon$ and $b - a \leq 1$ we obtain

$$b - \epsilon \leq b + 2\epsilon - (a - \epsilon)(b - a) \leq b + 3\epsilon - a(b-a),$$
$$a(b-a) \leq 4\epsilon, \; b - a \leq 4\epsilon/a.$$

Since $a > 0$ and $\epsilon > 0$ is arbitrary, it follows that $b - a = 0$. This proves the existence of the limit of $P_{ii}^{(n)}$ equal to $b = a$. Lemma 2a directly implies that this limit is $1/M_{ii}$.

Proof of Theorem 4a. Consider, together with the given Markov chain, a new chain determined by the elementary transition probabilities

$$\bar{p}_{ij} = P_{ij}^{(d)},$$

where d is the period of the class considered. Clearly, for all states of our class

$$\overline{P}_{ij}^{(n)} = P_{ij}^{(nd)}.$$

With respect to the new Markov chain our class of states forms one single subclass. Therefore, according to Lemma 4b,

$$\lim_{n\to+\infty} P_{ii}^{(nd)} = \lim_{n\to+\infty} \overline{P}_{ii}^{(n)} = \frac{1}{\overline{M}_{ii}} = \frac{d}{M_{ii}}.$$

Thus Theorem 4a is proved for $i = j$. To prove it for the general case, let q be the minimal number of steps in which we can pass from E_i to E_j (clearly $q \equiv \gamma - \beta \,(\mathrm{mod}\, d)$). Then,

$$P_{ij}^{(nd+q)} = \sum_{m=0}^{n} K_{ij}^{(md+q)} P_{jj}^{(nd-md)}.$$

In this case,

$$\sum_{m=1}^{\infty} K_{ij}^{(md+q)} = 1,$$

and $P_{jj}^{(nd-md)}$ tends to d/M_{jj} for constant m and $n \to +\infty$. This implies that

$$\lim_{n\to+\infty} P_{ij}^{(nd+q)} = d/M_{jj}.$$

An important supplement to Theorem 4a is the following.

Theorem 4b. *In a positive class the sum of the limits P_j over all states in a subclass is 1 for every subclass.*

Theorem 4b follows directly from the next lemma:

Lemma 5. *In a positive class there exists for any $\epsilon > 0$ a finite system of states $E_{j_1}, E_{j_2}, \ldots, E_{j_k}$ such that for any E_i from the same class and for all sufficiently large n,*

$$\sum_{s=1}^{k} P_{ij_s}^{(n)} > 1 - \epsilon.$$

Proof. Choose i_0 arbitrarily. According to Lemma 3 there exists an m such that for any n the probability that, starting from E_{i_0}, we visit E_{i_0} at least once between the nth and the $(n+m)$th step is greater than $1 - \epsilon/3$.

Clearly, we can always choose a system of states $E_{j_1}, E_{j_2}, \ldots, E_{j_k}$ such that for any $r \le m$,

$$\sum_{s=1}^{k} P_{i_0 j_s}^{(r)} > 1 - \frac{\epsilon}{3}.$$

We will prove that this system of states satisfies the conditions of the Lemma. For this we take some fixed i and choose q such that

$$\sum_{t=1}^{q} K_{ii_0}^{(t)} > 1 - \frac{\epsilon}{3}.$$

Now let $n > m + q$. Set $n = q' + m$, $q' > q$. With probability greater than $1 - \epsilon/3$, starting from E_i we visit E_{i_0} in the first q' steps. No matter at what step $\leq q'$ we first visit E_{i_0}, with probability greater than $1 - \epsilon/3$ we return to E_{i_0} between the q'th and the $(q' + m)$th step. If this happens at some $(q' + m - r)$th step, then with probability greater than $1 - \epsilon/3$ we are in one of the chosen states E_{j_s} after $q' + m$ steps. Thus with probability greater than $(1 - \epsilon/3)^3 > 1 - \epsilon$, having started from E_i we arrive at one of our chosen states E_{j_s} in $n = q' + m$ steps. This proves the Lemma and with it Theorem 4b.

Remark. Theorem 4a implies that not only does $\pi_{ii}^{(n)}$ tend to $1/M_{ii}$ (Lemma 2a), but also that for any E_j from the same class as E_i, the $\pi_{ji}^{(n)}$ tend to the same limit. By Theorem 4b, the sum $\sum(1/M_{ii})$ taken over the states of one subclass is equal to $1/d$ (where d is the period of the class), and the same sum taken over all the states of one class is equal to 1.

§6. Asymptotic behaviour of probabilities in other cases

We disregard combinations of states E_i and E_j for which $P_{ij}^{(n)} = 0$ for all n, due to the considerations of §2, and note that if E_j is inessential, then always $P_{ij}^{(n)} \to 0$ as $n \to +\infty$. We consider here the most difficult case: E_i is inessential, while E_j is essential and belongs to a certain class $S^{(\alpha)}$. For an inessential state E_i let $N_i^{(\alpha)}$ be the probability that, having started from E_i, we visit sooner or later one of the states of $S^{(\alpha)}$. Clearly,

$$\sum_{\alpha} N_i^{(\alpha)} \leq 1,$$

since, having once arrived in one of the states of the class $S^{(\alpha)}$ it is impossible to leave this class. In the case when we enter the class $S^{(\alpha)}$ at the initial state E_i, we denote by n_0 the number of steps before the first visit to one of the states E_j of the class $S^{(\alpha)}$ and by β_0 the number of the subclass $S_{\beta_0}^{(\alpha)}$ to which this first state E_j belongs. Now let $N_{i,\gamma}^{(\alpha)}$ be the probability that, given the initial state E_i, we visit the class $S^{(\alpha)}$ so that $n_0 \equiv \beta_0 + \gamma \pmod{d(\alpha)}$. Clearly,

$$\sum_{\gamma=1}^{d(\alpha)} N_{i,\gamma}^{(\alpha)} = N_i^{(\alpha)}.$$

The following result can be proved.

Theorem 5. *In the case of an inessential E_i and an essential E_j from a subclass $S_\gamma^{(\alpha)}$, the probability $P_{ij}^{(n)}$ tends to $N_{i,\beta-\gamma}^{(\alpha)} P_j$ as $n \to +\infty$ and runs through the values $n \equiv \beta \pmod{d(\alpha)}$.*

Thus, we see that for fixed i and j the dependence of the probability $P_{ij}^{(n)}$ on n in all cases (essential or inessential) is asymptotically periodic. The averages $\pi_{ij}^{(n)}$ however, always have certain limits π_{ij} as $n \to +\infty$.

Komarovka, 22 December 1936

References

1. A. Kolmogoroff, 'Anfangsgründe der Theorie der Markoffschen Ketten mit unendlich vielen möglichen Zuständen', *Mat. Sb.* **1** (1936), 607–610.
2. W. Doeblin, 'Sur les chaînes discrètes de Markoff', *C. R. Acad. Sci. Paris* **203** (1936), 24–26.
3. A. Khintchine, 'Sur la loi des grands nombres', *C. R. Acad. Sci. Paris* **188** (1928), 477–479.

24. ON THE REVERSIBILITY OF THE STATISTICAL LAWS OF NATURE *

1. Statement of the problem

Consider an n-dimensional manifold R. Let $f(t,x,y)dy_1dy_2\ldots dy_n$ be the probability of transition in time $t>0$ from a point x to a point η with coordinates η_i, $i=1,2,\ldots,n$, such that $y_i<\eta_i<y_i+dy_i$. Assume that $f(t,x,y)$ is differentiable up to a certain sufficiently high order and satisfies the following conditions:

$$f(t,x,y)\geq 0, \tag{1}$$

$$\iint\limits_R\cdots\int f(t,x,y)dy_1dy_2\ldots dy_n=1, \tag{2}$$

$$f(s+t,x,y)=\iint\limits_R\cdots\int f(s,x,z)f(t,z,y)dz_1dz_2\ldots dz_n, \tag{3}$$

$$\iint\limits_G\cdots\int f(t,x,y)dy_1dy_2\ldots dy_n\to 1 \quad \text{as } t\to 0, \tag{4}$$

where x is an interior point of a domain G (cf. [1], [2]). Given $f(t,x,y)$, $p(x)$ determines a *stationary probability distribution* compatible with $f(t,x,y)$ if and only if

$$p(x)\geq 0 \tag{5}$$

$$\iint\limits_R\cdots\int p(x)dx_1dx_2\ldots dx_n=1, \tag{6}$$

$$p(y)=\iint\limits_R\cdots\int p(x)f(t,x,y)dx_1dx_2\ldots dx_n. \tag{7}$$

A stationary distribution is called *ergodic* if $f(t,x,y)\to p(y)$ as $t\to\infty$ for any x and y. Formula (7) implies that an ergodic stationary distribution is always the only stationary distribution, that is, if there exists an ergodic stationary distribution $p_0(x)$, then no other ergodic stationary distribution $p(x)$ other than $p_0(x)$ can exist. Note that if R is closed, then the existence of

* 'Zur Umkehrbarkeit der statistischen Naturgesetze', *Math. Ann.* **113** (1937), 766–772.

an ergodic (and, consequently, the only possible) stationary distribution directly follows from the single condition that $f(t,x,y) > 0$ for any x and y and sufficiently large t (cf. [2],§5).

We suppose that $f(t,x,y)$, as well as a certain stationary distribution $p(x)$, are given beforehand. Assume, moreover, that $p(x) > 0$ for any x. In this case, knowing the position of y at the end of the time interval of duration t, we can also determine the conditional distribution of the probabilities of the initial position of x (for given y). Denoting by $h(t,x,y)$ the density of this conditional probability distribution, we have

$$h(t,x,y)p(y) = p(x)f(t,x,y). \tag{8}$$

Clearly, (8) uniquely defines $h(t,x,y)$.

The question on the reversibility of the laws of nature mentioned in the title of this paper[1] can be formulated in the following way: *under what conditions does the following relation hold*:

$$h(t,x,y) = f(t,y,x)\ ? \tag{9}$$

This paper deals with the special case in which the function $f(t,x,y)$ satisfies the following Fokker-Planck equations:

$$\frac{\partial f}{\partial t} = \sum_i A^i(x)\frac{\partial f}{\partial x^i} + \sum_i\sum_j B^{ij}(x)\frac{\partial^2 f}{\partial x^i \partial x^j}, \tag{10}$$

$$\frac{\partial f}{\partial t} = -\sum_i \frac{\partial}{\partial y^i}\{A^i(y)f\} + \sum_i\sum_j \frac{\partial^2}{\partial y^i \partial y^j}\{B^{ij}(y)f\} \tag{11}$$

(see [2]).

For simplicity we will further confine ourselves to the case of a closed manifold R. In this case, (11) implies that every stationary distribution $p(y)$ satisfies the equation

$$-\sum_i \frac{\partial}{\partial y^i}\{A^i(y)p\} + \sum_i\sum_j \frac{\partial^2}{\partial y^i \partial y^j}\{B^{ij}(y)p\} = 0. \tag{12}$$

[1] In [3] this problem is discussed for the case of Markov chains with a finite number of states; cf. also [4].

In addition to the propositions given above we further assume that the quadratic form B^{ij} is everywhere strictly positive definite. The case of a degenerate form B^{ij} is very interesting for many physical questions (see, for example, [5]), but we shall not consider it here. The assumption of positive definiteness of B^{ij} implies that $f(x,y,t) > 0$ for any x, y and $t > 0$, and hence, the existence of a unique stationary distribution $p(x)$, where $p(x) > 0$ for any x. This unique stationary distribution $p(x)$ is the unique solution of (12) such that

$$\iint_R \cdots \int p(x)dx_1 dx_2 \ldots dx_n = 1. \tag{13}$$

Under these assumptions, necessary and sufficient conditions for (9) to hold can be directly expressed in terms of properties of the coefficients A^i and B^{ij} of the equations (10) and (11). Namely, *under the above assumptions, (9) holds if and only if the vector $\alpha(y)$ defined below, whose components are expressed in terms of $A^i(y)$ and $B^{ij}(y)$ is the gradient of a scalar potential.* If $B^{ij}(y) = \delta^{ij}$, this condition becomes especially simple, since then $\alpha(y)$ is none other than the vector with components $A^i(y)$.

Let us once again state our assumptions:

1) the function $f(t,x,y)$ is differentiable a sufficient number of times and satisfies (1)–(4), (10), (11);

2) the manifold R is closed;

3) the form $B^{ij}(y)$ is positive definite.[2]

2. Invariant form of the Fokker-Planck equations

Now we consider a more general case than that discussed in §1. Namely, we now assume that the transitions from x to y in the time between the moments s and $t > s$ have probability distribution described by the probability density $f(s,t,x,y)$. In this case $f(s,t,x,y)$ must satisfy the following conditions (cf. [1], [2]):

$$f(s,t,x,y) \geq 0, \tag{14}$$

$$\iint_R \cdots \int f(s,t,x,y)dy_1 dy_2 \ldots dy_n = 1, \tag{15}$$

[2] As already mentioned, this implies that $p(x) > 0$ for any x.

$$f(s,t,x,y) = \iint_R \cdots \int f(s,u,x,z)f(u,t,z,y)dz_1dz_2\dots dz_n, \tag{16}$$

$$s < u < t,$$

$$\iint_G \cdots \int f(s,t,x,y)dy_1dy_2\dots dy_n \to 1 \quad \text{as } t \to s, \tag{17}$$

with x belonging to the interior of G.

In this case the Fokker-Planck equations take the form:

$$\frac{\partial f(s,t,x,y)}{\partial s} = -\sum_i A^i(s,x)\frac{\partial}{\partial x^i}f(s,t,x,y)-$$

$$-\sum_i\sum_j B^{ij}(s,x)\frac{\partial^2}{\partial x^i\partial x^j}f(s,t,x,y), \tag{18}$$

$$\frac{\partial f(s,t,x,y)}{\partial t} = -\sum_i \frac{\partial}{\partial y^i}\{A^i(t,y)f(s,t,x,y)\}+$$

$$+\sum_i\sum_j \frac{\partial^2}{\partial y^i\partial y^j}\{B^{ij}(t,y)f(s,t,x,y)\}. \tag{19}$$

The coefficients $B^{ij}(s,x)$ form a contravariant tensor of rank two, whereas the coefficients $A^i(s,x)$ transform according to the following more complex law:

$$\overline{A}^i = \bar{x}^i_k A^k + \frac{\partial^2 \bar{x}^i}{\partial x^k \partial x^m}B^{km}.$$

(From now on the summation sign is omitted.)

Let us assume that the quadratic form $B^{ij}(s,x)$ is positive definite everywhere and for all s, and choose it to be the principal metric form on R (note that our metric depends on s). We then set

$$\alpha^i(s,x) = A^i(s,x) - \Gamma^{ik}_k(s,x), \tag{20}$$

where Γ^{ik}_k is the Christoffel symbol corresponding to $B^{ij}(s,x)$. The contravariant vector α^i coincides with A^i in every geodesic coordinate system (at x). In a geodesic coordinate system chosen at the point x, (18) may be written in the following way:

$$\partial f/\partial s = \alpha^i(s,x)\Delta^{(x)}_i f - \Delta^{(x)} f, \tag{21}$$

where (x) indicates the argument over which the derivative is taken, Δ_i is the covariant derivative, and Δ is the Laplace operator, that is, $\Delta = \Delta_i\Delta^i$.

The latter equation, by virtue of its invariance, must hold also in any other coordinate system. When the coordinates x change, $f(s,t,x,y)$ does not vary; if, however, the coordinates y change, this function transforms as a scalar density.

Now set

$$f(s,t,x,y) = \sqrt{|B_{ij}(t,y)|}\phi(s,t,x,y). \tag{22}$$

Clearly, $\phi(s,t,x,y)$ is now invariant with respect to changes in both x and y; it also satisfies (21):

$$\partial\phi/\partial s = -\alpha^i(s,x)\Delta_i^{(x)}\phi - \Delta^{(x)}\phi. \tag{23}$$

As for (19), it takes the following form:

$$\frac{\partial\phi}{\partial t} = -\Delta_i^{(y)}\{\alpha^i(t,y)\phi\} + \Delta^{(y)}\phi + \frac{1}{2}\frac{\partial \log|B_{ij}(t,y)|}{\partial t}\phi. \tag{24}$$

3. Solution of the problem

Now let us return to the case when the conditions indicated at the end of §1 hold. Then the quadratic form $B^{ij}(x)$ is time-independent and depends only on the coordinates of x on R. We set

$$\alpha^i(x) = A^i(x) - \Gamma_k^{ik}(x), \tag{25}$$

$$f(t,x,y) = \sqrt{|B_{ij}(y)|}\phi(t,x,y), \tag{26}$$

$$p(x) = \sqrt{|B_{ij}(x)|}\pi(x), \tag{27}$$

$$h(t,x,y) = \sqrt{|B_{ij}(x)|}\psi(t,x,y). \tag{28}$$

Then to find the conditions when (9) holds, we merely have to the find conditions under which the equivalent relation

$$\psi(t,x,y) = \phi(t,y,x) \tag{29}$$

holds. In view of the results of §2, $\phi(t,x,y)$ and $\pi(x)$ satisfy the equations

$$\partial\phi/\partial t = \alpha^i(x)\Delta_i^{(x)}\phi + \Delta^{(x)}\phi, \tag{30}$$

$$\partial\phi/\partial t = -\Delta_i^{(y)}\{\alpha^i(y)\phi\} + \Delta^{(y)}\phi, \tag{31}$$

$$-\Delta_i^{(x)}\{\alpha^i(x)\pi\} + \Delta^{(x)}\pi = 0. \tag{32}$$

Since, moreover,

$$\psi(t,x,y) = \phi(t,x,y)\pi(x)/\pi(y),$$

(29) can be rewritten as

$$\phi(t,x,y)\pi(x) = \phi(t,y,x)\pi(y). \tag{33}$$

Assume that (29) (and hence (33)) holds. Interchanging x and y and using (31), we obtain the following equation for $\phi(t,y,x)$:

$$\partial\phi/\partial t = -\Delta_i\{\alpha^i\phi\} + \Delta\phi. \tag{34}$$

(Hereafter all the derivatives are taken either with respect to t or x, but never with respect to y, and α^i depends only on x.) The same equation must be satisfied by $\psi(t,x,y)$ and therefore by $\psi(t,x,y)\pi(y) = \phi(t,x,y)\pi(x)$:

$$\partial(\phi\pi)/\partial t = -\Delta_i\ (\alpha^i\phi\pi) + \Delta(\phi\pi), \tag{35}$$

or

$$\pi\frac{\partial\phi}{\partial t} = -\pi\alpha^i\Delta_i\phi - \phi\Delta_i(\alpha^i\pi) + \pi\Delta\phi + 2(\Delta^i\pi)(\Delta_i\phi) + \phi\Delta\pi. \tag{36}$$

Multiplying (32) by $-\phi$ and (30) by $-\pi$ and adding the resulting inequalities to (36), we have

$$0 = -2\pi\alpha^i\Delta_i\phi + 2(\Delta^i\pi)(\Delta_i\phi),$$

or

$$(\Delta_i\phi)(-\alpha^i\pi + \Delta^i\pi) = 0. \tag{37}$$

It can easily be seen that $\Delta_i\phi$ does not vanish identically (that is, for any $t > 0$) in any domain of R. Therefore, on an everywhere dense set of points x and hence, by continuity, everywhere,

$$-\alpha^i\pi + \Delta^i\pi = 0, \quad -\alpha_i\pi + \Delta_i\pi = -\alpha_i\pi + \partial\pi/\partial x^i = 0, \tag{38}$$

that is,

$$\alpha_i = \partial(\log\pi)/\partial x^i. \tag{39}$$

Putting $\log\pi = P$ we see that P is a potential of α. This proves the necessity of the condition stated in §1.

Assume now that, on the contrary, there exists P such that

$$\alpha_i = \partial P/\partial x^i;$$

we must prove that in this case (29) holds.

It can easily be seen that e^P satisfies (32). Since, however, π is the only solution of (32) such that

$$\iint_R \cdots \int \pi\sqrt{|B_{ij}(x)|}dx_1dx_2\dots dx_n = 1,$$

we clearly have

$$\pi = (1/J)e^P, \tag{40}$$

where

$$J = \iint_R \cdots \int e^P\sqrt{|B_{ij}(x)|}dx_1dx_2\dots dx_n.$$

Now (40) successively implies (39), (38), (37), (36) and (35), while (35) implies that $\psi(t,x,y)$ satisfies

$$\partial\psi/\partial t = -\Delta_i(\alpha^i\psi) + \Delta\psi, \tag{41}$$

that is, the equation that was earlier denoted by (34) holds for $\phi(t,y,x)$ as well. The initial values for $\phi(t,y,x)$ and $\psi(t,x,y)$ clearly coincide at $t = 0$. Hence, $\phi(t,y,x) = \psi(t,x,y)$, which proves our theorem.[3]

Moscow, 1 July 1936

References

1. A.N. Kolmogoroff, 'Über die analytischen Methoden in der Wahrscheinlichkeitsrechnung', *Math. Ann.* **104**:3 (1931), 415–458.
2. A.N. Kolmogoroff, 'Zur Theorie der stetigen zufälligen Prozesse', *Math. Ann.* **108**:1 (1933), 149–160.
3. A.N. Kolmogoroff, 'Zur Theorie der Markoffschen Ketten', *Math. Ann.* **112**:1 (1936), 155–160.
4. B. Hostinsky and J. Potoček, 'Chaînes de Markoff inverses', *Bull. Intern. Acad. Tchèque Sci.* **36** (1935), 64–67.
5. A.N. Kolmogoroff, 'Zufällige Bewegungen', *Ann. of Math.* **35**:1 (1934), 116–117.

[3] Cf. the uniqueness proof in [2]. The existence of continuous derivatives of the coefficients A^i and B^{ij} follows from the existence of sufficiently high derivatives of f. The condition of continuity of (29) in [2] for the case of a closed manifold R is equivalent to our condition (4).

25. SOLUTION OF A BIOLOGICAL PROBLEM *

R.A. Fisher [1] gave an interesting application of iteration theory to the laws of breeding a new gene in an unbounded population. Recently J.F. Steffenson [2] gave a detailed exposition of the probability that all offspring from an individual die out. Both these questions are, however, mathematically identical. After giving a brief statement of the problem and recapitulating the known results, [1] we will give several additions to the results by Fisher and Steffenson in certain directions.

1. Mathematical statement of the problem

The following assumptions are essential for our mathematical discussion. Let $F_0, F_1, \dots, F_n$, be a sequence of generations and let F_0 have K_0 individuals with a feature M (M-individuals); we assume that crossing between M-individuals is impossible; each M-individual from F_{n+1} is assumed to be an offspring from a certain M-individual in F_n. We are given the probabilities p_k that an M-individual from generation F_n has exactly k M-offspring in F_{n+1}, and these probabilities are independent of the fate of other branches of M-offspring.

Our goal is to determine the probabilities $P_k^{(n)}$ $(n \geq 1)$ that the number K_n of M-individuals in the nth generation is equal to k. Here $P_0^{(n)}$ is the probability that all the M-offspring die out. Clearly, $P_0^{(n)}$ can only increase with n. The main result of this paper is an asymptotic formula for $P_0^{(n)}$ for large n.

We assume that the three first factorial moments

$$a = p_1 + 2p_2 + \dots + kp_k + \dots$$
$$b = 2p_2 + 6p_3 + \dots + k(k-1)p_k + \dots$$
$$c = 6p_3 + 24p_4 + \dots + k(k-1)(k-2)p_k + \dots$$

are finite.

2. Biological explanations

K_n denotes the number of M-individuals in the nth generation that have reached a certain age. Consequently, the probabilities p_k, which are assumed

---

* *Izv. NII Mat. Mekh. Tomsk. Univ.* **2**:1 (1938), 7–12 (in Russian).

[1] References to previous works can be found in Steffenson's paper.

to be prescribed, take into account the effect of selection. If, for instance, a grown-up M-individual always has a very large number m of M-offspring in the next generation which, however, have a very insignificant probability a/m of reaching the age set for determining K_n, then we obtain

$$p_k = C_m^k(1-a/m)^{m-k}(a/m)^k,$$

or, approximately, by Poisson's formula

$$p_k = \frac{a^k}{k!}e^{-a}. \tag{1}$$

In this case it is easy to compute that

$$b = a^2, \quad c = a^3.$$

Most important for the fate of an M-generation is the first moment a, that is, the average number of M-offspring of an M-individual in the next generation. If K_n is very large, then we have approximately

$$K_{n+1} = aK_n.$$

We should distinguish three cases which are essentially different from each other: a greater than, smaller than, or equal to 1.

If $a < 1$, then it is clear that the M-offspring will finally disappear. In what follows we will show that this also holds when $a = 1$, $b > 0$. If $a > 1$, then the probability P_0 of the M-offspring dying out is $\neq 1$ ($P_0 = \lim P_0^{(n)}$ as $n \to \infty$). The probability that the offspring do not die but propagate unrestrictedly is $1 - P_0$.

The exception $a = 1$, $b = 0$ is only possible when $p_1 = 1$, $p_k = 0$ $(k \neq 1)$. In this case, clearly K_n always equals K_0 and consequently, $P_0^{(n)}$ is always 0.

A specific problem that led to Fisher's studies is the following: in a very large stationary population consisting only of individuals of BB type there appears a small number K_0 of individuals of Bb type (generation F_0). If in the further generations $F_1, F_2, \dots$ Bb-offspring remain comparatively small in number, then $Bb \times Bb$ crossing is virtually excluded. Under this assumption our scheme can be applied to Bb-offspring.

If $a = 1$, then Bb-individuals are as viable as BB-individuals. However, due to random fluctuations of K_n they should surely die out (here the case $a =$

1, $b = 0$ is impossible). Bb-individuals certainly disappear if $a < 1$. If $a > 1$, then there is a positive probability $1 - P_0$ that Bb-individuals will propagate unrestrictedly ($a > 1$ corresponds to the assumption that Bb-individuals are more viable than BB-individuals). In the real case, however, even for $a = 1$ random fluctuations of K_n could be so substantial that the crossing $Bb \times Bb$ attains a non-zero probability. Then one must also take into account the viability of bb-individuals.

3. The general theory

The main result of Fisher and Steffenson consists in reducing the computation of the probabilities $P_k^{(n)}$ to determining the coefficients of known power series. For this purpose they assume that

$$q(x) = p_0 + p_1 x + p_2 x^2 + \ldots + p_k x^k + \ldots, \tag{2}$$

$$Q^{(n)}(x) = P_0^{(n)} + P_1^{(n)} x + P_2^{(n)} x^2 + \ldots + P_k^{(n)} x^k + \ldots \tag{3}$$

Since $p_k \leq 1$, $P_k^{(n)} \leq 1$, it follows that $q(x)$ and $Q^{(n)}(x)$ are analytic functions for $|x| < 1$. Let $q^n(x)$ be the nth iterate of $q(x)$:

$$q^1(x) = q(x), \quad q^{n+1}(x) = q(q^n(x)).$$

The basic formula of the the whole theory is: [2]

$$Q^{(n)}(x) = [q^n(x)]^{K_0}. \tag{4}$$

In particular,

$$P_0^{(n)} = Q^{(n)}(0)$$

and consequently,

$$P_0^{(n)} = [q^n(0)]^{K_0}, \tag{5}$$

$$P_0 = \lim P_0^{(n)} = \lim [q^{(n)}(0)]^{K_0} = \lambda^{K_0}, \ n \to \infty. \tag{6}$$

The limit

$$\lambda = \lim q^n(0), \ n \to \infty,$$

[2] See [2]. The reader can easily reproduce the proof.

can be determined using general iteration theory. It can be proved that λ equals the minimal non-negative root of the equation[3]

$$q(\lambda) - \lambda = 0. \tag{7}$$

This root does exist, since

$$q(1) = p_0 + p_1 + \ldots + p_k + \ldots = 1, \quad q(1) - 1 = 0.$$

It can further be proved that for $a > 1$ the required smallest root λ is smaller than 1. Therefore in this case we have a positive probability

$$1 - P_0 = 1 - \lambda^{K_0}$$

that the M-offspring propagate unrestrictedly. If, on the other hand, $a \leq 1$, (and if $a = 1$, then b is also greater than zero), then $\lambda = 1$ and consequently, $P_0 = 1$. In this case M-offspring must necessarily die out. Therefore it is desirable to study the asymptotic behaviour of the probability

$$R^{(n)} = 1 - P_0^{(n)}$$

that the offspring will survive up to the nth generation. Neither Fisher nor Steffenson did this under general assumptions.

4. Asymptotic behaviour of $R^{(n)}$ for $a \leq 1$

Let

$$r_n = 1 - q^n(0).$$

In our case r_n converges to zero as $n \to \infty$. Hence, as $n \to \infty$ we have asymptotically

$$R^{(n)} = 1 - [q^n(0)]^{K_0} = 1 - (1 - r_n)^{K_0} \sim K_0 r_n. \tag{8}$$

It remains to study the asymptotic behaviour of r_n. First note that

$$q'(1) = a,$$

---

[3] See [2].

while for $0 \le x \le 1$,

$$0 \le q''(x) \le q''(1) = b, \quad 0 \le q'''(x) \le q'''(1) = c.$$

Now let $a < 1$. Expanding $q(1-y)$ in a Taylor series we obtain

$$q(1-y) = q(1) - q'(1)y + O(y^2) = 1 - ay + O(y^2).$$

Set $y = r_{n-1} = 1 - q^{n-1}(0)$. Then

$$r_n = 1 - q^n n(0) = 1 - q[q^{n-1}(0)] = 1 - q(1 - r_{n-1}) = ar_{n-1} + O(r_{n-1}^2),$$
$$r_n/r_{n-1} = a[1 + O(r_{n-1})]. \tag{9}$$

As can easily be seen, (9) implies that asymptotically

$$r_n \sim Ca^n, \tag{10}$$

where C is a suitable constant. Finally, by (8),

$$R^{(n)} \sim CK_0a^n. \tag{11}$$

Thus we see that for $a < 1$ the probability $R^{(n)}$ decreases asymptotically as a geometric progression with common ratio a.

Let $a = q'(1) = 1$. Then

$$q(1-y) = q(1) - q'(1)y + \tfrac{1}{2}q''(1)y^2 + O(y^3) = 1 - y + \tfrac{1}{2}by^2 + O(y^3),$$
$$r_n = 1 - q^n(0) = 1 - q(1 - r_{n-1}) = r_{n-1} - \tfrac{1}{2}br_{n-1}^2 + O(r_{n-1}^3),$$
$$r_n/r_{n-1} = 1 - \tfrac{1}{2}br_{n-1} + O(r_{n-1}^2). \tag{12}$$

It follows from (12) that asymptotically (for $b > 0$)

$$r_n \sim 2/nb. \tag{13}$$

Finally, by (8),

$$R^{(n)} \sim 2K_0/nb. \tag{14}$$

Thus we see that for $a = 1$, $R^{(n)}$ tends to zero considerably more slowly. Formula (14) shows that for $a = 1$ and large n the probability $R^{(n)}$ of survival until the nth generation is inversely proportional to the second moment b. In particular, if the probabilities p_k satisfy the Poisson formula (1), then $b = a^2 = 1$ and consequently,

$$R^{(n)} \sim 2K_0/n.$$

This formula has been given by Fisher. As a second example we consider the case when every Bb-individual gives exactly two offspring after the crossing $Bb \times BB$. Then $p_0 = \frac{1}{4}$, $p_1 = \frac{1}{2}$, $p_2 = \frac{1}{4}$, $p_k = 0$ $(k > 2)$, which implies that $b = \frac{1}{2}$ and $R^{(n)} = 4K_0/n$.

5. An approximate formula

Now let us return to the case $a > 1$. Assume that

$$\omega = a - 1$$

is small in absolute value compared to $b > 0$. Then

$$\mu = 1 - \lambda$$

is also small, and we can approximately set

$$q(\lambda) = q(1-\mu) \sim 1 - q'(1)\mu + \tfrac{1}{2}q''(1)\mu^2 = 1 - a\mu + \tfrac{1}{2}b\mu^2.$$

Equation (7) gives

$$q(\lambda) - \lambda \sim 1 - a\mu + \tfrac{1}{2}b\mu^2 - 1 + \mu = -\omega\mu + \tfrac{1}{2}b\mu^2 = 0,$$

$$\mu \sim 2\omega/b, \tag{15}$$

$$\lambda \sim 1 - 2\omega/b. \tag{16}$$

If ω is small compared to K_0, then (15) and (16) imply approximately

$$1 - P_0 \sim 2K_0\omega/b,$$

that is, the probability $1 - P_0$ of unrestricted propagation of M-offspring is directly proportional to the selection coefficient and inversely proportional to the second moment b.

References

1. R.A. Fisher, *The genetic theory of natural selection*, Oxford, 1930.
2. J.F. Steffenson, 'Deux problèmes du calcul des probabilités', *Ann. Inst. H. Poincaré* **3** (1933), 331–334.

26. ON A NEW CONFIRMATION OF MENDEL'S LAWS *

In the discussion on genetics that took place in the autumn of 1939 much attention was paid to checking whether or not Mendel's laws were really true. In the basic discussion on the validity of the entire concept of Mendel, it was quite reasonable and natural to concentrate on the simplest case, which, according to Mendel, results in splitting in the ratio 3:1. For this simplest case of crossing $Aa \times Aa$, with the feature A dominating over the feature a, it is well known that Mendel's concept leads to the conclusion that in a sufficiently numerous progeny (no matter whether it consists of one family or involves many separate families resulting from various pairs of heterozygous parents of type Aa) the ratio between the number of individuals with the feature A (that is, the individuals of the type AA or Aa) to the number of individuals with the feature a (aa type) should be close to the ratio 3:1. T.K. Enin [1], [2], N.I. Ermolaeva [3] and E. Kol'man [4] have concentrated on checking this simplest consequence of Mendel's concept. However, Mendel's concept not only results in this simplest conclusion on the approximate ratio 3:1 but also makes it possible to predict the average deviations from this ratio. Owing to this it is the statistical analysis of deviations from the ratio 3:1 that gives a new, more subtle and exhaustive way of proving Mendel's ideas on feature splitting. In this paper we will try to indicate what we think to be the most rational methods of such checking and to illustrate these methods on the material of the paper by N.I. Ermolaeva [3]. In contrast to the opinion of Ermolaeva herself, this material proved to be a brilliant new confirmation of Mendel's laws. [1]

§1. General considerations on the role of a random event in hereditary phenomena

Let us first assume a viewpoint independent of Mendelism. Suppose that the crossing of two individuals α and β give a progeny of n individuals. In this case, usually each of the individuals α and β produces a number of gametes which is much greater than the number of offspring n. Let α produce gametes

$$\alpha_1,\ \alpha_2, \ldots, \alpha_{k_1}$$

---

* *Dokl. Akad. Nauk SSSR* **27** (1940), 38–42 (in Russian).

[1] My attention was drawn to the work of N.I. Ermolaeva by A.S. Serebrovskii.

and β gametes

$$\beta_1, \beta_2, \ldots, \beta_{k_2}.$$

It depends on many factors which exactly of the large number k_1, k_2 of possible gamete pairs will be actually used for producing the progeny. We divide these factors into internal ones, determined by biological properties of the individuals α and β and gametes α_i and β_i, and external ones independent of these biological properties. Take, for example, pollination of plants. Which pollen grains will get onto the stigma and which will be lost ? What will be the distribution of the pollen grains on the stigma ? All this depends on innumerable factors that are extraneous with respect to the biological laws of behaviour. In studying the biological laws of behaviour, these fertilization factors should be considered as random and treated using the techniques of probability theory.

We can choose n pairs

$$(\alpha_{i_1}, \beta_{j_1}),\ (\alpha_{i_2}, \beta_{j_2}),\ \ldots,\ (\alpha_{i_n}, \beta_{j_n})$$

in $s = \frac{k_1!k_2!}{n!(k_1-n)!(k_2-n)!}$ different ways. In accordance with the above our further study stems from the assumption that for each of these possible choices there is only a certain probability of its actual fulfillment as determined by the biological factors.

The derivation of Mendel's laws is based on the simplest assumption that the probabilities corresponding to any of these s possible choices are equal (and, consequently, all equal to $1/s$) (see §2). From the biological viewpoint this assumption implies the same viability of gametes, the absence of selective fertilization, and equal viability (at least up to the moment of counting the offspring) of the individuals resulting from any pair combination of gametes α_i, β_i. For simplicity, we call this the independence hypothesis (the probability of obtaining some set of gametes used for producing the progeny is assumed to be independent of the biological features of these gametes).

Like any other hypothesis on the independence of some phenomenon of certain other ones, our hypothesis taken as an absolute dogma not allowing any corrections is wrong: there are a number of well-known examples of deviations from this hypothesis; some are quantitatively insignificant, while others are quite considerable.

It is quite clear that a viewpoint that totally rejects the role of biological external random events in the selection of the gametes that take part in fertilization would be equally without foundation.

It can be seriously disputed, however, which of the following two viewpoints is correct.

1. In most cases the independence hypothesis can well serve as a first approximation to the real state of affairs (the proponents of Mendel's and Morgan's genetics).

2. Selective fertilization and unequal viability always play so decisive a role that the considerations based on the independence hypothesis are fruitless for biology (Acad. T.D. Lysenko's school).

§2. Splitting in the ratio 3:1.

Let us now return to the more special Mendelian assumption for the case of crossing $Aa \times Aa$ and domination of the feature A. In this case we assume that each of the parents forms as many type A gametes as those of type a, gamete pairs of AA and Aa types give offspring with feature A and gamete pairs of aa type, offspring with feature a. These assumptions, together with the assumption that k_1 and k_2 are much greater than n, and the independence hypothesis, imply the following:

I. The probability that there are exactly m individuals with feature A in a group of n offspring (all the rest of them having feature a) is

$$P_n(m) = \frac{n!}{m!(n-m)!}\left(\frac{3}{4}\right)^m \left(\frac{1}{4}\right)^{n-m}. \tag{1}$$

Now assume that a large number r of $Aa \times Aa$ crossings is made, and each crossing resulted in a family consisting of

$$n_1, n_2, \ldots, n_r$$

individuals, of which

$$m_1, m_2, \ldots, m_r$$

respectively have the feature A. The question is how one can best check whether or not this result is in agreement with Mendel's assumption.

If the number of individuals in every family is very small (for example, less than 10), then it is reasonable to check (1) directly using Pearson's χ^2-criterion.

If each of the families is sufficiently numerous, then it is better to apply another method. In this case (1) implies the following:

II. The normalized deviations

$$\Delta = (m/n - 3/4) : \sigma_n,$$

where

$$\sigma_n = (\sqrt{3}/4)\sqrt{n},$$

approximately obey the Gaussian law with variance 1, that is, the probability of

$$\Delta \leq x$$

is approximately equal to

$$P(x) = \frac{1}{\sqrt{2\pi}} \int_{-\infty}^{x} e^{-\xi^2/2} d\xi. \tag{2}$$

Here $\sigma_n = (\sqrt{3}/4)\sqrt{n}$ is the mean square deviation of the frequency m/n from 3/4. We see that this mean square deviation is proportional to $1/\sqrt{n}$. Hence it is only for very large families that Mendel's theory predicts that m/n is very close to 3/4. For example, we can affirm that

$$|m/n - 3/4| < 0.01,$$

with probability 0.99 only when $n > 12,000$.

On the other hand, by examining a large number of families of medium size we can achieve a much finer verification of Mendel's assumptions by considering the distribution of the deviations Δ. In particular, (2) implies that the probability of $|\Delta| \leq 1$ is approximately 0.68. Hence, according to Mendel's theory, among a sufficiently large number of families there should be about 68% with $\Delta \leq 1$ and about 32% with $|\Delta| > 1$. In Ermolaeva's work only a series of 98 families with splitting in the colouring of a flower and leaf axil (Table 4 from [3]) and the series of 123 families with splitting in the colouring of cotyledons (Table 6 from [3]) can be used to verify this consequence of Mendel's theory. The rest of the series contain too few families (or groups of families) to provide a reliable verification. The results are given in the following table. [2]

[2] The summary Table 1 in Ermolaeva's paper [3] presents some figures other than those given here, since she takes into account certain families (2 from the first series and 4 from the second series) that, for reasons we do not know, were not included in her Tables 4 and 6. Our summary Table includes only the families represented in Tables 4 and 6. However, the conclusions given below remain the same if drawn from the data given in the summary Table 1 of Ermolaeva.

No. of families	Splitting (in colour) of the flower and leaf axil		Splitting (in colour) of cotyledons		Theoretical, %
total	98	100%	123	100%	100
with $\|\Delta\| \leq 1$	66	67%	85	69%	68
with $\|\Delta\| > 1$	32	33%	38	31%	32

For this number of families in the series the correspondence with the theory should be considered quite good. Due to some strange misunderstanding, Ermolaeva herself claims in her work that the presence of a noticeable percentage of families with $|\Delta| > 1$ disproves Mendel's theory.

In a similar way we could verify the coincidence of the percentage of families for which $|\Delta| \leq a$ for $a \neq 1$. For example, the theory predicts that approximately 50% of the families should have $|\Delta| \leq 0.674$. However, it is best just to check whether the actually observed distribution of the deviations Δ is close to the theoretical Gaussian distribution. To do this, we draw on the same plot the theoretical curve $y = P(x)$ in accordance with (2) and the empirical step curve

$$y = q(x)/r,$$

where r denotes the total number of families in a given series and $q(x)$ is the number of families in the series for which $\Delta \leq x$. The results of this verification for the two series from Ermolaeva's paper are shown in Figs. 1 and 2. It is clear that in both cases the coincidence between the theoretical and empirical curves is sufficiently good. To evaluate whether the observed discrepancy between these two curves is admissible for the given size of the series we should use the formulas derived earlier (see [5]). Using these formulas for the cases given in Figs. 1 and 2, we find that

$$\lambda_1 = 0.82, \quad \lambda_2 = 0.75,$$

$$\Phi(\lambda_1) = 0.49, \quad \Phi(\lambda_2) = 0.37,$$

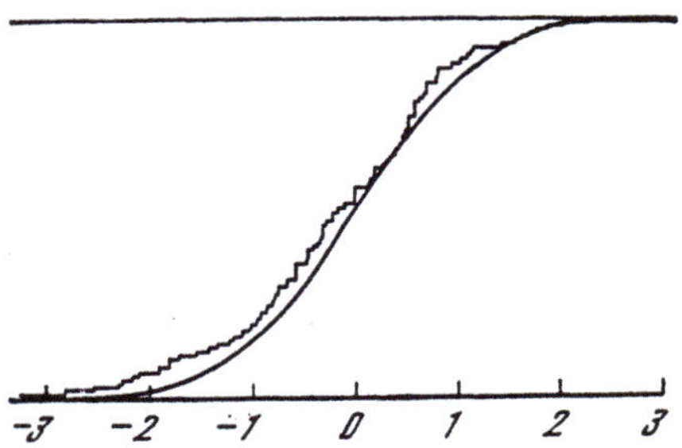

Fig. 1

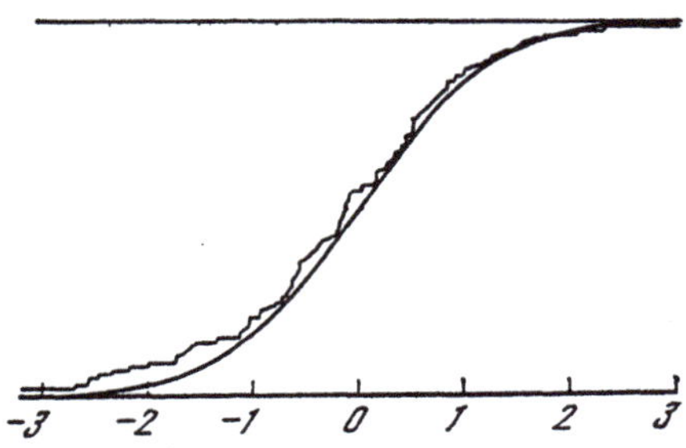

Fig. 2

which is quite satisfactory.

So we see that according to Mendel's theory, for the given family sizes we could not have expected the frequencies m/n for the individual families to be closer to their mean value 3/4 than those obtained by Ermolaeva.

If, for some sufficiently large series of families the deviations of m/n from 3/4 had been systematically smaller than required by the theory, then this would have disproved the applicability of the above assumptions to this series to the same extent as systematic overestimation of theoretically predicted deviations would do. Hints to the fact that the frequencies m/n are systematically too close to 3/4 can be found in T.K. Enin's work [1]. However, the material of this work is not broad enough (25 families as compared with two series of 98 and 123 families in Ermolaeva's work), and give reason for other doubts (the author himself thinks that they are not sufficiently homogeneous). Therefore we will not discuss them here.

Kol'man's paper referred to in the beginning of this note does not contain any new facts; it only analyses Enin's data and is based on a complete misunderstanding of the circumstances set forth in this paper.

20 February 1940

References

1. T.K. Enin, *Dokl. Akad. Nauk SSSR* **24** (1939), 176–178 (in Russian).
2. T.K. Enin, *Dokl. VASKhNIL* **6** (1939), (in Russian).
3. N.I. Ermolaeva, *Yarovizatsiya* **2**:23 (1939), 79–86 (in Russian).
4. E. Kol'man, *Yarovizatsiya* **3**:24 (1939), 70–73 (in Russian).
5. V. Romanovskii, *Mathematical statistics*, Moscow-Leningrad, 1939 (in Russian).

27. STATIONARY SEQUENCES IN HILBERT SPACE*

Introduction

Definition 1. A sequence $\{x(t)\}$ of elements in a complex Hilbert space H, where t runs through all integers from $-\infty$ to $+\infty$, is called ***stationary*** if the scalar products

$$B_{xx}(k) = (x(t+k), x(t)) \tag{0.1}$$

do not depend on t.

Definition 2. Two stationary sequences $\{x(t)\}$ and $\{y(t)\}$ are called ***jointly stationary*** if the scalar products

$$B_{yx}(k) = (y(t+k), x(t)) \tag{0.2}$$

do not depend on t.

The definitions of $B_{yx}(k)$ and $B_{xx}(k)$ immediately imply that

$$B_{xy}(k) = \overline{B}_{yx}(-k), \tag{0.3}$$

$$B_{xx}(k) = \overline{B}_{xx}(-k). \tag{0.4}$$

This kind of stationary and jointly stationary sequence is of great importance in probability theory and mathematical statistics. In terms of probability theory they are studied in detail in H. Wold's book [1] and H. Cramér's paper [2].

In §3 and §7 the main results of Wold and Cramér are given in terms of the geometry of Hilbert space.

All the new problems that are studied and solved in this paper have also arisen from probability theory and mathematical statistics. The application of our results to problems of extrapolation and interpolation of stationary series of random variables is illustrated in detail in my paper [7].

The following theorem shows a very simple connection between stationary and jointly stationary sequences and unitary operators.

Theorem 1. *Let the sequences*

$$\{x_1(t)\}, \{x_2(t)\}, \dots, \{x_n(t)\}$$

* *Byull. Moskov. Gos. Univ. Mat.* **2**:6 (1941), 1–40 (in Russian).

be stationary and pairwise jointly stationary, and let $H_{x_1 x_2 \dots x_n}$ *be the smallest closed linear subspace of* H *containing all the elements of these sequences. Then the equations*

$$Ux_\mu(t) = x_\mu(t+1), \quad \mu = 1, 2, \dots, n, \ -\infty < t < +\infty,$$

uniquely define a unitary operator

$$U = U_{x_1 x_2 \dots x_n},$$

mapping $H_{x_1 x_2 \dots x_n}$ *onto itself.*

This theorem allows us to obtain a number of basic properties of stationary and jointly stationary sequences as direct consequences of the spectral theory of unitary operators. The same is true for all the theorems of §§3–6.

The results of §§8–10 are more original and basically new; §§1–2 are of an auxiliary nature.

Theorem 1 follows directly from Definitions 1 and 2 and Lemma 1 proved in §1.

§1. Two lemmas

Let T be an operator defined on a certain subset M of H and taking values in H. The operator T is called an *isometry* if for any $x, y \in M$,

$$(x, y) = (Tx, Ty). \tag{1.1}$$

Let H_M be the smallest closed linear subspace of H containing M. Then

Lemma 1. *The operator* T *can be extended as an isometry onto the whole of* H_M. *This extension is unique.*

For the proof we choose an everywhere dense sequence of elements of M,

$$z_1, z_2, \dots, z_n, \dots. \tag{1.2}$$

We eliminate from this sequence all the elements z_n that depend linearly on the preceding z_k. Then we have a sequence

$$z_{n_1}, \ z_{n_2}, \dots, \ z_{n_k}, \dots , \tag{1.3}$$

for which the set R of all elements z of the form

$$z = a_1 z_{n_1} + a_2 z_{n_2} + \ldots + a_p z_{n_p} \tag{1.4}$$

contains all elements of the sequence (1.2) and is dense everywhere in H_M. Since the elements of R can be uniquely represented in the form (1.4) (by the linear independence of the elements in (1.3)), the formula

$$T^* z = a_1 T z_{n_1} + a_2 T z_{n_2} + \ldots + a_p T z_{n_p} \tag{1.5}$$

uniquely defines an operator T^* on R such that

$$(z', z'') = (T^* z', T^* z'')$$

for any z' and z'' in R, that is, T^* is an isometry on R. It can be extended by continuity to an isometry onto the whole of H_M.

Since all the elements of (1.2) are contained in R, (1.5) implies that for all elements of (1.2),

$$T^* z_n = T z_n;$$

by continuity this also holds for the whole set M.

Thus, the existence of the required extension is proved.

By (1.5) any other isometric extension of T coincides with T^* on R. By continuity this coincidence also holds on the whole of H_M. Consequently, the isometric extension of T onto the whole space H_M is unique, which proves the Lemma.

Lemma 2. *A sequence*

$$u_1, u_2, \ldots, u_n, \ldots,$$

satisfying the conditions

$$(u_m, u_n) = c_{mn}; \quad m, n, = 1, 2, \ldots, \tag{1.6}$$

exists on the space H if and only if for any $k, m_1, m_2, \ldots, m_k$ the matrix $\|c_{m_i m_j}\|$ is hermitian and non-negative, that is, if[1]

$$S = \sum_{i,j=1}^{k} c_{m_i m_j} \xi_i \bar{\xi}_j \geq 0 \tag{1.7}$$

[1] Here and in what follows, $a \geq 0$ means that a is real and non-negative.

for any complex numbers $\xi_1, \xi_2, \ldots, \xi_k$.

To prove the necessity of (1.7) consider an element

$$u = \xi_1 u_{m_1} + \xi_2 u_{m_2} + \xi_k u_{m_k}$$

of H. It is easy to compute that

$$\|u\|^2 = (u, u) = S. \tag{1.8}$$

Since always $\|u\|^2 \geq 0$, (1.8) implies (1.7).

We now prove the sufficiency of (1.7). In accordance with (1.7), assume that for any n,

$$S_n = \sum_{i,j=1}^{n} c_{ij} \xi_i \bar{\xi}_j \geq 0.$$

We find a transformation

$$\eta_i = \sum_{j=1}^{n} a_{ij}^{(n)} \xi_j$$

that reduces S_n to the form

$$S_n = \sum_{i=1}^{m} |\eta_i|^2.$$

Choose in H an orthonormal system of vectors

$$z_1^{(n)}, z_2^{(n)}, \ldots, z_m^{(n)}$$

and set

$$u_i^{(n)} = \sum_{j=1}^{m} a_{ji}^{(n)} z_j, \quad i = 1, 2, \ldots, n.$$

Then it can easily be computed that

$$(u_i^{(n)}, u_j^{(n)}) = c_{ij}.$$

Having done this for all positive integers n, we construct a sequence

$$u_1, u_2, \ldots, u_n, \ldots$$

using the following inductive process:

1) We set

$$u_1 = u_1^{(1)};$$

2) assuming $u_1, u_2, \ldots, u_N$ to be already known and to satisfy the requirement (1.6) for all $m, n \leq N$, we set

$$T_N u_n^{N+1} = u_n, \quad n = 1, 2, \ldots, N.$$

The operator T_N is an isometry on the set $M_N = \{u_1^{(N+1)}, \ldots, u_N^{(N+1)}\}$.

By Lemma 1 it can be extended to an isometry on H_{M_N}. Since H_{M_N} and $T_{M_N} H_{M_N}$ are finite dimensional, T_N can be extended to an isometry on the whole of H (see, for example, Theorem 2.49 from M. Stone's book [3]). Having chosen such an extension, we set

$$u_{N+1} = T_N u_{N+1}^{(N+1)}.$$

Clearly, the elements $u_1, u_2, \ldots, u_{N+1}$ satisfy (1.6) for all $m, n \leq N+1$.

§2. Operator calculus in H

In accordance with Theorem 1 we denote by H_x the minimal closed linear subspace of H containing all the elements of a stationary sequence $\{x(t)\}$ and consider the unitary operator U_x defined by the formula

$$U_x x(t) = x(t+1). \tag{2.1}$$

Let

$$U_x = \int_{-\pi}^{\pi} e^{i\lambda} dE_x(\lambda) \tag{2.2}$$

be the spectral representation [2] of U_x. For $-\pi \leq \lambda \leq +\pi$ we set

$$F_{xx}(\lambda) = \mu |E_x(\lambda) x(0)|^2. \tag{2.3}$$

---

[2] See M. Stone's book [3] on spectral representations of linear operators. In what follows we will only use spectral families $E(\lambda)$ corresponding to unitary operators for which $E(-\pi) = 0$ and $E(+\pi) = 1$. All the theorems from Stone's book used below (formulated there for the more general case of spectral families corresponding to spectral representations of self-adjoint operators of the form $A = \int_{-\infty}^{\infty} \lambda dE(\lambda)$) are applicable to them.

After Stone ([3], Chapter 6, §1), we denote by L_x^2 the class of all complex functions ϕ, measurable with respect to $F_{xx}(\lambda)$, defined on the segment $-\pi \leq \lambda \leq \pi$, and for which

$$\int_{-\pi}^{\pi} |\phi(\lambda)|^2 dF_{xx}(\lambda) \tag{2.4}$$

is finite. Two functions ϕ_1 and ϕ_2 from L_x^2 are considered to be identical if

$$\phi_1(\lambda) = \phi_2(\lambda)$$

almost everywhere with respect to $F_{xx}(\lambda)$.

To every function ϕ from L_x^2 there corresponds (see [3] Theorem 6.1) the operator

$$T(\phi) = \int_{-\pi}^{\pi} \phi(\lambda) dE_x(\lambda), \tag{2.5}$$

whose domain $\delta(\phi)$ consists of all the elements z of H_x for which the integral

$$\int_{-\pi}^{\pi} |\phi(\lambda)|^2 d|E(\lambda)z|^2 \tag{2.6}$$

is finite. The finiteness of (2.4) implies that $x(0)$ belongs to $\delta(\phi)$ for any ϕ from L_x^2. Therefore to each ϕ in L_x^2 there corresponds a certain element

$$z_\phi = T(\phi)x(0) = \int_{-\pi}^{\pi} \phi(\lambda) dE_x(\lambda)x(0) \tag{2.7}$$

of the space H_x.

These definitions allow us to state the following two Lemmas, which are almost immediate corrolaries of Theorems 6.1 and 6.2 of Stone [3].

Lemma 3. *The map $\phi \rightarrow T(\phi)$ is a one-to-one correspondence between L_x^2 and the class $\mathcal{T}_x$ of all operators T representable in the form (2.5), where ϕ belongs to L_x^2.*

Under this map

$$a\phi \rightarrow aT(\phi), \tag{3a}$$

$$\phi_1 + \phi_2 \rightarrow T(\phi_1) + T(\phi_2); \tag{3b}$$

if $\phi_1\phi_2$ belongs to L_x^2, then

$$\phi_1\phi_2 \rightarrow T(\phi_1)T(\phi_2) = T(\phi_2)T(\phi_1); \tag{3c}$$

$$e^{it\lambda} \rightarrow U_x^t; \tag{3d}$$

if $\chi_\lambda(\mu) = 1$ *for* $\mu \leq \lambda$ *and* $\chi_\lambda(\mu) = 0$ *for* $\mu > \lambda$, *then*

$$\chi_\lambda \to E_x(\lambda); \tag{3e}$$

Lemma 4. *The map*

$$\phi \to z_\phi$$

maps the class L_x^2 *in one-to-one fashion onto the whole space* H_x. *Under this map*

$$a\phi \to az_\phi, \tag{4a}$$

$$\phi_1 + \phi_2 \to z_{\phi_1} + z_{\phi_2}; \tag{4b}$$

if $\phi_1\phi_2$ *belongs to* L_x^2, *then*

$$\phi_1\phi_2 \to T(\phi_1)z_{\phi_2} = T(\phi_2)z_{\phi_1}, \tag{4c}$$

$$e^{it\lambda}\phi \to U_x^t z_\phi, \tag{4d}$$

$$\chi_\lambda(\phi) \to E_x(\lambda)z_\phi, \tag{4e}$$

$$(z_{\phi_1}, z_{\phi_2}) = \int_{-\pi}^{\pi} \phi_1(\lambda)\phi_2(\lambda)dF_{xx}(\lambda), \tag{4f}$$

$$(E_x(\lambda)z_{\phi_1}, z_{\phi_2}) = \int_{-\pi}^{\pi} \phi_1(\lambda)\phi_2(\lambda)dF_{xx}(\lambda). \tag{4g}$$

The properties (3a), (3b) and (3c) of the map $\phi \to T\phi$ follow from Theorem 6.2 of Stone [3]. The properties (3d) and (3e) can be derived by comparing (2.5) with the formulas

$$U_x^t = \int_{-\pi}^{\pi} e^{it\lambda}dE_x(\lambda), \tag{2.8}$$

$$E_x(\lambda) = \int_{-\pi}^{\lambda} dE_x(\mu) = \int_{-\pi}^{\pi} \chi_\lambda(\mu)dE_x(\mu). \tag{2.9}$$

The fact that the map $\phi \to T(\phi)$ is one-to-one follows from the fact that for $\phi_1 \neq \phi_2$ (in the sense of equality in L_x^2) the difference

$$T(\phi_1)x(0) - T(\phi_2)x(0) = \int_{-\pi}^{\pi} [\phi_1(\lambda) - \phi_2(\lambda)]dE_x(\lambda)x(0)$$

has norm

$$\int_{-\pi}^{\pi} |\phi_1(\lambda) - \phi_2(\lambda)|^2 d|E_x(\lambda)x(0)|^2 = \int_{-\pi}^{\pi} |\phi_1(\lambda) - \phi_2(\lambda)|^2 dF_{xx}(\lambda) \neq 0$$

and consequently

$$T(\phi_1)x(0) \neq T(\phi_2)x(0).$$

To prove Lemma 4 we note the following: the set $\mathfrak{M}$ of all elements of H_x representable in the form (2.7), where ϕ belongs to L_x^2, is, according to Theorem 6.2. of Stone [3], a closed linear subspace of H_x; since

$$x(t) = U_x^t x(0) = \int_{-\pi}^{\pi} e^{it\lambda} dE_x(\lambda)x(0) \tag{2.10}$$

and $e^{it\lambda}$ belongs to L_x^2, the set $\mathfrak{M}$ contains all the $x(t)$, which implies that

$$\mathfrak{M} = H_x.$$

In view of this remark, the fact that $\phi \to z_\phi$ is a one-to-one correspondence and its properties (4a), (4b) and (4f) follow directly from Theorem 6.2 of Stone [3].

The properties (4c), (4d) and (4e) of the correspondence $\phi \to z_\phi$ follow from the corresponding properties (3c), (3d) and (3e) of the map $\phi \to T(\phi)$. Finally, (4e) and (4f) imply (4g):

$$(E_x(\lambda)z_{\phi_1}, z_{\phi_2}) = \int_{-\pi}^{\pi} [\chi_\lambda(\mu)\phi_1(\mu)]\bar{\phi}_2(\mu)dF_{xx}(\mu) =$$

$$= \int_{-\pi}^{\lambda} \phi_1(\lambda)\bar{\phi}_2(\lambda)dF_{xx}(\lambda).$$

This proves Lemmas 3 and 4.

§3. Basic spectral properties

Theorem 2. *For any stationary sequence $\{x(t)\}$ the values $B_{xx}(k)$ can be represented as*

$$B_{xx}(k) = \int_{-\pi}^{\pi} e^{ik\lambda} dF_{xx}(\lambda), \tag{3.1}$$

where F_{xx} is a real continuous function, non-decreasing from the right, defined on the interval $-\pi \leq \lambda \leq \pi$, and such that

$$F_{xx}(-\pi) = 0. \tag{3.2}$$

The function $F_{xx}(\lambda)$ satisfying these conditions is defined uniquely by B_{xx}.

Here we note also that $F_{xx}(\lambda)$, which in what follows will be called the *spectral function* of the sequence $\{x(t)\}$, coincides with the function defined by (2.3), as will be clear from the text below.

Theorem 3. *If two stationary sequences $\{x(t)\}$ and $\{y(t)\}$ are jointly stationary, then*

$$B_{xy}(k) = \int_{-\pi}^{\pi} e^{ik\lambda} dF_{xy}(\lambda), \tag{3.3}$$

where $F_{xy}(\lambda)$ is an (in general complex) function continuous from the right, of bounded variation, defined on the interval $-\pi \leq \lambda \leq \pi$, for which

$$F_{xy}(-\pi) = 0. \tag{3.4}$$

The function $F_{xy}(\lambda)$ satisfying these conditions is uniquely defined by B_{xy}.

Since any stationary sequence is jointly stationary with respect to itself, Theorem 2 can be considered as a particular case of Theorem 3, except for the statement that $F_{xx}(\lambda)$ is real and non-decreasing.

To prove Theorem 3, we consider the spectral representation

$$U_{xy} = \int_{-\pi}^{\pi} e^{i\lambda} dE_{xy}(\lambda) \tag{3.5}$$

of the operator U_{xy} whose existence follows from Theorem 1 for

$$n = 2, \quad x_1 = x, \quad x_2 = y.$$

For $-\pi \leq \lambda \leq \pi$, we set

$$(E_{xy}(\lambda)x(0),\ y(0)) = F_{xy}(\lambda) \tag{3.6}$$

It is known (see Stone [3], p.177) that $F_{xy}(\lambda)$ is a function, continuous from the right and of bounded variation. The equality

$$E_{xy}(-\pi) = 0 \tag{3.7}$$

implies that

$$F_{xy}(-\pi) = 0.$$

Finally,

$$B_{xy}(k) = (x(k), y(0)) = (U^k x(0), y(0)) = \Big(\int_{-\pi}^{\pi} e^{ik\lambda} dE(\lambda)x(0), y(0)\Big) =$$

$$= \int_{-\pi}^{\pi} e^{ik\lambda} d(E(\lambda)x(0), y(0)) = \int_{-\pi}^{\pi} e^{ik\lambda} dF_{xy}(\lambda).$$

Thus, $F_{xy}(\lambda)$ defined by (3.6) satisfies all the conditions of the theorem.

For the particular case of one sequence $\{x(t)\} = \{y(t)\}$ the function

$$F_{xx}(\lambda) = (E_x(\lambda)x(0), x(0)) = |E_x(\lambda)x(0)|^2 \tag{3.8}$$

is real and non-decreasing (see Stone [3], p.189). This justifies the corresponding statement of Theorem 2.

Elementary properties of Fourier series for functions of bounded variation (see, for example, A. Zygmund [4], p.13) imply that

$$F_{xy}(\lambda) = \frac{1}{2\pi} W_{xy}(\lambda + 0) + C, \tag{3.9}$$

where

$$W_{xy}(\lambda) = B_{xy}(0)\lambda - \sum_{k \neq 0} \frac{B_{xy}(k)}{ik} e^{-ik\lambda}, \tag{3.10}$$

and the constant C can be found from the condition

$$F_{xy}(-\pi) = 0. \tag{3.11}$$

Formulas (3.9) and (3.10) show that $F_{xy}(\lambda)$ is uniquely defined by B_{xy}. Thus, Theorems 2 and 3 are completely proved.

Note also that (3.8) directly implies that

$$F_{yx}(\lambda) = \overline{F}_{xy}(\lambda). \tag{3.12}$$

Theorem 4 (H. Cramér). *If the sequences*

$$\{x_1(t)\}, \{x_2(t)\}, \ldots, \{x_n(t)\}$$

are stationary and jointly stationary, then for any $-\pi \leq \alpha \leq \beta \leq \pi$ *the increments*

$$\Delta F_{x_\mu x_\nu} = F_{x_\mu x_\nu}(\beta) - F_{x_\mu x_\nu}(\alpha) \tag{3.13}$$

form a non-negative hermitian matrix, that is,

$$\sum_{\mu,\nu=1}^{n} \Delta F_{x_\mu x_\nu} \xi_\mu \bar{\xi}_\nu \geq 0, \tag{3.14}$$

for any complex numbers $\xi_1, \xi_2, \ldots, \xi_n$.

Conversely, if the functions

$$F_{\mu\nu}(\lambda), \quad \mu, \nu = 1, 2, \ldots, n,$$

which are continuous from the right and are defined on $-\pi \leq \lambda \leq \pi$, *satisfy the conditions*

$$\sum_{\mu,\nu=1}^{n} \Delta F_{\mu\nu}\xi_\mu\bar{\xi}_\nu \geq 0, \tag{3.15}$$

$$F_{\mu\nu}(-\pi) = 0, \tag{3.16}$$

then there exists in H *a system of stationary and jointly stationary sequences*

$$\{x_1(t)\}, \{x_2(t)\}, \ldots, \{x_n(t)\}$$

for which

$$F_{x_\mu x_\nu}(\lambda) = F_{\mu\nu}(\lambda). \tag{3.17}$$

To prove this first part of the theorem we consider the spectral representation

$$U = \int_{-\pi}^{\pi} e^{i\lambda} dE(\lambda)$$

of the operator

$$U = U_{x_1 x_2 \ldots x_n}$$

of Theorem 1 and setting

$$z = \sum_{\mu=1}^{n} \xi_\mu x_\mu(0),$$

we consider on the interval $-\pi \leq \lambda \leq \pi$ the function

$$\phi(\lambda) = |E(\lambda)z|^2 = (E(\lambda)z, z) = \Big(E(\lambda)\sum_{\mu=1}^{n} \xi_\mu x_\mu(0), \sum_{\nu=1}^{n} \xi_\nu x_\nu(0)\Big) =$$

$$= \sum_{\mu,\nu=1}^{n} (\xi_\mu E(\lambda)x_\mu(0), \xi_\nu x_\nu(0)) = \sum_{\mu,\nu=1}^{n} \xi_\mu\bar{\xi}_\nu (E(\lambda)x_\mu(0), x_\nu(0)) =$$

$$= \sum_{\mu,\nu=1}^{n} F_{x_\mu x_\nu}(\lambda)\xi_\mu\xi_\nu.$$

Since $\phi(\lambda)$ is real and monotone non-decreasing (see Stone [3], p.189), it follows that for $\alpha < \beta$

$$\phi(\beta) - \phi(\alpha) = \Delta\phi = \sum_{\mu,\nu=1}^{n} \Delta F_{x_\mu x_\nu} \bar{\xi}_\mu \xi_\nu \geq 0,$$

as required.

Passing to the second part of the theorem, we set

$$B_{\mu\nu}(k) = \int_{-\pi}^{\pi} e^{ik\lambda} dF_{\mu\nu}(\lambda).$$

According to Lemma 2, in order that there exist elements $x_\mu(t)$ in H corresponding to any integer t and any $\mu = 1, 2, \ldots, n$ such that

$$(x_\mu(t'),\ x_\nu(t'')) = B_{\mu\nu}(t' - t''),$$

it suffices to show that

$$B = \sum_{p,q=1}^{r} B_{\mu_p \mu_q}(t_p - t_q)\xi_p \bar{\xi}_q \geq 0$$

for any r, μ_p, t_p, ξ_p. This is always the case since, setting

$$\xi_p(\lambda) = e^{it_p\lambda}\xi_p, \quad \zeta_\mu(\lambda) = \sum\nolimits^{(\mu)} \xi_p(\lambda),$$

where the sum $\sum^{(\mu)}$ is taken over all p for which $\mu_p = \mu$, we have, by (3.15),

$$B = \sum_{p,q=1}^{r} \int_{-\pi}^{\pi} e^{i(t_p - t_q)} dF_{\mu_p \mu_q}(\lambda)\xi_p \bar{\xi}_q = \sum_{p,q=1}^{r} \int_{-\pi}^{\pi} \xi_p(\lambda)\bar{\xi}_q(\lambda) dF_{\mu_p \mu_q}(\lambda) =$$

$$= \int_{-\pi}^{\pi} \sum_{\mu,\nu=1}^{n} \zeta_\mu(\lambda)\bar{\zeta}_\nu(\lambda) dF_{\mu\nu}(\lambda) \geq 0.$$

For the elements $x_\mu(t)$ of H, whose existence is thus established, we have

$$B_{x_\mu x_\nu}(k) = (x_\mu(t+k), x_\nu(t)) = B_{\mu\nu}(k).$$

Therefore the sequences

$$\{x_1(t)\}, \{x_2(t)\}, \ldots, \{x_n(t)\}$$

are stationary and jointly stationary. It is easy to see that in this case

$$F_{x_\mu x_\nu}(\lambda) = F_{\mu\nu}(\lambda).$$

Theorem 4 is completely proved.

For $n = 1$ the second part of Theorem 4 implies

Theorem 5. *For any real non-decreasing function $F(\lambda)$, continuous from the right, defined on $-\pi \le \lambda \le \pi$ and such that $F(-\pi) = 0$, there exists a stationary sequence $\{x(t)\}$ with*

$$F_{xx}(\lambda) = F(\lambda).$$

For $n = 2$ (3.15) is equivalent to the set of conditions

$$\Delta F_{x_1x_2} \ge 0, \quad \Delta F_{x_2x_2} \ge 0, \quad |\Delta F_{x_1x_2}|^2 \le \Delta F_{x_1x_1}\Delta F_{x_2x_2}$$

(see Cramér, [2], p. 227). In this case Theorem 4 gives

Theorem 6. *If the sequences $\{x(t)\}$ and $\{y(t)\}$ are stationary and jointly stationary, then for any $-\pi \le \alpha < \beta \le \pi$,*

$$|\Delta F_{xy}|^2 \le \Delta F_{xx}\Delta F_{yy}. \tag{3.18}$$

Conversely, if the functions $F_{11}(\lambda), F_{22}(\lambda)$ and $F_{12}(\lambda)$ defined on $-\pi \le \lambda \le \pi$ are continuous from the right and satisfy the conditions

$$\Delta F_{11} \ge 0, \quad \Delta F_{22} \ge 0, \quad |\Delta F_{12}|^2 \le \Delta F_{11}\Delta F_{22},$$
$$F_{11}(-\pi) = F_{22}(-\pi) = F_{12}(-\pi) = 0, \tag{3.19}$$

then there exist stationary and jointly stationary sequences $\{x(t)\}$ and $\{y(t)\}$ for which $F_{xx}(\lambda) = F_{11}(\lambda)$, $F_{yy}(\lambda) = F_{22}(\lambda)$, $F_{xy}(\lambda) = F_{12}(\lambda)$.

In what follows we require the following strengthened version of Theorem 6:

Theorem 7. *If $F_{11}(\lambda), F_{22}(\lambda)$ and $F_{12}(\lambda)$ satisfy all the conditions of the second part of Theorem 6, the stationary sequence $\{x(t)\}$ is such that*

$$F_{xx}(\lambda) = F_{11}(\lambda)$$

and the orthogonal complement $H \ominus H_x$ of H_x in H is infinite dimensional, then there exists a sequence $\{y(t)\}$ in H, stationary and jointly stationary with $\{x(t)\}$, for which

$$F_{yy}(\lambda) = F_{22}(\lambda), \quad F_{xy}(\lambda) = F_{12}(\lambda).$$

To prove Theorem 7, we note that, by Theorem 6, the conditions assumed with respect to $F_{11}(\lambda)$, $F_{22}(\lambda)$ and $F_{12}(\lambda)$ imply the existence of stationary and jointly stationary sequences $\{x^*(t)\}$ and $\{y^*(t)\}$ for which

$$F_{x^*x^*}(\lambda) = F_{11}(\lambda), \quad F_{y^*y^*}(\lambda) = F_{22}(\lambda), \quad F_{x^*y^*}(\lambda) = F_{12}(\lambda).$$

For all integers t we set

$$Tx^*(t) = x(t).$$

It can easily be verified that T is an isometry on the set $\{x^*(t)\}$. By Lemma 1, T extends as an isometry onto the whole space H_{x^*}. Then clearly,

$$TH_{x^*} = H_x.$$

Since the orthogonal complement $H \ominus H_x$ is infinite dimensional, T can be extended as an isometry from H_x onto the whole space H_{xy} (see Stone [3], Theorem 2.49). Having chosen such an extension, we set

$$Ty^*(t) = y(t).$$

It is easy to see that the sequence $\{y(t)\}$ has all the properties required by Theorem 7.

§4. Subordinate sequences

Definition 3. A stationary sequence $\{y(t)\}$ is said to be *subordinate* to a stationary sequence $\{x(t)\}$ if $\{x(t)\}$ and $\{y(t)\}$ are jointly stationary and all the elements of $\{y(t)\}$ belong to H_x.

Clearly, if $\{y(t)\}$ is subordinate to $\{x(t)\}$, then

$$H_{xy} = H_x, \tag{4.1}$$

$$U_{xy} = U_x. \tag{4.2}$$

Using the notions introduced in §2 we will prove the following theorem.

Theorem 8. *To every sequence $\{y(t)\}$ subordinate to $\{x(t)\}$ there corresponds a unique function $\phi_y^{(x)}(\lambda)$ of class L_x^2 for which*

$$F_{yy}(\lambda) = \int_{-\pi}^{\lambda} |\phi_y^{(x)}(\lambda)|^2 dF_{xx}(\lambda), \tag{4.3}$$

$$F_{yx}(\lambda) = \int_{-\pi}^{\lambda} \phi_y^{(x)}(\lambda) dF_{xx}(\lambda). \tag{4.4}$$

The map

$$\{y(t)\} \to \phi_y^{(x)}(\lambda)$$

is a one-to-one correspondence from the class Y_x of all sequences $\{y(t)\}$ subordinate to $\{x(t)\}$ onto the class of L_x^2 functions $\phi(\lambda)$. For two sequences $\{y_1(t)\}$ and $\{y_2(t)\}$ from Y_x we have

$$F_{y_1 y_2}(\lambda) = \int_{-\pi}^{\lambda} \phi_{y_1}^{(x)}(\lambda)\overline{\phi_{y_2}^{(x)}(\lambda)} dF_{xx}(\lambda). \tag{4.5}$$

The assertion of Theorem 8 that the function $\phi_y^{(x)}(\lambda)$ is unique is to be understood in the sense of identity of functions $\phi(\lambda)$ from the class L_x^2 given in §2. The one-to-one correspondence follows from (4.4) by virtue of the equality

$$\phi_y^{(x)}(\lambda) = dF_{yx}(\lambda)/dF_{xx}(\lambda). \tag{4.6}$$

We will prove the existence of functions $\phi_y^{(x)}(\lambda)$ for any sequence $\{y(t)\}$ from the class Y_x using Lemma 4 in §2. If $\{y(t)\}$ belongs to Y_x, then $y(0)$ belongs to H_x, and according to Lemma 4, there exists a function $\phi_y^{(x)}(\lambda)$ of class L_x^2 such that

$$y(0) = z_{\phi_y^{(x)}}. \tag{4.7}$$

By (3.6) and (4g) of Lemma 4 we have

$$F_{y_1 y_2}(\lambda) = (E_{y_1 y_2}(\lambda) y_1(0), y_2(0)) = (E_x(\lambda) y_1(0), y_2(0)) =$$

$$= \int_{-\pi}^{\lambda} \phi_{y_1}^{(x)}\overline{\phi_{y_2}^{(x)}}(\lambda) dF_{xx}(\lambda)$$

for any sequences $\{y_1(t)\}$ and $y_2(t)\}$ in Y_x, which proves (4.5). In the particular case $y_1 = y$, $y_2 = x$ we obtain (4.4), while for $y_1 = y_2 = x$ we obtain (4.3).

The one-to-one correspondence between $\{y(t)\}$ and $\phi_y^{(x)}(\lambda)$ also follows from Lemma 4.

An important addition to Theorem 8 is

Theorem 9. *If two sequences $\{x(t)\}$ and $\{y(t)\}$ are stationary and jointly stationary, then $\{y(t)\}$ is subordinate to $\{x(t)\}$ if and only if there exists a function $\phi(\lambda)$ in L_x^2 for which*

$$F_{yy}(\lambda) = \int_{-\pi}^{\lambda} |\phi(\lambda)|^2 dF_{xx}(\lambda), \tag{4.8}$$

$$F_{yx}(\lambda) = \int_{-\pi}^{\lambda} \phi(\lambda) dF_{xx}(\lambda). \tag{4.9}$$

The necessity of the condition follows from Theorem 8. Let us prove its sufficiency. If there exists $\phi(\lambda)$ in L_x^2 satisfying (4.8) and (4.9), then by Theorem 8 there exists a sequence $\{y^*(t)\}$ subordinate to $\{x(t)\}$ for which

$$\phi_{y^*}^{(x)}(\lambda) = \phi(\lambda).$$

Therefore

$$F_{y^*y^*}(\lambda) = F_{yy}(\lambda), \quad F_{y^*x}(\lambda) = F_{yx}(\lambda).$$

By Lemma 1 we can determine an isometry T on H_{xy^*} such that for any integer t,

$$Tx(t) = x(t), \quad Ty^*(t) = y(t).$$

Clearly

$$T(H_{xy^*}) = H_{xy}, \quad T(H_x) = H_x. \tag{4.10}$$

Since $\{y^*(t)\}$ is subordinate to $\{x(t)\}$,

$$H_{xy^*} = H_x. \tag{4.11}$$

Equalities (4.10) and (4.11) imply that

$$H_{xy} = H_x,$$

which means that $\{y(t)\}$ is subordinate[3] to $\{x(t)\}$.

[3] This conclusion makes it clear that $\{y(t)\}$ coincides with $\{x(t)\}$.

Definition 4. Stationary sequences $\{x(t)\}$ and $\{y(t)\}$ are called *equivalent* if each of them is subordinate to the other, that is, if they are jointly stationary and if

$$H_x = H_y.$$

Theorem 10. *A sequence $\{y(t)\}$ subordinate to $\{x(t)\}$ is equivalent to $\{x(t)\}$ if and only if $\phi_y^{(x)}(\lambda)$ is non-vanishing almost everywhere with respect to $F_{xx}(\lambda)$. If this is the case, then*

$$\phi_x^{(y)}(\lambda) = 1/\phi_y^{(x)}(\lambda) \tag{4.12}$$

almost everywhere with respect to $F_{xx}(\lambda)$ and $F_{yy}(\lambda)$.

1°. We prove the necessity of the condition. Assume that $\{y(t)\}$ is subordinate to $\{x(t)\}$, and $\{x(t)\}$ is subordinate to $\{y(t)\}$. Then it follows from (4.3) and the formula

$$F_{xx}(\lambda) = \int_{-\pi}^{\lambda} |\phi_x^{(y)}(\lambda)|^2 dF_{yy}(\lambda) \tag{4.13}$$

obtained by interchanging the indices, that $F_{xx}(\lambda)$ and $F_{yy}(\lambda)$ are absolutely continuous with respect to each other. Therefore the notions "almost everywhere with respect to F_{xx}" and "almost everywhere with respect to F_{yy}" coincide. By (4.3) and (4.13),

$$F_{xx}(\lambda) = \int_{-\pi}^{\lambda} |\phi_x^{(y)}(\lambda)|^2 dF_{yy}(\lambda) = \int_{-\pi}^{\lambda} |\phi_x^{(y)}(\lambda)|^2 |\phi_y^{(x)}(\lambda)|^2 dF_{xx}(\lambda)$$

and hence, almost everywhere with respect to F_{xx},

$$|\phi_x^{(y)}(\lambda)|^2 |\phi_y^{(x)}(\lambda)|^2 = 1. \tag{4.14}$$

Formula (4.14) shows that $\phi_y^{(x)}(\lambda) \neq 0$ almost everywhere with respect to F_{xx}, as required.

2°. We now prove the sufficiency of the condition. Assume that $\{y(t)\}$ is subordinate to $\{x(t)\}$ and $\phi_y^{(x)}(\lambda) \neq 0$ almost everywhere with respect to F_{xx}.

Since, according to (4.3), $F_{yy}(\lambda)$ is absolutely continuous with respect to $F_{xx}(\lambda)$, the function $\phi_y^{(x)}(\lambda)$ is also absolutely continuous almost everywhere with respect to $F_{yy}(\lambda)$. Therefore we can write

$$\int_{-\pi}^{\lambda} \frac{dF_{yy}(\lambda)}{|\phi_y^{(x)}(\lambda)|^2} = \int_{-\pi}^{\lambda} \frac{|\phi_y^{(x)}(\lambda)|^2 dF_{xx}(\lambda)}{|\phi_y^{(x)}(\lambda)|^2} = \int_{-\pi}^{\lambda} dF_{xx}(\lambda) = F_{xx}(\lambda), \tag{4.15}$$

$$\int_{-\pi}^{\lambda} \frac{dF_{yy}(\lambda)}{\phi_y^{(x)}(\lambda)} = \int_{-\pi}^{\lambda} \frac{|\phi_y^{(x)}(\lambda)|^2 dF_{xx}(\lambda)}{\phi_y^{(x)}(\lambda)} = \int_{-\pi}^{\lambda} \overline{\phi_y^{(x)}} dF_{xx}(\lambda) =$$

$$= \overline{\int_{-\pi}^{\lambda} \phi_y^{(x)}(\lambda) dF_{xx}(\lambda)} = \overline{F_{yx}(\lambda)} = F_{xy}(\lambda). \tag{4.16}$$

Formulas (4.15) and (4.16) show that the function

$$\phi(\lambda) = 1/\phi_y^{(x)}(\lambda) \tag{4.17}$$

satisfies the conditions

$$F_{xx}(\lambda) = \int_{-\pi}^{\lambda} |\phi(\lambda|^2 dF_{yy}(\lambda), \tag{4.18}$$

$$F_{xy}(\lambda) = \int_{-\pi}^{\lambda} \phi(\lambda) dF_{yy}(\lambda). \tag{4.19}$$

By Theorem 9 this implies that $\{x(t)\}$ is subordinate to $\{y(t)\}$, as required.

3°. By Theorem 8, (4.15) and (4.16) imply that

$$\phi_x^{(y)}(\lambda) = \phi(\lambda) = 1/\phi_y^{(x)}(\lambda).$$

§5. Decomposition of stationary sequences into orthogonal summands

Definition 5. Two sequences $\{y_1(t)\}$ and $\{y_2(t)\}$ are called *mutually orthogonal* if

$$(y_1(t+k),\ y_2(t)) = 0 \tag{5.1}$$

for all integers k and t.

Clearly, two stationary and jointly stationary sequences $\{y_1(t)\}$ and $\{y_2(t)\}$ are mutually orthogonal if and only if

$$B_{y_1 y_2}(k) \equiv 0, \tag{5.2}$$

or, equivalently, if

$$F_{y_1y_2}(\lambda) \equiv 0. \tag{5.3}$$

Theorem 11. *If two sequences $\{y_1(t)\}$ and $\{y_2(t)\}$ are stationary and mutually orthogonal, then*

a) *the sequence $\{x(t)\} = \{y_1(t) + y_2(t)\}$ is stationary;*

b) *$\{y_1(t)\}$ and $\{y_2(t)\}$ are jointly stationary both with respect to each other and with respect to $\{x(t)\}$:*

(c_1) $B_{xx}(k) = B_{y_1y_1}(k) + B_{y_2y_2}(k)$,

(c_2) $B_{y_1x}(k) = B_{y_1y_1}(k)$,

(c_3) $B_{y_2x}(k) = B_{y_2y_2}(k)$,

(d_1) $F_{xx}(\lambda) = F_{y_1y_1}(\lambda) + F_{y_2y_2}(\lambda)$,

(d_2) $F_{y_1x}(\lambda) = F_{y_1y_1}(\lambda)$,

(d_3) $F_{y_2x}(\lambda) = F_{y_2y_2}(\lambda)$.

Theorem 11 is proved by simple computations based on (0.2), (3.9)–(3.11).

In the conditions of Theorem 11, by (d_1) and the non-negativeness of the increments ΔF_{xx}, $\Delta F_{y_1y_1}$, $\Delta F_{y_2y_2}$ (for non-negative $\Delta\lambda$) we have:

$$|\Delta F_{y_1y_1}/\Delta F_{xx}| \le 1, \quad |\Delta F_{y_2y_2}/\Delta F_{xx}| \le 1. \tag{5.4}$$

Therefore the functions

$$\psi_{y_1}^{(x)}(\lambda) = \frac{dF_{y_1y_1}(\lambda)}{dF_{xx}(\lambda)}, \quad \psi_{y_2}^{(x)}(\lambda) = \frac{dF_{y_2y_2}(\lambda)}{dF_{xx}(\lambda)} \tag{5.5}$$

are bounded and uniquely defined (in the sense adopted in [3]) by the formulas

$$\begin{aligned} F_{y_1y_1}(\lambda) &= \int_{-\pi}^{\lambda} \psi_{y_1}^{(x)}(\lambda) dF_{xx}(\lambda), \\ F_{y_2y_2}(\lambda) &= \int_{-\pi}^{\lambda} \psi_{y_2}^{(x)}(\lambda) dF_{xx}(\lambda). \end{aligned} \tag{5.6}$$

Theorem 12. *Under the conditions of Theorem 11, either both sequences $\{y_1(t)\}$ and $\{y_2(t)\}$ are subordinate to $\{x(t)\}$ or neither is. The first of these two cases takes place if and only if*

$$\psi_{y_1}^{(x)}(\lambda)\psi_{y_2}^{(x)}(\lambda) = 0 \tag{5.7}$$

almost everywhere with respect to F_{xx}.

Proof. 1°. If $\{y_1(t)\}$ is subordinate to $\{x(t)\}$, then by virtue of the relation

$$y_2(t) = x(t) - y_1(t),$$

$\{y_2(t)\}$ is also subordinate to $x(t)$. Similarly, if $\{y_2(t)\}$ is subordinate to $\{x(t)\}$, then $\{y_1(t)\}$ is also subordinate to $\{x(t)\}$.

2°. If both sequences $\{y_1(t)\}$ and $\{y_2(t)\}$ are subordinate to $\{x(t)\}$, then, comparing (5.6) and (4.3), we see that almost everywhere with respect to F_{xx},

$$\psi_{y_1}^{(x)} = |\phi_{y_1}^{(x)}|^2, \quad \psi_{y_2}^{(x)} = |\phi_{y_2}^{(x)}|^2, \tag{5.8}$$

but if we take into account (d_2) and (d_3), then comparing (5.6) with (4.4), we see that almost everywhere with respect to F_{xx},

$$\psi_{y_1}^{(x)} = \phi_{y_1}^{(x)}, \quad \psi_{y_2}^{(x)} = \phi_{y_2}^{(x)}. \tag{5.9}$$

Comparing (5.8) and (5.9), we conclude that $\psi_{y_1}^{(x)}$ and $\psi_{y_2}^{(x)}$ are equal to 0 or 1 almost everywhere with respect to F_{xx}. Since, moreover, almost everywhere with respect to F_{xx}

$$\psi_{y_1}^{(x)} + \psi_{y_2}^{(x)} = \frac{dF_{y_1y_1}}{dF_{xx}} + \frac{dF_{y_2y_2}}{dF_{xx}} = \frac{dF_{xx}}{dF_{xx}} = 1, \tag{5.10}$$

it follows that (5.7) holds.

3°. Suppose that (5.7) holds. Then, by (5.10), $\psi_{y_1}^{(x)}$ and $\psi_{y_2}^{(x)}$ are equal to 0 or 1 almost everywhere with respect to F_{xx}.

Hence, almost everywhere with respect to F_{xx} we have

$$\psi_{y_1}^{(x)} = |\psi_{y_1}^{(x)}|^2, \quad \psi_{y_2}^{(x)} = |\psi_{y_2}^{(x)}|^2. \tag{5.11}$$

Formulas (5.11), (5.6), (d_2) and (d_3) imply that

$$\begin{aligned} F_{y_1y_1}(\lambda) &= \int_{-\pi}^{\lambda} |\psi_{y_1}^{(x)}(\lambda)|^2 dF_{xx}(\lambda), \\ F_{y_1x}(\lambda) &= \int_{-\pi}^{\lambda} \psi_{y_1}^{(x)}(\lambda) dF_{xx}(\lambda), \\ F_{y_2y_2}(\lambda) &= \int_{-\pi}^{\lambda} |\psi_{y_2}^{(x)}(\lambda)|^2 dF_{xx}(\lambda), \\ F_{y_2x}(\lambda) &= \int_{-\pi}^{\lambda} \psi_{y_2}^{(x)}(\lambda) dF_{xx}(\lambda). \end{aligned} \tag{5.12}$$

By Theorem 9, the existence of $\psi_{y_1}^{(x)}$ and $\psi_{y_2}^{(x)}$ satisfying (5.12) implies that the sequences $\{y_1(t)\}$ and $\{y_2(t)\}$ are subordinate to $\{x(t)\}$.

Theorem 13. *Suppose that for a stationary sequence $\{x(t)\}$, the orthogonal complement $H \ominus H_x$ of the space H_x in H is infinite dimensional. Then corresponding to any representation of $F_{xx}(\lambda)$ as a sum*

$$F_{xx}(\lambda) = F_1(\lambda) + F_2(\lambda) \tag{5.13}$$

of two real non-decreasing functions $F_1(\lambda)$ and $F_2(\lambda)$ continuous from the right and for which $F_1(-\pi) = F_2(-\pi) = 0$, there is at least one representation of $\{x(t)\}$ as

$$x(t) = y_1(t) + y_2(t), \tag{5.14}$$

where $\{y_1(t)\}$ and $\{y_2(t)\}$ are mutually orthogonal and have spectral functions

$$F_{y_1y_1}(\lambda) = F_1(\lambda), \quad F_{y_2y_2}(\lambda) = F_2(\lambda). \tag{5.15}$$

To prove this we set

$$F_{11}(\lambda) = F_{xx}(\lambda), \quad F_{22}(\lambda) = F_1(\lambda), \quad F_{12}(\lambda) = F_1(\lambda).$$

The functions $F_{11}(\lambda)$, $F_{22}(\lambda)$ and $F_{12}(\lambda)$ satisfy the conditions of the second part of Theorem 6. Therefore, by Theorem 7 there exists in H a sequence $\{y_1(t)\}$ that is stationary and jointly stationary with respect to $\{x(t)\}$ and is such that

$$F_{y_1y_1}(\lambda) = F_{22}(\lambda) = F_1(\lambda), \quad F_{xy_1}(\lambda) = F_{12}(\lambda) = F_1(\lambda).$$

Setting

$$y_2(t) = x(t) - y_1(t),$$

and using simple computations we see that $\{y_1(t)\}$ and $\{y_2(t)\}$ are mutually orthogonal and satisfy (5.15).

Theorem 12 gives conditions that should be imposed on $F_1(\lambda)$ and $F_2(\lambda)$ in the representation (5.13) in order to make $\{y_1(t)\}$ and $\{y_2(t)\}$ subordinate to $\{x(t)\}$, that is, to ensure that they lie in H_x. Under these additional conditions the requirement that the dimension of the complement $H \ominus H_x$ be infinite is superfluous,[4] and it is possible to strengthen the conclusion of Theorem 13 so that the representation (5.14) is uniquely[5] determined by (5.13).

We obtain the following theorem:

Theorem 14. *For any stationary sequence $\{x(t)\}$ and for each representation of the spectral function $F_{xx}(\lambda)$ in the form*

$$F_{xx}(\lambda) = F_1(\lambda) + F_2(\lambda),$$

where $F_1(\lambda)$ and $F_2(\lambda)$ are real non-decreasing functions of λ, continuous from the right with $F_1(-\pi) = F_2(-\pi) = 0$, and almost everywhere with respect to F_{xx},

$$\frac{dF_1}{dF_{xx}} + \frac{dF_2}{dF_{xx}} = 0, \tag{5.16}$$

there corresponds one representation $\{x(t)\}$ in the form

$$x(t) = y_1(t) + y_2(t),$$

for which $\{y_1(t)\}$ and $\{y_2(t)\}$ are mutually orthogonal and

$$F_{y_1y_1}(\lambda) = F_1(\lambda), \quad F_{y_2y_2}(\lambda) = F_2(\lambda).$$

The sequences $\{y_1(t)\}$ and $\{y_2(t)\}$ are subordinate to $\{x(t)\}$.

§6. Sliding summation

Definition 6. We say that a sequence $\{x(t)\}$ is obtained from a stationary sequence $\{u(t)\}$ by *sliding summation* if we can choose coefficients a_n such that

$$x(t) = \sum_{-\infty}^{+\infty} a_n u(t-n), \tag{6.1}$$

where the series on the right-hand side converges in norm.

---

[4] Indeed, consider an extension H^* of the space H for which $H^* \ominus H$ is infinite dimensional. By Theorem 13 the desired sequences $\{y_1(t)\}$ and $\{y_2(t)\}$ exist in H^* while under the additional condition of Theorem 12 they belong to H_x.

[5] This follows from the fact that according to Theorem 8, the functions $\phi_{y_1}^{(x)} = dF_1/dF_{xx}$ and $\phi_{y_2}^{(x)} = dF_2/dF_{xx}$ uniquely define sequences $\{y_1(t)\}$ and $\{y_2(t)\}$ subordinate to $x(t)$.

Clearly, a sequence $\{x(t)\}$ obtained by sliding summation from a stationary sequence $\{u(t)\}$ is always stationary and subordinate to $\{u(t)\}$. In this section we consider a particular case for which the converse statement is also true.

Definition 7. A sequence $\{u(t)\}$ of elements of H is called *fundamental* if it is stationary and satisfies the conditions

$$B_{uu}(0) = 1, \quad B_{uu}(k) = 0 \text{ for } k \neq 0. \tag{6.2}$$

By (3.9) and (3.10) it is easy to compute that for every fundamental sequence $\{u(t)\}$,

$$F_{uu}(\lambda) = (\lambda + \pi)/2\pi. \tag{6.3}$$

Conversely, every stationary sequence $\{u(t)\}$ with a spectral function of the form (6.3) is fundamental.

The elements of a fundamental sequence $\{u(t)\}$ form a complete orthonormal system in H_u. Therefore, each element z of H_u can be uniquely represented as

$$z = \sum_{t=-\infty}^{+\infty} c_t u(t), \tag{6.4}$$

where

$$\sum_{t=-\infty}^{+\infty} c_t^2 = \|z\|^2 < +\infty. \tag{6.5}$$

The coefficients c_t in (6.4) are determined by the equalities

$$c_t = (z, u(t)). \tag{6.6}$$

Theorem 15. *A stationary sequence $\{x(t)\}$ is obtained from a fundamental sequence $\{u(t)\}$ by sliding summation if and only if $\{x(t)\}$ is subordinate to $\{u(t)\}$. If this condition holds, then the unique representation of $\{x(t)\}$ in the form* (6.1) *is given by the formula*

$$x(t) = \sum_{n=-\infty}^{\infty} B_{xu}(n) u(t-n). \tag{6.7}$$

The necessity of the condition of the theorem has already been proved. If this condition holds, then every element $x(t)$ of the sequence $\{x(t)\}$ belongs to H_u and, by (6.4) and (6.6), is representable in the form

$$x(t) = \sum_{s=-\infty}^{\infty} (x(t), u(s)) u(s) = \sum_{s=-\infty}^{\infty} B_{xu}(t-s) u(s)$$

or, setting $t - s = n$, in the form (6.7).

Taking into account the uniqueness of the representation of any element in the form (6.4), we see that each representation of $\{x(t)\}$ in the form (6.1) is the same as (6.7).

Note also that a comparison of (6.7) with (6.4) and (6.5) yields

$$\sum_{n=-\infty}^{\infty} |B_{xu}(n)|^2 = \|x(t)\|^2 = B_{xx}(0). \tag{6.8}$$

Formulas (4.3), (4.4) and (6.3) imply that

$$F_{xx}(\lambda) = \frac{1}{2\pi} \int_{-\infty}^{\lambda} |\phi_x^{(x)}(\lambda)|^2 d\lambda, \tag{6.9}$$

$$F_{xu}(\lambda) = \frac{1}{2\pi} \int_{-\infty}^{\lambda} \phi_x^{(u)}(\lambda) d\lambda, \tag{6.10}$$

Formula (6.9) shows that $\phi_x^{(u)}(\lambda)$ is a square integrable function. It follows from (3.3) and (6.10) that

$$B_{xu}(n) = \frac{1}{2\pi} \int_{-\pi}^{\pi} e^{in\lambda} \phi_x^{(u)}(\lambda) d\lambda. \tag{6.11}$$

Thus the $B_{xu}(n)$ are the Fourier coefficients of $\phi_x^{(u)}(\lambda)$, that is,

$$\phi_x^{(u)}(\lambda) \sim \sum_{n=-\infty}^{\infty} B_{xu}(n) e^{-in\lambda}. \tag{6.12}$$

With regard to the above formulas we note that, by (6.3), the words "almost everywhere with respect to F_{uu}" mean "excluding a set of Lebesgue measure zero" and the class L_u^2 coincides with the class L^2 of square integrable functions in the usual sense of Lebesgue.

Theorem 16. *In order that a stationary sequence $\{x(t)\}$ can be obtained by sliding summation from some fundamental sequence it is necessary that the spectral function $F_{xx}(\lambda)$ be absolutely continuous.*

If $H \ominus H_x$ is infinite dimensional, then this condition is also sufficient.

The necessity of the condition follows from (6.9). To prove its sufficiency we consider an arbitrary function $\gamma(\lambda)$ in L^2 for which, almost everywhere on $-\pi \leq \lambda \leq \pi$,

$$|\gamma(\lambda)|^2 = 2\pi dF_{xx}/d\lambda \tag{6.13}$$

and define $F_{11}(\lambda), F_{22}(\lambda)$ and $F_{12}(\lambda)$ by the formulas

$$F_{11}(\lambda) = F_{xx}(\lambda), \quad F_{22}(\lambda) = \frac{\lambda+\pi}{2\pi}, \quad F_{12}(\lambda) = \frac{1}{2\pi}\int_{-\pi}^{\lambda} \gamma(\lambda)d\lambda.$$

It is easy to see that these functions satisfy the conditions of the second part of Theorem 6. By Theorem 7, if $H \ominus H_x$ is infinite dimensional, then there exists a sequence $\{u(t)\}$, stationary and jointly stationary with $\{x(t)\}$, for which

$$F_{uu}(\lambda) = F_{22}(\lambda) = (\lambda+\pi)/2\pi, \tag{6.14}$$

$$F_{xu}(\lambda) = F_{12}(\lambda) = \frac{1}{2\pi}\int_{-\pi}^{\lambda} \gamma(\lambda)d\lambda. \tag{6.15}$$

Formula (6.14) shows that $\{u(t)\}$ is a fundamental sequence. The absolute continuity of $F_{xx}(\lambda)$ and formula (6.13) imply that

$$F_{xx}(\lambda) = F_{11}(\lambda) = \frac{1}{2\pi}\int_{-\pi}^{\lambda} |\gamma(\lambda)|^2 d\lambda. \tag{6.16}$$

By Theorem 9, we conclude from (6.15) and (6.16) that $\{x(t)\}$ is subordinate to $\{u(t)\}$. By Theorem 15 this implies that $\{x(t)\}$ can be obtained from $\{u(t)\}$ by sliding summation.

Theorem 17. *A stationary sequence $\{x(t)\}$ can be obtained by sliding summation from a fundamental sequence subordinate to $\{x(t)\}$ if and only if $F_{xx}(\lambda)$ is absolutely continuous and the function*

$$f_{xx}(\lambda) = dF_{xx}(\lambda)/d\lambda \tag{6.17}$$

is almost everywhere positive on $-\pi \le \lambda \le \pi$.

If these conditions are satisfied, then:

A. *Every fundamental sequence from which $\{x(t)\}$ can be obtained by sliding summation is subordinate to $\{x(t)\}$.*

B. *The sequence $\{x(t)\}$ can be obtained by sliding summation from any fundamental sequence subordinate to it.*

C. *A sequence $\{u(t)\}$ subordinate to $\{x(t)\}$ is fundamental if and only if*

$$|\phi_u^{(x)}(\lambda)|^2 = \frac{1}{2\pi} f_{xx}(\lambda) \tag{6.18}$$

almost everywhere on $-\pi \leq \lambda \leq \pi$.

Regarding (6.18), note that if $F_{xx}(\lambda)$ is absolutely continuous and $f_{xx}(\lambda)$ is positive almost everywhere, then the class of sets of measure zero with respect to F_{xx} coincides with the class of sets of Lebesgue measure zero. Therefore, functions of class L_x^2, in particular $\phi_u^{(x)}$, are defined to within a set of measure zero in the sense of Lebesgue.

Proof of Theorem 17.

1°. The necessity of the absolute continuity has already been established in Theorem 16. Let us prove that f_{xx} must be positive almost everywhere. If a fundamental sequence $\{u(t)\}$ is subordinate to $\{x(t)\}$, then by Theorem 9,

$$\frac{\lambda+\pi}{2\pi} = F_{uu}(\lambda) = \int_{-\pi}^{\lambda} |\phi_u^{(x)}(\lambda)|^2 dF_{xx}(\lambda), \tag{6.19}$$

which implies after differentiation with respect to λ, that almost everywhere

$$1/2\pi = |\phi_u^{(x)}(\lambda)|^2 f_{xx}(\lambda). \tag{6.20}$$

It is clear from (6.20) that f_{xx} is positive almost everywhere.

2°. Now assume that F_{xx} is absolutely continuous and f_{xx} is positive almost everywhere. By Theorem 16, the absolute continuity of F_{xx} on its own implies that $\{x(t)\}$ can be obtained by sliding summation from a fundamental sequence $\{u(t)\}$ belonging to an extension H' of H_x. By (6.9), we have almost everywhere

$$|\phi_x^{(u)}(\lambda)|^2 = 2\pi f_{xx}(\lambda) \neq 0.$$

Hence by Theorem 10, $\{u(t)\}$ is subordinate to $\{x(t)\}$ (that is, actually belongs to H_x itself). Thus, we have proved the sufficiency of the conditions of the Theorem and with it, the additional assertion A.

A sequence $\{u(t)\}$ subordinate to $\{x(t)\}$ is fundamental if and only if

$$F_{uu}(\lambda) = \int_{-\pi}^{\lambda} |\phi_u^{(x)}(\lambda)|^2 dF_{xx}(\lambda) = \int_{-\pi}^{\lambda} |\phi_u^{(x)}(\lambda)|^2 f_{xx}(\lambda) d\lambda = \frac{\lambda+\pi}{2\pi},$$

that is, if almost everywhere

$$|\phi_u^{(x)}(\lambda)|^2 f_{xx}(\lambda) = 1/2\pi. \tag{6.21}$$

This proves C.

By (6.21) $|\phi_u^{(x)}(\lambda)|^2 \neq 0$ almost everywhere for any fundamental sequence $\{u(t)\}$ subordinate to $\{x(t)\}$. By Theorem 10 this implies that $\{x(t)\}$ is subordinate to $\{u(t)\}$, and hence, by Theorem 15, $\{x(t)\}$ is obtained from $\{u(t)\}$ by sliding summation. This proves B.

§7. Wold decomposition

For a stationary sequence $\{x(t)\}$ we denote by $H_x(t)$ the smallest closed linear subspace of H_x containing all the $x(s)$ for $s \leq t$, and let S_x be the intersection of all the $H_x(t)$. Clearly,

$$U_x^k H_x(t) = H_x(t+k), \tag{7.1}$$

$$U_x^k S_x = S_x. \tag{7.2}$$

Definition 8. A stationary sequence $\{x(t)\}$ is called ***singular*** if

$$S_x = H_x. \tag{7.3}$$

Clearly, for a singular sequence $\{x(t)\}$,

$$H_x(t) = H_x \tag{7.4}$$

for any t. Conversely, if (7.4) is true for some t, then $\{x(t)\}$ is singular. Indeed, if (7.4) holds for one t, then by (7.1), it also holds for all other t, whence it follows that $S_x = H_x$.

Let $s_x(t)$ be the projection of $x(t)$ onto S_x. It is easy to see that $\{s_x(t)\}$ is a singular stationary sequence subordinate to $\{x(t)\}$ for which

$$S_{s_x} = H_{s_x} = S_x. \tag{7.5}$$

The sequence $\{s_x(t)\}$ is called the *singular component* of $\{x(t)\}$.

Now assume that $\{x(t)\}$ is non-singular. Then $x(t)$ can be uniquely represented as

$$x(t) = \xi(t) + \Delta(t), \tag{7.6}$$

where $\xi(t)$ belongs to $H_x(t-1)$ and $\Delta(t) \neq 0$ is orthogonal to $H_x(t-1)$. Set

$$u_x(t) = \Delta(t)/||\Delta(t)||. \tag{7.7}$$

Clearly the elements

$$u_x(t),\ u_x(t-1),\ \ldots,\ u_x(t-n)\ldots$$

form a complete orthonormal system in the space

$$H_x(t) \ominus S_x.$$

Since $x(t)$ belongs to $H_x(t)$, it follows that $x(t)$ can be uniquely represented in the form

$$x(t) = s_x(t) + \sum_{n=0}^{\infty} c_n^{(x)} u_x(t-n). \tag{7.8}$$

Since $\{u_x(t)\}$ is stationary and subordinate to $\{x(t)\}$, the coefficients

$$c_n^{(x)} = (x(t), u_x(t-n)) = B_{xu}(n) \tag{7.9}$$

in (7.8) do not depend on t. It follows from (7.6) and (7.7) that

$$c_0 = \|\Delta(t)\| > 0. \tag{7.10}$$

The representation of a non-singular sequence in the form (7.8) is called the *Wold decomposition*. The above arguments make it clear that

$$s(t) = s_x(t), \quad u(t) = u_x(t), \quad c_n = c_n^{(x)}$$

have the following properties:

W_1) the sequence $\{s(t)\}$ is singular and subordinate to $\{x(t)\}$;

W_2) $\{u(t)\}$ is a fundamental sequence subordinate to $\{x(t)\}$;

W_3) $u(t)$ belongs to $H_x(t)$;

W_4) the sequences $s(t)$ and $u(t)$ are mutually orthogonal;

W_5) $c_0 > 0$.

The following theorem shows that the decomposition (7.8) is uniquely determined by these five properties.

Theorem 18. *If a stationary sequence $\{x(t)\}$ is represented as*

$$x(t) = s(t) + \sum_{n=0}^{\infty} c_n u(t-n), \tag{7.11}$$

and if $\{s(t)\}$ *and* $\{u(t)\}$ *and the coefficients* c_n *satisfy the conditions* W$_1$)–W$_5$), *then* $\{x(t)\}$ *is non-singular and*

$$s(t) = s_x(t), \quad u(t) = u_x(t), \quad c_n = c_n^{(x)}.$$

Proof. Since $x(t')$ and all the $u(t'-n)$ belong to $H_x(t')$ for $n = 0, 1, 2, \ldots$ (by W$_3$), then so does

$$s(t') = x(t') - \sum_{n=0}^{\infty} c_n u(t'-n).$$

Since $H_x(t')$ is contained in $H_x(t)$ for $t' \leq t$, all the $s(t')$ belong to $H_x(t)$ for $t' \leq t$. Hence $H_s(t)$ is contained in $H_x(t)$. Since $S_s = H_s(t)$ (by W$_1$)), S_s is contained in all $H_x(t)$ and hence in S_x. This implies that all the $s(t')$ belong to S_x and hence to all $H_x(t)$.

We rewrite (7.11) as

$$x(t) = \zeta(t) + c_0 u(t), \tag{7.12}$$

where

$$\zeta(t) = s(t) + \sum_{n=1}^{\infty} c_n u(t-n).$$

Since all the $s(t)$ and all the $u(t-n)$ for $n = 1, 2, 3, \ldots$ belong to $H_x(t-1)$, it follows that $\zeta(t)$ belongs to $H_x(t-n)$. Since for $t' < t$ all the $x(t')$ can be written as linear combinations of elements $s(t'')$ and $u(t'')$ with $t'' < t$ and $u(t)$ is orthogonal to all $s(t'')$ and $u(t'')$ for $t'' < t$, we find that $u(t)$ is orthogonal to all $x(t')$ for $t' < t$, and hence to the space $H_x(t-1)$.

Since $\zeta(t)$ belongs to $H_x(t-1)$ and $u(t)$ is orthogonal to $H_x(t-1)$, comparing (7.12) with (7.6) we see that

$$\zeta(t) = \xi(t), \quad c_0 u(t) = \Delta(t). \tag{7.13}$$

Comparing (7.13) with (7.7) and taking W$_5$) into account we obtain

$$u(t) = u_x(t). \tag{7.14}$$

By W$_2$) and W$_4$), (7.11) implies that

$$c_n = [x(t), u(t-n)]. \tag{7.15}$$

From (7.14) , (7.15) and (7.9) we obtain

$$c_n = c_n^{(x)}. \tag{7.16}$$

Comparing (7.8) with (7.11) and taking into account (7.15) and (7.16) we finally obtain

$$s(t) = s_x(t).$$

§8. Regular sequences

Theorem 19. *A stationary sequence $\{x(t)\}$ can be represented in the form*

$$x(t) = \sum_{n=0}^{\infty} c_n u(t-n), \tag{8.1}$$

where $\{u(t)\}$ is a fundamental sequence, if and only if

$$s_x(t) = 0. \tag{8.2}$$

If $x(t) = 0$, then the condition $s_x(t) = 0$ is fulfilled and the representation (8.1) is possible (with all $c_n = 0$). Now we have only to consider the case $x(t) \neq 0$. The sufficiency of the condition $s_x(t) = 0$ in this case is also clear: if $s_x(t) = 0$ and $x(t) = 0$, then the sequence $x(t)$ is non-singular and, according to (7.8) of the previous section, can be represented in the form

$$x(t) = \sum_{n=0}^{\infty} c_n^{(x)} u_x(t-n). \tag{8.3}$$

Let us prove the necessity of (8.2). For this purpose note that if there exists a representation (8.1), then the space $H_x(t)$ is contained in the space $H_u(t)$ and S_x in S_u. Therefore, if the equality

$$S_u = 0 \tag{8.4}$$

is true for every fundamental sequence, then $S_x = 0$ and hence $s_x(t) = 0$.

Definition 9. A stationary sequence $\{x(t)\}$ is said to be *regular* if

$$x(t) \neq 0, \quad s_x(t) = 0.$$

If $\{u(t)\}$ is a fundamental sequence, then (8.1) is a particular case of the representations studied in detail in §6. Therefore, in accordance with (6.7), (6.8) and (6.12),

$$B_{xu}(n) = c_n \text{ for } n \geq 0, \quad B_{xu}(n) = 0 \text{ for } n < 0, \tag{8.5}$$

$$\sum_{n=0}^{\infty} |c_n|^2 = |x(t)|^2 = B_{xx}(0), \tag{8.6}$$

$$\phi_x^{(u)}(\lambda) = \sum_{n=0}^{\infty} c_n e^{-in\lambda}. \tag{8.7}$$

By (8.6), the series

$$\Gamma_x^u(\zeta) = \sum_{n=0}^{\infty} c_n \zeta^n \tag{8.8}$$

represents an analytic function in the disk $|\zeta| < 1$. By (8.7), its boundary values on $|\zeta| = 1$ are given by the formula

$$\Gamma_x^u(e^{-i\lambda})) = \phi_x^{(u)}(\lambda). \tag{8.9}$$

The condition $x(t) \neq 0$ implies that $\Gamma_x^u(\zeta)$ cannot vanish identically. Therefore,[6] almost everywhere on $|\zeta| = 1$,

$$\phi_x^{(u)}(\lambda) \neq 0. \tag{8.10}$$

Since by (6.9)

$$\frac{dF_{xx}(\lambda)}{d\lambda} = f_{xx}(\lambda) = \frac{1}{2\pi} |\phi_x^{(u)}(\lambda)|^2 \tag{8.11}$$

almost everywhere, (8.10) implies that almost everywhere

$$f_{xx}(\lambda) > 0. \tag{8.12}$$

By Theorem 17 we conclude that $\{u(t)\}$ is subordinate to $\{x(t)\}$.

Theorem 20. *If a regular sequence $\{x(t)\}$ can be represented in the form* (8.1) *with a fundamental sequence $\{u(t)\}$, then $\{u(t)\}$ is subordinate to $\{x(t)\}$.*

---

[6] See, for example, I.I. Privalov's monograph [5], p.39.

A stationary sequence $\{x(t)\}$ subordinate to a fundamental sequence $\{u(t)\}$ can be represented in terms of $\{u(t)\}$ in the form (8.1) if and only if the function $\gamma_x^{(u)}(\zeta)$, defined for $|\zeta| = 1$ by

$$\gamma_x^{(u)}(e^{-i\lambda}) = \phi_x^{(u)}(\lambda), \tag{8.13}$$

coincides almost everywhere on the circle $|\zeta| = 1$ with the boundary values of the analytic function $\Gamma_x^{(u)}(\zeta)$, defined for $|\zeta| < 1$ by

$$\Gamma_x^{(u)}(\zeta) = \frac{1}{2\pi i}\int_{|\xi|=1} \frac{\gamma_x^{(u)}(\xi)d\xi}{\xi - \zeta}. \tag{8.14}$$

If this condition holds, then the coefficients c_n of (8.1) can be determined from (8.8).

The first part of the theorem, the necessity of the conditions in the second part of the theorem, and the assertion on the form of dependence between the coefficients c_n and the function $\Gamma_x^{(u)}(\zeta)$ have been proved above.

It remains to prove the sufficiency of the conditions of the second part of the theorem.

If these conditions hold, then for the boundary values of $\Gamma_x^{(u)}(\zeta)$ we have the Fourier series expansion

$$\gamma_x^{(u)}(e^{-i\lambda}) = \phi_x^{(u)}(\lambda) \sim \sum_{n=0}^{\infty} c_n e^{-in\lambda}. \tag{8.15}$$

Comparing (8.15) with (6.12) and (6.7) we conclude that $x(t)$ can be represented in the form (8.1) with coefficients c_n determined from (8.15) or, what is the same, from (8.8). This completes the proof of Theorem 20.

Theorem 21. *For any regular sequence $x(t)$, the function*

$$\Gamma_x(\zeta) = \Gamma_x^{(u_x)}(\zeta) = \sum_{n=0}^{\infty} c_n^{(x)}\zeta^n \tag{8.16}$$

does not vanish inside the disk $|\zeta| < 1$.

Assume that, on the contrary, $\Gamma_x(\zeta)$ vanishes at a ζ_0 with $|\zeta| < 1$. We construct a sequence $\{u(t)\}$ subordinate to $\{u_x(t)\}$ such that

$$\phi_u^{(u_x)}(\lambda) = \frac{e^{-i\lambda} - \zeta_0}{1 - \bar{\zeta}_0 e^{-i\lambda}}. \tag{8.17}$$

This sequence exists by Theorem 8 since the function

$$\phi(\lambda) = \frac{e^{-i\lambda} - \zeta_0}{1 - \bar{\zeta}_0 e^{-i\lambda}}$$

is bounded and is therefore of class $L^2_{u_x} = L^2$.

It is easily checked that for any λ

$$|\phi_u^{(u_x)}(\lambda)|^2 = 1. \tag{8.18}$$

Now

$$f_{u_x u_x}(\lambda) = \frac{d}{d\lambda} F_{u_x u_x}(\lambda) = \frac{d}{d\lambda} \frac{\lambda + \pi}{2\pi} = \frac{1}{2\pi}.$$

Therefore according to assertion C of Theorem 17, (8.18) implies that $\{u(t)\}$ is a fundamental sequence.

By (4.5),

$$F_{xx}(\lambda) = \int_{-\pi}^{\lambda} \phi_x^{(u_x)}(\lambda)\bar{\phi}_u^{(u_x)}(\lambda) dF_{u_x u_x}(\lambda) = \frac{1}{2\pi}\int_{-\pi}^{\lambda} \phi_x^{(u_x)}(\lambda)\bar{\phi}_u^{(u_x)}(\lambda) d\lambda. \tag{8.19}$$

It follows from (4.6) that almost everywhere

$$\phi_x^{(u)}(\lambda) = \frac{dF_{xu}(\lambda)}{dF_{uu}(\lambda)} = 2\pi \frac{dF_{xu}(t)}{d\lambda} = \phi_x^{(u_x)}(\lambda)\bar{\phi}_u^{(u_x)}(\lambda). \tag{8.20}$$

By (8.18),

$$\bar{\phi}_x^{(u_x)}(\lambda) = 1/\phi_u^{(u_x)}(\lambda).$$

Therefore (8.20) can be written as

$$\phi_x^{(u)}(\lambda) = \phi_x^{(u_x)}(\lambda)/\phi_u^{(u_x)}(\lambda). \tag{8.21}$$

Now, (8.9) and (8.16) imply that

$$\phi_x^{(u_x)}(\lambda) = \Gamma_x^{(u_x)}(e^{-i\lambda}) = \Gamma_x(e^{-i\lambda}). \tag{8.22}$$

From (8.21) , (8.22) and (8.17) we conclude that

$$\phi_x^{(u)}(\lambda) = \Gamma_x(e^{-i\lambda}) \frac{1 - \bar{\zeta}_0 e^{-i\lambda}}{e^{-i\lambda} - \zeta_0}. \tag{8.23}$$

It is easy to see that

$$\gamma_x^{(u)}(e^{-i\lambda}) = \phi_x^{(u)}(\lambda)$$

coincides almost everywhere with the boundary values of the function

$$\Gamma_x^{(u)}(\zeta) = \Gamma_x(\zeta)\frac{1-\bar{\zeta}_0\zeta}{\zeta-\zeta_0} \tag{8.24}$$

for $|\zeta| = 1$.

Since ζ_0 is a zero of $\Gamma_x(\zeta)$, it follows that $\Gamma_x^{(u)}(\zeta)$ is analytic for $|\zeta| < 1$. By Theorem 20 this implies that [7]

$$x(t) = \sum_{n=0}^{+\infty} c_n u(t-n), \tag{8.25}$$

where the coefficients c_n are determined from (8.8). The function

$$\gamma_u^{(u_x)}(e^{-i\lambda}) = \phi_u^{(u_x)}(\lambda) = \frac{e^{-i\lambda}-\zeta_0}{1-\bar{\zeta}_0 e^{-i\lambda}}$$

coincides almost everywhere on the circle $|\zeta| = 1$ with the boundary values of the function

$$\Gamma_u^{(u_x)}(\zeta) = \frac{\zeta-\zeta_0}{1-\bar{\zeta}_0\zeta}, \tag{8.26}$$

which is analytic in the disk $|\zeta| < 1$. Therefore, according to Theorem 20

$$u(t) = \sum_{n=0}^{\infty} d_n u_x(t-n), \tag{8.27}$$

where the d_n are the coefficients of the Taylor series of the function $\Gamma_u^{(u_x)}(\zeta)$. Formula (8.27) shows that $u(t)$ belongs to $H_{u_x}(t)$. Since by property W_3) of the sequence $\{u_x(t)\}$ (see §7), $H_{(u_x)}(t)$ lies in $H_x(t)$, we see that $u(t)$ also lies in $H_x(t)$.

This implies that for $n > 0$ all the $u(t-n)$ belong to $H_x(t-n)$. Therefore, comparing (8.25) and (7.6) we obtain

$$c_0 u(t) = \Delta(t),$$

and, by (7.7) and (7.10),

$$c_0 u(t) = c_0^{(x)} u_x(t).$$

[7] For Theorem 20 to be applicable it is essential that $\Gamma_x^{(u)}(\zeta)$ can be represented as a Cauchy integral in terms of the boundary values $\gamma_x^{(u)}(e^{-i\lambda})$. This holds, as we can see, for example, from Remark C on p. 94 of Riesz's paper [6], since, clearly $\Gamma_x^{(u)}(\zeta)$, as well as $\Gamma_x(\zeta)$, belong to the class H_1 in the sense of that paper.

It can easily be verified that the sequence $\{u(t)\}$ defined by the formula

$$u(t) = (c_0^{(x)}/c_0)u_x(t),$$

must correspond to the function

$$\phi_u^{(u_x)}(\lambda) = c_0^{(x)}/c_0. \tag{8.28}$$

The contradiction between (8.28) and (8.17) proves Theorem 21.

Theorem 22. *A stationary sequence $\{x(t)\}$ is regular if and only if the following conditions hold on $-\pi \leq \lambda \leq \pi$:*

1) *$F_{xx}(\lambda)$ is absolutely continuous;*
2) *$f_{xx}(\lambda)$ is positive almost everywhere;*
3) *$\log f_{xx}(\lambda)$ is summable.*

If all these conditions hold, then

$$\Gamma_x(\zeta) = \sqrt{2\pi}e^{Q_x(\zeta)}, \tag{8.29}$$

$$Q_x(\zeta) = \frac{a_0^{(x)}}{2} + \sum_{k=1}^{\infty}(a_k^{(x)} + ib_k^{(x)})\zeta^k, \tag{8.30}$$

where the coefficients a_k and b_k are determined from the expansion

$$\tfrac{1}{2}\log f_{xx}(\lambda) \sim \frac{a_0^{(x)}}{2} + \sum_{k=1}^{\infty}(a_k^{(x)}\cos k\lambda + b_k^{(x)}\sin k\lambda). \tag{8.31}$$

Since every regular sequence $\{x(t)\}$ can be obtained by sliding summation from a subordinate fundamental sequence $\{u_x(t)\}$ of it, the necessity of conditions 1 and 2 follows from Theorem 17. Let us prove the necessity of condition 3. If $\{x(t)\}$ is regular, then by Theorem 21, $\Gamma_x(\zeta)$ is analytic and does not vanish in the disk $|\zeta| < 1$. Denote by $Q_x(\zeta)$ the branch of $\log(\Gamma_x(\zeta)/\sqrt{2\pi})$ obtained by analytic continuation from the real value

$$\log\frac{\Gamma_x(0)}{\sqrt{2\pi}} = \log\frac{c_0^{(x)}}{\sqrt{2\pi}}.$$

Since $\Gamma_x(\zeta)$ has no zeros, the function $Q_x(\zeta)$ is uniquely defined at all points of the disk $|\zeta| < 1$. For the real part of $Q_x(\zeta)$ we have, by (8.29),

$$\operatorname{Re} Q_x(\zeta) = \log|\Gamma_x(\zeta)| - \tfrac{1}{2}\log 2\pi. \tag{8.32}$$

Let $\operatorname{Re}^+ Q_x(\zeta)$ be the function equal to $\operatorname{Re} Q_x(\zeta)$ for $\operatorname{Re} Q_x(\zeta) > 0$ and equal to 0 otherwise. Since (8.32) implies

$$\operatorname{Re}^+ Q_x(\zeta) < \log^+ |\Gamma_x(\zeta)| \leq |\Gamma_x(\zeta)|,$$

it follows that for any $\rho < 1$

$$\int_{-\pi}^{\pi} \operatorname{Re}^+ Q_x(\rho e^{-i\lambda}) d\lambda < \int_{-\pi}^{\pi} |\Gamma_x(\rho e^{-i\lambda})| d\lambda \leq \int_{-\pi}^{\pi} |\phi_x^{(u_x)}(\lambda)| d\lambda = K.$$

Since

$$\int_{-\pi}^{\pi} \operatorname{Re} Q_x(\rho e^{-i\lambda}) d\lambda = 2\pi \operatorname{Re} Q_x(0),$$

we find that

$$\int_{-\pi}^{\pi} |\operatorname{Re} Q_x(\rho e^{-i\lambda})| d\lambda = \int_{-\pi}^{\pi} [2 \operatorname{Re}^+ Q_x(\rho e^{-i\lambda}) - \operatorname{Re} Q_x(\rho e^{-i\lambda})] d\lambda \leq$$

$$\leq 2K - 2\pi \operatorname{Re} Q_x(0).$$

The latter inequality implies that the boundary values

$$\operatorname{Re} Q_x(e^{-i\lambda}) = \log\left|\frac{\Gamma_x(e^{-i\lambda})}{\sqrt{2\pi}}\right| = \log\left|\frac{\phi_x^{(u_x)}(\lambda)}{\sqrt{2\pi}}\right| =$$

$$= \tfrac{1}{2} \log\left|\frac{\phi_x^{(u_x)}(\lambda)}{\sqrt{2\pi}}\right|^2 = \tfrac{1}{2} \log f_{xx}(\lambda) \tag{8.33}$$

of $\operatorname{Re} Q_x(e^{-i\lambda})$ are integrable with respect to λ. This proves the necessity of the integrability of $\log f_{xx}(\lambda)$, that is, condition 3 of our theorem.

Before passing to the proof of the sufficiency of the conditions of the theorem, let us show that for any regular sequence $\{x(t)\}$ the function $\Gamma_x(\zeta)$ is determined by (8.29)–(8.31). Since, according to what was proved above, the function $\frac{1}{2} \log f_{xx}(\lambda)$ is integrable for a regular sequence $\{x(t)\}$, it can be expanded in a Fourier series (8.31). Formulas (8.33) and (8.31) imply that

$$\operatorname{Re} Q_x(e^{i\lambda}) \sim \frac{a_0^{(x)}}{2} + \sum_{k=1}^{\infty} (a_k^{(x)} \cos k\lambda - b_k^{(x)} \sin k\lambda). \tag{8.34}$$

Since

$$Q_x(0) = \log \frac{\Gamma_x(0)}{\sqrt{2\pi}} = \log \frac{c_0^{(x)}}{\sqrt{2\pi}} \tag{8.35}$$

is real, the imaginary part $\operatorname{Im} Q_x(e^{i\lambda})$ is given by

$$\operatorname{Im} Q_x(e^{i\lambda}) \sim \sum_{k=1}^{\infty}(a_k^{(x)} \sin k\lambda + b_k^{(x)} \cos k\lambda). \tag{8.36}$$

Formulas (8.34) and (8.36) imply (8.30). Finally, (8.29) follows from the definition of $Q_x(\zeta)$.

Now let us prove the sufficiency of the conditions of the theorem. For this we assume that these conditions hold, and we set

$$Q(\zeta) = \frac{a_0^{(x)}}{2} + \sum_{k=1}^{\infty}(a_k^{(x)} + ib_k^{(x)})\zeta^k, \tag{8.37}$$

$$\Gamma(\zeta) = \sqrt{2\pi} e^{Q(\zeta)}, \tag{8.38}$$

where the coefficients $a_k^{(x)}$ and $b_k^{(x)}$ are determined from the expansion (8.31). Then [8]

$$\log \frac{|\Gamma(\rho e^{-i\lambda})|^2}{2\pi} = 2 \operatorname{Re} Q(\rho e^{-i\lambda}) = a_0 + 2\sum_{k=1}^{\infty} \rho^k (a_k^{(x)} \cos k\lambda + $$

$$+ b_k^{(x)} \sin k\lambda) = \int_{-\pi}^{\pi} \log f_{xx}(\mu) P_\rho(\mu - \lambda) d\mu, \tag{8.39}$$

where

$$P_\rho(\theta) = \frac{1}{2\pi} \frac{1-\rho^2}{1+\rho^2 - 2\rho\cos\theta}. \tag{8.40}$$

The inequality between the geometric mean and the arithmetic mean [9] implies that

$$\frac{|\Gamma(\rho e^{-i\lambda})|^2}{2\pi} \le \int_{-\pi}^{\pi} f_{xx}(\mu) P_\rho(\mu - \lambda) d\mu. \tag{8.41}$$

Since $f_{xx}(\lambda)$ is integrable and $\int_{-\pi}^{\pi} P_\rho(\mu - \lambda) d\mu = 1$, by (8.41) we have

$$\frac{1}{2\pi} \int_{-\pi}^{\pi} |\Gamma(\rho e^{i\lambda}|^2 d\lambda = \int_{-\pi}^{\pi} f_{xx}(\mu) \int_{-\pi}^{\pi} P_\rho(\mu - \lambda) d\lambda d\mu = \int_{-\pi}^{\pi} f_{xx}(\lambda) d\lambda. \tag{8.42}$$

---

[8] See [5], p.15.

[9] We apply this inequality in the following form: if

$$m = \int_a^b P(x) f(x) dx, \quad \log s = \int_a^b P(x) \log f(x) dx, \quad \int_a^b P(x) dx = 1,$$
$$P(x) \ge 0, \ f(x) \ge 0,$$

then $s \le m$.

The boundedness of the integral on the left-hand side of (8.42) thus established guarantees[10] that $\Gamma(\zeta)$ can be represented as the Cauchy integral of its boundary values

$$\Gamma(e^{-i\lambda}) = \phi(\lambda).$$

Formulas (8.38) and (8.31) imply that

$$|\phi(\lambda)|^2 = |\Gamma(e^{-i\lambda})|^2 = 2\pi \exp(2\,\mathrm{Re}\, Q(e^{-i\lambda})) = 2\pi f_{xx}(\lambda). \tag{8.43}$$

Therefore,

$$\int_{-\pi}^{\pi} \frac{1}{|\phi(\lambda)|^2} dF_{xx}(\lambda) = 1$$

and consequently $1/\phi(\lambda)$ belongs to the class L_x^2. Let us find a sequence $\{u(t)\}$ subordinate to $\{x(t)\}$ for which

$$\phi_u^{(x)}(\lambda) = 1/\phi(\lambda).$$

Equation (8.43) and Theorem 17C imply that $\{u(t)\}$ is a fundamental sequence equivalent to $\{x(t)\}$. By Theorem 10,

$$\phi_x^{(u)}(\lambda) = 1/\phi_x^{(u)}(\lambda) = \phi(\lambda) = \Gamma(e^{-i\lambda}).$$

Thus, $\{u(t)\}$ satisfies all the conditions of the second part of Theorem 20. Hence $\{x(t)\}$ can be represented in terms of $\{u(t)\}$ in the form (8.1), which proves the sufficiency of the conditions of Theorem 22. It is easy to see that the sequence $\{u(t)\}$ so constructed in fact coincides with $\{u_x(t)\}$.

In conclusion we note the formula[11]

$$c_0^{(x)} = \sqrt{2\pi} e^Q x^{(0)} = \sqrt{2\pi} \exp \frac{a_0^{(x)}}{2} = \sqrt{2\pi} \exp\Big(\frac{1}{4\pi} \int_{-\pi}^{\pi} \log f_{xx}(\lambda) d\lambda\Big), \tag{8.44}$$

which follows from (8.30) and (8.35).

9. Spectral characterization of singular sequences and singular components

By the results in §7, every non-singular stationary sequence $\{x(t)\}$ can be represented as

$$x(t) = s_x(t) + r_x(t), \tag{9.1}$$

[10] See Remark C on p.94 of [6].

[11] This formula holds not only for regular, but also for all non-singular sequences, as will be clear from Theorem 23 of the following section.

where

$$r_x(t) = \sum_{n=0}^{\infty} c_n^{(x)} u_x(t-n). \tag{9.2}$$

Let us call $\{r_x(t)\}$ the *regular component*[12] of $\{x(t)\}$. Clearly, it is always subordinate to $\{x(t)\}$ and is regular.

For any stationary sequence we set

$$A_{xx}(\lambda) = \int_{-\pi}^{\lambda} f_{xx}(\lambda) d\lambda, \tag{9.3}$$

$$D_{xx}(\lambda) = F_{xx}(\lambda) - A_{xx}(\lambda). \tag{9.4}$$

We will distinguish the following three cases:

1) $f_{xx}(\lambda) = 0$ on a set of positive measure;

2) $f_{xx}(\lambda) = 0$ only on a set of measure zero but $\log f_{xx}(\lambda)$ is non-integrable;

3) $f_{xx}(\lambda) = 0$ only on a set of measure zero and $\log f_{xx}(\lambda)$ is integrable.

Theorem 23. *In cases* 1) *and* 2) *the sequence* $\{x(t)\}$ *is singular. In case* 3) $\{x(t)\}$ *is non-singular,*

$$F_{s_x s_x}(\lambda) = D_{xx}(\lambda), \quad F_{r_x r_x}(\lambda) = A_{xx}(\lambda) \tag{9.5}$$

and the coefficients $c_0^{(x)}$ *in* (9.2) *are determined from* (8.16), *where* $\Gamma_x(\zeta)$ *is defined by* (8.29)–(8.31).

Proof. Assume that $\{x(t)\}$ is non-singular. Then by the regularity of $\{r_x(t)\}$ and Theorem 22, $f_{r_x r_x}(\lambda)$ is positive almost everywhere and $\log f_{r_x r_x}(\lambda)$ is integrable. Since $\{s_x(r)\}$ and $\{r_x(t)\}$ are orthogonal to each other (see §7), Theorem 11 implies that

$$F_{xx}(\lambda) = F_{s_x s_x}(\lambda) + F_{r_x r_x}(\lambda) \tag{9.6}$$

and consequently, almost everywhere

$$f_{xx}(\lambda) = f_{s_x s_x}(\lambda) + f_{r_x r_x}(\lambda). \tag{9.7}$$

[12] Since the sequence identically equal to zero is singular, but not regular, the singular component of a regular sequence equals 0 and a singular sequence has no regular component.

It follows from (9.7) that almost everywhere

$$f_{xx}(\lambda) \geq f_{r_x r_x}(\lambda) > 0. \tag{9.8}$$

Therefore, every set of positive Lebesgue measure has positive measure also with respect to $F_{xx}(\lambda)$.

Since $\{s_x(t)\}$ and $\{r_x(t)\}$ are subordinate to $\{x(t)\}$, Theorem 12 implies that

$$\frac{dF_{s_x s_x}(\lambda)}{dF_{xx}(\lambda)} \frac{dF_{r_x r_x}(\lambda)}{dF_{xx}(\lambda)} = 0$$

almost everywhere with respect to $F_{xx}(\lambda)$, and hence, in the sense of Lebesgue.

Since by (9.8) almost everywhere

$$\frac{dF_{r_x r_x}(\lambda)}{dF_{xx}(\lambda)} = \frac{f_{r_x r_x}(\lambda)}{f_{xx}(\lambda)} > 0,$$

it follows that almost everywhere

$$\frac{dF_{s_x s_x}(\lambda)}{dF_{xx}(\lambda)} = \frac{f_{s_x s_x}(\lambda)}{f_{xx}(\lambda)} = 0,$$

that is,

$$f_{s_x s_x}(\lambda) = 0, \tag{9.9}$$

$$f_{xx}(\lambda) = f_{r_x r_x}(\lambda). \tag{9.10}$$

Formula (9.10) and the facts that $f_{r_x r_x}(\lambda)$ is positive almost everywhere and $\log f_{r_x r_x}(\lambda)$ is integrable imply that for a non-singular sequence case 3) always holds.

Let us show that, conversely, in case 3) $\{x(t)\}$ is non-singular. According to Theorem 14 in case 3) $\{x(t)\}$ can be uniquely represented as a sum

$$x(t) = s(t) + r(t)$$

of orthogonal sequences subordinate to $\{x(t)\}$ for which

$$F_{ss}(\lambda) = D_{xx}(\lambda), \quad F_{rr}(\lambda) = A_{xx}(\lambda).$$

By Theorem 22, $\{r(t)\}$ is regular. Therefore,

$$x(t) = s(t) + \sum_{n=0}^{\infty} c_n^{(r)} u_r(t-n). \tag{9.11}$$

Formula (9.11) shows that in this case $H_x(t)$ is contained in the space[13] $S_s \oplus H_r(t)$. Representing $x(t+1)$ as

$$x(t+1) = \alpha + \beta,$$

where

$$\alpha = c_0^{(r)} u_r(t+1), \quad \beta = s(t+1) + \sum_{n=1}^{\infty} c_n^{(r)} u_r(t-n+1),$$

it is easy to see that β belongs to $S_x \oplus H_{u_r}(t)$ and $\alpha \neq 0$ is orthogonal to this space. Therefore $x(t+1)$ does not lie in $S_x \oplus H_{u_r}(t)$ and hence does not lie in $H_x(t)$. This implies that $\{x(t)\}$ is non-singular.

We have proved (9.10) for any non-singular $\{x(t)\}$. Therefore $\Gamma_x(\zeta)$ determined in accordance with (8.29)–(8.31) coincides with $\Gamma_{r_x}(\zeta)$. To prove that the coefficients $c_n^{(x)}$ can be obtained from the expansion (8.16) of $\Gamma_x(\zeta)$, it only remains to establish that

$$c_n^{(x)} = c_n^{(r_x)}. \tag{9.12}$$

For this it suffices to show that (9.2) satisfies W_1)–W_5) (with zero singular component). Clearly $W_1), W_2), W_4)$ and W_5) hold. Let us prove that W_3) holds for (9.2). For this we note that by (9.8) and Theorem 17A, $\{u_x(t)\}$ is subordinate to $\{r_x(t)\}$. From (9.1) and (9.2) it is clear that $H_x(t)$ is contained in $S_x \oplus H_{r_x}(t)$. Since $u_x(t)$ belongs to $H_x(t)$ and is orthogonal to S_x (see §7), $u_x(t)$ belongs to $H_{r_x}(t)$, which means that (9.2) satisfies W_3).

For a non-singular sequence, (9.5) follows immediately from (9.1), (9.10) and the absolute continuity of $F_{r_x r_x}$ (the latter is ensured by the regularity of $\{r_x(t)\}$, see Theorem 16).

§10. Minimal sequences

For any stationary sequence $\{x(t)\}$ denote by $\widehat{H}_x(t)$ the smallest closed linear subspace of H_x containing all the $x(s)$ for $s \neq t$. Since

$$U_x \widehat{H}_x(t) = \widehat{H}_x(t+k), \tag{10.1}$$

only one of two cases is possible: either all the $\widehat{H}_x(t)$ coincide with H_x or all the $\widehat{H}_x(t)$ are different from H_x.

[13] $\mathfrak{G}_1 \oplus \mathfrak{G}_2$ denotes the minimal linear closed subspace of H containing both $\mathfrak{G}_1$ and $\mathfrak{G}_2$.

Definition 10. A stationary sequence is called *minimal* if

$$\widehat{H}_x(t) \neq H_x. \tag{10.2}$$

Each element $x(t)$ of the sequence $\{x(t)\}$ can be uniquely represented as a sum

$$x(t) = v(t) + \delta(t), \tag{10.3}$$

where $v(t)$ belongs to $\widehat{H}_x(t)$ and $\delta(t)$ is orthogonal to $\widehat{H}_x(t)$. Set

$$dx = |\delta(t)|. \tag{10.4}$$

Clearly, the condition

$$dx > 0 \tag{10.5}$$

is necessary and sufficient for $\{x(t)\}$ to be minimal.

Theorem 24. *A stationary sequence $\{x(t)\}$ is minimal if and only if $f_{xx}(\lambda)$ is positive almost everywhere and the integral*

$$\int_{-\pi}^{\pi} \frac{d\lambda}{f_{xx}(\lambda)} \tag{10.6}$$

is finite.

If these conditions hold, then

$$d_x^2 = (2\pi)^2 : \int_{-\pi}^{\pi} \frac{d\lambda}{f_{xx}(\lambda)}. \tag{10.7}$$

First we prove the sufficiency of the conditions of the theorem. Let

$$\phi(\lambda) = 1 : f_{xx}(\lambda) \Big(\int_{-\pi}^{\pi} \frac{d\lambda}{f_{xx}(\lambda)} \Big)^{1/2},$$

if $f_{xx}(\lambda)$ is defined and finite, and

$$\phi(\lambda) = 0,$$

if $f_{xx}(\lambda)$ is not defined (and is infinite). Then

$$\int_{-\pi}^{\pi} |\phi(\lambda)|^2 dF_{xx}(\lambda) = 1.$$

Hence $\phi(\lambda)$ belongs to the class L_x^2 and the equality

$$\phi_y^{(x)}(\lambda) = \phi(\lambda)$$

defines a certain sequence $\{y(t)\}$ subordinate to $\{x(t)\}$ for which, according to (4.4),

$$F_{yx}(\lambda) = \int_{-\pi}^{\lambda} \phi_y^{(x)}(\lambda) dF_{xx}(\lambda) = (\lambda + \pi) : \left(\int_{-\pi}^{\pi} \frac{d\lambda}{f_{xx}(\lambda)}\right)^{1/2}.$$

By (3.3), for $s \neq t$,

$$(y(t), x(s)) = B_{yx}(t-s) = \int_{-\pi}^{\pi} e^{i(t-s)\lambda} dF_{yx}(\lambda) = 0.$$

Hence the element $y(t)$ is orthogonal to the space $\widehat{H}_x(t)$. Since

$$|y(t)|^2 = B_{yy}(0) = \int_{-\pi}^{\pi} dF_{yy}(\lambda) = \int_{-\pi}^{\pi} |\phi_y^{(x)}(\lambda)|^2 dF_{xx}(\lambda) = 1 \neq 0,$$

the space H_x, which contains $y(t)$, cannot coincide with $H_x(t)$. This proves the minimality of $\{x(t)\}$.

To prove the necessity of the conditions of the theorem, assume that $\{x(t)\}$ is minimal. It is easy to see that $\{\delta(t)\}$ is stationary and subordinate to $\{x(t)\}$. Then

$$B_{\delta x}(k) = (\delta(t+k), x(t)) = 0 \text{ for } k \neq 0, \tag{10.8}$$
$$B_{\delta x}(0) = (\delta(t), x(t)) = |\delta(t)|^2 = d_\lambda^2. \tag{10.9}$$

By (3.9)–(3.11) this implies that

$$F_{\delta x}(\lambda) = d_x^2(\lambda + \pi)/2\pi. \tag{10.10}$$

At the same time, by (4.4),

$$F_{\delta x}(\lambda) = \int_{-\pi}^{\pi} \phi_\delta^{(x)}(\lambda) dF_{xx}(\lambda). \tag{10.11}$$

Formulas (10.10) and (10.11) imply that

$$\lambda + \pi = \frac{2\pi}{d_x^2} \int_{-\pi}^{\lambda} \phi_\delta^{(x)}(\lambda) dF_{xx}(\lambda) \tag{10.12}$$

and consequently, the derivative

$$\frac{d\lambda}{dF_{xx}(\lambda)} = \frac{2\pi}{d_x^2}\phi_\delta^{(x)}(\lambda) \tag{10.13}$$

is finite almost everywhere with respect to $F_{xx}(\lambda)$. Since by (10.12) λ is absolutely continuous with respect to $F_{xx}(\lambda)$, the derivative $d\lambda/dF_{xx}(\lambda)$ is finite almost everywhere with respect to λ, that is, in the sense of Lebesgue. Therefore the derivative

$$f_{xx}(\lambda) = dF_{xx}(\lambda)/d\lambda$$

is non-vanishing almost everywhere. Taking this into account, it follows from (10.13) that almost everywhere

$$\phi_\delta^{(x)}(\lambda) = \frac{d_x^2}{2\pi}\frac{d\lambda}{dF_{xx}(\lambda)} = \frac{d_x^2}{2\pi} : \frac{dF_{xx}(\lambda)}{d\lambda} = \frac{d_x^2}{2\pi f_{xx}(\lambda)}.$$

According to Theorem 8, the integral

$$\int_{-\pi}^{\pi} |\phi_\delta^{(x)}(\lambda)|^2 dF_{xx}(\lambda) = \int_{-\pi}^{\pi} \left|\frac{d_x^2}{2\pi f_{xx}(\lambda)}\right|^2 dF_{xx}(\lambda) = \frac{d_x^4}{(2\pi)^2}\int_{-\pi}^{\pi}\frac{d\lambda}{f_{xx}(\lambda)} \tag{10.14}$$

is finite. This proves the necessity of the conditions of the theorem. Noting that

$$d_x^2 = |\delta(t)|^2 = B_{\delta\delta}(0) = \int_{-\pi}^{\pi} dF_{\delta\delta}(\lambda) = \int_{-\pi}^{\pi} |\phi_\delta^{(x)}(\lambda)|^2 dF_{xx}(\lambda), \tag{10.15}$$

we obtain (10.7) from (10.14) and (10.15).

References

1. H. Wold, *A study in the analysis of stationary time series*, Uppsala, 1938.
2. H. Cramér, *Ann. of Math.* **41** (1940), 215–230.
3. M. Stone, 'Linear transformations in Hilbert space', *Amer. Math. Soc. Colloq. Publ.* **15** (1932).
4. A. Zygmund, *Trigonometric series*, Cambridge Univ. Press, 1979, 2 vols.
5. I.I. Privalov, *The Cauchy integral*, Saratov, 1919 (in Russian).
6. F. Riesz, *Math. Z.* **18** (1923), 87–95.
7. A.N. Kolmogorov, *Izv. Akad. Nauk SSSR Ser. Mat.* **5** (1941), 3–14 (in Russian) (No. 28 in this book).

28. INTERPOLATION AND EXTRAPOLATION OF STATIONARY RANDOM SEQUENCES *

Spectral conditions are established for the possibility of extrapolating and interpolating stationary random sequences by a sufficiently large number of terms with any prescribed accuracy.

Introduction

For each integer t $(-\infty < t < +\infty)$ let $x(t)$ be a real random variable whose square has finite expectation. The sequence of random variables $x(t)$ will be called stationary if the expectations[1]

$$m = \mathbf{E}x(t)$$

and

$$B(k) = \mathbf{E}[(x(t+k) - m)(x(t) - m)]$$

do not depend on t. Without loss of generality we may set

$$m = \mathbf{E}x(t) = 0. \tag{1}$$

Then

$$B(k) = \mathbf{E}[x(t+k)x(t)]. \tag{2}$$

Since

$$B(-k) = B(k), \tag{3}$$

it suffices to consider the second moments $B(k)$ only for $k \geq 0$.

The problem of linear extrapolation of a stationary sequence satisfying (1) is to select for given $n > 0$ and $m \geq 0$ real coefficients a_s for which the linear combination

$$L = a_1 x(t-1) + a_2 x(t-2) + \ldots + a_n x(t-n)$$

of random variables

$$x(t-1), x(t-2), \ldots, x(t-n)$$

* *Izv. Akad. Nauk SSSR Ser. Mat.* **5** (1941), 3–14.

[1] The expectation of a random variable y will be denoted by $\mathbf{E}y$.

gives the closest possible approximation to the random variable $x(t+m)$. It is natural to take the expectation

$$\sigma^2 = \mathsf{E}(x(t+m) - L)^2 = B(0) - 2\sum_{s=1}^{n} B(m+s)a_s + \sum_{p=1}^{n}\sum_{q=1}^{n} B(p-q)a_p a_q$$

as the measure of accuracy of such an approximation.

If the second moments $B(k)$ are known, then it is easy to find the coefficients a_s for which σ^2 takes the smallest value. This smallest value of σ^2 will be denoted by $\sigma_E^2\ (n,m)$.

Clearly, $\sigma_E^2\ (n,m)$ cannot increase with n. Therefore the limit

$$\lim_{n\to\infty} \sigma_E^2(n,m) = \sigma_E^2(m) \tag{4}$$

exists. To find this limit is the first problem we solve in this paper.

As for the interpolation problem, we will consider only the case of estimating $x(t)$ from

$$x(t+1),\ x(t+2),\ \ldots,\ x(t+n),$$
$$x(t-1),\ x(t-2),\ \ldots,\ x(t-n).$$

For this case, let $\sigma_I^2(n)$ be the minimal expectation

$$\sigma^2 = \mathsf{E}(x(t) - Q)^2,$$

where Q is a linear form

$$Q = a_1x(t+1) + a_2x(t+2) + \ldots + a_nx(t+n) +$$
$$+ a_{-1}x(t-1) + a_{-2}x(t-2) + \ldots + a_{-n}x(t-n)$$

with constant real coefficients a_s.

Since $\sigma_I^2(n)$ does not increase with n, the limit

$$\lim_{n\to\infty} \sigma_I^2(n) = \sigma_I^2 \tag{5}$$

exists.

Our second goal is to determine σ_I^2. The solution to both these problems was announced in my paper [1].[2] It uses notions relating to the spectral theory of stationary random processes.

[2] There is a misprint in formula (1) of [1]. The correct form of (1) is:

$$\lim_{n\to\infty} \sigma_I^2(n) = \pi : \int_0^{\pi} \frac{d\lambda}{s(\lambda)}.$$

The spectral theory of stationary processes was constructed by A.Ya. Khinchin [2] for the case of continuous variation of the time argument t. The theory for the case we are interested in now, that is, of a discrete stationary sequence, is given in detail in Wold's book [3]. Of principal importance here is the following [3]

Theorem 1. *For any stationary sequence $\{x(t)\}$ the second moments $B(k)$ can be represented in the form*

$$B(k) = \frac{1}{\pi}\int_0^\pi \cos k\lambda dW(\lambda), \tag{6}$$

where $W(\lambda)$ is the non-decreasing real function defined by the formula

$$W(\lambda) = B(0)\lambda + 2\sum_{k=1}^{\infty} \frac{B(k)}{k} \sin k\lambda. \tag{7}$$

The derivative

$$w(\lambda) = dW(\lambda)/d\lambda$$

of the non-decreasing function $W(\lambda)$ exists almost everywhere, is non-negative and summable. Since

$$\log w(\lambda) \le w(\lambda),$$

it follows from the summability of $w(\lambda)$ that the integral

$$P = \frac{1}{\pi}\int_0^\pi \log w(\lambda) d\lambda \tag{8}$$

is either finite or equal to $-\infty$.[4] Further, we prove the following:

Theorem 2. *If $P = -\infty$, then $\sigma_E^2(m) = 0$ for all $m \ge 0$. If the integral P is finite, then*

$$\sigma_E^2(m) = e^P(1 + r_1^2 + r_2^2 + \ldots + r_m^2), \tag{9}$$

where the r_i are determined from the formulas

$$e^{a_1\zeta + a_2\zeta^2 + \ldots} = 1 + r_1\zeta + r_2\zeta^2 + \ldots, \tag{10}$$

$$a_k = \frac{1}{\pi}\int_0^\pi \cos k\lambda \log w(\lambda) d\lambda. \tag{11}$$

Since $w(\lambda) \ge 0$, the integral

$$R = \frac{1}{\pi}\int_0^\pi \frac{d\lambda}{w(\lambda)} \tag{12}$$

is either finite or equal to $+\infty$.[5] Then we prove the following:

[3] See [3], §17.
[4] If $w(\lambda) = 0$ on a set of positive measure, we set $P = -\infty$.
[5] If $w(\lambda) = 0$ on a set of positive measure, we set $R = +\infty$.

Theorem 3. *If $R = +\infty$, then $\sigma_I^2 = 0$. If the integral is finite, then*

$$\sigma_I^2 = 1/R. \tag{13}$$

In my paper [4] I constructed a theory of stationary sequences of elements of a complex Hilbert space. In §1 of the present paper it is shown that stationary random sequences in the sense defined above can be considered as a special case of the stationary sequences considered in [4]. This allows us to obtain the above Theorems 1, 2 and 3 as simple corollaries of the results of [4].

In what follows, references to formulas indexed by two numbers (for example, (8.44)) apply to formulas from [4].

§1. Stationary random sequences and Hilbert space geometry

Our discussion will be based on the axioms and the construction of the basic notions suggested in my book [5], with one difference: our random variables can take not only real, but also complex values. [6]

Consider the set $\mathfrak{H}$ of all random variables of a certain Borel probability field (F, P) with finite expectation of the squared absolute value, regarding equivalent random variables (that is, random variables that differ from each other with probability zero) as identical. We introduce in $\mathfrak{H}$ the scalar product

$$(x, y) = \mathbf{E}(x\bar{y}). \tag{14}$$

The norm in $\mathfrak{H}$ is given by

$$\|x\|^2 = (x, x) = \mathbf{E}|x|^2. \tag{15}$$

Summation of elements in $\mathfrak{H}$ and their multiplication by a complex number is understood as usual.

It is easy to verify that under these definitions the set $\mathfrak{H}$ satisfies the axioms A, B and E from M. Stone's book [6], that is, all the axioms of an abstract unitary space.

Now let $\{x(t)\}$ be a stationary sequence of real random variables $x(t)$ of the probability field (F, P) in the sense adopted in the Introduction, satisfying

[6] A complex function $x(\xi)$ defined on the set E of elementary events ξ is called a random variable if for any choice of real numbers a and b the set of all ξ for which the real and imaginary parts of $x(t)$ satisfy $\operatorname{Re} x(\xi) < a$ and $\operatorname{Im} x(\xi) < b$ respectively, belongs to the system F.

the additional condition (1). Then, by (2) and the fact that $x(t)$ is real, we have

$$B(k) = \mathbf{E}[x(t+k)x(t)] = (x(t+k), x(t)).$$

Since, by definition, $B(k)$ does not depend on t, $\{x(t)\}$ is a stationary sequence of elements of $\mathfrak{H}$ in the sense of [4].

In [4] I consider stationary sequences in a Hilbert space, that is, in the space satisfying not only the Axioms A, B and E, but also C and D from [6]. However, this restriction is inessential. Indeed, denoting by H_x the smallest closed linear subspace of $\mathfrak{H}$ containing all elements of $\{x(t)\}$, it is easy to show that H_x is separable, that is, satisfies Axiom D. A separable unitary space is either a Hilbert space itself (that is, satisfies not only Axioms A, B and E, but also Axiom C), or is finite dimensional, and in this latter case can be extended to a Hilbert space H.

Thus, all the results obtained in [4] may be applied to $\{x(t)\}$, setting

$$B(k) = B_{xx}(k) = (x(t+k), x(t)). \tag{16}$$

§2. Proof of Theorem 1

For the case of real random variable $x(t)$, by (3) and (16) we have

$$B_{xx}(-k) = B_{xx}(k). \tag{17}$$

Therefore from (3.10) we obtain

$$W_{xx}(\lambda) = B_{xx}(0)\lambda - \sum_{k\neq 0} \frac{B_{xx}(k)}{ik} e^{-ik\lambda} =$$

$$= B_{xx}(0)\lambda + 2\sum_{k=1}^{\infty} \frac{B_{xx}(k)}{k} \sin k\lambda. \tag{18}$$

It follows from (18) that

$$W_{xx}(-\lambda) = -W_{xx}(\lambda). \tag{19}$$

Finally, from (3.1), (3.9) and (19) we obtain

$$B(k) = B_{xx}(k) = \int_{-\pi}^{+\pi} e^{ik\lambda} dF_{xx}(\lambda) =$$

$$= \frac{1}{2\pi}\int_{-\pi}^{+\pi} e^{ik\lambda} dW_{xx}(\lambda) = \frac{1}{\pi}\int_0^{\pi} \cos k\lambda dW_{xx}(\lambda). \tag{20}$$

It is clear from (18) and (3.9) and Theorem 2 of [4] that $W_{xx}(\lambda)$ is a non-decreasing real function. Together with (18) and (20) this shows that the function

$$W(\lambda) = W_{xx}(\lambda)$$

satisfies the requirements of Theorem 1.

§3. $\sigma_E^2(m)$ in the general case

After [4], we denote by $H_x(t-1)$ the smallest closed linear subspace of H_x containing the elements

$$x(t-1), x(t-2), \ldots, x(t-n), \ldots .$$

For any $m \geq 0$ the element $x(t+m)$ can be uniquely represented as

$$x(t+m) = \xi(t-1, m) + \Delta(t-1, m) \tag{21}$$

where $\xi(t-1,m)$ belongs to $H_x(t-1)$ and $\Delta(t-1,m)$ is orthogonal to $H_x(t-1)$.

It is easy to show that in the case of a stationary sequence of real random variables [7]

$$\sigma_E^2(m) = \|\Delta(t-1, m)\|^2. \tag{22}$$

In the general case of stationary sequences in the sense of [4] we take (22) as the definition of $\sigma_E^2(m)$.

If $\{x(t)\}$ is singular, then $H_x(t-1) = H_x$ and hence

$$\sigma_E^2(m) = 0. \tag{23}$$

---

[7] As is known, $\|\Delta(t-1,m)\|$ equals "the distance" of the point $x(t+m)$ from the space $H_x(t-1)$, that is, the greatest lower bound of the distances $\|x(t+m)-y\|$ for all y in $H_x(t-1)$. Since the elements of the form

$$L = a_1 x(t-1) + a_2 x(t-2) + \ldots + a_n x(t-n)$$

are everywhere dense in $H_x(t-1)$, $\|\Delta(t-1,m)\|$ also equals the greatest lower bound of the distances

$$\|x(t+m) - L\| = \sqrt{\mathsf{E}|x(t+m) - L|^2}. \tag{*}$$

If all the $x(s)$ are real random variables, then the greatest lower bound of (*) does not change when we consider only real coefficients a_k, in which case it clearly coincides with $\sigma_E(m)$

If $\{x(t)\}$ is non-singular, then, by (7.8),

$$x(t+m) = s_x(t+m) + \sum_{n=0}^{\infty} c_n^{(x)} u_x(t+m-n). \tag{24}$$

Since $s_x(t+m)$ and $u_x(t+m-n)$ belong to $H_x(t-1)$ for $n > m$ and $u_x(t+m-n)$ are orthogonal to $H_x(t-1)$ for $n \leq m$, comparing (21) and (24) we obtain

$$\Delta(t-1, m) = c_0^{(x)} u_x(t+m) + c_1^{(x)} u_x(t+m-1) + \ldots + c_m^{(x)} u_x(t). \tag{25}$$

Since the elements $u_x(t+i)$ are pairwise orthogonal and normalized, (25) implies that

$$\sigma_E^2(m) = \|\Delta(t-1)\|^2 = (c_0^{(x)})^2 + (c_1^{(x)})^2 + \ldots + (c_m^{(x)}). \tag{26}$$

§4. $\sigma_E^2(m)$ in the real case

It is easy to derive Theorem 2 from (26) for stationary sequences of real random variables. This is carried out in this section.

By (3.9) we have

$$f_{xx}(\lambda) = \frac{dF_{xx}(\lambda)}{d\lambda} = \frac{1}{2\pi}\frac{dW_{xx}(\lambda)}{d\lambda} = \frac{1}{2\pi} w(\lambda). \tag{27}$$

Formula (19) implies that

$$w(-\lambda) = w(\lambda). \tag{28}$$

Using (27) and (28) we obtain

$$\int_{-\pi}^{+\pi} \log f_{xx}(\lambda) d\lambda = 2\int_0^{\pi} log w(\lambda) d\lambda - 2\pi \log 2\pi. \tag{29}$$

Together with Theorem 23 from [4], (29) shows that the equality

$$P = \frac{1}{\pi}\int_0^{\pi} \log w(\lambda) d\lambda = -\infty$$

is necessary and sufficient for the singularity of the sequence $\{x(t)\}$. We have already seen in §3 that in this and only in this case

$$\sigma_E^2(m) = 0. \tag{30}$$

If $\{x(t)\}$ is non-singular, then (8.44) and (29) imply that

$$(c_0^{(x)})^2 = 2\pi \exp\Big(\frac{1}{2\pi}\int_{-\pi}^{+\pi} \log f_{xx}(\lambda)d\lambda\Big) = \exp\Big(\frac{1}{\pi}\int_0^{\pi} \log w\lambda\, d\lambda\Big) = e^P. \quad (31)$$

Under the same assumption, namely if $\{x(t)\}$ is non-singular, [8] (8.31), (27) and (28) imply that

$$\begin{aligned} a_k^{(x)} &= \frac{1}{2\pi}\int_{-\pi}^{+\pi} \cos kx \log f_{xx}(\lambda)d\lambda = \frac{1}{\pi}\int_0^{\pi} \cos k\lambda \log w(\lambda)d\lambda \\ b_k^{(x)} &= \frac{1}{2\pi}\int_{-\pi}^{+\pi} \sin kx \log f_{xx}(\lambda)d\lambda = \frac{1}{2\pi}\int_{-\pi}^{+\pi} \sin k\lambda \log w(\lambda)d\lambda = 0. \end{aligned} \quad (32)$$

Formulas (8.16), (8.29), (8.30) and (32) imply that

$$\sum_{n=0}^{\infty} c_n^{(x)}\zeta^n = \Gamma_x(\zeta) = \Gamma_x(0)\frac{\Gamma_x(\zeta)}{\Gamma_x(0)} = \Gamma_x(0)e^{Q_x(\zeta)-Q_x(0)} =$$

$$= c_0^{(x)} \exp\Big(\sum_{k=1}^{\infty} a_k^{(x)}\zeta^k\Big). \quad (33)$$

Setting

$$\exp\Big(\sum_{k=1}^{\infty} a_k^{(x)}\zeta^k\Big) = 1 + r_1\zeta + r_2\zeta^2 + \dots, \quad (34)$$

and comparing (33) and (34), we have

$$c_n^{(x)}/c_0^{(x)} = r_n. \quad (35)$$

Formulas (35) and (26) imply that

$$\sigma_E^2(m) = (c_0^{(x)})^2(1 + r_1^2 + r_2^2 + \dots + r_m^2). \quad (36)$$

Formulas (30), (31) and (36) now complete the proof of Theorem 2.

§5. Definition of σ_I^2

After [4], we denote by $\widehat{H}_x(t)$ the smallest closed linear subspace of H_x containing the elements

$$x(t+1), x(t+2), \dots, x(t+n), \dots,$$
$$x(t-1), x(t-2), \dots, x(t-n), \dots .$$

[8] As indicated in Theorem 23 from [4], (8.16), (8.29)–(8.31) are applicable to any non-singular sequence.

The element $x(t)$ can be uniquely represented in the form (10.3),

$$x(t) = \nu(t) + \delta(t),$$

where $\nu(t)$ belongs to $\widehat{H}_x(t)$ and $\delta(t)$ is orthogonal to $\widehat{H}_x(t)$. It is easy to show that for a stationary sequence of real random variables

$$\sigma_1^2 = \|\delta(t)\|^2 = d_x^2. \tag{37}$$

By Theorem 24 from [4],

$$d_x^2 = (2\pi)^2 : \int_{-\pi}^{+\pi} \frac{d\lambda}{f_{xx}(\lambda)}, \tag{38}$$

where $d_x^2 = 0$ if the integral in the denominator on the right-hand side of (38) is infinite. Formulas (37), (38), (27) and (28) imply that

$$\sigma_I^2 = d_x^2 = \pi : \int_0^{\pi} \frac{d\lambda}{w(\lambda)} = \frac{1}{R},$$

which proves Theorem 3.

Steklov Mathematical Institute, 26 November 1940

References

1. A. Kolmogoroff, 'Sur l'interpolation et extrapolation des suites stationnaires', *C. R. Acad. Sci. Paris* **208** (1939), 2043–2045.
2. A. Khintchine, 'Korrelationstheorie der stationären stochastischen Prozesse', *Math. Ann.* **109** (1934), 604–615.
3. H. Wold, *A study in the analysis of stationary time series*, Uppsala, 1938.
4. A.N. Kolmogorov, 'Stationary sequences in Hilbert space', *Byull. Moskov. Gos. Univ. Mat.* **2**:6 (1941), 1–40 (in Russian) (No. 27 in this volume).
5. A.N. Kolmogorov, *Basic notions of probability theory*, ONTI, Moscow, 1936.
6. M. Stone, 'Linear transformations in Hilbert space', *Amer. Math. Soc. Colloq. Publ.* **15** (1932).

29. ON THE LOGARITHMIC NORMAL DISTRIBUTION OF PARTICLE SIZES UNDER GRINDING *

In a recent paper [1] N.K. Razumovskii indicates many cases when the logarithms of particle sizes (gold grits in gold placers, rock particles under grinding, etc.) obey approximately the Gauss distribution law. The aim of this paper is to give a fairly general scheme of the random process of particle grinding, for which in the limit (when grinding does not stop) the Gauss law for the logarithms of particle sizes can be established theoretically. Perhaps similar considerations will help to explain also why the Gauss distribution is applicable to the logarithms of mineral contents in separate samples (Razumovskii's paper is mainly devoted to this question).

Let us study the general number of particles $N(t)$ and their distribution in size at successive times $t = 0, 1, 2, \dots$.

Let $N(r,t)$ denote the number of particles with sizes $\rho \leq r$ at time t (in what follows it is immaterial whether ρ denotes diameter, weight or any other characteristic of a particle's size, provided the size of each particle obtained after grinding a particle of size r does not exceed r).

We denote by $Q(k)$ the expectation of the number of particles of size $\rho \leq kr$ formed during the period from t to $t+1$ from one particle which at time t had size r. We set

$$A = \frac{1}{Q(1)} \int_0^1 \log k \, dQ(k), \tag{1}$$

$$B^2 = \frac{1}{Q(1)} \int_0^1 (\log k - A)^2 dQ(k). \tag{2}$$

Under certain assumptions given below it can be proved that for sufficiently large t the ratio

$$N(e^x, t)/N(t) \tag{3}$$

is arbitrarily close to

$$\frac{1}{\sqrt{2\pi t} B} \int_{-\infty}^{x} \exp\Big\{-\frac{(\xi - At)^2}{2B^2 t}\Big\} d\xi \tag{4}$$

with probability arbitrarily close to 1.

* *Dokl. Akad. Nauk SSSR* **31** (1941), 99–101.

The most essential assumption used to derive this relation is that the probability that a particle is ground into a certain number of parts of certain relative sizes per unit time does not depend on the size of the initial particle.

To formulate rigorously the required assumptions we introduce our notation. Let p_n be the probability of obtaining exactly n particles from one particle during the period between t and $t+1$ and let

$$F_n(a_1, a_2, \dots, a_n) = \mathbf{P}\{k_1 \le a_1, k_2 \le a_2, \dots, k_n \le a_n\}$$

be the conditional distribution law for the ratios $k_i = r_i/r$ of the sizes of the resulting n particles to the size of the original particle. The n particles resulting from grinding are supposed to be enumerated in increasing order of size: $r_1 \le r_2 \le r_3 \le \dots \le r_n$.

In accordance with this, $F_n(a_1, a_2, \dots, a_n)$ is defined only for $0 \le a_1 \le a_2 \le \dots \le a_n \le 1$.

Clearly,

$$Q(k) = \sum_{n=1}^{\infty} p_n\{F_n(k,1,1,\dots,1,1) + F_n(k,k,1,\dots,1,1) + \dots \\ \dots + F_n(k,k,k,\dots,k,1) + F_n(k,k,k,\dots,k,k)\}.$$

We assume that

a) the probabilities p_n and the distributions F_n do not depend on the absolute size of a particle, on its prehistory (that is, what kind of previous grinding it resulted from), and on the fate of other particles;

b) the expectation $Q(1)$ of the total number of particles obtained between t and $t+1$ from one particle, is finite and greater than 1;

c) the integral

$$\int_0^1 |\log k|^3 dQ(k) \tag{5}$$

is finite;

d) at initial time $t = 0$ there is a certain number of particles $N(0)$ with arbitrary size distribution $N(r, 0)$.

Under these assumptions:

1) The expectation of the total number of particles $N(t)$ at the moment t is

$$\overline{N}(t) = N(0)Q^t(1); \tag{6}$$

2) for sufficiently large t, (3) is arbitrarily close to the ratio

$$\frac{\overline{N}(e^x,t)}{\overline{N}(t)} = \frac{\overline{N}(e^x,t)}{N(0)Q^t(1)} = T(x,t) \tag{7}$$

of the corresponding expectations[1] with probability arbitrarily close to 1.

In view of all these comments our problem reduces to estimating $T(x,t)$. Assumptions a) and b) imply that

$$\overline{N}(r,t+1) = \int_0^1 \overline{N}\Big(\frac{r}{k},t\Big)dQ(k). \tag{8}$$

Setting

$$Q(k) = Q(1)S(\log k), \tag{9}$$

from (7) and (8) we obtain

$$T(x,t+1) = \int_{-\infty}^{0} T(x-\xi,t)dS(\xi). \tag{10}$$

It is easy to deduce from (7) and (9) that $S(x)$ and $T(x,0) = N(e^x,0)/N(0)$ satisfy all the requirements for distribution functions.[2] In view of the recurrence relation (10), the same is true for the functions[3] $T(x,t)$ for any integer $t > 0$. Condition (c) implies that

$$\int_{-\infty}^{0} |x|^3 dS(x) \tag{11}$$

is finite. By (10) and (11) and *Lyapunov's theorem* we have as $t \to \infty$,

$$T(x,t) \to \frac{1}{\sqrt{2\pi t}B}\int_{-\infty}^{x} \exp\Big\{-\frac{(\xi-At)^2}{2B^2t}\Big\}d\xi \tag{12}$$

---

[1] This point should not be dismissed as being trivial. For $N(t)$ and $N(r,t)$ taken separately, the corresponding statement, that is, that the ratios $N(t)/\overline{N}(t)$ and $N(r,t)/\overline{N}(r,t)$ are close to 1, would be wrong.

[2] In this case we set for $x > 0$,

$$S(x) = S(0) = 1.$$

Therefore in all integrals involving dS the upper limit 0 can be replaced by $+\infty$ without changing the value of the integral.

[3] In our problem $S(x)$ and $T(x,t)$ are not probability distributions, but rather express certain expectations. This does not prevent us from invoking Lyapunov's theorem, considered as a theorem in pure analysis.

uniformly with respect to x, where

$$A = \int_{-\infty}^{0} x dS(x), \quad B^2 = \int_{-\infty}^{0} (x - A)^2 dS(x)$$

are clearly the same as the A and B defined by (1) and (2).

It would be interesting to study mathematical schemes in which the rate of grinding decreases (or increases) with the decrease of particle size. Then it would be natural to consider first the cases when the grinding rate is proportional to a certain power of particle size. If this power is non-zero, then the logarithmic normal law is no longer applicable.

Steklov Mathematical Institute, 17 December 1940

References

1. N.K. Razumovskii, *Dokl. Akad. Nauk SSSR* **28**:8 (1940) (in Russian).

30. JUSTIFICATION OF THE METHOD OF LEAST SQUARES*

From the purely practical viewpoint, the standard literature on the method of least squares has one essential drawback: it does not give any indications on the use of the Student distribution and χ^2 laws for evaluating the reliability of results obtained (for computing the probability that the errors exceed certain limits). On the other hand, the use of the Gauss law for a small number of observations results in a very high, and in practice, quite noticeable, overestimation of this reliability (see below §§9, 10).

In addition, the standard handbooks do not explain adequately the techniques involved in the method of least squares; this is in evidence in the teaching courses in universities and pedagogical institutes where the students are supposed to have a good knowledge of linear algebra. The point is that usually all the basic results of the method of least squares are obtained in a very clumsy *purely computational way*, whereas the use of proper *general methods* of modern linear algebra (for example, the notion of orthogonality) gives the same results much more transparently. The presentation is most lucid when it is performed with the help of notions of n-dimensional vector geometry.

The aim of this paper is to indicate by means of an example of a very simple problem on the method of least squares, under the assumption that all observations are equally valid, how one may obviate both these drawbacks. The reader is supposed to have a knowledge of linear algebra in geometric vector presentation and the fundamentals of probability theory. Greek characters, excluding π and Γ, denote random variables.

I. LIST OF MAIN RESULTS REQUIRING JUSTIFICATION

§1. Statement of the problem

We assume that the variables

$$y, x_1, x_2, \ldots, x_n$$

are related by a homogeneous linear dependence

$$y = \sum_{j=1}^{n} a_j x_j. \tag{I}$$

* *Uspekhi Mat. Nauk* 1:1 (1946), 57–70 (in Russian).

The coefficients a_j are unknown. To determine them, we find by experiment the values

$$y_r = \sum_{j=1}^{n} a_j x_{jr}, \quad r = 1, 2, \dots, N. \tag{II}$$

We assume that the a_j are uniquely determined by the values of x_{jr} and y_r, that is, the rank of the matrix rank $\|x_{jr}\|$ is greater than or equal to n. This implies that $N \geq n$. In the experimental determination of the y_r certain errors are inevitable. Instead of the true values of y_r we experimentally obtain the values

$$\eta_r = y_r + \Delta_r. \tag{III}$$

Given x_{ir} and η_r, we have to determine the best rational approximate values α_i of the coefficients a_i.

§2. Dogmatic presentation of Gauss's results

Let

$$\eta_r^* = \sum_{j=1}^{n} \alpha_j x_{jr}, \tag{IV}$$

$$\epsilon_r = \eta_r - \eta_r^*. \tag{V}$$

According to Gauss, the α_j are determined from the condition

$$[\epsilon\epsilon] = \sum_{r=1}^{N} \Big(\eta_r - \sum_{j=1}^{n} \alpha_j x_{jr}\Big)^2 = \min. \tag{VI}$$

Here, and in what follows, we use Gauss's notation

$$[ab] = \sum_{r=1}^{N} a_r b_r. \tag{*}$$

Requirement (VI) is equivalent to the system of equations

$$\sum_{j=1}^{n} [x_i x_j] \alpha_j = [x_i \eta], \quad i = 1, 2, \dots, n. \tag{VII}$$

From these "normal equations" α_i is found.

The resulting approximate values of α_i do not contain any "systematic error", that is,

$$\mathsf{E}\alpha_j = a_j. \tag{VIII}$$

The accuracy of these approximations is determined by the formula for the mean square error, the variance $\mathbf{D}(\alpha_j)$:

$$\mathbf{D}\alpha_j = \mathbf{E}\{\alpha_j - a_j\}^2 = q_{jj}s^2, \tag{IX}$$

where s is the mean square error of the experimental values of y_r (it is assumed to be independent of the number r) and q_{ir} is determined from the equations

$$\sum_{j=1}^{n}[x_i x_j]q_{jk} = e_{ik}; \ i,k = 1,2,\ldots,n. \tag{X}$$

Here

$$e_{ik} = \begin{cases} 0 & \text{for } i \neq k, \\ 1 & \text{for } i = k \end{cases}$$

Formula (IX) is a particular case (for $i = j$) of the formula

$$\mathbf{E}\{(\alpha_i - a_i)(\alpha_j - a_j)\} = q_{ij}s^2. \tag{XI}$$

Practical applications of this formula are discussed in §11.

If s is not known *a priori*, it is usually taken to be approximately equal to

$$\sigma = \sqrt{[\epsilon\epsilon]/(N-n)}. \tag{XII}$$

This stratagem is based on the fact that

$$\mathbf{E}\sigma^2 = s^2, \tag{XIII}$$

$$\mathbf{D}\sigma^2 = \mathbf{E}\{\sigma^2 - s^2\}^2 = 2s^4/(N-n). \tag{XIV}$$

The latter formula shows that for large $N-n$ the ratio $\sigma : s$ is indeed close to 1 with probability close to 1. Together with formula (IX) this makes it possible to consider $q_{jj}\sigma^2$ as an approximate value of $\mathbf{D}\alpha_j$:

$$\mathbf{D}\alpha_j \sim q_{jj}\sigma^2. \tag{XV}$$

Remark. If $N = n$, then the system of equations

$$\eta_r = \sum_{j=1}^{n}\alpha_j x_{jr}; \quad r = 1,2,\ldots,N = n$$

is solvable and (VI) implies that

$$[\epsilon\epsilon] = 0.$$

In this case the denominator on the right-hand side of (XII) vanishes and σ is undefined.

§3. Continuation of the dogmatic presentation of Gauss's results

Formulas (VIII), (IX), (XIII) and (XIV) give but a rough indication of the size of the error resulting from the replacement of a_j by α_j and s by σ. The final solution of this problem should have consisted in producing distribution laws for the deviations $\alpha_j - a_j$ and $\sigma - s$. Gauss did this (under the assumption that the errors Δ_r are independent and obey the Gauss distribution law with mean equal to zero) for $\alpha_j - a_j$. Namely, he found that

$$(\alpha_j - a_j)/s$$

has the normal distribution

$$\mathbf{P}\left\{\frac{\alpha_j - a_j}{\sqrt{q_{jj}}s}\right\} = \frac{1}{\sqrt{2\pi}}\int_{-\infty}^{t} e^{-t^2/2}dt = G(t). \tag{XVI}$$

§4. More recent results

The deviations of a_j from α_j can be evaluated more accurately with the help of the distribution law

$$\mathbf{P}\{(N-n)\sigma^2/s^2 < h^2\} = H_{N-n}(h^2), \tag{XVII}$$

where

$$H_m(h^2) = \frac{1}{2^{(m-2)/2}\Gamma(m/2)}\int_0^h h^{m-1}e^{-h^2/2}dh. \tag{XVIII}$$

Tables of the function H_m and its inverse are widely used. In §10 we point out a possible modification in these tables which seems to be desirable from the viewpoint of practical application of the method of least squares.

The error resulting from replacing a_j by α_j for an unknown s is estimated using the theorem stating that

$$\tau = (\alpha_j - a_j)/\sqrt{q_{jj}}\sigma$$

obeys the Student distribution law

$$\mathbf{P}\left\{\frac{\alpha_j - a_j}{\sqrt{q_{jj}}\sigma} < t\right\} = S_{N-n}(t), \tag{XIX}$$

where[1]

$$S_m(t) = \frac{\Gamma((m+1)/2)}{\sqrt{m\pi}\Gamma(m/2)} \int_{-\infty}^{t} \left(1 + \frac{t^2}{m}\right)^{-(m+1)/2} dt. \tag{XX}$$

II. THE PROOFS

§5. Formula VII

The fact that requirement (VI) is equivalent to the system of normal equations (VII) is quite elementary and can be established in many different ways. We will obtain it by vector methods for the sake of uniformity of style.

For this purpose we note that from the viewpoint of modern linear algebra the Gaussian brackets $[bc]$ are just the *scalar product* of two n-dimensional vectors

$$b = (b_1, b_2, \ldots, b_n), \quad c = (c_1, c_2, \ldots, c_n).$$

Regarding

$$y_r,\ x_{ir},\ \eta_r,\ \Delta_r,\ \eta_r^*,\ \epsilon_r$$

as the components of the N-dimensional vectors

$$y,\ x_i,\ \eta,\ \Delta,\ \eta^*,\ \epsilon,$$

we express (II)–(V) in the form

$$y = \sum_{j=1}^{n} a_j x_j, \tag{1}$$

$$\eta = y + \Delta, \tag{2}$$

$$\eta^* = \sum_{j=1}^{n} \alpha_j x_j, \tag{3}$$

$$\eta = \eta^* + \epsilon = \sum_{j=1}^{n} \alpha_j x_j + \epsilon. \tag{4}$$

[1] At the end of this paper I give the table of the function inverse to $S_m(t)$ which is a reduced version of the table published by E.N. Pomerantseva. I believe it satisfies the basic needs for a practical use of the method of least squares (see §9).

Since the choice of the α_i is at our disposal, (3) merely asserts that

$$\eta^* \in L, \tag{5}$$

where

$$L = L(x_1, x_2, \ldots, x_n)$$

is the linear subspace generated in the N-dimensional vector space V^N by the x_i.

In the language of vectors, requirement (VI),

$$[\epsilon\epsilon] = \min \tag{6}$$

together with (5) means the following:

In (4) η^* *is the orthogonal projection of* η *onto* L *and* ϵ *is the complementary vector orthogonal to* L.

This implies that

$$[\epsilon x_i] = 0, \ i = 1, 2, \ldots, n. \tag{7}$$

By (7), taking the vector product of (4) and x_i we obtain

$$\sum_j [x_i x_j]\alpha_j = [x_i \eta], \ i = 1, 2, \ldots, n. \tag{8}$$

These are the normal equations (VII) for determining α_j. The corresponding determinant

$$G = \|[x_i x_j]\| \tag{9}$$

is the Gram determinant of the vectors $x_1, x_2, \ldots, x_n$. Since, by the assumption of §1, $\|x_{jr}\|$ has rank n, the vectors x_i are linearly independent and $G \neq 0$. Therefore (8) uniquely determine the α_j.

§6. Formulas (VIII)–(XII)

We need certain prerequisites from probability theory in order to justify formulas (VIII)–(XII). These are as follows:

A) *The true errors* Δ_r *are random variables;*

B) $\mathsf{E}\Delta_r = 0$,

C) $\mathsf{E}\Delta_r^2 = s^2$, *where* s^2 *is finite and does not depend on* r*;*

D) $\mathsf{E}\{\Delta_r \Delta_{r'}\} = 0$ *for* $r \neq r'$.

We define a system of vectors

$$u_1, u_2, \ldots, u_n,$$

in L that are δ_n-orthogonal to the system $x_1, x_2, \ldots, x_n$; this means that

$$[x_i u_j] = e_{ij}, \tag{10}$$

$$u_j \in L. \tag{11}$$

Since each of the systems $\{x_1, x_2, \ldots, x_n\}$ and $\{u_1, u_2, \ldots, u_n\}$ is a basis in L, it follows that

$$u_i = \sum_{k=1}^{n} q_{ik} x_k, \quad x_i = \sum_{k=1}^{n} c_{ik} u_k.$$

Taking the scalar product of the first of these equalities and u_j, and the scalar product of the second and x_j, and using (10) we obtain

$$q_{ij} = [u_i u_j], \quad c_{ij} = [x_i x_j], \tag{12}$$

that is,

$$u_i = \sum_{j=1}^{n} q_{ij} x_j = \sum [u_i u_j] x_j, \tag{13}$$

$$x_i = \sum_{j=1}^{n} [x_i x_j] u_j. \tag{14}$$

By (13)–(14), $\|q_{ij}\|$ is the inverse to $\|[x_i x_j]\|$, that is, the q_{ij} are indeed determined from (X).

Taking the scalar product of (1) and u_i we obtain

$$a_i = [y u_i]. \tag{15}$$

Similarly, (3) implies that

$$\alpha_i = [\eta^* u_i]. \tag{16}$$

But since ϵ is orthogonal to L, we have

$$[\epsilon u_i] = 0$$

and (16), together with (4), implies that

$$\alpha_i = [\eta u_i]. \tag{17}$$

Together with (2), formulas (15) and (17) give

$$\alpha_i = a_i + [\Delta u_i], \tag{18}$$

whence from B), C) and D) we obtain

$$\mathsf{E}\alpha_i = a_i, \tag{19}$$

$$\mathsf{E}\{(\alpha_i - a_i)(\alpha_j - a_j)\} = \mathsf{E}\{[\Delta u_i][\Delta u_j]\} =$$

$$= \sum_{s=1}^{N}\sum_{s'=1}^{N} \mathsf{E}\{\Delta_s \Delta_{s'}\} u_{is} u_{js'} = s^2 \sum_{s=1}^{n} u_{is} u_{js} = [u_i u_j] s^2,$$

that is, in view of (12),

$$\mathsf{E}\{(\alpha_i - a_i)(\alpha_j - a_j)\} = q_{ij} s^2. \tag{20}$$

Formulas (19) and (20) are none other than formulas (VIII) and (XI). As has already been mentioned, (XI) gives (IX) for $i = j$.

We set

$$\Delta^* = \eta^* - y. \tag{21}$$

Formulas (2), (4) and (21) imply that

$$\Delta = \Delta^* + \epsilon. \tag{22}$$

Since Δ^* belongs to L together with η^* and y, and ϵ is orthogonal to L, it follows that (22) is the decomposition of Δ into its orthogonal projection onto L and the orthogonal complement to this projection.

In our N-dimensional vector space we choose an orthogonal basis

$$b_r = (b_{r1}, b_{r2}, \dots, b_{rN}), \quad r = 1, 2, \dots, N,$$

$$[b_r b_{r'}] = e_{rr'},$$

such that the first n vectors belong to L and the remaining $N-n$ are orthogonal to L.

Setting

$$\tilde{\Delta}_r = [\Delta b_r], \tag{23}$$

we obtain

$$\Delta = \sum_{r=1}^{N} \tilde{\Delta}_r b_r, \tag{24}$$

$$\Delta^* = \sum_{r=1}^{n} \tilde{\Delta}_r b_r, \tag{25}$$

$$\epsilon = \sum_{r=n+1}^{N} \tilde{\Delta}_r b_r. \tag{26}$$

Clearly, by C) and D)

$$\mathsf{E}\{\tilde{\Delta}_r \tilde{\Delta}_{r'}\} = \mathsf{E}\Big(\sum_{k=1}^{N} \Delta_k e_{rk} \sum_{k'=1}^{N} \Delta_{k'} e_{r'k'}\Big) =$$

$$= \sum_{k=1}^{N} \sum_{k'=1}^{N} \mathsf{E}\{\Delta_k \Delta_{k'}\} e_{rk} e_{r'k'} = s^2 \sum_{k=1}^{N} e_{rk} e_{r'k'} = \begin{cases} s^2 & \text{for } r = r', \\ 0 & \text{for } r \neq r'. \end{cases} \tag{27}$$

Formula (26) and B) imply that

$$\mathsf{E}\epsilon = 0 \tag{28}$$

and (26) and (27) imply that

$$\mathsf{E}[\epsilon\epsilon] = (N-n)s^2. \tag{29}$$

Formula (29) and (XII) (this formula merely serves as a definition of σ) immediately imply (XIII).

§7. The hypothesis of the Gaussian distribution and the independence of the true errors

For our further derivations, the assumptions B), C) and D) given at the beginning of §6 should be replaced by the following stronger ones:

G) *The errors Δ_r obey the Gauss distribution*

$$\mathsf{P}\{\Delta_r < t\} = \frac{1}{\sqrt{2\pi}s} \int_{-\infty}^{t} e^{-t^2/2s^2}\, dt,$$

where s^2 does not depend on r;

U) *The errors $\Delta_1, \Delta_2, \ldots, \Delta_N$ are N mutually independent random variables.*

It is well known that G) implies B) and C), and B) together with U) implies D).

Assumptions G) and U) imply that the system $\Delta_1, \Delta_2, \ldots, \Delta_N$ of random variables obeys the N-dimensional distribution law with probability density

$$f(t_1, t_2, \ldots, t_N) = \prod_{r=1}^{N} \Bigl(\frac{1}{\sqrt{2\pi}s} e^{-t_r^2/2s^2} \Bigr), \tag{30}$$

or, in vector notation,

$$f(t) = (\sqrt{2\pi}s)^{-N} e^{-[tt]/2s^2}. \tag{31}$$

Formula (31) gives the probability density of a random vector Δ with

$$dv = dt_1 dt_2 \ldots dt_N \tag{32}$$

as the volume element.

In the coordinate system

$$\tilde{t}_r = [tb_r], \tag{33}$$

$$\tilde{\Delta}_r = [\Delta b_r], \tag{34}$$

where the basis vectors b_r satisfy the usual orthogonality condition

$$[b_r b_{r'}] = e_{rr'}, \tag{35}$$

the volume element (32) retains the form

$$dv = d\tilde{t}_1 d\tilde{t}_2 \ldots d\tilde{t}_N. \tag{32'}$$

Therefore in this new coordinate system $f(t)$ can also be considered as the probability density of Δ. In other words, the probability density of the random variables $\tilde{\Delta}_1, \tilde{\Delta}_2, \ldots, \tilde{\Delta}_N$ is of the form

$$f(\tilde{t}_1, \tilde{t}_2, \ldots, \tilde{t}_N) = \prod_{r=1}^{N} \Bigl(\frac{1}{\sqrt{2\pi}s} e^{-\tilde{t}_r^2/2s^2} \Bigr). \tag{30'}$$

Formula (30′) shows that the random variables $\tilde{\Delta}_r$ are independent and obey the same distribution law

$$\mathsf{P}\{\tilde{\Delta}_r < t\} = \frac{1}{\sqrt{2\pi}s}\int_{-\infty}^{t} e^{-t^2/2s^2}\,dt, \tag{36}$$

as do the Δ_r.

It is important to be aware of the fact that the independence of $\tilde{\Delta}_r$ is derived from both G) and U). This conclusion is no longer the case when the Gaussian distribution for the errors of the first kind is replaced by some other one.

§8. Proof of formulas (XIV)–(XX)

By (18) α_i is a linear function (with non-random coefficients) of the random variables Δ_r. Therefore G) and U) imply that α_i obeys a Gaussian probability distribution. Normalizing this distribution in accordance with (VIII) and (IX) we obtain (XVI).

Now we derive the distribution law for

$$\chi^2 = [\epsilon\epsilon]/s^2 = (N-n)\sigma^2/s^2. \tag{37}$$

For this we note that, by (26),

$$[\epsilon\epsilon] = \sum_{r=n+1}^{N} \tilde{\Delta}_r^2, \tag{38}$$

$$\chi^2 = \sum_{r=n+1}^{N} \left(\frac{\tilde{\Delta}_r}{s}\right)^2. \tag{39}$$

The independence of the random variables $\tilde{\Delta}_r$ implies that of the random variables $\tilde{\Delta}_r/s$. By (36) each of these latter variables obeys the normal distribution law with probability density $\frac{1}{\sqrt{2\pi}}e^{-t^2/2}$. Therefore their $(N-n)$-dimensional distribution law is characterized by the probability density

$$(2\pi)^{-(N-n)/2}\exp\Big\{-\tfrac{1}{2}\sum_{r=n+1}^{N} t_r^2\Big\}.$$

The probability of the inequality

$$\chi < h$$

is given by integrating this density over the $(N-n)$-dimensional volume for which

$$R^2 = \sum_{r=n+1}^{N} t_r^2 < h^2.$$

Thus we obtain [2]

$$\mathbf{P}\{\chi < h\} = (2\pi)^{-(N-n)/2} \iint \ldots \int_{R^2<h^2} e^{-R^2/2} dt_{n+1} dt_{n+2} \ldots dt_N =$$

$$= \frac{1}{2^{(N-n-2)/2}\Gamma((N-n)/2)} \int_0^h h^{N-n-1} e^{-h^2/2} dh. \tag{40}$$

This is formula (XVIII).

From (40) it follows that

$$\mathbf{E}\chi^2 = N - n, \tag{41}$$

$$\mathbf{D}\chi^2 = \mathbf{E}\{\chi^2 - (N-n)\}^2 = 2(N-n). \tag{42}$$

By (37), formulas (41) and (42) are equivalent to (XIII) and (XIV).

It only remains to consider the distribution law for

$$\tau_j = (\alpha_j - a_j)/\sqrt{q_{jj}}\sigma. \tag{43}$$

For this it is helpful to write τ_j in the form

$$\tau_j = \gamma_j/\chi, \tag{44}$$

where

$$\gamma_j = \sqrt{N-n}(\alpha_j - a_j)/\sqrt{q_{jj}}s. \tag{45}$$

We now note that by (15), (16), (21) and (25),

$$\alpha_j - a_j = [\Delta^* u_j] = \sum_{r=1}^{n} \tilde{u}_{jr} \tilde{\Delta}_r, \tag{46}$$

[2] For the derivation of (40) it should be noted that the area of the surface of a sphere of radius R in m-dimensional space is

$$\frac{2\pi^{m/2}}{\Gamma(m/2)} R^{m-1}.$$

where

$$\tilde{u}_{jr} = [u_j b_r]. \tag{47}$$

Comparing (46) with (37)–(38) we see that the $\tilde{\Delta}_r$ entering in the expression for $\alpha_j - a_j$ are only those $\tilde{\Delta}_r$ that do not occur in the expression for χ^2. Therefore, the independence of the $\tilde{\Delta}_r$ implies that $\alpha_j - a_j$ is independent of χ. By (45), γ_j and χ are also independent. In accordance with (XVI), the probability density of γ_j is

$$\frac{1}{\sqrt{2\pi(N-n)}} e^{-c^2/2(N-n)}.$$

Together with (40) this implies that the 2-dimensional probability density of γ_j and χ is

$$\frac{1}{\sqrt{\pi(N-n)} \cdot 2^{(N-n-1)/2}\Gamma((N-n)/2)} h^{N-n-1} \exp\Big\{-\frac{c^2}{2(N-n)} - \frac{h^2}{2}\Big\}.$$

Integrating this probability density over the region where $c/h < t$, we obtain

$$\mathbf{P}\{\tau_j < t\} = \mathbf{P}\Big\{\frac{\gamma_j}{\chi} < t\Big\} =$$

$$= \frac{1}{\sqrt{\pi(N-n)} 2^{(N-n-1)/2}\Gamma((N-n)/2)} \times$$

$$\times \iint h^{N-n-1} \exp\Big\{-\frac{c^2}{2(N-n)} - \frac{h^2}{2}\Big\} dc\,dh. \tag{48}$$

We now change the variables c and h to s and u via the formulas

$$s = \frac{c}{h}, \quad u = \frac{c^2}{2(N-n)} + \frac{h^2}{2}.$$

Since

$$\frac{\partial(s,u)}{\partial(c,h)} = 1 + \frac{s^2}{N-n}, \quad h^2 = \frac{2u}{1 + s^2/(N-n)},$$

(48) implies that

$$\mathbf{P}\{t_j < t\} =$$

$$= \frac{1}{\sqrt{\pi(N-n)}\Gamma((N-n)/2)} \int_0^\infty u^{\frac{N-n-1}{2}} e^{-n} du \int_0^t \Big(1 + \frac{s^2}{N-n}\Big)^{-\frac{N-n+1}{2}} ds =$$

$$= \frac{\Gamma((N-n+1)/2)}{\sqrt{\pi(N-n)}\Gamma((N-n)/2)} \int_0^t \Big(1 + \frac{t^2}{N-n}\Big)^{-(N-n+1)/2} dt. \tag{49}$$

Formulas (XIX) and (XX) are nothing but another way of expressing (49).

III. REMARKS ON THE PRACTICAL SIGNIFICANCE OF IMPROVING THE STANDARD ACCOUNT OF THE METHOD OF LEAST SQUARES

§9. On the use of the Student distribution law

In accordance with the general principles of modern mathematical statistics[3] the two values α_j' and α_j'' such that for any acceptable hypothesis concerning the unknown values a_j we have

$$\mathsf{P}\{\alpha_j' \leq a_j \leq \alpha_j''\} \geq \omega \tag{50}$$

will be called the "confidence limits" for a_j corresponding to the "confidence coefficient" ω.

If s is known and only the $a_1, a_2, \dots, a_n$ are unknown, then by (XVI), the formulas

$$\alpha_j' = \alpha_j - t\sqrt{q_{jj}}\,s, \quad \alpha_j'' = \alpha_j + t\sqrt{q_{jj}}\,s, \tag{51}$$

satisfy this requirement, where t is defined by the equation

$$G(t) = (1+\omega)/2. \tag{52}$$

The values of t corresponding to various ω are given in the second last line ($m = \infty$) of Table 1.

If s is known, then in accordance with (XV), it is customary to consider σ as an approximate value of s. However, for the confidence limits

$$\alpha_j' = \alpha_j - t\sqrt{q_{jj}}\,\sigma, \quad \alpha_j'' = \alpha_j + t\sqrt{q_{jj}}\,\sigma \tag{53}$$

to satisfy (50), t should be defined not by (52) but, in accordance with (XIX), (XX), by

$$S_{N-n}(t) = (1+\omega)/2. \tag{54}$$

The values of t corresponding to various ω according to (54) for different values of $N - n = m$ are given in the same Table 1.

---

[3] See [1] or [2].

Table 1. Values of t satisfying the equation $S_m(t) = P$

m	$P = 0.75$	$P = 0.90$	$P = 0.95$	$P = 0.975$	$P = 0.99$	$P = 0.995$	$P = 0.9975$	$P = 0.999$	$P = 0.9995$
1	1.000	3.078	6.314	12.076	31.821	63.657	127.321	318.309	636.619
2	0.816	1.886	2.920	4.303	6.965	9.925	14.089	22.327	31.600
3	0.765	1.638	2.353	3.182	4.541	5.841	7.453	10.214	12.922
4	0.741	1.533	2.132	2.776	3.747	4.604	5.597	7.173	8.610
5	0.727	1.476	2.015	2.571	3.365	4.032	4.773	5.893	6.869
6	0.718	1.440	1.943	3.447	3.143	3.707	4.317	5.208	5.959
7	0.711	1.415	1.895	2.365	2.998	3.499	4.029	4.785	5.408
8	0.706	1.397	1.860	2.306	2.896	3.355	3.833	4.501	5.041
9	0.703	1.383	1.833	2.262	2.821	3.250	3.690	4.297	4.781
10	0.700	1.372	1.812	2.228	2.764	3.169	3.581	4.144	4.587
11	0.697	1.363	1.796	2.201	2.718	3.106	3.497	4.025	4.437
12	0.69	1.356	1.782	2.179	2.681	3.055	3.428	3.930	4.318
13	0.694	1.350	1.771	2.160	2.650	3.012	3.372	3.852	4.221
14	0.692	1.345	1.761	2.145	2.624	2.977	3.326	3.787	4.140
15	0.691	1.341	1.753	2.131	2.602	2.947	3.286	3.733	4.073
16	0.690	1.337	1.746	2.120	2.583	2.921	3.252	3.686	4.015
17	0.689	1.333	1.740	2.110	2.567	2.898	3.222	3.646	3.965
18	0.688	1.330	1.734	2.101	2.552	2.878	3.197	3.610	3.922
19	0.688	1.328	1.729	2.093	2.539	2.861	3.174	3.579	3.883
20	0.687	1.325	1.725	2.086	2.528	2.845	3.153	3.552	3.849
21	0.686	1.323	1.721	2.080	2.518	2.831	3.135	3.527	3.819
22	0.696	1.321	1.717	2.074	2.508	2.819	3.119	3.505	3.792
23	0.685	1.319	1.714	2.069	2.500	2.807	3.104	3.486	3.768
24	0.685	1.318	1.711	2.064	2.492	2.797	3.092	3.467	3.745
25	0.684	1.316	1.708	2.060	2.485	2.787	3.078	3.450	3.725
26	0.684	1.315	1.706	2.056	2.479	2.779	3.067	3.435	3.707
27	0.684	1.314	1.703	2.052	2.473	2.771	3.057	3.421	3.690
28	0.683	1.313	1.701	2.048	2.467	2.763	3.047	3.408	3.674
29	0.683	1.311	1.699	2.045	2.462	2.756	3.038	3.396	3.659
30	0.683	1.310	1.697	2.042	2.457	2.750	3.030	3.386	3.646
∞	0.67449	1.21855	1.64485	1.95996	2.32634	2.57582	2.80703	3.09023	3.29053
ω	0.5	0.8	0.9	0.95	0.98	0.99	0.995	0.998	0.999

As $m \to \infty$, t determined from (54) converges to t determined from (52) with s replaced by σ, but this convergence is quite slow, especially for ω close to 1. For example, by (52), $t = 3.090$ for $\omega = 0.998$, which approximately corresponds to the classical "three sigma" rule, while (54), for example, yields $t = 5.959$ for $m = 6$.

§10. Estimation of s from σ.

The equality

$$\mathsf{P}\{k_1\sigma \le s \le k_2\sigma\} = \omega \tag{55}$$

is satisfied most simply by choosing k_1 and k_2 so that

$$\mathsf{P}\{s < k_1\sigma\} = (1-\omega)/2, \quad \mathsf{P}\{k_2\sigma \le s\} = (1+\omega)/2. \tag{56}$$

In accordance with (XVII), (XVIII) this is achieved if k_1 and k_2 are determined from

$$H_{N-n}\Big(\frac{N-n}{k_1}\Big) = \frac{1+\omega}{2}, \quad H_{N-n}\Big(\frac{N-n}{k_2}\Big) = \frac{1-\omega}{2}. \tag{57}$$

From the viewpoint of the method of least squares, it would appear desirable to have a table which would give k_1 and k_2 directly from $m = N - n$ and ω. This is done in our Table 2 for $\omega = 0.98$.

The sharp asymmetry between the upper and lower bounds of the estimate

$$k_1\sigma \le s \le k_2\sigma$$

is worth noting. For example, for $\omega = 0.98$ and $m = 3$,

$$k_1 = 0.514, \quad k_2 = 5.111,$$

that is, σ twice as large as s are found only with probability 0.01 but with probability 0.01 s is 5 times larger than σ. Therefore for small m it would be unreasonable to use confidence limits for s of the form $(1 \pm k')\sigma$, because then one would often have to take the lower bound negative ! Only for very large m can we assume that

$$k_1 \sim 1 - t/\sqrt{2m}, \quad k_2 \sim 1 + t\sqrt{2m}, \tag{58}$$

where t is determined from (52) by the Gaussian law.

Table 2. Values of k satisfying the equation $H_m(m/k) = \mathbf{P}(k\sigma \leq s\} = P$.

m	P		m	P		m	P	
	0.01	0.99		0.01	0.99		0.01	0.99
1	0.388	79.750	8	0.631	2.204	40	0.792	1.343
2	0.466	9.975	9	0.644	2.076	50	0.810	1.297
3	0.514	5.111	10	0.657	1.977	60	0.824	1.265
4	0.549	3.669	15	0.708	1.694	70	0.835	1.241
5	0.576	3.003	20	0.730	1.556	80	0.844	1.222
6	0.597	2.623	25	1.751	1.473	90	0.852	1.207
7	0.615	2.377	30	0.768	1.416	100	0.858	1.195

§11. Remarks on the value of the matrix $\|q_{ij}\|$

The value of the covariance matrix

$$S_{ij} = \mathbf{E}\{(\alpha_i - a_i)(\alpha_j - a_j)\} = q_{ij}s^2 \tag{59}$$

gives much more exhaustive information on the nature of the errors

$$\alpha_i - a_i$$

than the values of its diagonal elements

$$\mathbf{D}(\alpha_i) = S_{ii} \tag{60}$$

alone. For example, to determine the mean square error made when replacing a_i by α_i in the expression

$$b = c_1 a_1 + c_2 a_2 + \ldots + c_n a_n,$$

where the coefficients c_i are given, knowledge of the variances $\mathbf{D}(\alpha_i)$ is not enough, but it is quite sufficient to know the covariance matrix S_{ij}. Setting

$$\beta = c_1 \alpha_1 + c_2 \alpha_2 + \ldots + c_n \alpha_n,$$

we have

$$\mathsf{E}\{\beta - b\}^2 = \sum_i \sum_j S_{ij} c_i c_j .$$

Usually the diagonal elements of the matrix $\|q_{ij}\|$ are introduced as the reciprocals of the "weights"

$$p_{ii} = 1/q_{ii}.$$

This points to the essential role of the entire matrix $\|q_{ij}\|$.

References

1. Yu. Neiman, 'Statistical estimation in problems of classical probability theory', *Uspekhi Mat. Nauk* **10** (1944), 207 (in Russian).
2. A.N. Kolmogorov, 'Determination of the centre of scattering and the measurement accuracy from a limited number of observations', *Izv. Akad. Nauk SSSR Ser. Mat.* **6** (1942), 3–32 (in Russian).

31. A FORMULA OF GAUSS IN THE METHOD OF LEAST SQUARES *

Jointly with A.A. Petrov and Yu.M. Smirnov

This paper refines certain estimates from Gauss's theory of the method of least squares.

In §39 of "Theoria combinationis observationum erroribus minimis obnoxiae", Gauss computes the mean square error made in replacing $\mu\mu$ by

$$\frac{M}{\pi - \rho} = \frac{\lambda\lambda + \lambda\lambda' + \lambda''\lambda'' + \dots}{\pi - \rho}.$$

Gauss shows that this mean square error equals

$$\frac{\nu^4 - \mu^4}{\pi - \rho} - \frac{\nu^4 - 3\mu^4}{(\pi - \rho)^2}[\rho - \sum(a\alpha + b\beta + c\gamma + \dots)^2]. \tag{1}$$

In §40, Gauss derives from (1) simple estimates of the mean square error. He derives these estimates from the inequalities

$$\rho\rho/\pi \leq \sum(a\alpha + b\beta + c\gamma + \dots)^2 < \pi. \tag{2}$$

Gauss overlooked the fact that the upper estimate in (2) can be essentially sharpened, and (2) may be replaced by

$$\rho\rho/\pi \leq \sum(a\alpha + b\beta + c\gamma + \dots)^2 \leq \rho. \tag{3}$$

Therefore the conclusions of §40 are unexpectedly weak: the lower estimate suggested by Gauss for the least square error is sometimes even negative.

Our aim is to prove the inequalities (3) and to show that they cannot be improved.

§1. Statement of the problem

Let us first recapitulate the problem in modern notation close to that in Kolmogorov's paper [1].

Let y_r, a_i and x_{ir} where $r = 1, 2, \dots, N$; $i = 1, 2, \dots, n$; $n < N$, satisfy the N equations

$$y_r = \sum_{i=1}^{n} a_i x_{ir}. \tag{4}$$

* *Izv. Akad. Nauk SSSR Ser. Mat.* **11**:6 (1947), 561–566.

The x_{ir} are assumed to be exact and instead of y_r we are given

$$\eta_r = y_r + \Delta_r, \tag{5}$$

where the Δ_r are independent random variables such that

$$\mathsf{E}(\Delta_r) = 0, \quad \mathsf{E}(\Delta_r^2) = s^2, \quad \mathsf{E}(\Delta_r^4) = f^4 \tag{6}$$

(E denotes the expectation). Let the rank of the matrix $\|x_{ir}\|$ be n.

Under the above assumptions the method of least squares recommends that one takes the α_i determined from the condition

$$\sum_{r=1}^{N}(\eta_r - \sum_{i=1}^{n} \alpha_i x_{ir})^2 = \min \tag{7}$$

as approximate values of the unknown a_i.

As is known, this problem has the unique solution

$$\alpha_i = \sum_{r=1}^{N} u_{ir}\eta_r. \tag{8}$$

The coefficients u_{ir} in (8) have the following geometric meaning: in N-dimensional space the vectors

$$u_i = (u_{i1}, u_{i2}, \dots, u_{iN}), \; i = 1, 2, \dots, n,$$

form a biorthogonal system with the vectors

$$x_i = (x_{i1}, x_{i2}, \dots, x_{iN}), \; i = 1, 2, \dots, n,$$

which means that the vectors $u_1, u_2, \dots, u_n$ are uniquely determined by the following conditions: they belong to the linear n-dimensional subspace L spanned by $x_1, x_2, \dots, x_n$ and satisfy the biorthogonality conditions

$$\sum_{r=1}^{N} x_{ir} u_{jr} = e_{ij} = \begin{cases} 1 & \text{for } i = j, \\ 0 & \text{for } i \neq j. \end{cases} \tag{9}$$

Setting

$$\sigma^2 = \frac{1}{N-n} \sum_{r=1}^{N} \epsilon_r^2, \tag{10}$$

where

$$\epsilon_r = \eta_r - \sum_{i=1}^{n} \alpha_i x_{ir}, \tag{11}$$

we have

$$\mathbf{E}\sigma^2 = s^2. \tag{12}$$

Gauss's problem mentioned in the beginning of our paper is to determine

$$\mathbf{D}\sigma^2 = \mathbf{E}\{\sigma^2 - s^2\}^2. \tag{13}$$

In our notation Gauss's formula (1) is expressed in the form

$$\mathbf{D}\sigma^2 = \frac{f^4 - s^4}{N - n} - \frac{f^4 - 3s^4}{(N-n)^2}(n - \Omega), \tag{14}$$

where

$$\Omega = \sum_{r=1}^{N}\Big(\sum_{i=1}^{n} x_{ir}u_{ir}\Big)^2. \tag{15}$$

§2. Estimation of Ω

Expression (15) has a simple geometric meaning. Namely, if we denote by $e_r^* = (e_{r1}^*, e_{r2}^*, \dots, e_{rN}^*)$ the projection of the unit coordinate vector e_r onto L, then

$$\Omega = \sum_{r=1}^{N} |e_r^*|^4 = \sum_{r=1}^{N}\Big[\sum_{k=1}^{N}(e_{rk}^*)^2\Big]^2. \tag{16}$$

Indeed, the projection z^* of any vector z onto L can be written in the form

$$z^* = \sum_{i=1}^{n} c_i x_i.$$

Computing the scalar product (zu_i) we obtain, by (9):

$$(zu_i) = (z^* u_i) = c_i.$$

If $z = e_r$, then the latter formula gives $c_i = u_{ir}$. Therefore

$$e_r^* = \sum_{i=1}^{n} u_{ir} x_i, \tag{17}$$

or, in coordinate form,

$$e^*_{rk} = \sum_{i=1}^{n} u_{ir} x_{ik}. \tag{18}$$

Similarly we obtain

$$e^*_r = \sum_{i=1}^{r} x_{ir} u_i, \tag{19}$$

$$e^*_{rk} = \sum x_{ir} u_{ik}. \tag{20}$$

Formulas (18) and (20) imply that

$$|e^*_r|^2 = \sum_{k=1}^{N} (e^*_{rk})^2 = \sum_{k=1}^{N} \sum_{i=1}^{n} \sum_{j=1}^{n} (u_{ir} x_{ik} x_{jr} u_{jk}) = \sum_{i=1}^{n} u_{ir} x_{ir}. \tag{21}$$

Formulas (21) together with (15) immediately give (16).

Now let us introduce in N-dimensional space a new orthogonal coordinate system, placing the first n axes in the space L, and choosing the other $N-n$ axes to be orthogonal to this space.

In this new coordinate system let the vectors e_r be given by

$$e_r = (\Omega_{r1}, \Omega_{r2}, \dots, \Omega_{rN}).$$

Clearly, in this new coordinate system

$$e^*_r = (\Omega_{r1}, \Omega_{r2}, \dots, \Omega_{rn}, 0, 0, \dots, 0).$$

Therefore,

$$\Omega = \sum_{r=1}^{N} \Big(\sum_{k=1}^{n} \Omega^2_{rk} \Big)^2. \tag{22}$$

The matrix $\|\Omega_{rk}\|$ is orthogonal. It is easy to see that the statement of our problem does not impose any other restrictions on the form of this matrix: under an appropriate choice of the matrix $\|x_{ir}\|$ of rank n we can make $\|\Omega_{rk}\|$ an arbitrary orthogonal matrix.

Thus, the problem of estimating the possible values of Ω can be put as follows: *which values can* (22) *take for* $n < N$, *for an orthogonal* $N \times N$ *matrix?*

Since

$$\sum_{k=1}^{n} \Omega^2_{rk} \le \sum_{k=1}^{N} \Omega^2_{rk} = 1,$$

it follows that always

$$\Big(\sum_{k=1}^{n} \Omega_{rk}^2\Big)^2 \le \sum_{k=1}^{n} \Omega_{rk}^2.$$

Therefore (22) implies

$$\Omega \le \sum_{r=1}^{N}\sum_{k=1}^{n} \Omega_{rk}^2 = n.$$

The inequality

$$\Omega \le n \tag{23}$$

thus established turns into an equality if $\|\Omega_{rk}\|$ is the identity matrix

$$\Omega_{rk} = e_{rk}.$$

On the other hand, since always

$$\sum_{r=1}^{N} c_r^2 \ge \frac{1}{N}\Big(\sum_{r=1}^{N} c_r\Big)^2,$$

(22) implies that

$$\Omega \ge \frac{1}{N}\Big(\sum_{r=1}^{N}\sum_{k=1}^{n} \Omega_{rk}^2\Big)^2 = \frac{n^2}{N}.$$

In the resulting inequality

$$\Omega \ge n^2/N \tag{24}$$

an equality can be attained for any N and $n < N$, as has been established by A.I. Mal'tsev [2] at my request.

§3. Conclusions

In §2 it was established that

$$n^2/N \le \Omega \le n \tag{25}$$

and both bounds in (25) can be attained. To estimate $\mathbf{D}\sigma^2$ we obtain by (14)

$$\begin{aligned} &\frac{f^4 - s^4}{N-n} - \frac{n}{N}\Big(\frac{f^4 - 3s^4}{N-n}\Big) \le \mathbf{D}\sigma^2 \le \frac{f^4 - s^4}{N-n}, \quad \text{if } f^4 - 3s^4 \ge 0,\\ &\frac{f^4 - s^4}{N-n} \le \mathbf{D}\sigma^2 \le \frac{f^4 - s^4}{N-n} + \frac{n}{N}\Big(\frac{3s^4 - f^4}{N-n}\Big), \quad \text{if } f^4 - 3s^4 \le 0, \end{aligned} \tag{26}$$

and the bounds in (26) can be attained for any N and $n < N$ in certain particular cases.

If $f^4 - 3s^4 = 0$, then from (26) we obtain the formula

$$\mathbf{D}\sigma^2 = 2s^4/(N-n), \tag{27}$$

which is well known for the case of the normal Gaussian distribution of the errors Δ_r.

If $n/N \to 0$ and $f^4 - s^4 > 0$, then asymptotically

$$\mathbf{D}\sigma^2 \sim (f^4 - s^4)/(N-n). \tag{28}$$

In fact, in Gauss's memoir only the asymptotic estimates

$$(2f^4 - 4s^4)/(N-n) \text{ and } 2s^4/(N-n)$$

are given for $\mathbf{D}\sigma^2$ as $n/N \to 0$.

Note also that in the degenerate case $f^4 = s^4$ (as is well known, the case $f^4 < s^4$ is impossible) one obtains from (26) the formula

$$\mathbf{D}\sigma^2 \leq 2ns^4/N(N-n). \tag{29}$$

5 May 1947

References

1. A.N. Kolmogorov, 'A justification of the method of least squares', *Uspekhi Mat. Nauk* **1**:1 (1946), 57–70 (in Russian).
2. A.I. Mal'tsev, 'A remark on the work by A.N. Kolmogorov, A.A. Petrov and Yu.M. Smirnov "A Gauss formula from the method of least squares"', *Izv. Akad. Nauk SSSR Ser. Mat.* **11** (1947), 567–568 (in Russian).

32. BRANCHING RANDOM PROCESSES *

Jointly with N.A. Dmitriev

§1. Statement of the problem

Consider a set of objects (for example, molecules) of n types $T_1, T_2, \ldots, T_n$ and assume that with probability $P_k^\alpha(t_1, t_2) = P(T_k \to S_\alpha | t_1, t_2)$, during the time interval (t_1, t_2) one object of type T_k turns into the set

$$S_\alpha = \alpha_1 T_1 + \alpha_2 T_2 + \ldots + \alpha_n T_n,$$

consisting of α_1 objects of the first type, α_2 objects of the second type, α_i objects of type i, etc. A random process consisting of this kind of transformation is called a *branching process* if the probabilities $P_k^\alpha(t_1, t_2)$ are uniquely determined by the times $t_1 < t_2$, the number k of the original type, $k = 1, 2, \ldots, n$, and the n-dimensional vector $\alpha = (\alpha_1, \alpha_2, \ldots, \alpha_n)$ with integer coefficients, $\alpha_i = 0, 1, 2, \ldots$.

It is essential here that we assume the probabilities to be *independent of*:

1) how the original object of type T_k appeared; it is only assumed that it exists at time t_1;

2) the fate of other objects of types $T_1, T_2, \ldots, T_n$ different from T_k at time t_1 and the objects originating from them at times $t > t_1$.

This probability-theoretic scheme finds numerous applications in biology, chemistry and elementary particle physics. In particular, in chemistry it can can be used to describe the initial stages of the most varied chemical reactions. At the initial stage of a chemical reaction the concentration of certain types $T_1', T_2', \ldots, T_m'$ of molecules may be considered high but approximately constant, whereas the concentrations of other types $T_1'', T_2'', \ldots, T_n''$ are variable but very low. Under these assumptions meeting between two types T'' is virtually impossible, while the results of meeting between one molecule of type T'' with one or several molecules of the types T' approximately obey the above requirements as regards the number of resulting molecules of types T''.

In chemical and physical questions it is natural to use a "continuous time" version of this scheme, assuming that the probabilities $P_k^\alpha(t_1, t_2)$ are differentiable with respect to t_1 and t_2. The differential equations derived under

---

* *Dokl. Akad. Nauk SSSR* **56**:1 (1947), 7–10.

this assumption in §3 of this paper for the special case of "monomolecular" reactions ($P_k^\alpha(t_1, t_2) > 0$) only for $\alpha_1 + \alpha_2 + \ldots + \alpha_n = 1$) have been given before by M.A. Leontovich [1]. In biological questions another, "discrete time" approach is more natural, where the "time" t takes only integer values and indicates the number of the generation. In this version the results given below were obtained for the case $n = 1$ by R.A. Fisher [2]. Fisher's studies were continued by J.F. Steffenson [3] and one of the authors of this paper [4].

§2. The basic functional equation

By the general principles of probability theory, the $P_k^\alpha(t_1, t_2)$ satisfy the conditions

$$P_k^\alpha(t_1, t_2) \geq 0, \tag{I}$$

$$\sum_\alpha P_k^\alpha(t_1, t_2) = 1. \tag{II}$$

Taking these probabilities to be defined for any $t_1 \leq t_2$, it is natural to assume that for $t_1 = t_2$,

$$P_k^\alpha(t, t) = E_k^\alpha = \begin{cases} 1 & \text{if } \alpha_k = 1, \alpha_i = 0,\ i \neq k, \\ 0 & \text{otherwise.} \end{cases} \tag{III}$$

Finally, in view of the above assumption, for any

$$t_1 \leq t_2 \leq t_3,$$

the $P_k^\alpha(t_1, t_2)$ satisfy the equation

$$P_k^\beta(t_1, t_3) = \sum_\alpha P_k^\alpha(t_1, t_2) P_\alpha^\beta(t_2, t_3), \tag{IV}$$

where

$$P_\alpha^\beta(t_1, t_2) = P(S_\alpha \to S_\beta | t_1, t_2)$$

is the probability that the set

$$S_\alpha = \alpha_1 T_1 + \alpha_2 T_2 + \ldots + \alpha_n T_n$$

is transferred into the set

$$S_\beta = \beta_1 T_1 + \beta_2 T_2 + \ldots + \beta_n T_n$$

during the time interval (t_1, t_2).

The probabilities $P_\alpha^\beta(t_1, t_2)$ can be expressed in terms of the probabilities $P_k^\alpha(t_1, t_2)$ as

$$P_\alpha^\beta(t_1, t_2) = \sum \prod_{i=1}^{n} \prod_{s=1}^{\alpha_i} P_i^{\beta(i,s)}(t_1, t_2), \tag{IVa}$$

where the sum is taken over all sets of vectors

$$\beta(i,s) = (\beta_1(i,s), \beta_2(i,s), \dots, \beta_n(i,s))$$

with non-negative integer coefficients $\beta_k(i,s)$ such that [1]

$$\sum_{i=1}^{n} \sum_{s=1}^{\alpha_i} \beta(i,s) = \beta. \tag{IVb}$$

Formulas (I)–(IV) are a complete translation of the notion of "branching random processes" defined in §1 in terms of probability theory, into purely analytical terms.

Formulas (IV) and (IVa) imply that the $P_\alpha^\beta(t_1, t_2)$ satisfy the main equation of Markov processes,

$$P_\alpha^\gamma(t_1, t_3) = \sum_\beta P_\alpha^\beta(t_1, t_2) P_\beta^\gamma(t_2, t_3). \tag{1}$$

Equation (1) follows immediately from the probabilistic assumptions of §1. In essence, (IV) is but a particular case of (1). This shows that "branching random processes" are, in essence, merely a particular case of Markov processes with a countable set of states. However, for this particular case we have a more effective analytical apparatus than the one that can be developed for the general case of Markov processes with a countable number of states. For this we introduce the generating functions

$$F_k(t_1, t_2; x_1, x_2, \dots, x_n) = \sum_\alpha P_k^\alpha(t_1, t_2) x_1^{\alpha_1} x_2^{\alpha_2} \dots x_n^{\alpha_n}. \tag{2}$$

It is convenient to express the n functions F_k as a single vector-valued function

$$F(t_1, t_2; x) = (F_1(t_1, t_2; x), F_2(t_1, t_2; x), \dots, F_n(t_1, t_2; x))$$

[1] For $\alpha_i = 0$ the corresponding product in (IVa) and (IVb) is set to be 1, and the corresponding sum 0.

of the vector argument $x = (x_1, x_2, \ldots, x_n)$. The reason for introducing the function $F(t_1, t_2; x)$ is that by means of this function (IV) can be expressed in the following way:

$$F(t_1, t_3; x) = F(t_1, t_2; F(t_2, t_3; x)). \tag{A}$$

Apart from the basic functional equation (A) for $F(t_1, t_2; x)$ (III) also gives the boundary value

$$F(t, t; x) = x \tag{B}$$

Naturally $F(t_1, t_2; x)$ is considered to be defined only for $t_1 \leq t_2$. The argument x has a purely formal meaning, although in any case, (I) and (II) imply that $F_k(t_1, t_2; x)$ is defined and analytic with respect to the arguments $x_1, x_2, \ldots, x_n$ for $|x_i| < 1$, $s = 1, 2, \ldots, n$.

Especially interesting is the case of processes homogeneous in time, that is, processes satisfying the condition

$$P_k^\alpha(t_1, t_2) = P_k^\alpha(t_2 - t_1). \tag{3}$$

In this case for

$$F_k(t; x) = \sum_\alpha P_k^\alpha(t) x_1^{\alpha_1} x_2^{\alpha_2} \ldots x_n^{\alpha_n} \tag{4}$$

we obtain, instead of (A) and (B):

$$F(t + \tau; x) = F(t; F(\tau; x)), \tag{A$_1$}$$

$$F(0; x) = x. \tag{B$_1$}$$

§3. Differential equations for the case of continuous time

We now assume that

$$P_k^\alpha(t, t + \Delta) = E_k^\alpha + \Delta p_k^\alpha(t) + o(\Delta), \tag{V}$$

where $o(\Delta)$ is an infinitesimal with respect to Δ. In this case it is natural to introduce the generating functions

$$f_k(t; x) = \sum_\alpha p_k^\alpha(t) x_1^{\alpha_1} x_2^{\alpha_2} \ldots x_n^{\alpha_n}. \tag{5}$$

Then for $|x_i| < r < 1,\ i = 1, 2, \ldots, n$, we obtain

$$F(t, t+\Delta; x) = x + \Delta f(t; x) + o(\Delta). \tag{6}$$

Setting $t_1 = t',\ t_2 = t' + \Delta,\ t_3 = t''$ in (A) and passing to the limit as $\Delta \to 0$, we obtain for $t' < t''$ the following differential equation for $F(t', t''; x)$:

$$\partial F/\partial t' = -f(t'; F). \tag{C}$$

This differential equation together with the boundary value (B) can serve as a basis for computing this type of processes. Indeed, $p_k^\alpha(t)$ can usually be determined directly from the conditions of the problem, while the purpose of the mathematical theory is to determine the probabilities $P_k^\alpha(t_1, t_2)$, and sometimes, their asymptotic behaviour as $t_2 \to +\infty$. Since f_k can easily be found from the given p_k^α and the desired P_k^α are obtained from the power series expansions of the F_k in the x_i, to determine the probabilities P_k^α and study their asymptotic behaviour we only have to solve (C) with the boundary value (B) and investigate the asymptotic behaviour of the solutions.

For a process homogeneous in time the probability densities p_k^α are constants and the $F_k(t; x_1, x_2, \ldots, x_n)$ are related to the functions

$$f_k(x_1, x_2, \ldots, x_n) = \sum_\alpha p_k^\alpha x_1^{\alpha_1} x_2^{\alpha_2} \ldots x_n^{\alpha_n} \tag{7}$$

by the equations

$$\frac{dF_1}{f_1(F_1, F_2, \ldots, F_n)} = \frac{dF_2}{f_2(F_1, F_2, \ldots, F_n)} = \ldots = \frac{dF_n}{f_n(F_1, F_2, \ldots, F_n)} = dt, \tag{C}$$

which has to be solved for $t > 0$, regarding x_i as constants, and with the initial value

$$F_k(0; x_1, x_2, \ldots, x_n) = x_k. \tag{B'}$$

This method is often much more efficient than directly dealing with the infinite systems of differential equations of Markov processes with a countable set of states derived from (1) under assumptions similar to (V). Here we will confine ourselves to one simple example of using this method (the same problem has been solved by N. Arley [5] using infinite systems):

$$n = 1:$$

$$p_1^0 = a, \quad p_1^1 = -(a+b), \quad p_1^2 = b, \quad p_1^r = 0 \text{ for } r > 2,$$

$$f(x) = a - (a+b)x + bx^2 = (1-x)(a-bx).$$

$$\frac{dF}{(1-F)(a-bF)} = dt, \quad F = \frac{a + ce^{(a-b)t}}{b + ce^{(a-b)t}},$$

$$F(0;x) = \frac{a+c}{b+c} = x, \quad c = \frac{bx-a}{1-x},$$

$$F(t;x) = \frac{a(1-x) + (bx-a)e^{(a-b)t}}{b(1-x) + (bx-a)e^{(a-b)t}} = P_1^0(t) + P_1^1(t)x + P_1^2(t)x^2 + \dots$$

$$P_1^0(t) = \frac{a}{b}\,\frac{1 - e^{(a-b)t}}{1 - (a/b)e^{(a-b)t}},$$

$$P_1^k(t) = \left(1 - \frac{a}{b}\right)^2 \frac{(1 - e^{(a-b)t})^{k-1}}{[1 - (a/b)e^{(a-b)t}]^{k+1}} e^{(a-b)t},$$

where $k = 1, 2, \dots$.

20 February 1947

References

1. M.A. Leontovich, 'Basic equations of kinetic gas theory from the point of view of the theory of random processes', *ZhETF*, **5** (1935), 211–231 (in Russian).
2. R.A. Fisher, *The genetic theory of natural selection*, Oxford, 1930.
3. J.F. Steffenson, 'Deux problèmes du calcul des probabilités', *Ann. Inst. H. Poincaré* **3** (1933), 331–334.
4. A.N. Kolmogorov, 'Solution of a biological problem', *Izv. NII Mat. Mekh., Tomsk Univ.* **2**:1 (1938), 7–12 (No. 25 in this volume).
5. N. Arley, On the theory of stochastic processes and their application to the theory of cosmic radiation, Dissertation, Copenhagen, 1943.

33. COMPUTATION OF FINAL PROBABILITIES FOR BRANCHING RANDOM PROCESSES *

Jointly with B.A. Sevastyanov

The terminology and notation in this paper are close to those in [1]. In what follows, we consider discrete schemes homogeneous in time t: t runs only through $1, 2, 3, \ldots$ and can be interpreted as the "generation number" of the particle in question. Accordingly,

$$P_k^\alpha(t) = P(T_k \to \alpha_1 T_1 + \alpha_2 T_2 + \ldots + \alpha_n T_n | t) \tag{1}$$

is the probability that one particle of type T_k gives α_1, $\alpha_2, \ldots$, α_n particles of the types $T_1, T_2, \ldots, T_n$, respectively in t generations. All further computations are based on the generating functions

$$F_k(1; x) = f_k(x) = \sum_\alpha p_k^\alpha x_1^{\alpha_1} x_2^{\alpha_2} \ldots x_n^{\alpha_n} \tag{2}$$

of the probabilities

$$p_k^\alpha = P_k^\alpha(1) \tag{3}$$

of the various transitions during one generation. Using these probabilities, we define by induction the generating functions

$$F_k(t; x) = \sum_\alpha P_k^\alpha(t) x_1^{\alpha_1} x_2^{\alpha_2} \ldots x_n^{\alpha_n} \tag{4}$$

for all positive integers t:

$$F_k(t+1; x) = f_k\{F_1(t; x),\ F_2(t; x), \ldots, F_n(t; x)\}. \tag{5}$$

In accordance with the general theory given in [1],

$$p_k^{(0,0,\ldots,0)} = f_k(0, 0, \ldots, 0) \tag{6}$$

denotes the probability of one particle of type T_k dying in one generation without producing any new particles of the considered types $T_1, T_2, \ldots, T_n$. In what follows it is useful to eliminate this possibility and to consider only schemes in which

$$f_k(0, 0, \ldots, 0) = 0. \tag{7}$$

* *Dokl. Akad. Nauk SSSR* **56**:8 (1947), 783–786.

If (7) does not hold in a certain scheme, it is always possible to introduce an additional ghost type T_{n+1} and consider that the death of a particle of any other type means its transition into a particle of type T_{n+1}, while the particles of type T_{n+1} remain unchanged. Analytically this is equivalent to transition from the initial system of generating functions $f_1, f_2, \dots, f_n$ to the new system

$$\begin{aligned}\bar{f}_k(x_1, \dots, x_n, x_{n+1}) = f_k(x_1, \dots, x_n) + \\ +(x_{n+1} - 1) f_k(0, 0, \dots, 0), \quad k = 1, 2, \dots, n, \\ \bar{f}_{n+1}(x_1, \dots, x_n, x_{n+1}) = x_{n+1}. \end{aligned} \tag{8}$$

In what follows we assume that (7) is already satisfied for the initial system of types $T_1, T_2, \dots, T_n$ and functions $f_1, f_2, \dots, f_n$.

A group of types $T_{k_1}, T_{k_2}, \dots, T_{k_m}$ is called *closed* if a particle of any other type produces only particles belonging to the types of the group. The system of all types $T_1, T_2, \dots, T_n$ is decomposable if it can be divided into two closed groups. It is natural to confine ourselves (and we shall do so) to indecomposable systems.

A group of types is called *final* if a) it is closed; b) each particle of any type in the group always produces exactly one particle; and c) it does not contain any smaller group with the properties a) and b).

It is easy to see that two final groups do not have common elements. Therefore, in general, the entire system $T_1, T_2, \dots, T_n$ consists of a certain number of final groups $\Psi_r = \{T_{r1}, T_{r2}, \dots, T_{rn_r}\}$, $r = 1, 2, \dots, s$, and a certain number of types $T_{01}, T_{02}, \dots, T_{0n_0}$ which do not belong to final groups. Clearly, $n_0 + n_1 + \dots + n_s = n$.

The process is considered to be completed if there only remain particles of the types from final groups. This way of understanding is quite natural, since the progeny of a particle of a type belonging to a final group consists, in any future generation, of one particle of a type belonging to the same final group, and transitions from one type to another within a final group are governed by the well known law of Markov chains in their simplest form, corresponding to the assumption that all "states" are "essential" and form one "class" (see, for example, [2]).

Denote by

$$q_k^\beta = P\{T_k \to \beta_1 \Psi_1 + \beta_2 \Psi_2 + \dots + \beta_s \Psi_s | \infty\} \tag{9}$$

the probability that the evolution of the progeny from one particle of type T_k eventually terminates in the final state, where β_r particles belonging to the types of each of the final groups Ψ_r remain;

$$Q_k = \sum_{\beta} q_k^{\beta} \leq 1 \tag{10}$$

is the total probability that the evolution of the progeny from one particle eventually terminates (in the above sense).

We introduce the generating functions

$$\phi_k(u_1, u_2, \dots, u_s) = \sum_{\beta} q_k^{\beta} u_1^{\beta_1} u_2^{\beta_2} \dots u_s^{\beta_s}. \tag{11}$$

In accordance with the double index of particle types we sometimes denote by ϕ_{rm} the functions among the ϕ_k corresponding to the type $T_k = T_{rm}$. Since $q_k^{\beta} \geq 0$, by (10), the functions (11) are in every case defined and analytic with respect to all variables for

$$0 \leq u_r < 1, \quad r = 1, 2, \dots, s. \tag{12}$$

When some of the variables attain the value 1, analyticity may be lost, but continuity remains. In particular,

$$\phi_k(1, 1, \dots 1) = Q_k. \tag{13}$$

It is easy to derive from probability considerations the following relation, which is of basic importance to us:

$$\phi_k = f_k(\phi_1, \phi_2, \dots, \phi_n), \quad k = 1, 2, \dots, n. \tag{14}$$

Moreover, it can be derived from the definition of a final group that

$$\phi_{rm} = u_r, \; m = 1, 2, \dots, n_r, \quad r = 1, 2, \dots, s. \tag{15}$$

Equations of the system (14) for which the number k corresponds to the type of a certain final group are corollaries of (15). Therefore, finally to determine ϕ_k we have the system of equations

$$\begin{aligned} \phi_{0m} &= f_{0m}(\phi_1, \phi_2, \dots, \phi_n), \quad m = 1, 2, \dots, n_0; \\ \phi_{rm} &= u_r, \quad m = 1, 2, \dots, n_r, \quad r = 1, 2, \dots, s. \end{aligned} \tag{16}$$

The junior author of this paper has proved the following theorem on uniqueness of the solution of (16).

Theorem. *System* (16) *together with the restrictions* $0 \leq \phi_k < 1$, $k = 1, 2, \ldots, n$, *uniquely determines the values of* $\phi_1, \phi_2, \ldots, \phi_n$ *for any given* $0 \leq u_r < 1$, $r = 1, 2, \ldots, s$.

Remark 1. The proof will be published elsewhere. It is based on properties of the f_k which follow from the assumption that the systems of types are indecomposable, and from the definition of final groups, assumption (7) and the relations

$$p_k^\alpha \geq 0, \ \sum_\alpha p_k^\alpha = 1. \tag{17}$$

These relations, however, are not fully used: analyticity of f_k is inessential in the proof.

Remark 2. All these considerations are applicable also to computing final probabilities q_k^β for branching processes with continuous time. It suffices to count not over time literally, but over "generations" of particles, introducing for the types that do not undergo any further transformation additional ghost transformations of their particles into themselves (with probability 1). This remark will be clarified with an example (see below).

Remark 3. In most applications to chemical chain reactions each final group consists of one type (final reaction products). In this case the theory simplifies.

Example. Let two positive transition probability densities be given in a scheme with continuous time and two types T_1 and T_2 (see [1],§3); we set

$$\bar{p}_1^{(2,0)} = \bar{p}(T_1 \to 2T_1), \quad \bar{p}_1^{(0,1)} = \bar{p}(T_1 \to T_2),$$

and let all other transitions be forbidden (in particular, particles of type T_2 do not turn into their own type). Counting over generations, we set

$$p_1^{(2,0)} = P\{T_1 \to 2T_1\} = \frac{\bar{p}_1^{(2,0)}}{\bar{p}_1^{(2,0)} + \bar{p}_1^{(0,1)}} = p,$$

$$p_1^{(0,1)} = P\{T_1 \to 2T_2\} = \frac{\bar{p}_1^{(0,1)}}{\bar{p}_1^{(2,0)} + \bar{p}_1^{(0,1)}} = 1 - p,$$

and additionally

$$p_2^{(0,1)} = P\{T_2 \to T_2\} = 1$$

(here the p_k^α have the meaning of probabilities introduced in this paper). In our example there is one final group of one type $\Psi_1 = \{T_2\}$. For $\phi_1(u_1), \phi_2(u_1)$ we obtain

$$\phi_1 = p\phi_1^2 + (1-p)\phi_2, \quad \phi_2 = u_1.$$

Solving these equations, we formally obtain

$$\phi_1 = \frac{1}{2p}(1 \pm \sqrt{1 - 4p(1-p)u_1}). \tag{18}$$

When $0 \le u_1 < 1$, $0 \le \phi_1 < 1$ there remains only one branch of the curve (18) (with minus sign).

Let us complete the computations for $p = \frac{1}{2}$. In this case we obtain for the coefficients of the expansion $\phi_1(u_1) = q_1^{(0)} + q_1^{(1)}u_1 + q_1^{(2)}u_1^2 + \dots$

$$q_1^{(0)} = 0, \quad q_1^{(1)} = \frac{1}{2}, \quad q_1^{(2)} = \frac{1}{8}, \dots, q_1^{(m)} = \frac{1 \cdot 3 \cdot 5 \dots (2m-3)}{2^m \cdot m!}, \quad m \ge 2,$$

that is, asymptotically

$$q_1^{(m)} \sim (\tfrac{1}{2}\sqrt{\pi})m^{-3/2}. \tag{19}$$

Note that although

$$\phi_1(1) = \sum_m q_1^{(m)} = Q_1 = 1,$$

that is, the process will inevitably terminate, the expectation $\mathbf{E}_1 = \sum_m mq_1^{(m)}$ of the number of particles of type T_2 obtained from one particle of type T_1 is infinite. This causes a peculiar phenomenon of non-stability of the number of particles of type T_2 generated by a given, though perhaps very large, number of particles of type T_1. To make this clear, we denote by μ_n the number of particles of type T_2 generated by n particles of type T_1. Clearly, $\mu_n = \kappa_1 + \kappa_2 + \dots + \kappa_n$ where κ_i denotes the number of particles of type T_2 generated by the ith particle of type T_1. The variables κ_i are independent and have probability distribution $\mathbf{P}\{\kappa_i = k\} = q_1^{(k)}$.

This together with (19) implies that (see [3]) the distribution law for

$$\xi_n = \mu_n/n^2 \tag{20}$$

is $S_n(x) = \mathbf{P}\{\xi_n < x\}$, which tends to a certain limit distribution law for the quantity

$$S(x) = \int_0^x s(x)dx. \tag{21}$$

The limit distribution law (21) can be found from the logarithm of its characteristic function,

$$\log \chi(t) = \log \int_0^\infty s(x)e^{itx}dx = \frac{1}{2\sqrt{\pi}} \int_0^\infty (e^{iut} - 1)\frac{du}{u^{3/2}}. \tag{22}$$

Thus for large n, μ_n is of order n^2, but the ratio μ_n/n^2 varies from case to case.

Moscow, 12 April 1947

References

1. A.N. Kolmogorov and N.A. Dmitriev, 'Branching random processes', *Dokl. Akad. Nauk SSSR* **56**:1 (1947), 7–10 (in Russian) (No. 32 in this volume).
2. A.N. Kolmogoroff, 'Anfangsgründe der Theorie der Markoffschen Ketten mit unendlich vielen möglichen Zustanden', *Mat. Sb.* **1**:4 (1936), 607–610.
3. B.V. Gnedenko, 'On the theory of domains of attraction of stable laws', *Uch. Zap. Moskov. Gos. Univ.* **30** (1939), 61–82 (in Russian).

34. STATISTICAL THEORY OF OSCILLATIONS WITH CONTINUOUS SPECTRUM *

1. The subject discussed in this sketch can be characterized in very general terms not connected with any special field of mechanics or physics. For simplicity, we confine ourselves to considering only changes in time of a certain finite number of variables

$$\xi_1(t), \xi_2(t), \dots, \xi_s(t).$$

Naturally the argument t (time) is assumed to be real, while $\xi_r(t)$, in general, will be complex. If the $\xi_r(t)$ are changing periodically with period ω, then the process of their variation is usually studied by means of Fourier series

$$\xi_r(t) \sim \sum_n \alpha_r^{(n)} e^{i2\pi nt/\omega}. \tag{1}$$

Owing to this, an oscillatory variation of $\xi_r(t)$ with period ω can be expanded in harmonic oscillations of period

$$\omega, \omega/2, \omega/3, \omega/4, \dots.$$

A natural generalization of periodic oscillations are almost periodic oscillations, of the form

$$\xi_r(t) \sim \sum_n \alpha_r^{(n)} e^{i\lambda_n t}. \tag{2}$$

Here the periods

$$\omega_n = 2\pi/|\lambda_n|,$$

are, in general, incommensurable. Naturally, a further generalization of (2) is to pass from sums with respect to a sequence of particular discrete frequencies λ_n to integration with respect to a continuous frequency λ ,

$$\xi_r(t) \sim \int_{-\infty}^{\infty} \phi_r(\lambda) e^{i\lambda t} d\lambda, \tag{3}$$

or, in a completely general form combining the discrete (2) and continuous (3) cases, to represent oscillatory processes by Stieltjes integrals

$$\xi_r(t) \sim \int_{-\infty}^{\infty} e^{i\lambda t} d\Phi_r(\lambda). \tag{4}$$

---

* In: *Collected papers on the 30th anniversary of the Great October Socialist Revolution*, Vol. 1, Akad. Nauk SSSR, Moscow-Leningrad, 1947, pp. 242–252.

However, if we restrict ourselves to the methods of classical analysis, then the above approach meets with an essential difficulty. The classical Fourier integrals (3) can only represent functions that tend, in the mean, to constants as $t \to \infty$. Integrals (4) can also represent only functions that can be decomposed into two components,

$$\xi(t) = \eta(t) + \zeta(t),$$

where $\eta(t)$ is almost periodic (that is, has discrete spectrum), and $\zeta(t)$ tends, in the mean, to a constant as $t \to \infty$. Thus, the classical tool of Fourier integrals for the case of a continuous spectrum leads only to damped oscillations. This is what is said in textbooks on physics that want to retain mathematical rigour.[1]

Nevertheless, the idea that even in the non-damped case, oscillations with continuous spectrum are possible and can be expressed by integrals of the form (3) is maintained by many, sometimes quite outstanding, researchers in mechanics and physics. Studies of this kind, despite the lack of mathematical rigour, in many cases bring about correct and useful results.

In essence, the prerequisites for a rigorous mathematical justification of the theory of non-damped oscillations with continuous spectrum are contained in the spectral theory of operators.[2] A mathematically rigorous spectral theory for functions $\xi(t)$ of a real argument t which was directly oriented to the mathematical justification of physical notions on non-damped oscillations with continuous spectrum was first created by Norbert Wiener[3] in 1925. Wiener's concept, however, does not bring us to either of the representations (3) or (4). Only the spectral theory of stationary random processes suggested by A.Ya. Khinchin in 1934 [1] completely clarified the above questions. In particular, it leads to a mathematically rigorous and, at the same time, physically reasonable justification of representing non-damped statistically stationary oscillations in the form (4). Apparently, this justification is general enough for practical applications. Still, even in Khinchin's concept, the representation of non-damped oscillations in the form (3) has no rigorous mathematical justification. I believe that such a state of affairs should be considered final: there seems

---

[1] See, for example, [15], §54.

[2] See the presentation of this theory in the papers by A.I. Plesner and V.A. Rokhlin [16], [17].

[3] A systematical presentation of Wiener's theory can be found in Chapter 4 of his book [18]; see also [19].

to be no reasonable way to ascribe definite derivatives $\phi_r(\lambda)$ to the functions $\Phi_r(\lambda)$ in the case of continuous spectrum.

The purpose of this paper is to review certain Soviet and foreign studies developing the concept of Khinchin.

2. All the very simple and complete results stated below are obtained by a radical change in viewpoint: the $\xi_r(t)$ are considered as random variables in the sense of probability theory. Thus, the main object of study is not a certain individual oscillatory process, but the law of the probability distribution in the function space of various possible versions of such a process.

Formally, the theory may be presented starting from the following definitions.[4]

An s-dimensional *random process* is a collection of complex random variables $\xi_r(t)$ given for all real t and $r = 1, 2, \dots, s$.

It is assumed that the expectations

$$\mathbf{E}|\xi_r(t)|^2$$

are finite and that the process is stochastically continuous, which means that

$$\mathbf{E}|\xi_r(t+\Delta) - \xi_r(t)|^2 \to 0$$

as $\Delta \to 0$.

A random process $\{\xi_1(t), \xi_2(t), \dots, \xi_s(t)\}$ is called *stationary* if for any $t, t_1, t_2, \dots, t_n$, the sn-dimensional distribution law of sn random variables

$$\begin{array}{l} \xi_1(t+t_1),\ \xi_1(t+t_2), \dots,\ \xi_1(t+t_n), \\ \xi_2(t+t_1),\ \xi_2(t+t_2), \dots,\ \xi_2(t+t_n), \\ \dots\dots\dots\dots\dots\dots\dots \\ \xi_s(t+t_1),\ \xi_s(t+t_2), \dots,\ \xi_s(t+t_n), \end{array}$$

does not depend on the shifts in t.

Stationarity implies, in particular, that the expectations[5]

$$B_{qr}(\tau) = \mathbf{E}\{\xi_q(t+\tau)\bar{\xi}_r(t)\} \qquad (5)$$

[4] The notions of probability theory used in what follows are rigorously introduced, for example, in my book [20].

[5] In (5), $\bar{\xi}$ denotes the complex conjugate of ξ.

do not depend on t. The independence itself of the expectations (5) on t is called ***stationarity in the wider sense.*** The account given here and in the next sections is based solely on the hypothesis of stationarity in the wider sense.

The main result of A.Ya. Khinchin, in the form given by Harald Cramér [2], is as follows:

The functions $B_{qr}(\tau)$ can be represented as

$$B_{qr}(\tau)=\int_{-\infty}^{\infty} e^{i\lambda\tau}\,dF_{qr}(\lambda), \tag{6}$$

where for any $\lambda_1<\lambda_2$ the matrix $\|F_{qr}(\Delta_\lambda)\|$ of the increments

$$F_{qr}(\Delta_\lambda)=F_{qr}(\lambda_2)-F_{qr}(\lambda_1)$$

is hermitian and non-negative.

To understand the meaning of Khinchin's spectral functions we must consider the discrete case, when (6) turns into the sums

$$B_{qr}(\tau)=\sum_n a_{qr}^{(n)}e^{i\lambda_n\tau}. \tag{7}$$

This case was studied by E.E. Slutskii [3]. He showed that under the assumption (7) there exists an expansion

$$\xi_r(t)\sim\sum_n \alpha_r^{(n)}e^{i\lambda_n\tau} \tag{8}$$

where the $\alpha_r^{(n)}$ are certain random variables uniquely [6] determined by the given $\xi_r(t)$. In this case,

$$a_{qr}^{(n)}=\mathsf{E}\{\alpha_q^{(n)}\bar{\alpha}_r^{(n)}\}. \tag{9}$$

In particular,

$$a_{rr}^{(n)}=\mathsf{E}|\alpha_r^{(n)}|^2. \tag{9'}$$

It is natural to assume that in the general case we also have

$$F_{qr}(\Delta_\lambda)=\mathsf{E}\{\Phi_q(\Delta_\lambda)\overline{\Phi}_r(\Delta_\lambda)\}, \tag{10}$$

$$F_{rr}(\Delta_\lambda)=\mathsf{E}|\Phi_r(\Delta_r)|^2, \tag{10'}$$

[6] Uniqueness is to within equivalent variables, that is, variables that are equal with *probability* 1.

where the $\Phi_r(\Delta_\lambda)$ are the increments of the spectral function occurring in the representation of $\xi_r(t)$ in the form (4).

This assumption appears to be correct. The further development of Khinchin's theory of stationary random processes follows almost automatically from the reduction of it to the spectral theory of one-parameter groups of unitary operators, as is indicated in [4]–[7] and presented in the following section.

3. The random variables $\xi_r(t)$ may be considered as elements of a Hilbert space, in which the scalar product is given by the formula

$$(\xi, \eta) = \mathbf{E}(\xi\bar{\eta}). \tag{11}$$

It can easily be shown that any stationary process $\{\xi_1(t), \xi_2(t), \dots, \xi_s(t)\}$ in the corresponding Hilbert space generates a one-parameter group of unitary operators $\{U^\tau\}$ satisfying the relation

$$U^\tau \xi_r(t) = \xi_r(t+\tau) \tag{12}$$

for all real t and τ and all $r = 1, 2, \dots, s$. As is well known, the operators U^τ can be represented as

$$U^\tau = \int_{-\infty}^{\infty} e^{i\lambda\tau} dE(\lambda), \tag{13}$$

where $E(\lambda)$ is the *resolution of the identity*. Setting

$$\Phi_r(\lambda) = E(\lambda)\xi_r(0), \tag{14}$$

we obtain (provided the integration sign is properly understood) the basic formula (4). A probabilistic interpretation of (4) has been given in subsequent papers by H. Cramér [8], M. Loève [9], A. Blanc-Lapierre and R. Fortet [10], [11] in a manner more general than in [6] and [7]. To understand the real meaning of the basic spectral functions $\Phi_r(\lambda)$ it is natural to consider their increments

$$\Phi_r(\Delta_\lambda) = \Phi_r(\lambda_2) - \Phi_r(\lambda_1)$$

and jumps

$$\alpha_r(\lambda) = \Phi_r(\lambda + 0) - \Phi_r(\lambda - 0)$$

at distinct points λ. Naturally, $\Phi_r(\Delta_\lambda)$ and $\alpha_r(\lambda)$ are *random variables*, as is $\xi_r(t)$, but, unlike $\xi_r(t)$, they do not depend on the time t. If for some λ the jump $\alpha_r(\lambda)$ is non-zero, then $\xi_r(t)$ contains a strictly periodic component

$$\xi_r(t, \lambda) = \alpha_r(\lambda) e^{i\lambda t}. \tag{15}$$

In particular, $\alpha_r(0)$ is the average (over time) of

$$\alpha_r(0) = \lim_{T\to\infty} \frac{1}{2T} \int_{-T}^{T} \xi_r(t)dt. \tag{16}$$

The real meaning of the increments $\Phi_r(\Delta_\lambda)$ for ***sufficiently short*** intervals Δ_λ is similar. Namely, the component

$$\xi_r(t, \Delta_\lambda) = \int_{\Delta_\lambda} e^{i\lambda t} d\Phi(\lambda), \tag{17}$$

corresponding to the interval Δ_λ of the spectrum can be arbitrary closely approximated by

$$\Phi_r(\Delta_\lambda)e^{i\lambda t}, \tag{17'}$$

where λ is an arbitrary point of Δ_λ, on any finite interval $-T \le t \le +T$ of the time axis, provided that Δ_λ is sufficiently short. It is natural, however, that in the case of a continuous spectrum the components

$$\xi_r(t, \Delta_\lambda)$$

are not strictly periodic for any finite fixed interval Δ_λ (unless they are identically zero). Their time behaviour on short intervals Δ_λ is similar to the oscillations of a pendulum with weak damping generated by chaotically distributed small random pushes.

The components $\xi_r(t, \Delta_\lambda)$ for non-intersecting intervals Δ'_λ and Δ''_λ are not correlated to each other, that is,

$$\mathbf{E}\{\xi_q(t, \Delta'_\lambda)\bar{\xi}_r(t, \Delta''_\lambda)\} = 0, \tag{18}$$

if Δ'_λ and Δ''_λ do not intersect. For one interval Δ_λ,

$$\mathbf{E}\{\xi_q(t, \Delta_\lambda)\bar{\xi}_r(t, \Delta_\lambda)\} = F_{qr}(\Delta_\lambda). \tag{19}$$

In particular,

$$\mathbf{E}|\xi_r(t, \Delta_\lambda)|^2 = F_{rr}(\Delta_\lambda), \tag{19'}$$

that is, $F_{rr}(\Delta_\lambda)$ is just the average value (with respect to probability) of the square of the spectral component $\xi_r(t, \Delta_\lambda)$ of $\xi_r(t)$.

4. The use of the abstract tool of operators in Hilbert space could have produced the impression on the reader that the spectral components $\xi_r(t, \Delta_\lambda)$

are some mathematical fiction far from possible direct experiment. This is not true: with any desired degree of approximation they can be isolated from any statistically stationary oscillatory process by means of appropriate *filters*. Namely, with any desired accuracy a device can, in principle, be constructed which associates with any given random stationary function $\xi_1(t)$ the random stationary function

$$\xi_2(t) = \int_0^\infty \xi_1(t-\tau)s(\tau)d\tau, \tag{20}$$

where $s(\tau)$ is subject to the sole requirement that

$$\int_0^\infty |s(\tau)|^2 d\tau < \infty.$$

If we set

$$s(\tau) = \frac{1}{2\pi i}\frac{e^{i\lambda_2(\tau-T)} - e^{i\lambda_1(\tau-T)}}{\tau-T}, \tag{21}$$

then for sufficiently large positive T

$$\xi_2(t+T) = \frac{1}{2\pi i}\int_{-T}^\infty \xi_1(t-\tau)\frac{e^{i\lambda_2\tau} - e^{i\lambda_1\tau}}{\tau}d\tau$$

is arbitrarily close to

$$\xi_1(t, \Delta_\lambda) = \frac{1}{2\pi i}\int_{-\infty}^\infty \xi_1(t-\tau)\frac{e^{i\lambda_2\tau} - e^{i\lambda_1\tau}}{\tau}d\tau$$

with probability close to 1.

Note that (20) can be approximately realized in the form

$$L\xi_2 = \xi_1, \tag{22}$$

where

$$L = c_0 + c_1\frac{d}{dt} + c_2\frac{d^2}{dt^2} + \ldots + c_n\frac{d^n}{dt^n} \tag{23}$$

and all the eigenvalues of L have negative real part. Under these assumptions, if $\xi_1(t)$ is statistically stationary, then $\xi_2(t)$ tends to a statistically stationary function as $t \to +\infty$. This statistically stationary limit behaviour of $\xi_2(t)$ is determined by the formula

$$\xi_2(t) = \int_{-\infty}^\infty \frac{e^{i\lambda t}}{K(i\lambda)}d\Phi_1(\lambda), \quad K(z) = c_0 + c_1 z + \ldots + c_n z^n. \tag{24}$$

Formula (24) is of interest in its own right and not merely for the experimental determination of the spectral components $\xi_r(t, \Delta_\lambda)$. It implies that

$$\Phi_2(\lambda) = \int_{-\infty}^{\lambda} \frac{d\Phi_1(\lambda)}{K(i\lambda)}, \tag{25}$$

$$F_{22}(\lambda) = \int_{-\infty}^{\lambda} \frac{dF_{11}(\lambda)}{|K(i\lambda)|^2}. \tag{26}$$

Formulas (24), (25) and (26) generalize the usual resonance theory to the case of an arbitrary statistically stationary external force. Usually, a non-statistical presentation of this theory can be applied only to the case when the external force has a discrete spectrum, that is, when $\xi_1(t)$ is of the form (2). In physical literature, however, for the case of continuous spectrum this kind of reasoning is widely used, though, of course, without rigorous mathematical justification.

5. In applications, for the case of a continuous spectrum, we usually have

$$F_{qr}(\lambda) = \int_{-\infty}^{\lambda} f_{qr}(\lambda)d\lambda, \tag{27}$$

that is, (6) may be replaced by

$$B_{qr}(\tau) = \int_{-\infty}^{\infty} e^{i\lambda\tau} f_{qr}(\lambda)d\lambda. \tag{28}$$

This passage to spectral densities $f_{qr}(\lambda)$ enables us to do without Stieltjes integrals in the part of the theory that does not involve $\Phi_r(\lambda)$.

The simplest and most important in applications is the case when the distribution laws of any finite number of variables $\xi_r(t)$ are Gaussian. In this case so are the $\Phi_r(\Delta_\lambda)$. Under the simplest and most typical case of a continuous spectrum (the applicability of (27) and (28) and Gaussian distributions) the random functions $\Phi_r(\lambda)$ are not differentiable and the passage from (4) to the formulas of type (3) seems impossible. On the intervals of the λ-axis where the spectral density $f_{rr}(\lambda)$ is continuous and non-zero, the nature of variation of $\Phi_r(\lambda)$ is the same as the time dependence of the coordinates of a Brownian particle, when neglecting inertial forces, that is, if for small $\Delta = \lambda_2 - \lambda_1$ the increments of $\Phi_r(\lambda)$ are of order $\sqrt{\Delta}$. Here is a new case where continuous nowhere differentiable functions of Weierstrass type intrude into mathematical physics. For Brownian motion, more refined arguments taking into account the inertial forces bring us back to differentiable functions; here non-differentiability is inherent in the idea of a continuous spectrum.

Beyond the mechanics and physics of oscillations, the ideas of spectral analysis of statistically stationary process are applied mainly in the slightly different form of the spectral theory of stationary sequences (for example, in meteorology). Apart from the papers [4], [5] mentioned above, the book of the Swedish mathematician Herman Wold [12] is devoted to this theory. Important additions to Wold's theory were made by V.N. Zasukhin [13]. The clear understanding of the fact that the existence of a spectrum follows automatically from statistical stationarity and does not necessarily indicate that the process being studied results from superposition of strictly periodic components was very important for a critical reconsideration of the so-called periodographics, which claim to play a major role in meteorology and even in economy. E.E. Slutsky's papers in this field (especially [14]) have become classical in statistics.

References

1. A. Khintchine, 'Korrelationstheorie der stationären stochastischen Prozesse', *Math. Ann.* **109**:4 (1934), 604–615.
2. H. Cramér, 'On the theory of stationary random processes', *Ann. of Math.* **41**:1 (1940), 215–230.
3. E. Slutsky, 'Sur les fonctions aléatoires presque périodiques et sur la décomposition des fonctions aléatoires stationnaires en composantes', *Actual. Sci.* **738** (1938), 35–55.
4. A.N. Kolmogorov, 'Interpolation and extrapolation of stationary random sequences', *Izv. Akad. Nauk SSSR Ser. Mat.* **5**:1 (1941), 3–14 (in Russian) (No. 28 in this volume).
5. A.N. Kolmogorov, 'Stationary sequences in Hilbert space', *Byull. Moskov. Gos. Univ. Ser. Mat.* **2**:6 (1941), 1–40 (in Russian) (No. 27 in this volume).
6. A.N. Kolmogorov, 'Curves in Hilbert space invariant with respect to a one-parameter group of motions', *Dokl. Akad. Nauk SSSR* **26**:1 (1940), 6–9 (in Russian).
7. A.N. Kolmogorov, 'Wiener's spiral and some other interesting curves in Hilbert space', *Dokl. Akad. Nauk SSSR* **26** (1940), 115–118 (in Russian).
8. H. Cramér, 'On harmonic analysis in certain functional spaces', *Ark. Mat. Astron. Fys. B* **28**:12 (1942), 1–7.

9. M. Loève, 'Analyse harmonique générale d'une fonction aléatoire', *C. R. Acad. Sci. Paris* **220**:12 (1945), 380–382.
10. A. Blanc-Lapierre and R. Fortet, 'Sur la décomposition spectrale de fonction aléatoire stationnaire d'ordre deux', *C. R. Acad. Sci. Paris* **222**:9 (1946), 467–468.
11. A. Blanc-Lapierre and R. Fortet, 'Résultats sur la décomposition spectrale des fonctions aléatoires stationnaires d'ordre 2', *C. R. Acad. Sci. Paris* **222**:13 (1946), 713–714.
12. H. Wold, *A study in the analysis of stationary time series*, Uppsala, 1938; 2nd. ed. 1954.
13. V.N. Zasukhin, 'On the theory of multidimensional stationary random processes', *Dokl. Akad. Nauk SSSR* **33**:5 (1941), 435–437.
14. E.E. Slutsky, 'Summation of random causes as a source of cyclic processes', *Voprosy Kon"yunktury* **3**:1 (1927), 34–64 (in Russian). English transl: *Econometrica* **5**:2 (1937), 105–146.
15. E.V. Shpol'skii, *Atomic physics*, Gostekhizdat, Moscow-Leningrad, 1944 (in Russian).
16. A.I. Plesner, 'Spectral theory of linear operators', *Uspekhi Mat. Nauk* **9** (1941), 3–124 (in Russian).
17. A.I. Plesner and V.A. Rokhlin, 'Spectral theory of linear operators', *Uspekhi Mat. Nauk* **1**:1 (1946), 192–216 (in Russian).
18. N. Wiener, *The Fourier integral and certain of its applications*, New York, 1933.
19. N. Wiener, 'The harmonic analysis of irregular motion', *J. Math. Phys.* **5** (1926), 99–122, 158–191.
20. A.N. Kolmogorov, *Basic notions of probability theory*, ONTI, Moscow-Leningrad, 1936; 2nd ed. Nauka, Moscow, 1974 (in Russian).

35. ON SUMS OF A RANDOM NUMBER OF RANDOM TERMS *

Jointly with Yu.V. Prokhorov

Wald and a number of other American authors have given interesting theorems concerning the sums

$$\zeta_\nu = \xi_1 + \xi_2 + \ldots + \xi_\nu$$

of the first ν random variables from an infinite sequence

$$\xi_1, \xi_2, \ldots, \xi_n, \ldots$$

where the number ν of terms is a random variable (see [1]–[3], where references to earlier literature can be found). In their method of proof these theorems go back to the work of one of the authors of the present paper [4], where for estimating the probability

$$\mathbf{P}\{\max_{1\leq n\leq N} |\zeta_n - A_n| \geq h\}$$

he considered sums ζ_ν with index ν equal to the first number n for which

$$|\zeta_n - A_n| \geq h.$$

The inequality proved in [4] (see also [5], p.154) can easily be derived from Theorem 5 of the present paper.

Further we give very simple proofs for theorems of Wald type relating to the first and second moments. Our conditions for the applicability of basic identities are somewhat broader than those of Wald and Wolfowitz. Our generalization of the conditions for the applicability of these identities is important for certain applications.

In what follows, ν denotes a random variable that can take only non-negative integer values, and

$$n = 0, 1, 2, 3, \ldots.$$

The event that $\nu = n$ will be denoted by

$$S_n = \{\nu = n\},$$

* *Uspekhi Mat. Nauk* **4**:4 (1949), 168–172.

and the probability of this event by

$$p_n = \mathbf{P}(S_n).$$

Moreover, we set

$$P_n = \mathbf{P}\{\nu = n\} = \sum_{m=n}^{\infty} p_m.$$

Expectations of random variables,

$$\mathbf{E}(\eta) = \int_U \eta d\mathbf{P}$$

will be understood in the sense of the abstract Lebesgue integral over the set of elementary events U. Accordingly, the expectations, when they exist, are always finite, and the existence of $\mathbf{E}(\eta)$ implies the existence of $\mathbf{E}(|\eta|)$. The conditional probability distributions and conditional expectations are understood in the sense explained in [6].

Of basic importance for all theorems of Wald type is the assumption

(w) *For $n > m$ the random variable ξ_n and the event S_m are independent.*

According to [6], (w) means that for $n > m$ the conditional distribution of ξ_n under the condition S_m coincides with the unconditional distribution

$$P_{\xi_n}(A|S_m) = P_{\xi_n}(A).$$

Theorem 1. *If condition* (w) *holds and the expectations*

$$\mathbf{E}(\nu) \textit{ and } \mathbf{E}(\xi) = a, \quad \mathbf{E}(|\xi_n|) = c,$$

exist, where a and c do not depend on n, then the expectation of ζ_ν exists and equals

$$\mathbf{E}(\zeta_\nu) = a\mathbf{E}(\nu). \tag{1}$$

Since

$$\mathbf{E}(\nu) = \sum_{n=1}^{\infty} p_n n = \sum_{n=1}^{\infty} P_n,$$

Theorem 1 is an obvious consequence of the following more general statement:

Theorem 2. *If condition* (w) *holds, the expectations*

$$\mathbf{E}(\xi_n) = a_n, \quad \mathbf{E}(|\xi_n|) = c_n$$

exist and the series

$$\sum_{n=1}^{\infty} P_n c_n \tag{I}$$

converges, then the expectation of ζ_ν *exists and equals*

$$\mathbf{E}(\zeta_\nu) = \sum_{n=1}^{\infty} p_n A_n, \quad A_n = \mathbf{E}(\zeta_n) = a_1 + a_2 + \ldots + a_n. \tag{II}$$

Proof of Theorem 2. In view of (I), the series on the right-hand side of (II) converges absolutely. Applying the Abel transformation to it, we obtain

$$\sum_{n=1}^{\infty} p_n A_n = \sum_{n=1}^{\infty} P_n a_n. \tag{2}$$

We recall that

$$P_n = \mathbf{P}\{\nu \geq n\}.$$

The event $\{\nu \geq n\}$ is complementary to the event $\{\nu < n\}$ which, by condition (w), is independent of ξ_n . But the independence of an event implies the independence of its complementary event. Therefore, denoting by $\mathbf{E}(\eta|A)$ the conditional expectation of η under condition A, we obtain

$$\sum_{n=1}^{\infty} p_n A_n = \sum_{n=1}^{\infty} \mathbf{P}\{\nu \geq n\} a_n = \sum_{n=1}^{\infty} \mathbf{P}\{\nu \geq n\} \mathbf{E}(\xi_n | \nu_n \geq n) =$$

$$= \sum_{n=1}^{\infty} \int_{\{\nu \geq n\}} \xi_n d\mathbf{P} = \sum_{n=1}^{\infty} \sum_{m=n}^{\infty} \int_{S_m} \mathbf{P}. \tag{3}$$

Since the fact that ξ_n is independent of the event $\{\nu \geq n\}$ implies that $|\xi_n|$ is also independent of this event, it follows from (I) that

$$\sum_{n=1}^{\infty} \sum_{m=n}^{\infty} \left| \int_{S_m} \xi_n d\mathbf{P} \right| \leq \sum_{n=1}^{\infty} \sum_{m=n}^{\infty} \int_{S_m} |\xi_n| d\mathbf{P} = \sum_{n=1}^{\infty} \int_{\{\nu \geq n\}} |\xi_n| d\mathbf{P} =$$

$$= \sum_{n=1}^{\infty} \mathbf{P}\{\nu \geq n\} \mathbf{E}(|\xi_n| \,|\, |\nu \geq n) =$$

$$= \sum_{n=1}^{\infty} \mathbf{P}\{\nu \geq n\} \mathbf{E}(|\xi_n|) = \sum_{n=1}^{\infty} P_n c_n < \infty. \tag{4}$$

Since (4) is finite, we can change the order of summation:

$$\sum_{n=1}^{\infty}\sum_{m=n}^{\infty}\int_{S_m}\xi_n d\mathbf{P}=\sum_{m=1}^{\infty}\sum_{n=1}^{m}\int_{S_m}\xi_n d\mathbf{P}=\sum_{m=1}^{\infty}\int_{S_m}\zeta_m d\mathbf{P}. \tag{5}$$

Since $\zeta_m = \zeta_\nu$ holds on S_m,

$$\sum_{m=1}^{\infty}\int_{S_m}\zeta_m d\mathbf{P}=\int_U \zeta_\nu d\mathbf{P}=\mathbf{E}(\zeta_\nu). \tag{6}$$

As for the latter equality, we note that the "necessary event" U is the union of the events S_m, including S_0. However, the omission of the term with $m=0$ on the left-hand side of (6) is inessential, since, in accordance with the accepted rule, we consider the sum ζ_0 of an empty set of terms to be identically equal to zero.

Comparing (3), (5) and (6), we obtain (II) as required.

When considering the second moments we will assume that

$$\xi_n = (\xi_n^1, \xi_n^2)$$

is a vector with two components ξ_n^1 and ξ_n^2. Condition (w) is now understood in the sense that for $n > m$ the two-dimensional conditional distribution of ξ_n under the condition S_m coincides with the unconditional distribution of the same vector. In addition to (w) we also assume the condition

(z) *The vectors $\xi_1, \xi_2, \xi_3, \ldots$ are independent.*

By contrast, the dependence among the components of the same vector can be arbitrary. We now consider the particular case when $\xi_n^1 \equiv \xi_n^2$:

Theorem 3. *If* (w) *and* (z) *hold and the expectations*

$$\mathbf{E}(\nu^{3/2}),\quad \mathbf{E}(\xi_n^i),\quad \mathbf{E}\{(\xi_n^i - a^i)(\xi_n^j - a^j)\} = b^{ij},\ i,j=1,2,$$

exist, where a^i and b^{ij} do not depend on n, then

$$\mathbf{E}\{(\zeta_\nu^1 - \nu a^1)(\zeta_\nu^2 - \nu a^2)\} = b^{12}\mathbf{E}(\nu) \tag{III}$$

exists.

Naturally, in (III) ζ_ν^i denotes

$$\zeta_\nu^i = \xi_1^i + \xi_2^i + \ldots + \xi_\nu^i;\ i=1,2.$$

Theorem 3 can easily be derived from the more general Theorem 4 if we note that the finiteness of

$$\mathbf{E}(\nu^{3/2}) = \sum_{n=1}^{\infty} p_n n^{3/2} = \sum_{n=1}^{\infty} P_n(n^{3/2} - (n-1)^{3/2})$$

is equivalent to the convergence of $\sum_{n=1}^{\infty} P_n\sqrt{n}$.

In the statement of Theorem 4 we use the notation

$$A_n^i = a_1^i + a_2^i + \ldots + a_n^i,$$

$$B_n^{ij} = b_1^{ij} + b_2^{ij} + \ldots + b_n^{ij}.$$

Theorem 4. *If* (w) *and* (z) *hold, the expectations*

$$\mathbf{E}(\xi_n^i) = a_n^i, \quad \mathbf{E}\{(\xi_n^i - a_n^i)(\xi_n^j - a_n^j)\} = b_n^{ij}; \ i,j = 1,2,$$

exist and the series

$$\sum_{n=1}^{\infty} P_n\left(\sqrt{b_n^{11}B_n^{22}} + \sqrt{b_n^{22}B_n^{11}}\right) \tag{7}$$

converges, then

$$\mathbf{E}\{(\zeta_\nu^1 - A_\nu^1)(\zeta_\nu^2 - A_\nu^2)\} = \sum_{n=1}^{\infty} p_n B_n^{12}. \tag{IV}$$

exists.

Proof of Theorem 4. By passing from the variables ξ_n^i to ${}^*\xi_n^i = \xi_n^i - a_n^i$, we can reduce the general case to the case

$$\mathbf{E}(\xi_n^i) = a_n^i = 0 \tag{8}$$

We consider this case in what follows. Setting

$$\phi_n = \xi_n^1\xi_n^2 + \xi_n^1\zeta_{n-1}^2 + \xi_n^2\zeta_{n-1}^1, \tag{9}$$

we have identically

$$\chi_n = \zeta_n^1\zeta_n^2 = \phi_1 + \phi_2 + \ldots + \phi_n. \tag{10}$$

From (8) (9) and the independence of ξ_n of ζ_{n-1} (which follows from (z)) we obtain

$$\mathbf{E}(\phi_n) = \mathbf{E}(\xi_n^1\xi_n^2) = b_n^{12}, \quad \mathbf{E}(\chi_n) = B_n^{12}. \tag{11}$$

Under condition (8) the equality (IV) to be proved turns into

$$\mathsf{E}(\chi_\nu) = \sum_{n=1}^{\infty} p_n B_n^{12}. \tag{12}$$

Comparing (10), (11) and (12) with the statement of Theorem 2, we see that (12) is simply (II) applied to the sums

$$\chi_\nu = \phi_1 + \phi_2 + \ldots + \phi_\nu.$$

It remains to prove that the conditions of Theorem 2 hold. Special computations are required to verify (I) which, applied to the sums χ_ν, is written in the form

$$\sum_{n=1}^{\infty} P_n \mathsf{E}(|\phi_n|) < \infty. \tag{13}$$

Using the Cauchy-Bunyakovskii-Schwarz inequality we obtain from (9) the estimate

$$\begin{aligned} \mathsf{E}(|\phi_n|) &\leq \mathsf{E}(|\xi_n^1 \xi_n^2|) + \mathsf{E}(|\xi_n^2 \zeta_{n-1}^1|) \leq \\ &\leq \sqrt{b_n^{11} b_n^{22}} + \sqrt{b_n^{11} B_n^{22}} + \sqrt{b_n^{22} B_n^{11}} \leq 2\left(\sqrt{b_n^{11} B_n^{22}} + \sqrt{b_n^{22} B_n^{11}}\right) \end{aligned} \tag{14}$$

for $\mathsf{E}(|\phi_n|)$. Comparing (7) and (14) we see that (13) holds, which completes the proof of Theorem 4.

In the particular case $\xi_n^1 \equiv \xi_n^2$ we can abandon the vector notation and write

$$\xi_n^1 = \xi_n^2 = \xi_n, \quad A_n = a_1 + a_2 + \ldots + a_n,$$
$$B_n = b_1 + b_2 + \ldots + b_n.$$

Then from Theorem 4 we obtain

Theorem 5. *If* (w) *and* (z) *hold for a sequence of random variables ξ_n, the expectations $\mathsf{E}(\xi_n) = a_n$, $\mathsf{E}(\xi_n - a_n)^2 = b_n$ exist and the series $\sum_{n=1}^{\infty} P_n \sqrt{b_n B_n}$ converges, then*

$$\mathsf{E}(\zeta_\nu - A_\nu)^2 = \sum_{n=1}^{\infty} p_n B_n \tag{V}$$

exists.

References

1. A. Wald, 'On cumulative sums of random variables', *Ann. Math. Statist.* **15** (1944), 283–296.
2. A. Wald, 'Differentiation under the expectation sign of the fundamental identity in sequential analysis', *Ann. Math. Statist.* **17** (1946).
3. J. Wolfowitz, 'The efficiency of sequential estimates and Wald's equation for sequential processes', *Ann. Math. Statist.* **18** (1947), 215–231.
4. A.N. Kolmogoroff, 'Über die Summen durch den Zufall bestimmter unabhängiger Grössen', *Math. Ann.* **99** (1928), 309–319.
5. S.N. Bernshtein, *Probability theory*, GTTI, Moscow-Leningrad, 1946 (in Russian).
6. A.N. Kolmogorov, *Fundamentals of probability theory*, ONTI, Moscow-Leningrad, 1936 (in Russian).

36. A LOCAL LIMIT THEOREM FOR CLASSICAL MARKOV CHAINS*

This paper studies the limiting distributions of the number of visits to various states for a Markov chain with a constant transition probability matrix.

§1. Statement of the problem and review of results

Consider a classical Markov chain, that is, a random Markov process with discrete time, a finite number of states ($s > 1$)

$$e_1, e_2, \ldots, e_s$$

and constant probabilities

$$p_\alpha^\beta = \mathbf{P}\{\epsilon(t+1) = e_\beta | \epsilon(t) = e_\alpha\} \tag{1.1}$$

of transition from a state $\epsilon(t) = e_\alpha$ at time t to a state $\epsilon(t+1) = e_\beta$ at time $t+1$. Naturally, we assume that the usual conditions hold:

$$p_\alpha^\beta \geq 0, \quad \sum_\beta p_\alpha^\beta = 1. \tag{1.2}$$

If the states e_α are the unit vectors

$$\begin{aligned} e_1 &= (1, 0, \ldots, 0), \\ e_2 &= (0, 1, \ldots, 0), \\ &\cdots\cdots\cdots\cdots \\ e_s &= (0, 0, \ldots, 1) \end{aligned} \tag{1.3}$$

in s-dimensional coordinate vector space, then the components

$$\mu^1, \mu^2, \ldots, \mu^s$$

of the vector

$$\mu(t) = \epsilon(1) + \epsilon(2) + \ldots + \epsilon(t) \tag{1.4}$$

are equal to the number of visits to the states

$$e_1, e_2, \ldots, e_s$$

* *Izv. Akad. Nauk SSSR Ser. Mat.* **13** (1949), 281–300.

at times $t' = 1, 2, \ldots, t$ respectively. We will adhere to these vector representations throughout. We suppose that at time $t = 0$ $\mu(t)$ is equal to the zero vector

$$\mu(0) \equiv 0 = (0, 0, \ldots, 0). \tag{1.5}$$

The only possible values for $\mu(t)$ are the vectors $m = (m^1, m^2, \ldots, m^s)$ with integer components such that

$$\overline{m} = \sum_{\beta} m^{\beta} = t. \tag{1.6}$$

For any integer vector m with $\overline{m} \geq 0$ we set

$$W_{\alpha}(m) = \mathbf{P}\{\mu(\overline{m}) = m | \epsilon(0) = e_{\alpha}\}. \tag{1.7}$$

Thus the function $W_{\alpha}(m)$ of the vector argument m comprises the conditional distributions for $\epsilon(0) = e_{\alpha}$ of all vectors $\mu(t)$ for $t = 1, 2, 3, \ldots$. This is possible because, since by virtue of the relation

$$\overline{\mu}(t) = \sum_{\beta} \mu^{\beta}(t) = t, \tag{1.8}$$

these distributions are at most $(s-1)$-dimensional. Naturally, the sum of the probabilities $W_{\alpha}(m)$ pertaining to one variable $\mu(t)$ is 1:

$$\sum_{\overline{m}=t} W_{\alpha}(m) = 1. \tag{1.9}$$

Our entire analysis is devoted to the clarification of the limiting behaviour of the probabilities $W_{\alpha}(m)$ as $\overline{m} \to \infty$. The study of all limit distributions of sums of random variables "linked into a chain" (according to Markov's terminology) can easily be reduced to this problem.

The most noteworthy is the case when, in terms of [1], all states e_{α} form one class, that is, when the following condition holds

(A) *For any two states e_{α} and e_{β} there exists a sequence of states $(e_{\alpha}, e_{\gamma_1}, e_{\gamma_2}, \ldots, e_{\gamma_k}, e_{\beta})$ along which all the transition probabilities $p_{\alpha}^{\gamma_1}, p_{\gamma_1}^{\gamma_2}, \ldots, p_{\gamma_{k-1}}^{\gamma_k}, p_{\gamma_k}^{\beta}$ are positive.*

The most general case can be reduced to the case (A). This is done in §7.

The following Lemmas 1–3 hold only under condition (A). In the statement of these lemmas we denote by $\mathbf{E}_{\alpha}$ the conditional expectations under the hypothesis $\epsilon(0) = e_\alpha$. Although these lemmas are proved in [1]–[3], Lemmas 2 and 3 are again proved in §2 of the present paper.

Lemma 1. *The system of equations*

$$\sum_\alpha q^\alpha p_\alpha^\beta = q^\beta \ (\beta = 1, 2, \ldots, s), \quad \bar{q} \equiv \sum_\alpha q^\alpha = 1 \tag{1.10}$$

has a unique solution.

Lemma 2.

$$A_\gamma^\alpha(t) \equiv \mathsf{E}_\gamma \mu^\alpha(t) = tq^\alpha + O(1) \tag{1.11}$$

as $t \to \infty$.

Lemma 3. *The second moments*

$$B_\gamma^{\alpha\beta}(t) \equiv \mathsf{E}_\gamma\{(\mu^\alpha(t) - A_\gamma^\alpha(t))(\mu^\beta(t) - A_\gamma^\beta(t))\} \tag{1.12}$$

are of the form

$$B_\gamma^{\alpha\beta}(t) = tb^{\alpha\beta} + O(1), \tag{1.13}$$

as $t \to \infty$, *where the constants* $b^{\alpha\beta}$ *are determined solely from the matrix of initial probabilities* p_α^β.

It follows from the general properties of second moments of any system of random variables that $\|b^{\alpha\beta}\|$ is symmetric and the corresponding quadratic form

$$b(x) = \sum_{\alpha,\beta} b^{\alpha\beta} x_\alpha x_\beta \tag{1.14}$$

is non-negative. Then it can easily be derived from (1.8) that

$$\sum_\gamma b^{\alpha\gamma} = \sum_\gamma b^{\gamma\beta} = 0. \tag{1.15}$$

Formula (1.15) implies that

$$b(x) = b(x - \bar{x}). \tag{1.16}$$

Therefore it suffices to consider the form $b(x)$ in the $(s-1)$-dimensional space N of vectors x with

$$\bar{x} = 0. \tag{1.17}$$

It is natural to consider the case when

(B) *the form $b(x)$ is positive in the space N of vectors x for which $\bar{x} = 0$*

as the general "non-degenerate" case of our problem.

By (1.15) the determinant of the form $b(x)$ is always 0;

$$|b^{\alpha\beta}| = 0, \tag{1.18}$$

and all the principal minors of the matrix $\|b^{\alpha\beta}\|$ are equal to one another:

$$\Delta_{11} = \Delta_{22} = \ldots = \Delta_{ss} = \Delta. \tag{1.19}$$

The fact that $b(x)$ is non-negative implies, of course, that always

$$\Delta \geq 0; \tag{1.20}$$

condition (B) is equivalent to the requirement that

$$\Delta > 0. \tag{1.21}$$

Under condition (B) the form $b(x)$ has an inverse form $c(x)$ in the space N which can, for example, be expressed as

$$c(x) = -\frac{1}{\Delta}\begin{vmatrix} b^{11} & b^{12} & \ldots & b^{1,s-1} & x^1 \\ b^{21} & b^{22} & \ldots & b^{2,s-1} & x^2 \\ \ldots & \ldots & \ldots & \ldots & \ldots \\ b^{s-1,1} & b^{s-1,2} & \ldots & b^{s-1,s-1} & x^{s-1} \\ x^1 & x^2 & \ldots & x^{s-1} & 0 \end{vmatrix} \tag{1.22}$$

or in a similar alternative way by choosing some other index γ instead of $s-1$.

Setting

$$p(x) = \frac{1}{\sqrt{s(2\pi)^{s-1}\Delta}} e^{-1/2c(x)}, \tag{1.23}$$

$$\xi(t) = (\mu(t) - tq)/\sqrt{t}, \tag{1.24}$$

we can state a "non-degenerate" integral limit theorem.

Theorem 1. *Under conditions* (A) *and* (B), *for any γ and any rectifiable domain G of the space N,*

$$\mathbf{P}\{\xi(t) \in G \mid \epsilon(0) = e_\gamma\} \to \int_G p(x)dx \tag{1.25}$$

as $t \to \infty$, where

$$dx = \sqrt{s}dx_1 dx_2 \dots dx_{s-1} \tag{1.26}$$

denotes the volume element in N.

This theorem is proved in §4.

However, to formulate the simplest local limit theorem, we must make a certain additional analysis of possible directions of transition from one state to another.

A *chain* is a sequence of states

$$(e_{\gamma_0}, e_{\gamma_1}, \dots, e_{\gamma_k}), \tag{1.27}$$

for which all transition probabilities

$$p_{\gamma_0}^{\gamma_1}, p_{\gamma_1}^{\gamma_2}, p_{\gamma_2}^{\gamma_3}, \dots, p_{\gamma_{k-1}}^{\gamma_k}$$

are positive. A chain (1.27) is called a *cycle* if

$$\gamma_0 = \gamma_k. \tag{1.28}$$

We then say that an s-dimensional vector z is *cyclic* if there exists a cycle (1.27) such that

$$z = e_{\gamma_1} + e_{\gamma_2} + \dots + e_{\gamma_k}. \tag{1.29}$$

Finally, the *fundamental lattice* is the set of all vectors m representable as

$$m = a_1 z_1 + a_2 z_2 + \dots + a_n z_n, \tag{1.30}$$

where $z_1, z_2, \dots, z_n$ are cyclic vectors and $a_1, a_2, \dots, a_n$ are arbitrary integers (the number of terms n is also arbitrary). We denote the fundamental lattice by Z. Clearly Z consists exclusively of vectors with integer components and forms a group with respect to addition. We now state our additional condition.

(C) *the fundamental lattice Z coincides with the set Q of all integer vectors of s-dimensional coordinate space.*

The results of §§6, 7 imply that both conditions (A) and (C) are necessary for the limiting behaviour of the probabilities $W_\gamma(m)$ to be independent of the indices γ. Thus, the case when both conditions hold is the only case when we can count on obtaining a local limit theorem in an ideally simple formulation. This makes us regard the results of §3 and §5 as definitive in a certain sense:

Corollary 3 of Lemma 10 (§3). *Conditions* (A) *and* (C) *imply* (B).

Theorem 3 (§5). *If*[1] (A) *and* (C) *hold, then whatever the choice of indices* γ,

$$W_\gamma(m) \sim \frac{1}{\sqrt{(2\pi\overline{m})^{s-1}\Delta}} e^{-c(x)/2} \tag{1.31}$$

as $\overline{m} \to \infty$, *where*

$$x = (m - \overline{m}g)/\sqrt{\overline{m}}. \tag{1.32}$$

A very interesting question is the following: are (B) and (C) equivalent if (A) holds? If the answer is yes, then the conditions of applicability of the Local Theorem 3 and the Integral Theorem given above would coincide.[2]

In §6 we give a full analysis of the complications which arise when (C) does not hold while (A) still holds. As mentioned earlier, in §7 we consider the case when (A) does not hold. The results of this section are stated in a somewhat more complicated way, though the question on the limiting behaviour of the probabilities $W_\gamma(m)$ is essentially solved for the most general case as fully as for the case when (A) and (C) hold.

§2. Doeblin's method

Local theorems, which are the main subject of this paper, will be derived in §5 and §6 using a certain strengthening of the method developed by Doeblin for

---

[1] Unlike (1.23), in (1.31) there is no factor s under the root sign, since in the $(s-1)$-dimensional space of vectors m with given $\overline{m}$ (for example, in the space N) integer points are distributed with density $1/\sqrt{s}$. Note also that in the text of §5 the statement of Theorem 3 differs from that given here in that the way in which the probabilities tend to their asymptotic expression is indicated more precisely.

[2] Let L be the linear closure of the fundamental lattice Z, that is, the set of all vectors representable in terms of cyclic ones by (1.30) with arbitrary real coefficients a_k. In §3 (see Corollary 2 of Lemma 10) it is proved that under condition (A), requirement (B) is equivalent to (B′):
(B′) The space L coincides with the entire s-dimensional vector space R.
The question whether (B) and (C) are equivalent would be solved if we could show that
(*) under condition (A) the fundamental lattice Z always coincides with the set of all integer points from L.
Assumption (*) is very likely to be true. If it were proved, this would lead to a certain improvement in the results of §6.
(Assumption (*) was later proved by Rosenknop in Moscow and Chulanovskii in Leningrad.)

proving the integral limit theorem for the case of an infinite number of states [4]. To make clear how this method was developed, in this section we briefly present Doeblin's method in its original form, which is only suitable for proving integral theorems.

The presentation of this paragraph will be brief, since the results given here are essentially known. Here, as in §§3–6, (A) will be assumed to hold without any special indication. Moreover, throughout this section γ will be assumed to be fixed and

$$\epsilon(0) = e_\gamma. \tag{2.1}$$

Under this hypothesis the conditional probabilities and expectations will be denoted by

$$\mathbf{P}\{A|\epsilon(0) = e_\gamma\} = \mathbf{P}_\gamma(A), \quad \mathbf{E}\{\xi|\epsilon(0) = e_\gamma\} = \mathbf{E}_\gamma(\xi).$$

Let

$$0 = \tau(0) < \tau(1) < \tau(2) < \ldots < \tau(n) < \ldots$$

be the sequence of all times t when the state e_γ is observed. For $n \geq 1$ we set

$$\delta(n) = \mu(\tau(n)) - \mu(\tau(n-1)), \tag{2.2}$$

$$\lambda(n) = \delta(1) + \delta(2) + \ldots + \delta(n) = \mu(\tau(n)). \tag{2.3}$$

For $n = 0$ set

$$\lambda(0) = 0. \tag{2.4}$$

The components $\delta^\alpha(n)$ of $\delta(n)$ denote the number of visits to the state e_α at times t satisfying the inequalities

$$\tau(n-1) < t \leq \tau(n),$$

that is, between the $(n-1)$th and the nth returns to the original state e_γ (including the very moment of the nth return). Clearly, we always have

$$\delta^\gamma(n) = 1, \tag{2.5}$$

while for the sums $\lambda(n)$ we have

$$\lambda^\gamma(n) = n. \tag{2.6}$$

Doeblin's method is based on a simple observation: the random vectors $\delta(n)$ are independent and identically distributed. Because of this, the limit theorems established for sums of independent summands may be applied to the sums $\lambda(n)$. Relation (2.3) allows transition from the sums $\lambda(n)$ to the sums $\mu(t)$. Further we give two versions of this transition: one is based on Lemma 6 and gives most accurate results for estimating the second moments $B_\gamma^{\alpha\beta}(t)$, while the other is based on Lemma 7 and is convenient for deriving integral limit theorems. In [1] I proved

Lemma 4. *The variables $\delta^\alpha(n)$ have the finite expectations*

$$a_\gamma^\alpha = \mathbf{E}_\gamma \delta^\alpha(n) = q^\alpha / q^\gamma. \tag{2.7}$$

Since the terms $\delta(n)$ are independent

$$\mathbf{E}_\gamma \lambda^\alpha(n) = n a_\gamma^\alpha, \tag{2.8}$$

and since

$$\tau(n) = \bar{\lambda}(n), \tag{2.9}$$

(2.7) and (2.8) imply that

$$\mathbf{E}_\gamma \tau(n) = \frac{n}{q^\gamma} \sum q^\alpha = \frac{n}{q^\gamma}. \tag{2.10}$$

By the methods of [1] it is easy to prove

Lemma 5. *There exist constants C and $D > 0$ such that for any k and for $\bar{\delta}(n) = \tau(n) - \tau(n-1)$ the following inequality holds:*

$$\mathbf{P}\{\bar{\delta}(n) \geq k\} \leq C e^{-kD}. \tag{2.11}$$

Lemma 5 implies

Corollary. *The second moments*

$$b_\gamma^{\alpha\beta} = \mathbf{E}_\gamma \{(\delta^\alpha(n) - a_\gamma^\alpha)(\delta^\beta(n) - a_\gamma^\beta)\} \tag{2.12}$$

are finite.

By analogy with (2.8), for the second moments $\lambda^\alpha(n)$ we obtain

$$\mathbf{E}_\gamma\{(\lambda^\alpha(n) - na^\alpha_\gamma)(\lambda^\beta(n) - na^\beta_\gamma)\} = nb^{\alpha\beta}_\gamma. \tag{2.13}$$

Let τ_t denote the smallest of the numbers $\tau(n)$ that are $\geq t$ and let $\nu(t)$ be the corresponding number n.

Setting

$$\lambda(t) = \lambda(\nu_t) = \mu(\tau_t) = \sum_{n=1}^{\nu_t} \delta(n), \tag{2.14}$$

we obtain

$$\lambda^\gamma_t = \nu_t, \tag{2.15}$$

$$\bar{\lambda}_t = \tau_t. \tag{2.16}$$

In the same way as for Lemma 5 we prove

Lemma 6. *There exist constants C and $D > 0$ such that for any k,*

$$\mathbf{P}_\gamma\{|\tau_t - t| \geq k\} \leq Ce^{-kD}. \tag{2.17}$$

Since always

$$|\mu^\alpha(t) - \mu^\alpha(t')| \leq |t - t'|, \tag{2.18}$$

(2.17) implies that

$$\mathbf{P}_\gamma\{|\lambda^\alpha_t - \mu^\alpha(t)| \geq k\} \leq Ce^{-kD}. \tag{2.19}$$

The well known Wald identities (see [5] concerning the conditions of their applicability) are applicable to the following sums with random upper limit ν_t:

$$\lambda_t = \sum_{n=1}^{\nu_t} \delta(n).$$

From these identities we obtain

$$\mathbf{E}_\gamma \lambda^\alpha_t = a^\alpha_\gamma \mathbf{E}_\gamma(\nu_t), \tag{2.20}$$

$$\mathbf{E}_\gamma\{(\lambda^\alpha_t - \nu_t a^\alpha_\gamma)(\lambda^\beta_t - \nu_t a^\beta_\gamma)\} = b^{\alpha\beta}_\gamma \mathbf{E}_\gamma(\nu_t). \tag{2.21}$$

From (2.16) and (2.20) we obtain

$$\mathbf{E}_\gamma \tau_t = \mathbf{E}_\gamma(\nu_t)\sum_\alpha a_\gamma^\alpha = \frac{\mathbf{E}_\gamma(\nu_t)}{q^\gamma}. \tag{2.22}$$

By Lemma 6,

$$\mathbf{E}_\gamma \tau_t = t + O(1) \tag{2.23}$$

as $t \to \infty$. Therefore, from (2.22) and (2.20) we obtain as $t \to \infty$

$$\mathbf{E}_\gamma \nu_t = q^{\gamma t} + O(1), \tag{2.24}$$

$$\mathbf{E}_\gamma \lambda_t^\alpha = q_t^\alpha + O(1). \tag{2.25}$$

Using (2.19) we infer from (2.25) that

$$\mathbf{E}_\gamma \mu^\alpha(t) = q^\alpha t + O(1). \tag{2.26}$$

Thus we obtain a new proof of Lemma 2 stated in §1.

Noting that identically

$$\lambda_t^\alpha - \tau_t q^\alpha = (\lambda_t^\alpha - \nu_t a_\gamma^\alpha) - q^\alpha \sum_\phi (\lambda_t^\phi - \nu_t a_\gamma^\phi), \tag{2.27}$$

and using (2.21), we then obtain

$$\mathbf{E}_\gamma\{(\lambda_t^\alpha - \tau_t q^\alpha)(\lambda_t^\beta - \tau_t q^\beta)\} =$$

$$= \mathbf{E}_\gamma(\nu_t)[b_\gamma^{\alpha\beta} - \sum_\phi (q^\beta b_\gamma^{\alpha\phi} + q^\alpha b_\gamma^{\phi\beta}) + \sum_{\phi,\psi} q^\alpha q^\beta b_\gamma^{\phi\psi}]. \tag{2.28}$$

Using Lemma 6 we can change the left-hand side of (2.28) to

$$B_\gamma^{\alpha\beta}(t) = \mathbf{E}_\gamma\{(\mu^\alpha(t) - A_\gamma^\alpha(t))(\mu^\beta(t) - A_\gamma^\beta(t))\} \tag{1.12}$$

with accuracy to within $O(1)$ as $t \to \infty$. Together with (2.24) this leads to the formulas

$$B_\gamma^{\alpha\beta}(t) = tb^{\alpha\beta} + O(1), \tag{1.13}$$

where

$$b^{\alpha\beta} = q^\nu[b_\gamma^{\alpha\beta} - \sum_\phi (q^\beta b_\gamma^{\alpha\phi} + q^\alpha b_\gamma^{\phi\beta}) + \sum_{\phi,\psi} q^\alpha q^\beta b_\gamma^{\phi\psi}]. \tag{2.29}$$

Thus we have proved Lemma 3 from §1, at the same time showing the connection between the coefficients $b^{\alpha\beta}$ and the moments $b^{\alpha\beta}_{\gamma}$ (the assertion of independence of $b^{\alpha\beta}$ of the index γ should be proved separately, but because of (A), this is quite simple).

Note here the conversion[3] of (2.29) (although we shall not need it in what follows):

$$b^{\alpha\beta}_{\gamma} = \frac{1}{q^{\gamma}}(b^{\alpha\beta} - a^{\alpha}_{\gamma}b^{\gamma\beta} - a^{\beta}_{\gamma}b^{\alpha\gamma} + a^{\alpha}_{\gamma}a^{\beta}_{\gamma}b^{\gamma\gamma}). \tag{2.30}$$

We introduce the vectors

$$\eta(n) = \sqrt{\frac{q^{\gamma}}{n}}[\lambda(n) - \bar{\lambda}(n)q] = \frac{1}{\sqrt{n}}\sum_{k=1}^{n}\Delta(k), \tag{2.31}$$

where

$$\Delta(n) = \sqrt{q^{\gamma}}[\delta(n) - \bar{\delta}(n)q]. \tag{2.32}$$

Using (2.7), (2.12) and (2.29) we obtain

$$\mathbf{E}_{\gamma}\Delta^{\alpha}(n) = \mathbf{E}_{\gamma}\eta^{\alpha}(n) = 0, \tag{2.33}$$

$$\mathbf{E}_{\gamma}\Delta^{\alpha}(n)\Delta^{\beta}(n) = \mathbf{E}_{\gamma}\eta^{\alpha}(n)\eta^{\beta}(n) = b^{\alpha\beta}. \tag{2.34}$$

Since the $\Delta(n)$ are independent and identically distributed, by (2.33) and (2.34) the vectors $\eta(n)$ have a Gaussian distribution as $n \to \infty$ with matrix of second moments $||b^{\alpha\beta}||$. Applying to the sums

$$\sum_{k=n'}^{n''}\Delta(k)$$

a well known strengthening of Chebyshev's inequality (see [2], p.154), it is easy to prove the following

Lemma. *If a random variable ν_n taking only positive integer values satisfies the condition*

$$|\nu_n - n| < C\sqrt{n}$$

[3] Formula (2.30) can be proved directly in a way similar to that given for (2.29) starting not from (2.27) but from the identity

$$\lambda^{\alpha}_{t} - \nu_t a^{\alpha}_{\gamma} = (\lambda^{\alpha}_{t} - tq^{\alpha}) - a^{\alpha}_{\gamma}(\lambda^{\gamma}_{t} - tq^{\gamma}),$$

which is trivial in view of (2.15).

as $n \to \infty$, *where* C *is a constant, then for any* $h > 0$

$$\mathbf{P}_\gamma\{|\eta(\nu_n) - \eta(n)| > h\} \to 0, \tag{2.35}$$

as $n \to \infty$.

In order to pass from the vectors $\eta(n)$ to the vectors

$$\xi(t) = (\mu(t) - tq)/\sqrt{t} \tag{1.24}$$

introduced in §1 we can use

Lemma 7. *If* n_t *is the integer part of* tq^γ, *then for any* $H > 0$,

$$\mathbf{P}_\gamma\{|\xi(t) - \eta(n_t)| > H\} \to 0 \tag{2.36}$$

as $t \to \infty$.

In the proof of Lemma 7 we can pass from $\xi(t)$ to $\eta(t)$ via the vectors $\xi(\tau_t)$ and

$$\eta(\nu_t) = \sqrt{q^\gamma \tau_t/\gamma_t}\xi(\tau_t). \tag{2.37}$$

To estimate ν_t we must use the relation

$$\mathbf{E}_\gamma(\nu_t - \tau_t q^\gamma)^2 = b^{\gamma\gamma}t + O(1), \tag{2.38}$$

obtained from (2.24) and (2.28) by setting $\alpha = \beta = \gamma$ in (2.28).

By Lemma 6, (2.28) implies that for sufficiently large C and t the probability

$$\mathbf{P}_\gamma\{|\nu_t - n_t| > C\sqrt{n_t}\}$$

can be made arbitrarily small. This enables us to use (2.35), setting $h = H/3$ and to write

$$\mathbf{P}_\gamma\{|\eta(\nu_t) - \eta(n_t)| > \tfrac{1}{3}H\} \to 0 \text{ as } t \to \infty. \tag{2.39}$$

Lemma 6, (2.38), the definition of n_t and (2.37) imply that

$$\mathbf{P}_\gamma\{|\xi(\tau_t) - \eta(\nu_t)| > \tfrac{1}{3}H\} \to 0 \text{ as } t \to \infty. \tag{2.40}$$

Finally, by Lemma 6,

$$\mathbf{P}_\gamma\{|\xi(t) - \xi(\tau_t)| > \tfrac{1}{3}H\} \to 0 \text{ as } t \to \infty. \tag{2.41}$$

Combining (2.39)–(2.41) we obtain (2.36).

An application of Lemma 7 to the derivation of integral theorems will be given in §4.

§3. Cyclic vectors and the fundamental lattice

Apart from the probabilities $W_\gamma(m)$ we shall also need the probabilities

$$W_\gamma^\alpha(m) = \mathbf{P}_\gamma\{\mu(\overline{m}) = m,\ \epsilon(\overline{m}) = e_\alpha\}. \tag{3.1}$$

$W_\gamma^\alpha(m)$ is the conditional probability under the hypothesis $\epsilon(0) = e_\gamma$ of the combination of the two events:

1) visiting the states e_{β_1} ($\beta = 1, 2, \ldots, s$) m^β times at the times $t = 1, 2, \ldots, \overline{m}$ respectively; and

2) visiting the state e_α at the final moment $t = \overline{m}$.

For $m = 0$ we set

$$W_\gamma^\alpha(0) = \begin{cases} 1 & \text{for } \gamma = \alpha, \\ 0 & \text{for } \gamma \neq \alpha. \end{cases} \tag{3.2}$$

Clearly, the probabilities $W_\gamma(m)$ can be expressed in terms of the probabilities $W_\gamma^\alpha(m)$ by the formula

$$W_\gamma(m) = \sum_\alpha W_\gamma^\alpha(m). \tag{3.3}$$

The probabilities $W_\gamma^\alpha(m)$ are of special interest when the upper and lower indices are equal. They can be used to express the probability distributions for the vectors $\lambda(n)$ considered in the previous section,

$$\mathbf{P}_\gamma\{\lambda(m^\gamma) = m\} = W_\gamma^\gamma(m). \tag{3.4}$$

Since always

$$\lambda^\gamma(n) = n, \tag{3.5}$$

the distribution of $\lambda(n)$ under the hypothesis $\epsilon(0) = e_\gamma$ is completely determined by the values $W_\gamma^\gamma(m)$ with $m^\gamma = n$ and for any $n \geq 0$

$$\sum_{m^\gamma = n} W_\gamma^\gamma(m) = 1. \tag{3.6}$$

Clearly, for any cyclic vector z for which $z^\gamma > 0$, we have

$$W_\gamma^\gamma(z) > 0. \tag{3.7}$$

Clearly, the converse is also true: if for some γ (3.7) holds, then z is cyclic.[4]

Especially important for us are the cyclic vectors with $z^\gamma = 1$. They are the only values of vectors $\delta(n)$ which (naturally, provided $\epsilon(0) = e_\alpha$) have positive probability. They are considered in

Lemma 8. *The minimal additive vector group containing all cyclic vectors with $z^\gamma = 1$ coincides with the fundamental lattice Z.*

To prove this lemma it suffices to establish that any cyclic vector z can be represented as a linear combination with integer coefficients of cyclic vectors z with $z^\gamma = 1$. To this end we consider three cases:

1) if $z^\gamma = 1$, then our assertion is already proved;

2) if $z^\gamma > 1$, then the cycle generating z can be divided into z^γ cycles, from the visit to e_γ up to the next (along the cycle) subsequent visit to e_γ, and then z can be represented as

$$z = z_1 + z_2 + \ldots + z_{z^\gamma},$$

where the z_k correspond to the partial cycles;

3) if $z_\gamma = 0$, that is, the z-generating cycle

$$(e_{\gamma_0}, e_{\gamma_1}, \ldots, e_{\gamma_0})$$

does not cointain e_γ at all, then by (A), we can find chains

$$(e_\gamma, e_{\alpha_1}, e_{\alpha_2}, \ldots, e_{\alpha_i}, e_{\gamma_0}), \quad (e_{\gamma_0}, e_{\beta_1}, e_{\beta_2}, \ldots, e_{\beta_j}, e_\gamma)$$

so that they can contain e_γ as indicated: the first chain as first element, and the second chain as last element. Then the chains

$$(e_{\gamma_0}, e_{\gamma_1}, \ldots, e_{\gamma_0}, e_{\beta_1}, \ldots, e_{\beta_j}, e_\gamma, e_{\alpha_1}, \ldots, e_{\alpha_i}, e_{\gamma_0}),$$
$$(e_{\gamma_0}, e_{\beta_1}, \ldots, e_{\beta_j}, e_\gamma, e_{\alpha_1}, \ldots, e_{\alpha_i}, e_{\gamma_0})$$

are cycles, and for the corresponding cyclic vectors z_1 and z_2 we obtain

$$z_1 - z_2 = z, \quad z_1^\gamma = 1, \; z_2^\gamma = 1.$$

---

[4] Note here also the interesting identity (although we shall not need it in what follows):

$$z^\beta W_\alpha^\alpha(z) = z^\alpha W_\beta^\beta(z)$$

This completes the proof of our lemma.

Denote by $\mathfrak{L}$ the smallest linear subspace of s-dimensional vector space containing all cyclic vectors (and consequently, the fundamental lattice Z). The dimension r of this space is called the *rank* of the Markov chain considered.

Denote by $\mathfrak{L}_0$ the intersection of $\mathfrak{L}$ with the $(s-1)$-dimensional space N of vectors x with $\bar{x}=0$. Since $\mathfrak{L}$ is not contained in N (there exist cyclic vectors z with $\bar{z}>0$), it follows that $\dim \mathfrak{L} = r-1$.

Lemma 9. *The space $\mathfrak{L}$ is equal to the linear span of the values of the vectors $\delta(n)$ which have (under the hypothesis $\epsilon(0)=e_\gamma$) positive probability, while the space $\mathfrak{L}_0$ is equal to the linear span of the possible values of the vectors*

$$\Delta(n)=\sqrt{q^\gamma}[\delta(n)-\bar{\delta}(n)q]. \tag{2.32}$$

The first part of Lemma 9 follows directly from Lemma 8. The second part can be proved in the following way:

1. Lemma 4 and the first part of Lemma 9 imply that q belongs to $\mathfrak{L}$.

2) It can easily be shown that $\bar{\Delta}(n)=0$. Therefore the possible values of $\Delta(n)$ belong to $\mathfrak{L}$.

3) Since the possible values of

$$\delta(n)=\frac{1}{\sqrt{q^\gamma}}\Delta(n)+\bar{\delta}(n)q$$

generate the whole space $\mathfrak{L}$ obtained by attaching to $\mathfrak{L}_0$ a vector q that does not belong to $\mathfrak{L}_0$, the possible values of $\Delta(n)$ generate the entire space $\mathfrak{L}_0$.

Since

$$\mathsf{E}_\gamma \Delta(n)=0, \tag{2.33}$$

$$\mathsf{E}_\gamma \Delta^\alpha(n)\Delta^\beta(n)=b^{\alpha\beta}, \tag{2.34}$$

it follows almost immediately from Lemma 9 that

Lemma 10.

$$b(x)\begin{cases} >0 & \text{for } x\in\mathfrak{L}_0,\\ =0 & \text{for } x \text{ orthogonal to } \mathfrak{L}_0.\end{cases} \tag{3.8}$$

Lemma 10 brings us the following conclusions (some of them were already mentioned in §1):

Corollary 1. *The rank of* $||b^{\alpha\beta}||$ *is* $r-1$.

Corollary 2. *Condition* (B) *is equivalent to the condition*

(B$_1$) $r = s$.

Corollary 3. *Condition* (C) *implies* (B).

To conclude this section we will show that the determination of the fundamental lattice Z and the verification of (C) are purely arithmetic problems which allow simple algorithmic solutions. It can easily be seen that when determining the cycles it suffices to consider, instead of $||p_\alpha^\beta||$, the matrix $||\theta_\alpha^\beta||$ where

$$\theta_\alpha^\beta = \begin{cases} 1 & \text{for } p_\alpha^\beta > 0, \\ 0 & \text{for } p_\alpha^\beta = 0. \end{cases} \tag{3.9}$$

The sequence

$$(e_{\gamma_0}, e_{\gamma_1}, \ldots, e_{\gamma_{k-1}}, e_{\gamma_0})$$

is a cycle if and only if

$$\theta_{\gamma_0}^{\gamma_1} = \theta_{\gamma_1}^{\gamma_2} = \ldots = \theta_{\gamma_{k-1}}^{\gamma_0} = 1. \tag{3.10}$$

A cycle will be called *simple* if it contains each state at most once. Simple cyclic vectors corresponding to simple cycles are characterized by the fact that all their components are at most 1:

$$z^\alpha \leq 1. \tag{3.11}$$

Since there are a finite number of simple cycles and all of them can easily be found, the following lemma gives a very effective way for determining the fundamental lattice.

Lemma 11. *All the vectors in* Z *are representable in the form* (1.30) *with simple cyclic vectors* z_i.

For the proof it suffices to note that any cycle in which a certain state is found more than once can be divided into two cycles. Repeating such division, any cycle can be divided into simple cycles.

Let

$$f_1, f_2, \ldots, f_h$$

be the system of all simple cyclic vectors. Then the above considerations immediately give the following result:

Lemma 12. *The lattice Z is the minimal additive group generated by the vectors $f_1, f_2, \ldots, f_h$, while the space $\mathfrak{L}$ is the linear span of this system of vectors. The rank r of the Markov process equals the rank of the matrix*

$$\|f_g^\alpha\|,$$

and (C) *is equivalent to the condition*

(C$_1$) *The rank of $\|f_g^\alpha\|$ equals s, and the greatest common divisor of the sth order determinants that can be formed from its rows $(f_g^1, f_g^2, \ldots, f_g^s)$ is equal to* 1.

Example. Let $s = 5$ and

$$\|\theta_\alpha^\beta\| = \begin{Vmatrix} 0 & 1 & 0 & 1 & 1 \\ 0 & 0 & 1 & 0 & 0 \\ 1 & 0 & 0 & 1 & 0 \\ 0 & 1 & 0 & 0 & 1 \\ 0 & 1^* & 1 & 0 & 0 \end{Vmatrix}.$$

It can easily be verified that in this case the simple cycles consist only of three or four states. Here is the complete table of them

$$\begin{array}{ll} (e_1, e_2, e_3, e_1) & f_1 = (11100) \\ (e_1, e_5, e_3, e_1) & f_2 = (10101) \\ (e_2, e_3, e_4, e_2) & f_3 = (01110) \\ (e_3, e_4, e_5, e_3) & f_4 = (00111) \\ (e_1, e_4, e_2, e_3, e_1) & f_5 = (11110) \\ (e_1, e_4, e_5, e_3, e_1) & f_6 = (10111) \\ (e_1, e_5, e_2, e_3, e_1)^* & f_7 = (11101)^* \\ (e_2, e_3, e_4, e_5, e_2)^* & f_8 = (01111)^* \end{array}$$

The vectors f_g corresponding to the cycles are given in the right column. Since

$$\begin{vmatrix} f_1 \\ f_2 \\ f_3 \\ f_4 \\ f_5 \end{vmatrix} = \begin{vmatrix} 1 & 1 & 1 & 0 & 0 \\ 1 & 0 & 1 & 0 & 1 \\ 0 & 1 & 1 & 1 & 0 \\ 0 & 0 & 1 & 1 & 1 \\ 0 & 1 & 1 & 1 & 1 \end{vmatrix} = -1,$$

$r = 5$ and (C) holds.

If in the matrix $||\theta_\alpha^\beta||$ of this example $\theta_5^2 = 1$ is replaced by $\theta_5^2 = 0$, then only the first six of the simple cycles f_g will remain and it can be verified that the rank is lowered to $r = 4$.

§4. Integral limit theorems

By Lemma 10 the distribution of the vectors $\Delta(n)$ and $\eta(n)$ is totally concentrated on the space $\mathfrak{L}$. Since the form $b(x)$ on this space is positive, it follows that as $n \to \infty$, $\eta(n)$ obeys a Gaussian distribution which is non-degenerate on $\mathfrak{L}_0$ and corresponds to $b(x)$. We denote by $p(x)$ the probability density of this Gaussian distribution. Then by Lemma 7 we have the following result.

Theorem 2. *Under condition* (A), *for any* γ *and any domain* G *the intersection of whose boundary with* $\mathfrak{L}_0$ *has measure* 0 *in* $\mathfrak{L}_0$ *we have*

$$\mathbf{P}_\gamma\{\xi(t) \in G\} \to \int_{G\cap\mathfrak{L}_0} p(x)dx, \tag{4.1}$$

as $t \to \infty$, *where* dx *is the volume element in* $\mathfrak{L}_0$.

In the particular case when (B) holds, $\mathfrak{L}_0$ coincides with N and Theorem 2 gives Theorem 1 stated in §1.

§5. The basic identity and local limit theorem in the non-degenerate case

The version of Doeblin's method which allows us to reduce local limit theorems for Markov chains to local limit theorems for independent terms is based on the following identity:

$$W_\gamma(m) = \sum_{l^\gamma=0} W_\gamma^\gamma(m-l)W_\gamma(l). \tag{5.1}$$

The sum on the right-hand side extends over all integer vectors l with non-negative components satisfying the conditions

$$l^\alpha \leq m^\alpha, \tag{5.2}$$

$$l^\gamma = 0. \tag{5.3}$$

Identity (5.1) is self-evident from the probabilistic viewpoint.

It is easily checked that the sum of the probabilities $W_\gamma(l)$ over all vectors l satisfying (5.3) is given by

$$\sum_{l^\gamma=0} W_\gamma(l) = \mathbf{E}_\gamma \delta(n) = \frac{1}{q^\gamma}. \tag{5.4}$$

Under the additional restriction (5.2) this sum may become somewhat smaller, but because of the absolute convergence of (5.4) this reduced sum tends to $1/q^\gamma$ as the m^α tend to infinity.

If (A) and (C) hold, then by Lemma 8, the minimal additive group generated by the values of the vectors $\delta(n)$ with positive probability (under the hypothesis $\epsilon(0) = e_\gamma$) consists of all integer vectors, for any γ. For the differences of the values of the vectors $\delta(n)$ having positive probability, m_γ is always zero (since for the possible values of $\delta(n)$ themselves m_γ is always 1), that is, they belong to the group Q_γ of all integer vectors m with $m_\gamma = 0$. It is easy to see that in fact they generate the entire group: if the vectors

$$m_1 - m_0,\ m_2 - m_0, \ldots,\ m_k - m_0, \ldots,$$

where $m_0,\ m_1, \ldots,\ m_k, \ldots$ are all the possible values of $\delta(n)$, did not generate the whole group Q_γ, but only a proper subgroup of it, then after adding one vector m_0 they could not generate the whole group Q, whereas the group generated by the vectors

$$m_0,\ m_1 - m_0,\ m_2 - m_0, \ldots,\ m_k - m_0, \ldots$$

clearly coincides with the group generated by the vectors

$$m_0,\ m_1,\ m_2, \ldots,\ m_k, \ldots,$$

that is, in view of (C), with the whole group Q of integer vectors. As a result, we have the following:

Lemma 13. *Under conditions* (A) *and* (C) *the minimal group generated by the differences of the values of the vectors* $\delta(n)$ *coincides with the group* Q_γ *of all integer vectors* m *with* $m^\gamma = 0$.

From Lemma 4, the corollary of Lemma 5 and Lemma 13, and using the limit theorem [6] applied to the sums

$$\lambda(n) = \delta(1) + \delta(2) + \ldots + \delta(n),$$

we obtain

Lemma 14. *If* (A) *and* (C) *hold, then*

$$(m^\gamma)^{(s-1)/2} W_\gamma^\gamma(m) = p_\gamma(y) + o(1), \tag{5.5}$$

as $m^\gamma \to \infty$, *where*

$$y = \frac{m - m^\gamma q/q^\gamma}{\sqrt{m^\gamma}}, \tag{5.6}$$

$p_\gamma(y)$ *is the Gaussian probability density in the space* Q_γ *corresponding to zero mean values and covariance matrix* $\|b_\gamma^{\gamma\beta}\|$, *and the estimate of the remainder term is uniform for*

$$|m^\alpha - m^\gamma q^\alpha/q^\gamma| < C\sqrt{m^\gamma} \tag{5.7}$$

with any fixed C.

Noting that under condition (5.7) and as $m^\gamma \to \infty$,

$$m^\gamma \sim \overline{m} q^\gamma,$$

we can express (5.5) in the form

$$(\overline{m})^{(s-1)/2} W_\gamma^\gamma(m) = (q^\gamma)^{-(s-1)/2} p_\gamma(y) + o(1). \tag{5.8}$$

This estimate must now be applied to $W_\gamma^\gamma(m-l)$ in (5.1). Because of the convergence of (5.4), on the right-hand side of (5.1), as $\overline{m} \to \infty$ we can confine ourselves to the terms for which the ratios

$$(m^\alpha - l^\alpha)/m^\alpha$$

are arbitrarily close to 1. Then from (5.1), (5.4) and (5.8) we obtain

$$(\overline{m})^{(s-1)/2} W_\gamma(m) = (q^\gamma)^{-(s+1)/2} p_\gamma(y) + o(1) \tag{5.9}$$

as $\overline{m} \to \infty$, where (5.7) can be replaced by

$$|m^\alpha - \overline{m} q^\alpha| < C\sqrt{\overline{m}}. \tag{5.10}$$

Since under condition (5.10) and as $\overline{m} \to \infty$,

$$x = \frac{m - \overline{m} q}{\sqrt{\overline{m}}} \sim \frac{1}{\sqrt{q^\gamma}}(y - \bar{y} q), \tag{5.11}$$

$$y \approx \sqrt{q^\gamma}(x - x^\gamma q/q^\gamma), \tag{5.12}$$

(5.9) can be rewritten in the form

$$\overline{m}^{(s-1)/2} W_\gamma(m) = (p^\gamma)^{-(s+1)/2} p_\gamma[\sqrt{q^\gamma}(x - x^\gamma q/q^\gamma)] + o(1). \tag{5.13}$$

Formula (5.13) acts uniformly provided that (5.10) holds. Taking this into account and keeping in mind the integral Theorem 1, we obtain[5]

$$p_\gamma[\sqrt{q^\gamma}(x - x^\gamma q/q^\gamma)] = \sqrt{s}p(x). \tag{5.14}$$

Finally, the results of this section can be stated as follows:

Theorem 3. *If conditions* (A) *and* (C) *hold, then as* $\overline{m} \to \infty$ *and for any* γ

$$(\overline{m})^{(s-1)/2} W_\gamma(m) = \sqrt{s}p(x) + o(1) \tag{5.15}$$

uniformly under condition (5.10).

§6. The case when condition (C) does not hold

In this section we retain condition (A), but discard (C). Since now the fundamental lattice Z does not coincide with the lattice Q of all integer vectors, it is natural to consider residues of Q modulo Z. In more detail, this means the following. Two vectors m_1 and m_2 will be considered congruent modulo Z if

$$m_1 - m_2 \in Z. \tag{6.1}$$

All vectors with integer components are divided into classes of congruent vectors modulo Z. These classes are residue classes modulo Z.

Lemma 15. *Vectors* m *such that for fixed* α *and* β

$$W_\alpha^\beta(m) > 0 \tag{6.2}$$

are congruent modulo Z.

[5] Of course, (5.13) can be proved directly from the relations between the moments $b_\gamma^{\alpha\beta}$ and $b^{\alpha\beta}$. Then the proof of the local Theorem 3 stated below would become independent of integral theorems. Such a presentation, more consistent from the algebraic viewpoint, would however be somewhat cumbersome. The factor $\sqrt{s}$ in (5.14), as mentioned in §1, is connected with the fact that the integer points in N are distributed with density $1/\sqrt{s}$.

Lemma 15 can also be stated in the following way:

Lemma 15′. *All vectors m satisfying* (6.2) *belong to the same residue class modulo Z. This residue class will be denoted by D^β_α.*

To prove Lemma 15, we assume that $W^\beta_\alpha(m_1) > 0$ and $W^\beta_\alpha(m_2) > 0$. Then there exist chains

$$(e_\alpha, e_{\phi_1}, e_{\phi_2}, \ldots, e_{\phi_i} = e_\beta), \quad (e_\alpha, e_{\psi_1}, e_{\psi_2}, \ldots, e_{\psi_j} = e_\beta)$$

for which

$$e_{\phi_1} + e_{\phi_2} + \ldots + e_{\phi_j} = m_1, \quad e_{\psi_1} + e_{\psi_2} + \ldots + e_{\psi_j} = m_2.$$

By condition (A) there exists a chain $e_\beta, e_{\chi_1}, e_{\chi_2}, \ldots, e_{\chi_k} = e_\alpha$. Clearly,

$$(e_\alpha, e_{\phi_1}, e_{\phi_2}, \ldots, e_{\phi_i} = e_\beta, e_{\chi_1}, e_{\chi_2}, \ldots, e_{\chi_k} = e_\alpha),$$
$$(e_\alpha, e_{\psi_1}, e_{\psi_2}, \ldots, e_{\psi_j} = e_\beta, e_{\chi_1}, e_{\chi_2}, \ldots, e_{\chi_k} = e_\alpha)$$

are cycles. We denote by z_1 and z_2 the corresponding cyclic vectors. Since $m_1 - m_2 = z_1 - z_2 \in Z$, our lemma is proved.

Clearly, always [6]

$$D^\gamma_\gamma = Z \tag{6.3}$$
$$D^\beta_\alpha + D^\gamma_\beta = D^\gamma_\alpha. \tag{6.4}$$

In complete analogy with Lemma 14, the general local limit theorem [6] gives

Lemma 16. *If* (A) *holds and $r > 1$, then*

$$(m^\gamma)^{(r-1)/2} W^\gamma_\gamma(m) = \begin{cases} 0 & \text{for } m \notin Z, \\ \omega_\gamma p_\gamma(y) + o(1) & \text{for } m \in Z, \end{cases} \tag{6.5}$$

$$y = \frac{m - m^\gamma q/q^\gamma}{\sqrt{m^\gamma}}, \tag{5.6}$$

as $m^\gamma \to \infty$, the estimate of the remainder is uniform for

$$|m^\alpha - m^\gamma q^\alpha/q^\gamma| < C\sqrt{m^\gamma}, \tag{5.7}$$

[6] Residue classes are added in accordance with the usual algebraic rules.

$p_\gamma(y)$ is the Gaussian density corresponding to mean value 0 and matrix of second moments $\|b_\gamma^{\alpha\beta}\|$ in the space $\mathfrak{L}_\gamma$ of vectors $\gamma \in \mathfrak{L}$ with $y^\gamma = 0$, and ω_γ is the density of the points of Z in $\mathfrak{L}_\gamma$.

In the same way that we replaced (5.5) by (5.8), we can rewrite (6.5) in the form

$$(\overline{m})^{(r-1)/2} W_\gamma^\gamma(m) = (q^\gamma)^{-(r-1)/2} p_\gamma(y) + o(1). \tag{6.6}$$

For each residue class D in the quotient group Q modulo Z we denote by $f_\gamma(D)$ the sum of all the $W_\gamma(l)$ with $l^\gamma = 0$ and $l \in D$:

$$f_\gamma(D) = \sum_{l^\gamma=0} W_\gamma(l). \tag{6.7}$$

According to (5.4),

$$\sum_D f_\gamma(D) = \frac{1}{q^\gamma}. \tag{6.8}$$

Since

$$W_\gamma(l) = \sum_\alpha W_\gamma^\alpha(l), \tag{3.3}$$

it follows by Lemma 15 that $f_\gamma(D)$ is positive only for D that coincide with some D_γ^α, that is, only for a finite number of residue classes D.

Treating (6.6), (5.1) and (6.7) in the same way as (5.9), we see that, as $m^\gamma \to \infty$ when (5.7) holds and m lies in D,

$$(\overline{m})^{(r-1)/2} W_\gamma(m) = (q^\gamma)^{-(r-1)/2} \omega_\gamma f_\gamma(D) p_\gamma(y_*) + o(1), \tag{6.9}$$

where

$$y_* = \frac{m_* - m_*^\gamma q/q^\gamma}{\sqrt{m^\gamma}}, \tag{6.10}$$

$$m_* = m - u_D \tag{6.11}$$

and u_D is a fixed vector in D.

It is easy to show that $y_* \in \mathfrak{L}_\gamma$. It is necessary to replace y by y_*, since y may not belong to $\mathfrak{L}_\gamma$, in which case the density $p(y)$ is not defined for it.

In exactly the same way that we derived Theorem 3 in §5, we obtain from (6.9):

Theorem 4. *If* (A) *holds and* $r > 1$, *then as* $\overline{m} \to \infty$ *and for* m *in the residue class* D *we have*

$$(\overline{m})^{(r-1)/2} W_\gamma(m) = \omega f_\gamma(D) p(x_*) + o(1), \tag{6.12}$$

where[7]

$$x_* = (m_* - \overline{m}_* q)/\sqrt{\overline{m}}, \tag{6.13}$$

ω *is the density of the points of* Z *in the space* $\mathfrak{L}_0$, $p(x)$ *is the Gaussian density in* $\mathfrak{L}_0$ *corresponding to mean value* 0 *and matrix of second moments* $\|b^{\alpha\beta}\|$ *and the estimate* $o(1)$ *is uniform under the condition*

$$|m^\alpha - \overline{m} q^\alpha| < C\sqrt{\overline{m}}. \tag{5.10}$$

§7. The case when condition (A) does not hold

In the most general case the set of states $e_1, e_2, \dots, e_s$ splits into a certain number of "classes" $K_1, K_2, \dots, K_n$ of "essential" states and a set R of "inessential" states (see [1]). Condition (A) holds within each class K_i: transitions from a state $e_\alpha \in K_i$ to a state $e_\beta \in K_j$ for $i \neq j$ are impossible as are transitions from a state belonging to one of the K_i to a state in R; on the other hand, there is always the possibility of passing in a certain number of steps from a state $e_\alpha \in R$ to a state in at least one of the K_i. Clearly transitions of the latter kind are irreversible: having reached a state of class K_i, our system is not able to come out of the states of this class.

Let K be the union of all classes K_i. For $e_\alpha \in K$ we denote by Ω_α the set of integer vectors m with non-negative components satisfying the conditions

$$m^\alpha = 1, \quad m^\beta = 0 \text{ for } \beta \neq \alpha, \ \beta \in K. \tag{7.1}$$

Since from any state $e_\gamma \in R$ transition to some state $e_\beta \in K$ eventually occurs, the following lemma holds:

Lemma 17. *If* $e_\gamma \in R$, *then*

$$\sum_{\alpha \in K} \sum_{m \in \Omega_\alpha} W_\gamma^\alpha(m) = 1. \tag{7.2}$$

---

[7] The definition of x_* depends on the choice of the vectors u_D in (5.7), but since $f_\gamma(D) > 0$ only for a finite number of residue classes D, this arbitrariness is immaterial in the limit theorems.

Consider the following sets of vectors m with non-negative integer components:

M': the components m^α of the vector m vanish for all α such that $e^\alpha \in K$;

M'': there are components $m^\alpha > 0$ of m with e^α belonging to more than one class K_i;

M_i: there are components $m^\alpha > 0$ of m with $e^\alpha \in K_i$, but there are no components $m^\alpha > 0$ with $e^\alpha \in K_j$, $j \neq i$.

In the same way as Lemmas 5 and 6, we can prove

Lemma 18. *There exist constants C and $D > 0$ such that for any γ and any $m \in M'$,*

$$W_\gamma(m) < Ce^{-\overline{m}D}. \tag{7.3}$$

We can now easily complete the study of the limit behaviour of $W_\gamma(m)$ as $\overline{m} \to \infty$:

(I) If $m \in M'$ we can apply Lemma 18;

(II) If $m \in M''$, then for any γ

$$W_\gamma(m) = 0 \tag{7.4}$$

(III) If $m \in M^{(i)}$, then m can be uniquely represented in the form

$$m = m_1 + m_2, \tag{7.5}$$

where $m_1 \in M'$ and m_2 has components $m_2^\alpha > 0$ only for $e^\alpha \in K_i$. Then

$$W_\gamma(m) = \sum_{\alpha \in K_i, m^\alpha > 0} W_\gamma^\alpha(m_1 + e_\alpha) W_\alpha(m_2 - e_\alpha). \tag{7.6}$$

It is easy to study the limit behaviour of the probabilities $W_\alpha(m_2 - e_\alpha)$ using Theorem 4, since within the class K condition (A) holds.

15 March 1949

References

1. A.N. Kolmogorov, 'Markov chains with a countable number of possible states', *Byull. Moskov. Gos. Univ. Mat. Mekh.* **1**:3 (1937), 1–16 (in Russian) (No. 23 in this volume).

2. S.N. Bernshtein, *Probability theory*, 4th ed., GTTI, Moscow-Leningrad, 1946 (in Russian).
3. M. Fréchet, *Recherches théoriques modernes sur la théorie des probabilités*, Paris, 1937.
4. W. Doeblin, 'Sur deux problèmes de M. Kolmogoroff concernant les chaînes dénombrables', *Bull Soc. Math. France* **66** (1938), 210–220.
5. A.N. Kolmogorov and Yu.V. Prokhorov, 'On the sums of a random number of random terms', *Uspekhi Mat. Nauk* **4**:4 (1949), 168–172 (in Russian).
6. D.G. Meizler, O.S. Parasyuk and E.L. Rvacheva, 'A multidimensional local limit theorem in probability theory', *Dokl. Akad. Nauk SSSR* **60** (1948), 1127–1128 (in Russian).

37. SOLUTION OF A PROBABILISTIC PROBLEM RELATING TO THE MECHANISM OF BED FORMATION *

A.B. Vistelius and O.V. Sarmanov drew my attention to a mathematical problem that they were confronted with in the statistical study of bed thickness in geological sediments. The scheme of bed formation considered by them can be understood from the example given in Fig. 1. The horizontal axis represents time, and the vertical axis, the thickness of sediments and washouts. The nth bed is deposited over the time interval $(t_{n-1}; s_n)$ and its thickness (before subsequent washouts) is $\xi_n = g_n - h_{n-1}$. During time intervals $(s_n; t_n)$ washouts reach the depth $\eta_n = g_n - h_n$.

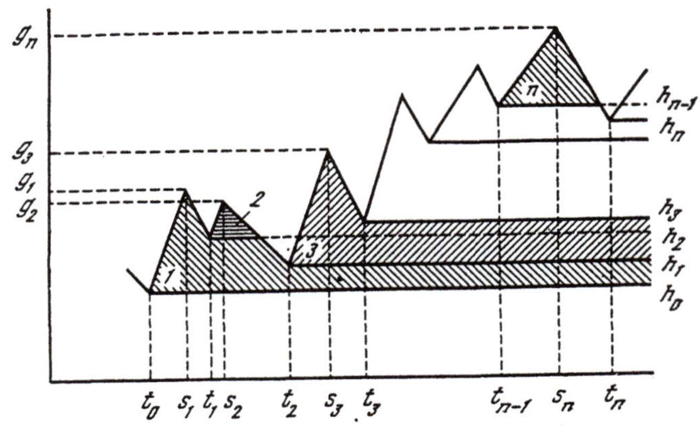

Fig. 1

As a result of these alternating sedimentations and washouts certain beds can be washed out several times (bed 1 is washed out twice during the first and the second washouts), and certain beds can even entirely disappear (bed 2 is washed out completely after the second washout). If we assume that the mean thickness of sediments is greater than the depth of washouts, then each bed would undergo a significant risk of washing out only during a small number of alternating sedimentations and washouts. Therefore we may speak of the probability of the "final" existence of a bed and of the conditional distribution of the probabilities of the "final" thickness of the remaining beds. It is these conditional distributions that we should compare with actually observed statistical distributions of bed thickness in a certain profile. According to the data

* *Dokl. Akad. Nauk SSSR* **65**:6 (1949), 793–796 (in Russian)

of A.B. Vistelius, these distributions are often highly asymmetric, as if "cut off" to the left of the zeroth abscissa.

Let $\delta_n = \xi_n - \eta_n = h_n - h_{n-1}$ be the difference between the initial thickness of the nth bed and the depth of the washout that occurs immediately after this bed has been formed. In our further mathematical constructions we will assume that an infinite sequence of random variables $\delta_1, \delta_2, \dots, \delta_n, \dots$ is given and we make the following assumptions: 1) the random variables δ_n are independent and have the same distribution law $\mathbf{P}\{\delta_n < x\} = G(x)$;

2) the expectation $M = \mathbf{E}\delta_n = \int_{-\infty}^{\infty} x dG(x)$ is positive;

3) the distribution of the δ_n is continuous, that is, it can be expressed in terms of a corresponding probability density $g(x)$ by the formula

$$G(x) = \int_{-\infty}^{x} g(x)dx.$$

The third assumption could have been omitted, thereby complicating further the analytical tools. But in view of the applied significance of the problem, we decided to make the presentation less cumbersome by avoiding the use of Stieltjes integrals. The second assumption guarantees[1] that the sums $\zeta_n^{(r)} = \delta_n + \delta_{n+1} + \dots + \delta_{n+r}$ tend to $+\infty$ as the second index tends to $+\infty$, therefore the greatest lower bounds $\phi_n = \inf(\zeta_n^{(0)}, \zeta_n^{(1)}, \dots, \zeta_n^{(r)}, \dots)$ are finite and will be attained at a certain (random) finite number r.

It can easily be seen that if $\phi_n \le 0$ the nth bed is washed out completely, while if $\phi_n > 0$, it retains some final thickness ϕ_n. Therefore the problem is reduced to determining the probability $p = \mathbf{P}\{\phi_n > 0\}$ and the conditional distribution of ϕ_n under the hypothesis $\phi_n > 0$. It is clear from assumption 1) that both p and this distribution do not depend on n.

Assumption 3) implies that the distribution of the random variables ϕ_n is continuous, that is, it can be characterized by a certain probability density $f(x)$. Clearly, $p = \int_0^\infty f(x)dx$ and the conditional distribution of ϕ_n under the

[1] By a well-known theorem of A.Ya. Khinchin, 1) and 2) imply that our sums obey the strong law of large numbers, that is,

$$\mathbf{P}\Big\{\lim_{r\to\infty} \frac{\zeta_n^{(r)}}{r+1} = M\Big\} = 1.$$

hypothesis $\phi_n > 0$ is given by the probability density

$$f^*(x) = f(x)/p \text{ for } x > 0, \quad f^*(x) = 0 \text{ for } x < 0. \tag{1}$$

To find $f^*(x)$ and p it suffices to know the function $s(x) = f(x)/p$. Then f^* is determined by (1) and p is computed by the formula

$$p = 1 : \int_{-\infty}^{\infty} s(x)dx. \tag{2}$$

Theorem. *Under the assumptions* 1), 2), 3) *the functional* $s(h) = f(x)/p$ $(-\infty < x < +\infty)$ *is the unique solution of the integral equation*

$$s(x) = g(x) + \int_{-\infty}^{0} g(x-y)s(y)dy. \tag{3}$$

Clearly (3) is equivalent to

$$f(x) = pg(x) + \int_{-\infty}^{0} g(x-y)f(y)dy. \tag{4}$$

The proof of (4) is as follows. From the definition of ϕ_n it can easily be derived that

$$\phi_n = \delta_n \text{ if } \phi_{n+1} > 0; \quad \phi_n = \delta_n + \phi_{n+1} \text{ if } \phi_{n+1} < 0. \tag{5}$$

By assumption 1), the quantities δ_n and ϕ_{n+1} are independent. Therefore, representing $f(x)$ as

$$f(x) = pf_1(x) + (1-p)f_2(x), \tag{6}$$

where $f_1(x)$ and $f_2(x)$ are the conditional densities for ϕ_n under the hypotheses $\phi_{n+1} > 0$ and $\phi_{n+1} < 0$ respectively and noting that the conditional density for ϕ_{n+1} under the hypothesis $\phi_{n+1} < 0$ is[2]

$$f^{**}(x) = f(x)/(1-p) \text{ for } x < 0; \quad f^{**}(x) = 0 \text{ for } x > 0 \tag{7}$$

[2] Here we set $p < 1$. When $p = 1$, $g(x) = 0$ for $x < 0$, $f(x) \equiv g(x)$ and (4) holds trivially.

we obtain

$$f_1(x) = g(x), \tag{8}$$

$$f_2(x) = \int_{-\infty}^{\infty} g(x-y)f^{**}(y)dy = \frac{1}{1-p}\int_{-\infty}^{0} g(x-y)f(y)dy. \tag{9}$$

Formula (4) now follows immediately from (6), (8) and (9).

To complete the proof of the theorem we now only need to establish the uniqueness of the solution to (3). Consider the iterated kernels

$$K_1(x,y) = g(x-y) \text{ for } y < 0; \quad K_1(x,y) = 0 \text{ for } y > 0;$$

$$K_r(x,y) = \int_{-\infty}^{\infty} K_1(x,z)K_{r-1}(z,y)dz.$$

All these kernels, as well as the functions

$$s_0(x) = g(x),$$

$$s_r(x) = \int_{-\infty}^{\infty} K_r(x,y)g(y)dy =$$

$$= \iint\cdots\int_{\substack{u_1+u_2+\ldots+u_k<0\\ k=1,2,\ldots,r}} g(u_1)\ldots g(u_r)g(x-u_1-\ldots-u_r)du_1\ldots du_r$$

are non-negative. A formal solution to (3) is of the form

$$s(x) = \sum_{r=0}^{\infty} s_r(x). \tag{10}$$

To see that it really is a solution, and the unique one, it suffices to prove the convergence of the series $\sum_{r=0}^{\infty} I_r$, where

$$I_r = \int_{-\infty}^{\infty} s_r(x)dx = \iint\cdots\int_{\substack{u_1+u_2+\ldots+u_k<0\\ k=1,2,\ldots,r}} g(u_1)\ldots g(u_r)du_1\ldots du_r. \tag{11}$$

The convergence of (11) can be derived from the probabilistic meaning of its terms. Denote by $A_n^{(r)}$ the event $\{\phi_n = \zeta_n^{(r)}\}$, which can alternatively be written in the form

$$A_n^{(r)} = \begin{cases} \delta_{n+k} + \delta_{n+k+1} + \ldots + \delta_{n+r} \le 0, & k = 1,2,\ldots,r. \\ \phi_{n+r+1} \ge 0. \end{cases}$$

Clearly,

$$\mathbf{P}\{A_n^{(r)}\} =$$
$$= \mathbf{P}\{\phi_{n+r+1} \geq 0\}\mathbf{P}\{\delta_{n+k} + \delta_{n+k+1} + \ldots + \delta_{n+r} \leq 0;\ k = 1, \ldots, r\} =$$
$$= pI_r.$$

Since the events $A_n^{(r)}$, $r = 1, 2, \ldots$, form (to within probability 0) a complete system of pairwise incompatible events, it follows that

$$\sum_0^\infty I_r = \frac{1}{p}\sum_r \mathbf{P}\{A_n^{(r)}\} = \frac{1}{p}. \tag{12}$$

This essentially solves the problem posed at the beginning of the article: formula (10) gives the function $s(x)$. Using this function we can compute $f^*(x)$:

$$f^*(x) = s(x) \text{ for } x > 0; \quad f^*(x) = 0 \text{ for } x < 0, \tag{13}$$

while the probability p is given by (12). We can estimate the remainder terms of the series (10) and (12). The use of these series for computing numerical results is somewhat cumbersome but entirely possible.

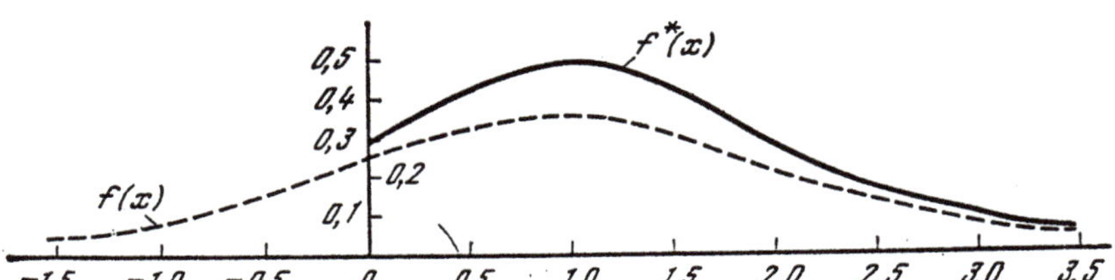

Fig. 2

Figure 2 plots $f(x)$ and $f^*(x)$ for the case

$$g(x) = \phi(x - a), \quad \phi(x) = \frac{1}{\sqrt{2\pi}}e^{-x^2/2}$$

for $a = 1$ (this corresponds to $p = 0.82$). The fact that $f^*(x)$ is obtained by "cutting off" $s(x)$ via formula (13) explains the qualitative features observed in reality.

2 March 1949

38. UNBIASED ESTIMATORS *

The article considers a number of problems on finding unbiased estimators $\phi(x_1, x_2, \dots, x_n)$ for various functionals $f(P)$, which depend on the distribution law P of observables $x_1, x_2, \dots, x_n$. Some of these problems are connected with questions of statistical control and inspection of industrial mass production.

§1. General definitions and theorems

Suppose that we are given a system $\mathfrak{P}$ of "admissible" probability distributions

$$P(A) = \mathbf{P}\{x \in A\}$$

for a random point x in a certain space X, and a functional $f(P)$ defined on $\mathfrak{P}$.

Definition 1. A function $\phi(x)$ defined on X is called an *unbiased estimator* for $f(P)$ if for any distribution P in $\mathfrak{P}$ [1]

$$\mathbf{E}^P \phi = f(P). \tag{1}$$

For example, if X is the n-dimensional space of points

$$x = (x_1, x_2, \dots, x_n),$$

and $\mathfrak{P}$ consists of distributions P given by the probability densities

$$p(x|a) = (2\pi)^{-n/2} \exp\Big\{-\tfrac{1}{2}\sum_{k=1}^{n}(x_k - a)^2\Big\},$$

then a well-known unbiased estimator for

$$f(P) = a$$

* *Izv. Akad. Nauk SSSR Ser. Mat.* **14**:4 (1950), 303–326.

[1] In (1) and in what follows $\mathbf{E}^P$ denotes the expectation corresponding to the distribution P:

$$\mathbf{E}^P = \int_X \phi(x) P(dx).$$

is the arithmetic mean

$$\bar{x} = (x_1 + x_2 + \ldots + x_n)/n,$$

while

$$\phi = \bar{x}^2 - 1/n$$

can serve as an unbiased estimator for

$$f(P) = a^2,$$

etc.

We shall see that in many important cases there are no unbiased estimators. In these cases the following definition is helpful.

Definition 2. Functions $\phi_+(x)$ and $\phi_-(x)$ are called the *upper* and *lower estimators* for $f(P)$ respectively if for any distribution P from $\mathfrak{P}$

$$\mathsf{E}^P \phi_+ \geq f(P), \tag{2}$$

$$\mathsf{E}^P \phi_- \leq f(P). \tag{3}$$

On the other hand, in most problems where unbiased estimators do exist, there are many of them. Thus in the example considered above, any linear form

$$\phi = c_1 x_1 + c_2 x_2 + \ldots + c_n x_n$$

with coefficients satisfying the condition

$$c_1 + c_2 + \ldots + c_n = 1$$

can serve as unbiased estimator for a.

The excessive diversity of unbiased estimators can be significantly decreased if we confine ourselves to unbiased estimators that are expressed in terms of properly chosen sufficient statistics of the problem. To formulate the appropriate general theorems, I shall have to give a somewhat generalized definition of a sufficient statistic. Since in this definition a sufficient statistic can be not only scalar, but also vector-valued, we do not need to introduce the notion of "a system of sufficient statistics": in our general definition such a system of statistics $(\chi_1, \chi_2, \ldots, \chi_s)$ can be considered as a single statistic

$$\chi = (\chi_1, \chi_2, \ldots, \chi_s).$$

The following, somewhat abstract formulations will be made more concrete in §2 and §3 which the reader should refer to should he find this formulation too complicated.

Let $\chi(x)$ be a function given on X and with values in some other space H. Consider the conditional probability distributions

$$P(A|h) = \mathbf{P}\{x \in A | \chi(x) = h\} \tag{4}$$

and the conditional expectations[2]

$$\mathbf{E}_h^P \phi = \int_X \phi(x) P(dx|h) \tag{5}$$

for a fixed value

$$\chi(x) = h.$$

Definition 3. A function $\chi(x)$ is called a *sufficient statistic* for the system of distributions $\mathfrak{P}$ if the conditional probability distributions $P(A|h)$ do not depend on the choice of P from $\mathfrak{P}$.

Since, in accordance with [1], the conditional probabilities $P(A|h)$ are only defined to within a set of values h taken by $\chi(x)$ of probability 0 (for the distribution P for x), the exact meaning of Definition 3 is as follows: $\chi(x)$ is called a *sufficient statistic* for $\mathfrak{P}$ if there exists a function $Q(A|h)$ of $A \subseteq X$ and $h \in H$ that for any P from $\mathfrak{P}$ can be taken for $P(A|h)$.

Theorem 1. *If $\chi(x)$ is a sufficient statistic for $\mathfrak{P}$, then for any $\phi(x)$ whose expectation $\mathbf{E}^P$ is finite for all $P \in \mathfrak{P}$, the conditional expectations $\mathbf{E}_h^P \phi$ do not depend on $P \in \mathfrak{P}$.*

As with Definition 3, the exact meaning of Theorem 1 needs to be clarified. It is as follows: for any function $\phi(x)$ whose expectation $\mathbf{E}^P \phi$ is finite for all P in $\mathfrak{P}$, there exists an $M(h)$ that for any $P \in \mathfrak{P}$ can be taken for $\mathbf{E}_h^P \phi$.

[2] The integrals in (5) should be understood in the sense of (10) and (11) of §4, Chapter 5 of my book [1].

To prove Theorem 1 it suffices to set

$$M(h) = \int_X \phi(x) Q(dx|h), \tag{6}$$

where $Q(A|h)$ is taken from the explanation to Definition 3 and the integral is understood in the sense of the footnote to (5).

The following two theorems give a generalization of Blackwell's results [3].

Theorem 2. *If $\phi(x)$ is an unbiased estimator for $f(P)$ and $\chi(x)$ a sufficient statistic for $\mathfrak{P}$, then*[3]

$$\phi^*(x) = \mathbf{E}_{\chi(x)}\phi(x) = M[\chi(x)] \tag{7}$$

is also an unbiased estimator for $f(P)$.

Theorem 3. *For any $P \in \mathfrak{P}$ for which the variance $\mathbf{D}^P\phi$ exists, under the conditions of Theorem 2 we have the inequality*

$$\mathbf{D}^P\phi^* \leq \mathbf{D}^P\phi. \tag{8}$$

To prove Theorems 2 and 3 it suffices to use the identities

$$\mathbf{E}(\mathbf{E}_\chi\phi) = \mathbf{E}\phi, \tag{9}$$

$$\mathbf{D}\phi = \mathbf{D}(\mathbf{E}_\chi\phi) + \mathbf{E}(\phi - \mathbf{E}_\chi\phi)^2, \tag{10}$$

which hold for any functions $\phi(x)$ and $\chi(x)$ provided that $\mathbf{E}\phi$ and $\mathbf{D}\phi$ exist.

Theorems 2 and 3 can be considered as a substantiation of the natural tendency to use only unbiased estimators that are expressed via sufficient statistics of the problem: Theorem 2 shows that in doing this we do not restrict the number of problems for which there exist unbiased estimators, and Theorem 3 demonstrates that when we pass from an unbiased extimator ϕ to the averaged estimator ϕ^* expressed in terms of the statistic χ, we can decrease the variance. It can be shown that the estimator ϕ^* is always "no worse" than the estimator ϕ generating it, even when using other methods for comparing the "quality" of estimators.

[3] The superscript P to $\mathbf{E}$ in (7) is omitted, since by Theorem 1 ϕ^* can be chosen independent of P.

Moreover, as we shall see, Theorem 2 gives an effective tool for finding unbiased estimators with low variances; often, by averaging "bad" but easily found estimators ϕ one can get "good" estimators ϕ^*. In a rather wide class of cases unbiased estimators expressed in terms of properly chosen sufficient statistics appear to be uniquely determined by the functional $f(P)$ to be estimated. This is discussed in an interesting work by Halmos [4] which represents one of a few attempts to approach unbiased estimators from the general viewpoint of sufficient statistics. Some results of Halmos will be given in §10 of this paper.

§2 Main formulas for the discrete case

If the set X of possible values of x is finite,

$$X = \{a_1, a_2, \ldots, a_n\}$$

and the distributions P^θ of the system $\mathfrak{P}$ are uniquely determined by one or several parameters θ,[4] then it is natural to introduce the probabilities

$$p_k(\theta) = \mathbf{P}^\theta\{x = a_k\},$$

represent the functional $f(P)$ by the function $f(\theta)$ and set

$$\phi_k = \phi(a_k)$$

for the possible values of the estimator ϕ. The condition that ϕ is an unbiased estimator of $f(\theta)$ is then expressed as

$$\sum_{k=1}^{n} \phi_k p_k = f(\theta). \tag{1}$$

Formula (1) shows that in the case when θ runs through an infinite number of values, only few functions $f(\theta)$, namely those expressed by linear forms (1), allow unbiased estimators ϕ. If the $p_k(\theta)$ are linearly independent, then for each function $f(\theta)$ which allows unbiased estimators there exists only one such estimator.

[4] In the case of several parameters $\theta_1, \theta_2, \ldots, \theta_n$, we shall denote by θ the vector $(\theta_1, \theta_2, \ldots, \theta_n)$.

If the possible values of x are enumerated by pairs of indices,

$$X = \{a_{hk}\},\ 1 \le h \le m,\ 1 \le k \le n_h,$$

then it is natural to give a similar enumeration to the probabilities $p_{hk}(\theta)$ and the values ϕ_{hk} of the estimator ϕ. Then the first index h will be a sufficient statistic of the problem if and only if $p_{hk}(\theta)$ can be written in the form

$$p_{hk}(\theta) = \bar{p}_h(\theta) q_{hk}, \tag{2}$$

where the q_{hk} do not depend on θ,

$$q_{hk} \ge 0, \quad \sum_{k=1}^{n_h} q_{hk} = 1,$$

$$\bar{p}_h(\theta) \ge 0, \quad \sum_{h=1}^{m} \bar{p}_h(\theta) = 1.$$

In case (2), in accordance with Theorems 2 and 3 we obtain from the unbiased estimator

$$\phi(a_{hk}) = \phi_{hk}$$

for $f(\theta)$, the unbiased estimator

$$\phi^*(a_{hk}) = \phi_h^* = \sum_{k=1}^{n_h} q_{hk}\phi_{hk} \tag{3}$$

with variance

$$\mathbf{D}^\theta \phi^* \le \mathbf{D}^\theta \phi$$

for any θ.

The proof of Theorems 2 and 3 which, in the general case, though extremely brief and simple, was based on a somewhat difficult general theory of conditional probabilities and expectations, now becomes, for the special case under consideration, purely arithmetic, since the expectations and variances under consideration can be expressed by the known formulas:

$$\mathbf{E}^\theta \phi = \sum_{h=1}^{m} \sum_{k=1}^{n_h} p_{hk}(\theta)\phi_{hk}, \quad \mathbf{E}^\theta \phi^* = \sum_{h=1}^{m} \bar{p}_h(\theta)\phi_h^*,$$

$$\mathbf{D}^\theta \phi = \sum_{h=1}^{m} \sum_{k=1}^{n_h} p_{hk}(\theta)(\phi_{hk} - \mathbf{E}^\theta \phi)^2,$$

$$\mathbf{E}^\theta \phi^* = \sum_{h=1}^{m} \bar{p}_h(\theta)(\phi_h^* - \mathbf{E}^\theta \phi^*)^2.$$

§3. Main formulas for the continuous case

Let the distributions P^θ from $\mathfrak{P}$ be given by probability densities

$$p(x|\theta).$$

In this case the condition for $\phi(x)$ to be an unbiased estimator for $f(\theta)$ is written in the form

$$\int_X \phi(x)p(x|\theta)dx = f(\theta), \tag{1}$$

that is, the determination of an unbiased estimator $\phi(x)$ for a given function $f(\theta)$ reduces to the solution of an integral equation.

Suppose further that a function $\chi(x)$ is given with values in a space H. This function divides X into subsets V_h on which χ takes constant values,

$$\chi(x) = h.$$

We assume that the volume element dx can be expressed in the form

$$dx = dx_1 dh$$

where dh is the volume element in H and dx_1 is a suitably defined volume element in V_h. Then χ is a sufficient statistic of the problem if and only if the probability densities $p(x|\theta)$ can be expressed in the form

$$p(x|\theta) = \bar{p}(h(x)|\theta)q(x), \tag{2}$$

where $q(x)$ does not depend on θ and

$$q(x) \geq 0, \quad \int_{V_h} q(x)dx_1 = 1,$$

$$\bar{p}(h) \geq 0, \quad \int_H \bar{p}(h|\theta)dh = 1.$$

In case (2), in view of Theorems 2 and 3, we can obtain from an unbiased estimator

$$\phi = \phi(x)$$

for $f(\theta)$, the unbiased estimator

$$\phi^* = \phi^*(x) = \int_{V_{\chi(x)}} \phi(x)q(x)dx_1 \tag{3}$$

with variance

$$\mathbf{D}^{\theta}\phi^{*} \leq \mathbf{D}^{\theta}\phi \tag{4}$$

for any θ, this variance depending only on $\chi(x)$:

$$\phi^{*} = s[\chi(x)].$$

§4. Unbiased estimators of random variables

In this section we further generalize the problem of §1; this generalization will be used in §5. Given a system $\mathfrak{P}$ of admissible probability distributions

$$P(A) = \mathbf{P}\{(x, y) \in A\}$$

for the pairs consisting of an "observable" point $x \in H$ and a random variable y that cannot be observed directly, a function $\phi(x)$ is called an unbiased estimator of y for all P in $\mathfrak{P}$ if

$$\mathbf{E}^{P}\phi = \mathbf{E}^{P}y. \tag{1}$$

The question of finding such unbiased estimators for random variables does not involve new difficulties, since it is equivalent to finding unbiased estimators for

$$f(P) = \mathbf{E}^{P}y;$$

only now, to characterize the accuracy of the estimator ϕ for y it is natural to consider the expression

$$\mathbf{E}^{P}(\phi - y)^{2}$$

instead of the variance $\mathbf{D}^{P}\phi$.

In the case when x and y are independent for any P in $\mathfrak{P}$ we have

$$\mathbf{E}^{P}(\phi - y)^{2} = \mathbf{D}^{P}\phi + \mathbf{D}^{P}y, \tag{2}$$

and since $\mathbf{D}^{P}y$ does not depend on the choice of ϕ, the problem of finding an unbiased estimator ϕ for y with least value of $\mathbf{E}^{P}(\phi-y)^{2}$ is equivalent to finding an unbiased estimator for $\mathbf{E}^{P}y$ with minimal variance $\mathbf{D}^{P}\phi$.

§5. Inspection by qualitative features based on a single sample: the case of destroying the product item while testing

In order to show, by means of a sufficiently specific example, the value of the theory of unbiased estimators in inspecting mass production based on selected data, we consider two problems from this field in this section and in §7. In this paper we shall not go into details on the practical application of the suggested method. In this section we assume that we destroy the industrial product item by testing and therefore, in principle, this testing can only be of a sample character. The inspection system under consideration is as follows.

Random sampling of n items is made from a batch containing N items. These n selected items are tested and the number x of "defective" items in the sample is found. If $x \leq c$, then the $N - n$ items not included in the sample are accepted. If $x \geq d = c + 1$, then the whole batch is rejected.

If there are y defective items in the batch before testing, then the number of accepted defective items will be

$$y^* = \begin{cases} y - x & \text{for } x \leq c, \\ 0 & \text{for } x \geq d. \end{cases}$$

We set

$$q = y/N, \quad q^* = y^*/N.$$

The aim of inspection is to guarantee that the q^* are small enough, without causing unnecessary, that is, for small q, too frequent rejection of the entire batch, and without excessively increasing the sample size n.

To satisfy all these requirements, appropriate values of n and c must be chosen. The main characteristic of the inspection system with given n and c is the conditional probability

$$L(q) = \mathbf{P}\{x \leq c|q\}$$

of obtaining $x \leq c$ for a given q, that is, of accepting the batch. To compute $L(q)$ we use the conditional probabilities

$$p_m(q) = \mathbf{P}\{x = m|q\} = \frac{n!}{m!(n-m)!} \frac{(N-n)!N^n}{N!} q\left(q - \frac{1}{N}\right) \times$$
$$\times \left(q - \frac{2}{N}\right) \cdots \left(q - \frac{m-1}{N}\right)(1-q)\left(1 - q - \frac{1}{N}\right) \cdots$$
$$\cdots \left(1 - q - \frac{n-m-1}{N}\right) \tag{1}$$

of obtaining $x = m$ for a given q. It can easily be seen that

$$L(q) = \sum_{m \leq c} p_m(q). \tag{2}$$

If it is necessary to reject batches with $q < q_0$ and to accept those with $q > q_0$, then the ideal operative characteristic for $L(q)$ would be the function

$$L(q) = \begin{cases} 1 & \text{for } q < q_0 \\ 0 & \text{for } q > q_0, \end{cases}$$

given in Figure 1. This form of $L(q)$ can be achieved only for $n = N$, which in case the tests are of a destructive character, would make the entire operation meaningless. However, we come close to the ideal characteristic in Figure 1 if we choose a sufficiently large c and take

$$n \sim c/q_0.$$

If N is very large, this does not lead to an excessive increase in the ratio n/N which, of course, should remain sufficiently small. Figure 2 gives operative characteristics for the case[5]

$$N = 10,000, \quad n = 1,000, \quad c = 20, \quad q_0 \sim c/n = 0.02.$$

Unfortunately, quite satisfactory operative characteristics even for N of the order 1,000–10,000 can often be obtained only at the expense of increasing the ratio n/N. Therefore, we often use inspection systems with operative characteristics that do not fall steeply enough as q grows.

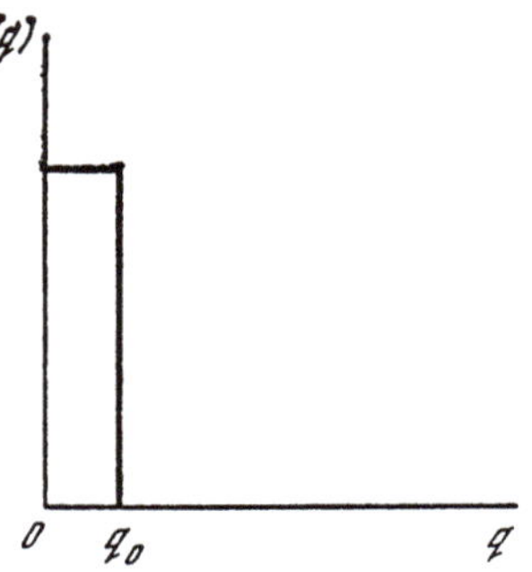

Fig. 1

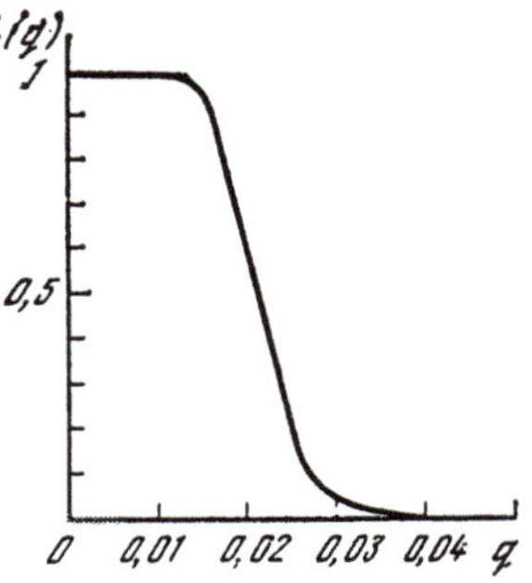

Fig. 2.

In these cases it becomes especially important to estimate the proportion of actually accepted defective items based on the test results. When production

[5] Sampling inspection. Statistical Research Group, Columbia Univ., 1948.

runs smoothly enough this proportion is often much smaller than the proportion of defective items q_0 which has been used for computing the inspection system.

Subsequent estimations of the proportion of defective items in the products to be checked based on test results and the proportion of defective items that were accepted using this inspection system, is certainly very important also in the case of quite satisfactory operative characteristics. This is the problem we will solve now.

Assume that the inspection system with given n and c is applied to a large number s of batches. The ratios q and q^* introduced above will be denoted by q_r and q_r^* for the rth batch. Moreover, s_m will be the number of batches with $x = m$ (where m is the number of defective items found in a batch). Clearly,

$$s = s_0 + s_1 + \ldots + s_n,$$

while

$$s' = s_0 + s_1 + \ldots + s_c$$

denotes the number of accepted batches. The total number of items in s batches is

$$R = sN,$$

while the number of accepted items is

$$R' = s'(N - n).$$

The total number of defective items is

$$Y = \sum_{r=1}^{s} y_r = N \sum_{r=1}^{s} q_r,$$

while the number of accepted defective items is

$$Y' = \sum_{r=1}^{s'} y_r' = N \sum_{r=1}^{s} q_r^*.$$

The success of the entire inspection procedure is measured by how much the ratio

$$q'_{\text{mean}} = Y'/R'.$$

is less than the ratio

$$q_{\text{mean}} = \frac{Y}{R} = \frac{1}{s} \sum_{r=1}^{s} q_r$$

Since

$$q'_{\text{mean}} = q^*_{\text{mean}} R/R',$$

where

$$q^*_{\text{mean}} = \frac{Y'}{R} = \frac{1}{s}\sum_{r=1}^{s} q^*_r,$$

while the ratio R/R' becomes known after acceptance and is close to 1 for any normal production run, we can take it that our problem is to estimate q_{mean} and q^*_{mean}.

If the number s of batches is large enough, and if $\phi(x)$ and $\phi^*(x)$ are unbiased estimators of q and q^* respectively with respect to x, then by the law of large numbers, q_{mean} and q^*_{mean} can be estimated using the approximate formulas

$$q_{\text{mean}} \sim \phi_{\text{mean}} = \frac{1}{s}\sum_{r=1}^{s} \phi(x_r), \tag{3}$$

$$q^*_{\text{mean}} \sim \phi^*_{\text{mean}} = \frac{1}{s}\sum_{r=1}^{s} \phi^*(x_r). \tag{4}$$

The accuracy of these formulas is determined by the standard methods, since the terms on the right-hand sides of the equalities

$$q_{\text{mean}} - \phi_{\text{mean}} = \frac{1}{s}\sum_{r=1}^{s} [q_r - \phi(x_r)],$$

$$q^*_{\text{mean}} - \phi^*_{\text{mean}} = \frac{1}{s}\sum_{r=1}^{s} [q^*_r - \phi^*(x_r)]$$

are independent. This is discussed at the end of the next section.

An unbiased estimator for q is well known:

$$\phi(x) = x/n. \tag{5}$$

It leads to the estimate

$$q_{\text{mean}} \sim \phi_{\text{mean}} = \frac{1}{ns}\sum_{m=1}^{n} m s_m \tag{6}$$

for q_{mean}.

Since always [6]

$$q^*_{\text{mean}} \le q_{\text{mean}}, \tag{7}$$

it follows that q_{mean} can also serve as an upper estimator (see Definition 2 of §1) for q^*_{mean}. However, this well-known estimator does not indicate the decrease in the number of defective items as a result of inspection. To obtain an idea about more efficient estimators of q^*_{mean} we consider the general question: which functions $f(x)$ have unbiased estimators with respect to the number of defective items x in a sample of n items. This question is solved very simply in accordance with §2. For any function $\phi(x)$,

$$f(q) = \mathbf{E}^q\phi(x) = \sum_{m=0}^{n} \phi(m)p_m(q) \tag{8}$$

is a polynomial in q of degree at most n. Since in this case the $n+1$ polynomials $p_m(q)$ are linearly independent, any polynomial $f(q)$ of degree at most n can be uniquely expressed in the form (8). In other words, *polynomials $f(q)$ of degree at most n, and only these polynomials, have unbiased estimators with respect to x, and such an estimator is uniquely determined by the polynomial $f(q)$.*

It can easily be verified that

$$Q(q) = \mathbf{E}^q q^* = \sum_{m\le c} \Big(q - \frac{m}{N}\Big)p_m(q) \tag{9}$$

is a polynomial of degree $n+1$. Therefore an accurate unbiased estimator $\phi^*(x)$ for q^* with respect to x *does not exist.* However, in the following section we shall see that the problem of finding such an estimator allows, under rather general assumptions, a very simple approximate solution. Moreover, the following two remarks are of interest.

1. For any $n' < n$ there exists an unbiased estimator for the number of defective items x in a sample of n items for q^* that would be obtained from inspection over samples of n' items (with any c).

2. By approximating $Q(q)$ by polynomials $Q_+(q)$ and $Q_-(q)$ of degree at most n such that

$$Q_-(q) \le Q(q) \le Q_+(q),$$

we can obtain upper and lower estimators $\phi^*_+(x)$ and $\phi^*_-(x)$ respectively for q^*.

---

[6] More precisely, $q^*_{\text{mean}} \le q_{\text{mean}} - \frac{1}{sN}\sum_{r=1}^{s} x_r$.

This method promises acceptable practical results for many cases in which the approximate formulas of the next section are inapplicable.

§6. The case $c \ll n \ll N$

If n is small compared to N, then we may set

$$q^* = \begin{cases} q & \text{for } x \le c, \\ 0 & \text{for } x \ge d, \end{cases}$$

in a sufficiently good approximation, and replace (1), (2), (9) of §5 by

$$p_m(q) = \frac{n!}{m!(n-m)!} q^m (1-q)^{n-m}, \tag{1}$$

$$L(q) = \sum_{m \le c} p_m(q), \tag{2}$$

$$\mathsf{E}^q q^* - Q(q) = qL(q). \tag{3}$$

Assuming that we can use (1) for the variance

$$\mathsf{D}^q \phi(x) = \mathsf{D}^q(x/n) = q(1-q)/n$$

of the estimator (5) of the previous section, we obtain the unbiased estimator

$$\psi^2 = \frac{x(n-x)}{n^2(n-1)}, \tag{4}$$

found by Girshick, Mosteller and Savage [5]. From this we obtain the unbiased estimator

$$\Delta^2 = \frac{1}{s^2 n^2 (n-1)} \sum_{m=1}^{n-1} m(n-m) s_m \tag{5}$$

for the variance of ϕ_{mean}.

For m small compared with n the probabilities $p_m(q)$ are non-zero only for small q, and (1) can be replaced by the simple approximate formula

$$p_m(q) = \frac{n^m}{m!} q^m e^{-nq}. \tag{6}$$

Since by (6),

$$q p_m(q) = \frac{m+1}{n} p_{m+1}(q),$$

under the assumption that c is small compared with n, (2) and (3) give

$$Q(q) = \frac{1}{n} \sum_{m \le d} m p_m(q). \tag{7}$$

Thus when we can use (3), (6) and (7), the function $Q(q)$ is a linear combination of the $p_m(q)$ and, in accordance with §2, we are able to construct an unbiased estimator for $Q(q)$ (and consequently, for q^*) with respect to x. This estimator has the form

$$\phi^*(x) = \begin{cases} x/n & \text{for } x \le d = c+1, \\ 0 & \text{for } x \ge d+1. \end{cases} \tag{8}$$

This gives the following unbiased estimator for q^*_{mean}:

$$q^*_{\text{mean}} \sim \phi^*_{\text{mean}} = \frac{1}{ns} \sum_{m \le d} m s^*_m. \tag{9}$$

It is instructive to compare (9) with formula (6) in §5. Comparing these formulas we see that for large s the difference

$$\phi_{\text{mean}} - \phi^*_{\text{mean}} = \frac{1}{ns} \sum_{m>d} m s_m \tag{10}$$

is approximately equal to $q_{\text{mean}} - q^*_{\text{mean}}$ or (when R/R' is close to 1) to $q_{\text{mean}} - q'_{\text{mean}}$, that is, the decrease of the proportion of defective items in the process of inspection.

We can quite effectively estimate the accuracy of the approximate estimator (9). The independence of the samples from different batches (this independence is implicit in the notion of a "random sample") implies that

$$\mathsf{E}(q^*_{\text{mean}} - \phi^*_{\text{mean}})^2 = \frac{1}{s^2} \sum_{r=1}^{s} \mathsf{E}[q^*_r - \phi^*(x_r)]^2. \tag{11}$$

Since

$$(q^*_{\text{mean}} - \phi^*_{\text{mean}})^2 = \begin{cases} (q - x/n)^2 & \text{for } x \le c, \\ (x/n)^2 & \text{for } x = d = c+1 \\ 0 & \text{for } x \ge d+1, \end{cases} \tag{12}$$

it follows that

$$\mathsf{E}^q(q^*_{\text{mean}} - \phi^*_{\text{mean}})^2 = \frac{1}{n^2} \sum_{m=1}^{d} m^2 p_m(q) + \frac{2q}{n} \sum_{m=1}^{c} m p_m(q) + q^2 \sum_{m=0}^{c} p_m(q)$$

or, after appropriate transformations,

$$\mathsf{E}^q(q^*_{\text{mean}} - \phi^*_{\text{mean}})^2 = \frac{1}{n^2} \Big[\sum_{m=1}^{d} m p_m(q) + d(d+1) p_{d+1}(q) \Big]. \tag{13}$$

Therefore an unbiased estimator of $\mathsf{E}^q(q^* - \phi^*)^2$ is

$$\psi_*^2(x) = \begin{cases} x/n^2 & \text{for } x \le d, \\ d(d+1)/n^2 & \text{for } x = d+1, \\ 0 & \text{for } x \ge d+2, \end{cases} \tag{14}$$

while

$$\Delta_*^2 = \frac{1}{s^2}\sum_{r=1}^{s}\psi^2(x_r) = \frac{1}{(ns)^2}\Big[\sum_{m=1}^{d} ms_m + d(d+1)s_{d+1}\Big] =$$
$$= \frac{\phi_{\text{mean}}^*}{ns} + \frac{d(d+1)}{(ns)^2}s_{d+1} \tag{15}$$

can serve as an unbiased estimator of $\mathsf{E}(q_{\text{mean}}^* - \phi_{\text{mean}}^*)^2$. For large s

$$\mathsf{E}(q_{\text{mean}}^* - \phi_{\text{mean}}^*)^2 \sim \Delta_*^2$$

and according to Lyapunov's theorem,

$$\mathsf{P}\{|q_{\text{mean}}^* - \phi_{\text{mean}}^*| \le t\Delta\} \sim \frac{2}{\sqrt{2\pi}}\int_0^t e^{-t^2/2}dt. \tag{16}$$

In conclusion, we consider a numerical example in which (6) of §5 and (4), (9), (15) and (16) of this section are applied. Let

$$N = 1000, \quad n = 50,$$
$$s = 200, \quad c = 1, d = 2.$$

m	0	1	2	3	4	5	6	7	8	9	10	11
s_m	143	27	12	9	3	1	2	1	1	–	–	1

$$\phi_{\text{mean}} = 0.0133, \quad \Delta_* \sim 0.0010,$$
$$\Delta \sim 0.0011, \quad q_{\text{mean}} \sim 0.0133 \pm 0.0011,$$
$$\phi_{\text{mean}}^* = 0.0051, \quad q_{\text{mean}}^* \sim 0.0051 \pm 0.0010,$$

or, under the 3σ-rule,

$$|q_{\text{mean}} - \phi_{\text{mean}}| < 3\Delta, \quad |q_{\text{mean}}^* - \phi_{\text{mean}}^*| < 3\Delta,$$
$$0.0100 < q_{\text{mean}} < 0.00166, \quad 0.0021 < q_{\text{mean}}^* < 0.0081.$$

I believe that for the chosen s, N, n and c the use of the formulas introduced under the assumptions $c \ll n \ll N$ and large s is already correct. This question, however, deserves a detailed study.

§7. Inspection by qualitative features based on a single sample: the case when the tests do not affect the product items

If a test does not affect the items, we may accept an inspection system different from that considered in §5.

A random sample of n items is made from a batch consisting of N items. The n items chosen are tested, and the number x of "defective" items in the sample is found. If $x \leq c$, then the x detected defective items are replaced by good ("non-defective") ones, and the whole batch is accepted. If $x \geq d = c+1$, then the whole batch is checked, all defective items thus detected are replaced by good ones, and only after that is the entire batch accepted.

As in §5,

$$q^* = \begin{cases} q - x/N & \text{for } x \leq c, \\ 0 & \text{for } x \geq d, \end{cases}$$

but now q^* has a simpler meaning: it is the proportion of defective items remaining in the batch after the above-described procedure. Formulas (1)–(9) of §5 hold. Here $Q(q)$ has the new meaning of expectation of the proportion of defective items in an accepted product * under the assumption that prior to inspection the proportion of defective items in each batch was equal to q. Therefore we can now make an *a priori* estimate of the worst possible average quality of a product accepted under this inspection system: whatever the distribution of defective items over the batches prior to inspection, in the product accepted after inspection the proportion of defective items is, on the average, at most

$$Q_L = \max_{0 \leq q \leq 1} Q(q). \tag{1}$$

In the setting of §5 this *a priori* estimate is impossible. In Figure 3 the function $Q(q)$ is plotted [7] for

$$N = 1000, \quad c = 1$$
$$n = 100, \quad d = 2.$$

The maximum $Q_L = 0.0035$ is attained here for $q = 0.009$.

In addition to the number of defective items x, in this new setting of simple inspection we can also observe the number

$$z = \begin{cases} 0 & \text{for } x \leq c, \\ qN - x & \text{for } x \geq d, \end{cases}$$

* Here a product can also be a set of batches (Translator's note).

[7] Grant, *Statistical quality control*, 1946, p.353.

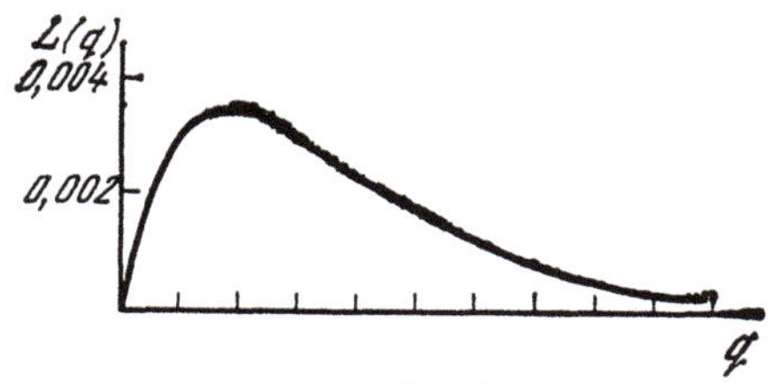

Fig. 3

of defective items found during overall checking of $N-n$ items not included in the sample, should such checking take place. This gives new possibilities for obtaining unbiased estimators of q and q^*.

It can easily be seen that the sum

$$u = x + z = \begin{cases} x & \text{for } x \leq c, \\ qN & x \geq d \end{cases}$$

is a sufficient statistic for the problem.

Now we shall see that *for any* $0 < c \leq n$ *and for any function* $f(q)$ *there exists a unique unbiased estimator* $\phi(u)$.

Indeed, the requirement

$$\mathbf{E}^q \phi(u) = \sum_{m \leq c} \phi(m) p_m(q) + \phi(qN)[1 - L(q)] = f(q) \tag{2}$$

comprises $N+1$ equations corresponding to the possible values of q:

$$q = 0, 1/N, 2/N, \ldots, 1.$$

Since for $q = 0, 1/N, \ldots, c/N$ the factor $1 - L(q)$ in (2) vanishes, for $u \leq c$ the function $\phi(u)$ is uniquely determined by the system of d equations

$$\sum_{u \leq c} \phi(u) p_u\left(\frac{m}{n}\right) = f\left(\frac{m}{N}\right), \quad m = 0, 1, \ldots, c. \tag{3}$$

For $u \geq d$, setting $q = u/N$, we obtain from (2)

$$\phi(u) = [f(u/N) - \sum_{m \leq c} \phi(m) p_m(u/N)] \,/\, [1 - L(u/N)]. \tag{4}$$

In particular, for $f(q) = Q(q)$ formulas (3), (4) give an unbiased estimator for $Q(q)$, that is, for q^*.

§8. Estimators of $f(a)$ for a normal distribution with given σ

Let $x_1, x_2, \ldots, x_n$ be independent and obey the normal distribution with probability density

$$p(x|a,\sigma) = \frac{1}{\sqrt{2\pi}\sigma} e^{-(x-a)^2/2\sigma^2}. \tag{1}$$

We assume that σ is known. Then

$$\bar{x} = (x_1 + x_2 + \ldots + x_n)/n \tag{2}$$

is a sufficient statistic for the problem. The probability density for $\bar{x}$ can be expressed in the form

$$p(\bar{x}|a,\sigma) = G(\bar{x} - a, T), \tag{3}$$

where

$$G(z,t) = \frac{1}{2\sqrt{\pi t}} e^{-z^2/4t}, \tag{4}$$

$$T = \sigma^2/2n. \tag{5}$$

In our case (for unbiased estimators of the form $\phi(x)$ for the function $f(a)$) the main equation (1) of §3 can be expressed in the form

$$\int_{-\infty}^{\infty} \phi(\bar{x}) G(\bar{x} - a, T) d\bar{x} = f(a). \tag{6}$$

For $t > -T$ we set

$$\phi(z,t) = \int_{-\infty}^{\infty} \phi(\zeta) G(\zeta - z, T - t) d\zeta. \tag{7}$$

Clearly,

$$\phi(z,0) = f(z). \tag{8}$$

Without neglecting solutions of the problem that can be of practical interest, we may confine ourselves to functions $\phi(z)$ such that

1) $\phi(z,t)$ is analytic for all real $z, t > -T$ with respect to the variables z and t;

2) $\phi(z,t)$ has the limits

$$\phi(z,-T) = \phi(z) \tag{9}$$

for almost all z.

As is known, for $t > -T$ the function $\phi(z,t)$ satisfies the heat equation

$$\partial\phi/\partial t = \partial^2\phi/\partial z^2 \tag{10}$$

and, under the above assumption, is uniquely defined for $t > -T$ by its values $f(z)$ at $t = 0$. By (9) this implies that $\phi(z)$ is uniquely determined by $f(z)$.[8]

Thus, *if the problem of finding unbiased estimators for $f(a)$ is solvable, then, under restrictions* 1) *and* 2) *its solution is unique.*

By (8)–(10) the problem of finding unbiased estimators for $f(a)$ is reduced to "the inverse heat conductivity problem", which is considered, for example, in [6]–[8]. In what follows we will note only the following concerning this. If at a certain time $T_0 > T$, $f(z)$ is already representable as

$$f(z) = \int_{-\infty}^{\infty} G(\zeta - z, T_0)\,dF(\zeta), \tag{11}$$

then the unbiased estimator $\phi(\bar{x})$ is given by the formula

$$\phi(z) = \int_{-\infty}^{\infty} G(\zeta - z, T_0 - T)\,dF(\zeta). \tag{12}$$

In particular, for the probability density itself,

$$p(x|a,\sigma) = G(x - a, \sigma^2/2) \tag{13}$$

at some fixed point x we obtain an unbiased estimator for $n > 1$ in the form

$$\phi_x(\bar{x}) = \frac{1}{\sqrt{2\pi}\sigma_0} e^{-(\bar{x}-x)^2/2\sigma_0^2}, \quad \sigma_0^2 = \left(1 - \frac{1}{n}\right)\sigma^2. \tag{14}$$

Integrating (14) for any fixed set A on the x-axis, we obtain for

$$P(A) = \mathbf{P}\{x \in A|a,\sigma\} = \int_A e^{-(x-a)^2/2\sigma^2}\,dx \tag{15}$$

the unbiased estimator[9]

$$\phi_A(\bar{x}) = \frac{1}{\sqrt{2\pi}\sigma_0} \int_A e^{-(x-\bar{x})^2/2\sigma_0^2}\,dx. \tag{16}$$

---

[8] Since by (3) $\bar{x}$ hits a set of measure zero with probability zero, the possible ambiguity of $\phi(z)$ on a set of measure zero does not bother us. In all similar cases it is legitimate to understand the uniqueness of a solution as uniqueness to within cases having probability zero for all admissible distributions P.

[9] It was Yu.V. Linnik who drew my attention to (14) and the possibility of obtaining (16) from it by integration. Before this remark by Yu.V. Linnik I derived estimators of type (16) for special kinds of sets A directly from (6).

Formula (16) is applicable for $n > 1$. When $n = 1$, we can use instead the unbiased estimator for $P(A)$ with respect to $\bar{x} = x_1$ of the form

$$\psi_A(x_1) = \begin{cases} 1 & \text{for } x_1 \in A, \\ 0 & \text{for } x_1 \overline{\in} A. \end{cases} \tag{17}$$

Similarly, for $n > 1$ we can also give for $P(A)$ an unbiased estimator:

$$\psi_A^*(x_1, x_2, \ldots, x_n) = \frac{1}{n} \sum_{m=1}^{n} \psi_A(x_m), \tag{18}$$

equal to $1/n$th of the number of points x_m belonging to the set A. This estimator has the advantage of being an unbiased estimator of $P(A)$ also in the case when the class $\mathfrak{P}$ of admissible distributions

$$P(A) = \mathbf{P}\{x \in A\}$$

is enlarged so as to include the class of all one-dimensional distributions $P(A)$. However, in the case of normal distributions with unknown a and given σ, the estimator ψ_A^* is far less efficient than ϕ_A.

§9. Estimators of $f(a, \sigma)$ for a normal distribution with unknown a and σ

We shall proceed from (1) of §8 and retain the assumption of independence of $x_1, x_2, \ldots, x_n$, but assume that both parameters are unknown. From the formula

$$p(x_1, x_2, \ldots, x_n | a, \sigma) = \frac{1}{(\sqrt{2\pi}\sigma)^n} \exp\left[-\frac{n}{2\sigma^2}\left\{(\bar{x} - a)^2 + s^2\right\}\right] \tag{1}$$

for the n-dimensional probability density it is clear that there are two sufficient statistics for the problem, namely $\bar{x}$ and

$$s^2 = \frac{1}{n} \sum_{m=1}^{n} (x_m - \bar{x})^2. \tag{2}$$

The two-dimensional probability density for $\bar{x}$ and s is expressed by

$$\bar{p}(\bar{x}, s | a, \sigma) = \frac{K_{n-2} n^{(n-1)/2} s^{n-2}}{(\sqrt{2\pi}\sigma)^n} \exp\left[-\frac{n}{2\sigma^2}\left\{(\bar{x} - a)^2 + s^2\right\}\right], \tag{3}$$

where

$$K_{n-2} = \frac{\pi n^{(n-1)/2}}{\Gamma((n-1)/2)} \tag{4}$$

is the $(n-2)$-dimensional volume of a sphere of radius 1 in $(n-1)$-dimensional space. The main equation (1) of §3 now takes the form

$$\int_{-\infty}^{\infty}\int_{-\infty}^{\infty} \phi(\bar{x}, s)\bar{p}(\bar{x}, s|a, \sigma)ds\, d\bar{x} = f(a, \sigma). \tag{5}$$

We will confine ourselves to finding, with the help of Theorem 2 of §1, the solution of (5) for the case

$$f(a, \sigma) = P(A) = \int_A p(x|a, \sigma)dx. \tag{6}$$

As already mentioned in §8, for $P(A)$ there exists an unbiased estimator

$$\psi_A(x_1) = \begin{cases} 1 & \text{for } x_1 \in A, \\ 0 & \text{for } x_2 \overline{\in} A. \end{cases}$$

To obtain an unbiased estimator for $P(A)$ of the form $\phi(\bar{x}, s)$ it remains to compute the integral

$$\psi_A(\bar{x}, s) = \int_{V_{\bar{x},s}} \psi_A(x_1) q\, dv, \tag{7}$$

where $V_{\bar{x},s}$ is the set of points $(x_1, x_2, \ldots, x_n)$ in n-dimensional space satisfying the equations

$$\frac{x_1 + x_2 + \ldots + x_n}{n} = \bar{x}, \quad \frac{1}{n}\sum_{m=1}^{n}(x_m - \bar{x})^2 = s^2,$$

which is an $(n-2)$-dimensional sphere of radius

$$\rho = \sqrt{n}s; \tag{8}$$

dv denotes the volume element in $V_{\bar{x},s}$, and q is the constant

$$q = \frac{1}{K_{n-2}(\sqrt{n}s)^{n-2}}. \tag{9}$$

For $n > 2$,

$$\phi_A(\bar{x}, s) = \int_A \phi_x dx, \tag{10}$$

where ϕ_x is an unbiased estimator with respect to $\bar{x}$ and s for the probability density $p(x|a,\sigma)$ at x. To determine ϕ_x we consider the volume of the annular zone

$$\phi_x dx/q$$

cut out from the sphere $V_{\bar{x},\sigma}$ by the hyperplanes

$$x_1 = x, \quad x_1 = x + dx.$$

When

$$\frac{n}{n-1}(x-\bar{x})^2 \geq \rho^2$$

this annulus disappears and

$$\phi_x = 0;$$

whereas for

$$\frac{n}{n-1}(x-\bar{x})^2 < \rho^2$$

its radius is

$$\rho' = \sqrt{\rho^2 - \frac{n}{n-1}(x-\bar{x})^2}, \tag{11}$$

its width is

$$\delta = \sqrt{\frac{n}{n-1}}\frac{\rho}{\rho'}dx$$

and its volume is

$$\frac{\phi_x dx}{q} = K_{n-3}(\rho')^{n-3}\delta = K_{n-3}\sqrt{\frac{n}{n-1}}\rho(\rho')^{n-4}dx.$$

Using (9)–(11) we finally obtain

$$\phi_x = \begin{cases} \frac{K_{n-3}}{K_{n-2}}\frac{1}{\sqrt{n-1}}\frac{1}{s}\left\{1 - \frac{1}{n-1}\left(\frac{x-\bar{x}}{s}\right)^2\right\}^{\frac{n-4}{2}}, & \text{if } \left|\frac{(x-\bar{x})}{s}\right| < \sqrt{n-1}, \\ 0 & \text{if } \left|\frac{(x-\bar{x})}{s}\right| > \sqrt{n-1}. \end{cases} \tag{12}$$

By introducing the function

$$f_n(t) = \begin{cases} C_n\left(1 - \frac{1}{n-1}t^2\right)^{\frac{n-4}{2}} & \text{for } |t| \leq \sqrt{n-1}, \\ 0 & \text{for } |t| \geq \sqrt{n-1}, \end{cases} \tag{13}$$

where

$$C_n = \frac{K_{n-3}}{K_{n-2}}\frac{\sqrt{n}}{n-1} = \frac{\Gamma((n-1)/2)}{\sqrt{2\pi}\Gamma((n-2)/2)\sqrt{n-1}}, \tag{14}$$

we can write ϕ_x as

$$\phi_x = \frac{1}{s} f_n\left(\frac{x-\bar{x}}{\sigma}\right). \tag{15}$$

As $n \to \infty$, the functions $f_n(t)$ converge to

$$f_\infty(t) = \frac{1}{\sqrt{2\pi}} e^{-t^2/2}. \tag{16}$$

§10. Unbiased estimators of characteristics of an arbitrary distribution based on independent observations

The examples that we have been considering have given the reader some idea of the variety of specific problems on finding unbiased estimators. It seems to me that this field of mathematical statistics deserves more attention than it has received so far. Concluding this article, which does not claim more than to demonstrate by way of examples the fact that this field is underdeveloped, I would like to point out a connection between the general approach to unbiased estimators and sufficient statistics given in §1, and the results of Halmos' work [4] quoted above.

Suppose that we are given a system of distributions $\mathfrak{P}_0$ in a certain space X_0. As admissible distributions in the space X of ordered systems

$$x = (x_1, x_2, \dots, x_n),$$

where $x_m \in X_0$ $(m = 1, 2, \dots, n)$, we will consider only those distributions that are obtained in the usual way from the assumption that all the x_m are independent and obey the same distribution $\theta \in \mathfrak{P}_0$.

The question: which functionals $f(\theta)$ defined on $\mathfrak{P}_0$ allow an unbiased estimator of the form

$$\phi(x_1, x_2, \dots, x_n),$$

is equivalent to the question: which functionals $f(\theta)$ defined on $\mathfrak{P}_0$ can be represented in the form

$$f(\theta) = \int_{X_0} \int_{X_0} \cdots \int_{X_0} \phi(x_1, x_2, \dots, x_n) \theta(dx_1) \theta(dx_2) \dots \theta(dx_n). \tag{1}$$

It is natural to call functionals of the form (1) *functionals of degree* $\leq n$. In many ways they are similar[10] to polynomials of degree $\leq n$. For example, it can easily be seen that they have the following property:

[10] It can easily be seen that any functional $f(\theta)$ that is representable in the form (1) for $n = n_1$ is also representable in the form (1) for $n = n_2 > n_1$. It is natural to call the functionals $f(\theta)$ representable in the form (1) for $n = n_0$ but not representable in the form (1) for any smaller n *functionals of degree* n_0.

(*) For any θ_1 and θ_2 the expression

$$F(\lambda) = f[\lambda\theta_1 + (1-\lambda)\theta_2]$$

may be expressed as a polynomial of degree $\leq n$ in λ for all λ such that

$$\theta = \lambda\theta_1 + (1-\lambda)\theta_2.$$

Under certain restrictions on $\mathfrak{P}_0$ the converse is true: (*) implies representability in the form (1), that is, (*) may be taken as a definition of functionals of degree $\leq n$ (cf. the definition of polynomial operations given in [9]).

This evident criterion for the existence of unbiased estimators for $f(\theta)$ has led Halmos to some interesting conclusions. For example, if $\mathfrak{P}_0$ consists of all distributions θ on the number axis for which the absolute moment

$$\beta_s(\theta) = \int_{-\infty}^{\infty} |t|^s \, d\theta$$

where s is a positive integer, is finite, then the central moment

$$\mu_s(\theta) = \int_{-\infty}^{\infty} [t - m(\theta)]^s \, d\theta, \quad m(\theta) = \int_{-\infty}^{\infty} t \, d\theta,$$

has an unbiased estimator of the form

$$\phi(x_1, x_2, \ldots, x_n)$$

if and only if

$$s \leq n.$$

(Unbiased estimators for $s \leq n$ are well-known, see [2],§27.6.)

It is then easy to see that, in accordance with the general definition of §1 for our problem of estimating $f(\theta)$ in terms of $x_1, x_2, \ldots, x_n$, there exists a sufficient statistic χ in the form of a system of values $(x_1, x_2, \ldots, x_n)$ considered independently of the order of the points x_m (but counting multiplicities, if certain x_m coincide). The functions $\phi(x_1, x_2, \ldots, x_n)$ expressed in terms of this sufficient statistic χ are, in fact, symmetric functions in $x_1, x_2, \ldots, x_n$. This evident remark brings us, by Theorems 2 and 3 of §1, to one of Halmos' results: if there exists an unbiased estimator of the form $\phi(x_1, x_2, \ldots, x_n)$ for $f(\theta)$, then for $f(\theta)$ there also exists a symmetric unbiased estimator $\phi^*(x_1, x_2, \ldots, x_n)$ with variance not exceeding the variance of $\phi(x_1, x_2, \ldots, x_n)$.

However, Halmos went even further than these evident results and proved that under certain not too restrictive limitations imposed on the system $\mathfrak{P}_0$, the symmetric unbiased estimator is *unique* (see [4]).

30 March 1950

References

1. A.N. Kolmogorov, *Basic notions of probability theory*, ONTI, Moscow-Leningrad, 1936.
2. H. Cramér, *Mathematical methods of statistics*, Princeton Univ. Press, 1946.
3. D. Blackwell, 'Conditional expectation and unbiased sequential estimation', *Ann. Math. Statist.* **18** (1947), 105–110.
4. P. Halmos, 'The theory of unbiased estimation', *Ann. Math. Statist.* **17** (1946), 34–43.
5. M.A. Girshick, F. Mosteller and L.J. Savage, 'Unbiased estimates for certain binomial sampling problems with applications', *Ann. Math. Statist.* **17** (1946), 13–23.
6. A.N. Tikhonov, 'Uniqueness theorems for the heat equation', *Dokl. Akad. Nauk SSSR* **1** (1935), 294–300 (in Russian).
7. A.N. Tikhonov, 'Uniqueness theorems for the heat equation', *Mat. Sb.* **42** (1935), 199–216 (in Russian).
8. P. Lévy, 'Sur un problème de calcul des probabilités lié à celui du refroidissement d'une barre homogène' *Ann. Scuola Norm. Super. Pisa Ser.* 2 **1** (1932), 283–296.
9. S. Mazur and W. Orlicz, 'Grundlegende Eigenschaften der polynomischen Operationen', *Stud. Math.* **5** (1934), 50–68, 179–189.
10. P. Halmos and L.J. Savage, 'Application of the Radon-Nikodým theorem to the theory of sufficient statistics', *Ann. Math. Statist.* **20** (1949), 225–241.

39. ON DIFFERENTIABILITY OF TRANSITION PROBABILITIES OF TIME-HOMOGENEOUS MARKOV PROCESSES WITH A COUNTABLE NUMBER OF STATES *

The transition probabilities $p_\alpha^\beta(t)$ we are interested in are defined for all real $t \geq 0$ and satisfy the relations

$$p_\alpha^\beta(t) \geq 0, \tag{I}$$

$$\sum_\beta p_\alpha^\beta(t) = 1, \tag{II}$$

$$p_\alpha^\beta(0) = e_\alpha^\beta = \begin{cases} 0 & \text{for } \alpha \neq \beta, \\ 1 & \text{for } \alpha = \beta, \end{cases} \tag{III}$$

$$\sum_\beta p_\alpha^\beta(t) p_\beta^\gamma(t') = p_\alpha^\gamma(t+t'). \tag{IV}$$

In addition to these relations of an algebraic character we will assume that the continuity condition holds:

$$p_\alpha^\beta(t) \to p_\alpha^\beta(0) = e_\alpha^\beta \quad \text{as } t \to 0. \tag{V}$$

Although (V) requires continuity of the functions $p_\alpha^\beta(t)$ only at $t = 0$, it can easily be seen that the conditions (I)–(V) imply that these functions are continuous for all $t \geq 0$.

In the case of a finite number of states, that is, when α, β and γ in (I)–(V) run only through $1, 2, \ldots, n$, (I)–(V) imply (see [2]) that the $p_\alpha^\beta(t)$ are differentiable for all $t \geq 0$ (naturally, only on the right for $t = 0$), that is, in particular, that the limits

$$a_\alpha = -\left[\frac{d}{dt} p_\alpha^\alpha(t)\right]_{t=0} = \lim_{t \to 0} \frac{1 - p_\alpha^\alpha(t)}{t}, \tag{1}$$

$$a_\alpha^\beta = \left[\frac{d}{dt} p_\alpha^\beta(t)\right]_{t=0} = \lim_{t \to 0} \frac{p_\alpha^\beta(t)}{t} \quad \text{for } \beta \neq \alpha, \tag{2}$$

exist. These limits satisfy the relations

$$a_\alpha = \sum_{\beta \neq \alpha} a_\alpha^\beta \tag{3}$$

* *Uch. Zap. Moskov. Gos. Univ. Mat.* **148**:4 (1951), 53–59.

and have the well-known probabilistic meaning of probability densities of leaving the αth state (for the a_α) and probability densities of transition from the αth state to the βth state (for the a_α^β). The transition probabilities $p_\alpha^\beta(t)$ themselves are uniquely determined by the transition densities a_α^β (in terms of which the a_α are defined by (3)) as solutions of any of the two systems of differential equations

$$\frac{d}{dt}p_\alpha^\beta(t) = -a_\alpha p_\alpha^\beta(t) + \sum_{\gamma\neq\alpha} a_\alpha^\gamma p_\gamma^\beta(t), \tag{4}$$

$$\frac{d}{dt}p_\alpha^\beta(t) = -p_\alpha^\beta(t)a_\beta + \sum_{\gamma\neq\beta} p_\alpha^\gamma(t)a_\gamma^\beta, \tag{5}$$

with initial values (III).

After my work [1] the systems (4) and (5) became the main apparatus for studying transition probabilities $p_\alpha^\beta(t)$ for a countable set of states as well, that is, when the indices α, β, γ in (I)–(V) run through the positive integers $1, 2, 3, \ldots$. However, in the general case even the preliminary questions of the existence of the limits (1) and (2), the differentiability of the transition probabilities $p_\alpha^\beta(t)$ for $t > 0$ and the applicability of formula (3) and the differential equations (4) and (5) remained open for a long time. Apparently, if we confine ourselves to the assumptions (I)–(V) and introduce no additional restrictions, then the case with a countable number of states is as follows:

(A) the limits a_α always exist, but there are some examples when they are equal to ∞;

(B) the a_α^β always exist for $\beta \neq \alpha$ and are finite;

(C) always

$$\sum_{\beta\neq\alpha} a_\alpha^\beta \leq a_\alpha, \tag{6}$$

but even for finite a_α there are examples when

$$\sum_{\beta\neq\alpha} a_\alpha^\beta < a_\alpha. \tag{7}$$

(D) the $p_\alpha^\beta(t)$ always have finite derivatives for $t \geq 0$;

(E) if all the a_α are finite, then (4) holds if the equations (3) hold;

(F) there are examples when the system (5) is not satisfied, even if all the a_α are finite and all the equations (3) hold;

(G) there are examples of two different systems of functions $p_\alpha^\beta(t)$ satisfying (I)–(V) with identical and finite a_α and a_α^β satisfying (3).

Of the above statements, only (D) remains conjectural. The others were partially proved by Doob (see [3], [4]), and are established completely in the present paper.

Proof of (A). The existence of finite or infinite limits was proved by Doob. We shall give another, quite elementary, proof (Doob's proof makes use of measures in function spaces).

Set

$$\tilde{a}_\alpha = \liminf_{t\to 0} \frac{1 - p_\alpha^\alpha(t)}{t}. \tag{8}$$

Clearly, $\tilde{a}_\alpha \geq 0$. If $\tilde{a}_\alpha = \infty$, then

$$\tilde{a}_\alpha = \lim_{t\to 0} \frac{1 - p_\alpha^\alpha(t)}{t} \tag{9}$$

and $a_\alpha = \tilde{a}_\alpha$ is the desired limit.

If $\tilde{a}_\alpha < \infty$, then for any $t > 0$,

$$(1 - p_\alpha^\alpha(t))/t \leq \tilde{a}_\alpha. \tag{10}$$

Indeed, take arbitrary $t > 0$ and $\epsilon > 0$ and choose h such that

$$(1 - p_\alpha^\alpha(h))/h \leq \tilde{a}_\alpha + \epsilon/2, \tag{11}$$

$$p_\alpha^\alpha(s) > 1 - (\epsilon/2)t \tag{12}$$

for $s < h$. We write t in the form

$$t = nh + s,$$

where $n \geq 0$ is an integer and $s < h$. Then

$$\begin{aligned} p_\alpha^\alpha(t) &\geq \{p_\alpha^\alpha(h)\}^n p_\alpha^\alpha(s) \geq \left\{1 - \left(\tilde{a}_\alpha + \frac{\epsilon}{2}\right)h\right\}^n \left(1 - \frac{\epsilon}{2}t\right) \geq \\ &\geq \left\{1 - n\left(\tilde{a}_\alpha + \frac{\epsilon}{2}\right)h\right\}\left(1 - \frac{\epsilon}{2}t\right) \geq \left\{1 - \left(\tilde{a}_\alpha + \frac{\epsilon}{2}\right)t\right\}\times \\ &\times\left(1 - \frac{\epsilon}{2}t\right) \geq 1 - (\tilde{a}_\alpha + \epsilon)t, \end{aligned}$$

$$(1 - p_\alpha^\alpha(t))/t \leq \tilde{a}_\alpha + \epsilon. \tag{13}$$

Since $\epsilon > 0$ is arbitrary, (13) implies (10). Formulas (10) and (8) imply (9). Thus the existence of the limit $a_\alpha^\alpha = -\tilde{a}_\alpha = -a_\alpha$ is also proved for finite $\tilde{a}_\alpha$.

Whether the case $a_\alpha = \infty$ is possible remains an open question in Doob's work. The affirmative answer is given by the following:

Lemma. *Suppose that for $\alpha \geq 2$ the $a_\alpha > 0$ are finite and are such that*

$$\sum_{\alpha=2}^{\infty} \frac{1}{a_\alpha} < \infty. \tag{14}$$

Then there exists a Markov process satisfying (I)–(V), *with given a_α for $\alpha \geq 2$,*

$$a_1 = \infty, \quad a_1^\beta = 1 \textit{ for } \beta \geq 2, \quad \left.\begin{matrix} a_\alpha^1 = a_\alpha \\ a_\alpha^\beta = 0 \end{matrix}\right\} \textit{ for } \alpha \geq 2,\ \alpha \neq \beta \geq 2.$$

The $p_\alpha^\beta(t)$ satisfying these requirements may be constructed in the following way. Setting

$$p_1^1(t) = \phi(t), \tag{15}$$

we assume that for $\beta \geq 2$,

$$\frac{d}{dt} p_1^\beta(t) = \phi(t) - a_\beta p_1^\beta(t),$$

which, together with

$$p_1^\beta(0) = 0,$$

gives

$$p_1^\beta(t) = \int_0^t \phi(s) e^{-a_\beta(t-s)} ds. \tag{16}$$

The requirement

$$\sum_\beta p_1^\beta(t) = p_1^1(t) + \sum_{\beta \geq 2} p_1^\beta(t) = 1$$

leads to the equation

$$\phi(t) + \int_0^t \phi(s) \sum_{\beta=2}^{\infty} e^{-a_\beta(t-s)} ds = 1. \tag{17}$$

Condition (14) implies that the kernel

$$k(\tau) = \sum_{\beta=2}^{\infty} e^{-a_\beta \tau}$$

has finite integral

$$\int_0^{\infty} k(\tau)d\tau = \sum_{\beta=2}^{\infty} \frac{1}{a_\beta}.$$

Therefore,[1] (17) has a continuous solution $\phi(t)$ for $t \geq 0$, which can be obtained in the usual way via the Fourier transformation. It can easily be shown that this solution is continuous and satisfies the conditions

$$\phi(0) = 1, \quad \phi'(0) = -\infty.$$

To complete our construction, it suffices to attach to (15) and (16) the $p_\alpha^\beta(t)$, for $\alpha \geq 2$ satisfying the equations[2]

$$\frac{d}{dt}p_\alpha^\alpha(t) = a_\alpha p_1^\alpha(t) - a_\alpha p_\alpha^\alpha(t),$$
$$\frac{d}{dt}p_\alpha^\alpha(t) = p_\alpha^1(t) - a_\alpha p_\alpha^\alpha(t),$$
$$\frac{d}{dt}p_\alpha^\beta(t) = p_\alpha^1(t) - a_\beta p_\alpha^\beta(t),$$

which can easily be integrated one by one (with initial values (III)). The probabilities $p_\alpha^\beta(t)$ so constructed satisfy all the requirements (I)–(V).

Proof of (B).

Lemma. *Suppose that for* $t \leq H$,

$$1 - p_\alpha^\alpha(t) \leq \epsilon, \quad 1 - p_\beta^\beta(t) \leq \epsilon. \tag{18}$$

Then for

$$nh \leq t \leq H,$$

where $n > 0$ *is an integer,*

$$p_\alpha^\beta(t) \geq np_\alpha^\beta(h)(1 - 3\epsilon). \tag{19}$$

---

[1] See E.C. Titchmarsh, *Introduction to the theory of Fourier integrals*, Oxford Univ. Press, Oxford, 1948, §11.5.

[2] Here the second and third equations are taken from (5), and the first one from (4).

Proof. Let[3]

$$P_k = \sum_{\substack{\gamma_i \neq \beta \\ i=1,2,\ldots,k-1}} p_\alpha^{\gamma_1}(h) p_{\gamma_1}^{\gamma_2}(h) \ldots p_{\gamma_i}^{\gamma_{i+h}}(h) \ldots p_{\gamma_{k-1}}^{\beta}(h),$$

$$P = \sum_{k=1}^{n} P_k,$$

$$Q_k = \sum_{\substack{\gamma_i \neq \beta \\ 1 \leq i \leq k-1}} p_\alpha^{\gamma_1}(h) p_{\gamma_1}^{\gamma_2}(h) \ldots p_{\gamma_i}^{\gamma_{i+1}}(h) \ldots p_{\gamma_{k-1}}^{\alpha}(h).$$

It can easily be seen that

$$1 - p_\alpha^\alpha(t) \geq p_\alpha^\beta(t) \geq \sum_{k=1}^{n} P_k p_\beta^\beta(t - kh) \geq (1 - \epsilon)P. \tag{20}$$

Formulas (18) and (20) imply that

$$P \leq \epsilon/(1 - \epsilon). \tag{21}$$

On the other hand,

$$Q_k \geq p_\alpha^\alpha(kh) - \sum_{i=1}^{k-1} P_i \geq p_\alpha^\alpha(kh) - P, \tag{22}$$

which, together with (18) and (21) gives

$$Q_k \geq 1 - \epsilon - \epsilon/(1 - \epsilon). \tag{23}$$

Finally,

$$p_\alpha^\beta(t) \geq \sum_{k=1}^{n} Q_{k-1} p_\alpha^\beta(h) p_\beta^\beta(t - kh), \tag{24}$$

which, together with (18) and (23), gives

$$p_\alpha^\beta(t) \geq n p_\alpha^\beta(h) \Big(1 - \epsilon - \frac{\epsilon}{1 - \epsilon}\Big)(1 - \epsilon) \geq n p_\alpha^\beta(h)(1 - 3\epsilon),$$

as required.

---

[3] To understand the meaning of inequalities (20), (22) and (24) the reader should remember the probabilistic meaning of P_k and Q_k. These inequalities, however, can easily be proved also in a purely algebraic way, based on (I)–(IV).

Now assume that the conditions of the lemma are satisfied for a given H and consider an $h \leq H$. Let n be the integer part of H/h and let $t = nh$. Then

$$n \geq H/2h$$

and, by (19),

$$1 \geq p_\alpha^\beta(t) \geq np_\alpha^\beta(h)(1-3\epsilon),$$

that is,

$$\frac{p_\alpha^\beta(h)}{h} \leq \frac{1}{nh(1-3\epsilon)} \leq \frac{2}{H(1-3\epsilon)}.$$

Hence

$$\tilde{a}_\alpha^\beta = \limsup_{t\to 0} \frac{p_\alpha^\beta(t)}{t} < \infty. \tag{25}$$

For any $t \leq H$ choose h such that

$$p_\alpha^\beta(h)/h \geq \tilde{a}_\alpha^\beta(1-\epsilon), \tag{26}$$

$$h/t \leq \epsilon, \tag{27}$$

and denote by n the integer part of t/h. Then

$$nh \geq t(1-\epsilon)$$

and, by (19), for $\epsilon < 1$,

$$\frac{p_\alpha^\beta(t)}{t} \geq \frac{nh}{t}\frac{p_\alpha^\beta(h)}{h}(1-3\epsilon) \geq \frac{p_\alpha^\beta(h)}{h}(1-3\epsilon)(1-\epsilon) \geq$$

$$\geq \tilde{a}_\alpha^\beta(1-4\epsilon)(1-\epsilon). \tag{28}$$

For any ϵ such that $1 > \epsilon > 0$ (28) is proved for $t \leq H$, where H is sufficiently small. This, together with (25), implies that

$$\lim_{t\to 0} \frac{p_\alpha^\beta(t)}{t} = \tilde{a}_\alpha^\beta = a_\alpha^\beta.$$

Proof of (C). Inequality (6) is elementary and can be found in the work of Doob [3], [4].

Here is an example where (7) holds for $\alpha = 1$, though all the a_α are finite.

1) For $\alpha \geq 3$ the probabilities $p_\alpha^\alpha(t)$, $p_\alpha^\beta(t)$ for $3 \leq \beta < \alpha$ and $p_\alpha^2(t)$ are determined from the differential equations

$$\frac{d}{dt} p_\alpha^\alpha(t) = -a_\alpha p_\alpha^\alpha(t),$$
$$\frac{d}{dt} p_\alpha^\beta(t) = a_{\beta+1} p_\alpha^{\beta+1}(t) - a_\beta p_\alpha^\beta(t),\ \beta = \alpha-1, \alpha-2, \ldots, 3,$$
$$\frac{d}{dt} p_\alpha^2(t) = a_3 p_\alpha^3(t)$$

(with the usual initial values (III)), where the $a_\alpha > 0$ are chosen so that

$$\sum_{\alpha=3}^{\infty} \frac{1}{a_\alpha} < \infty. \tag{29}$$

2) For $\alpha \geq 2$ and $\beta = 1$ or $\beta > \alpha$, $p_\alpha^\beta(t) \equiv 0$.

3) $p_2^2(t) \equiv 1$.

4) $p_1^1(t) = e^{-t}$.

5) For $\beta > 1$,

$$p_1^\beta(t) = \int_0^t e^{-s} p_*^\beta(s) ds,$$

where

$$p_*^\beta(t) = \lim_{\alpha \to \infty} p_\alpha^\beta(t).$$

Since for these $p_\alpha^\beta(t)$

$$a_1 = -\left[\frac{d}{dt} p_1^1(t)\right]_{t=0} = 1,$$
$$a_1^\beta = \left[\frac{d}{dt} p_1^\beta(t)\right]_{t=0} = 0 \quad \text{for all } \beta > 1,$$

it follows that

$$\sum_{\beta \neq 1} a_1^\beta < a_1.$$

It can be verified that (I)–(V) hold in this example.

Statement (E) was proved by Doob (see [4], VIII). For the above example equation (4) fails. Indeed, in this example $a_1 = 1$ and $a_1^\beta = 0$ for $\beta \neq 1$. Therefore for $\alpha = 1$, (4) are of the form

$$dp_1^\beta(t)/dt = -p_1^\beta(t)$$

and, together with $p_1^\beta(0) = 0$ yields

$$p_1^\beta(t) = 0$$

for all $\beta \neq 1$, which is absurd.

Proof of (F) *and* (G). This can be found in Doob's work (see [4], Theorem 2.2).

References

1. A.N. Kolmogorov, 'On analytic methods in probability theory', *Uspekhi Mat. Nauk* **5** (1938), 5–41.
2. W. Doeblin, 'Sur l'équation matricielle $A^{(t+s)} = A^{(t)}A^{(s)}$', *Bull. Sci. Math.* **62** (1938), 21–32.
3. J.L. Doob, 'Topics in the theory of Markoff chains', *Trans. Amer. Math. Soc.* **52** (1942), 37–64.
4. J.L. Doob, 'Markoff chains — denumerable case', *Trans. Amer. Math. Soc.* **58** (1945), 455–473.

40. A GENERALIZATION OF POISSON'S FORMULA FOR A SAMPLE FROM A FINITE SET *

A box contains N balls, M of which are white and $N-M$ black. If n balls are taken out of the box at random, then the probability of obtaining m white balls is given by

$$P_n(m/N, M) = C_M^m C_{N-M}^{n-m} / C_N^n. \tag{1}$$

Although (1) is elementary, computations based on it are difficult, and it is replaced by the approximate relations

$$P_n(m/N, M) \sim P_n^*(m/N, M), \tag{2}$$

where P^* is some expression that is easier to compute or use than the precise formula for P. In what follows we indicate conditions when (2) becomes precise in the limit, meaning that

$$\sum_{m=0}^{n} |P - P^*| \to 0. \tag{3}$$

Most often the Laplace-type approximation is used:

$$P_n^*(m/N, M) = \frac{1}{\sqrt{2\pi}\sigma} e^{-(m-a)^2/2\sigma^2}, \tag{4}$$

where

$$a = np, \tag{5}$$

$$\sigma^2 = np(1-p)(N-n)/(N-1), \tag{6}$$

$$p = M/N. \tag{7}$$

The most widely known conditions for the applicability of approximation (4) are as follows:[1]

$$\lambda = n/N < \lambda_0 < 1, \quad a = np \to \infty, \quad n(1-p) \to \infty, \tag{8}$$

where λ_0 is assumed to be constant.

---

* *Uspekhi Mat. Nauk* **6**:3 (1951), 133–134.

[1] For Bernoulli's scheme, cf. P.A. Kozulyaev, *Uch. Zap. Moskov. Gos. Univ. Mat.* **15** (1939), 179–182.

The condition $\lambda \leq \lambda_0 < 1$ is not restrictive since for $\lambda \geq \frac{1}{2}$ we can consider the "complementary" sample from the remaining

$$n' = N - n$$

balls, for which $\lambda' \leq \frac{1}{2}$.

Therefore in all cases when $n \to \infty$ it only remains to consider the cases of bounded $a = np$ and bounded $n(1-p)$. The second of these cases is reduced to the first one by calling white balls black and black balls white. The aim of the present note is to show that if

$$n \to \infty, \quad a = np \leq c \tag{9}$$

the following generalized Poisson approximation is correct:

$$P_n^*(m/N, M) = \frac{a(a-\lambda)(a-2\lambda)\dots(a-(m-1)\lambda)}{m!(1-\lambda)^m} e^{-\omega a}, \tag{10}$$

where

$$\omega = -[\log(1-\lambda)]/\lambda. \tag{11}$$

Approximation (10) is also correct under more general conditions. Namely, for its applicability in the sense of (3) it suffices that

$$p \to 0. \tag{12}$$

Using the easily verified identity

$$\frac{a(a-\lambda)(a-2\lambda)\dots(a-(m-1)\lambda)}{m!(1-\lambda)^m} e^{-\omega a} = C_M^m \lambda^m (1-\lambda)^{n-m} \tag{13}$$

and the well-known duality

$$P_n(m/N, M) = P_M(m/N, n) \tag{14}$$

the problem is reduced to proving that the binomial approximation

$$P_n(m/N, M) = C_n^m p^m (1-p)^{n-m} \tag{15}$$

is applicable in the sense of (3) under the condition

$$\lambda \to 0. \tag{16}$$

41. SOME RECENT WORK ON LIMIT THEOREMS IN PROBABILITY THEORY *

Introduction

In the middle of the 1940's it seemed that the topic of limit theorems of classical type (that is, problems of the limiting behaviour of distributions of sums of a large number of terms that are either independent or connected in Markov chains) was basically exhausted. The monograph written by me together with B.V. Gnedenko [1] was intended to summarize the results of the previous years. In reality, however, starting from the late 1940's more papers in these classical fields appeared. This can be explained by several circumstances. First, it became clear that from a practical viewpoint the accuracy of remainder terms obtained so far was far from sufficient. Secondly, certain problems that were solved earlier only under complicated and very restrictive conditions unexpectedly obtained very simple complete solutions (in the sense of necessary and sufficient conditions). These include, for example, the problem of "localization" of limit theorems, which turn out to have an exhaustive solution both for the case of identically distributed independent summands and for the case of the distribution of the number of separate states visited in a homogeneous Markov chain.

Naturally, these results have been a stimulus for seeking similar complete results for a number of other cases. Finally, the very statements of the problems have become more refined and transparent, owing to the introduction of suitable distances between distributions and the ideas of computing least upper bounds of residue terms borrowed from the theory of best approximation.

This paper briefly presents the results of some recent papers that illustrate these new tendencies. The results given in the monograph [1] are supposed to be known to the reader, though sometimes they are mentioned in order to make the presentation more coherent.

§1. Estimates of the proximity of distributions and various kinds of convergence of distributions

If ξ is a random variable of a certain set X, then the function

$$P(A) = \mathbf{P}\{\xi \in A\},$$

* *Vestn. Moskov. Gos. Univ.* **10** (1953), 29–38.

defined for an appropriate system of subsets A of X and giving the probability that ξ belongs to A, is called the probability distribution of the random variable ξ. When X is a metric space (for example, the number axis R^1, or n-dimensional coordinate space R^n), we assume that $P(A)$ is only defined on the B-sets. As is known (in the one-dimensional case), $P(A)$ is uniquely determined by the corresponding distribution function

$$F(x) = \mathbf{P}\{\xi < x\}.$$

It is most natural to consider two distributions P_1 and P_2 to be close if the probabilities $P_1(A)$ and $P_2(A)$ are close for all admissible A. This understanding of being close corresponds to the distance between distributions

$$\rho_1(P_1, P_2) = \sup_A |P_1(A) - P_2(A)|,$$

where the supremum is taken over all A for which the probabilities are defined. In the one-dimensional case, $\rho_1(P_1, P_2)$ is equal to half the total variation of the difference $F_1 - F_2$:

$$\rho_1(P_1, P_2) = \tfrac{1}{2}\operatorname{var}(F_1 - F_2).$$

Convergence of distributions P_n to the distribution P determined by the requirement

$$\rho_1(P_n, P) \to 0$$

will be called *convergence in variation*. This convergence is the strongest of all convergences with a direct probabilistic meaning: if two distributions are close enough in terms of ρ_1, then for all real problems dealing with a limited (though quite large) number of tests, the distributions P_1 and P_2 are virtually equivalent.[1]

In the one-dimensional case we are sometimes concerned only with the probabilities $P(\Delta)$ for intervals. A new distance $\rho'(P_1, P_2)$ might be defined by the formula

$$\rho'(P_1, P_2) = \sup_\Delta |P_1(\Delta) - P_2(\Delta)|,$$

[1] Let P^n be the distribution of a system $(\xi_1, \xi_2, \ldots, \xi_n)$ of independent random variables ξ_k each of which obeys the distribution P. Then it is easy to prove that $\rho_1(P_1^n, P_2^n) \le n\rho_1(P_1, P_2)$, which shows that when $n\rho_1(P_1, P_2)$ is sufficiently small, the probability of any given outcome of n independent tests under the distribution P_1 differs only slightly from the probability of the same outcome under the distribution P_2.

where the supremum is taken over all intervals Δ. However, it is easy to prove that

$$\sup|F_1 - F_2| \leq \rho'(P_1, P_2) \leq 2\sup|F_1 - F_2|.$$

Therefore the use of ρ' is basically equivalent to the use of

$$\rho_2(P_1, P_2) = \sup|F_1 - F_2|.$$

Clearly,

$$\rho_2(P_1, P_2) \leq \rho_1(P_1, P_2).$$

Convergence in the sense that

$$\rho_2(P_n, P) \to 0$$

will be called *strong convergence* of distributions. Strong convergence is weaker than the convergence in variation.

In practice, random variables are usually measurable and are needed only to within a certain accuracy. From this point of view the distances ρ_1 and ρ_2 introduced above sometimes overestimate the practical significance of the difference between two distributions. For example, for the degenerate distributions

$$\epsilon_x(A) = \begin{cases} 1, & \text{if } x \in A, \\ 0, & \text{if } x \notin A, \end{cases}$$

for $a \neq b$ we always have

$$\rho_1(\epsilon_a, \epsilon_b) = \rho_2(\epsilon_a, \epsilon_b) = 1,$$

whereas for small $b - a$ it is natural to consider the distributions ϵ_a and ϵ_b as being close to each other.

This remark can be made more precise as follows. Let $P^{(\sigma)}$ be the distribution of the variable

$$\eta = \xi + \delta,$$

where ξ obeys the distribution P and δ (the measurement error) is independent of ξ and obeys the normal distribution with mean 0 and variance σ^2. We introduce two new types of convergence by the requirements

a) $P_n \to P$, if for any $\sigma > 0$

$$\rho_1(P_n^{(\sigma)}, P^{(\sigma)}) \to 0;$$

b) $P_n \to P$, if for any $\sigma > 0$

$$\rho_2(P_n^{(\sigma)}, P^{(\sigma)}) \to 0.$$

These two, formally different, types of convergence appear in fact to be equivalent to the well known *weak convergence*[2] of distributions.

We confine ourselves to distributions on the straight line and somewhat arbitrarily introduce errors that obey normal probability distributions. One might think, however, that in a more general setting weak convergence is in good agreement with the idea of closeness of distributions which takes into account practical equivalence (or identity) of close points of the set X.

Weak convergence is weaker than convergence in variation (in any metric space X) and also weaker than strong convergence with respect to the distance ρ_2 (on the number axis). Weak convergence of distributions in any metric space can be defined by a properly introduced distance. Lévy's distance $L(P_n, P)$ (see [1], §9) can serve for this purpose on the number axis. However, all such distances are artificial and non-invariant with respect to the simplest transformations of X (for instance, shifts (translations) on the number axis). This distinguishes weak convergence from convergence in variation and strong convergence, defined above (for the number axis), which are connected with quite natural definitions of distance between distributions.[3]

§2. A catalogue of possible limiting distributions. Distribution families sufficient for uniform approximation

For a long time general limit theorems were being found "by feel". Only after the limiting distribution for a particular case was computed was the problem raised of finding general conditions for convergence to this distribution. The first example of a different approach I am aware of is Lévy's theorem on limiting distributions of the successive sums

$$\zeta_n = \xi_1 + \xi_2 + \ldots + \xi_n$$

[2] We say that in an arbitrary metric space X distributions P_n *weakly converge* to a distribution P if $\int_X f dP_n \to \int_X f dP$ for any bounded continuous function $f(x)$.

[3] Clearly, the distance ρ_1 is invariant with respect to all one-to-one transformations of the set X onto itself. The distance ρ_2 is invariant with respect to one-to-one transformations of the straight line onto itself.

of the terms of a sequence

$$\xi_1, \xi_2, \ldots, \xi_n$$

of independent identically distributed terms (see [1], §33). The theorem states that the distributions of the variables

$$(\zeta_n - A_n)/B_n$$

can only converge weakly to a stable distribution, and to any such distribution (under an appropriate choice of the distribution of the terms ξ_n). It was only much later that W. Doeblin and B.V. Gnedenko fully studied conditions under which this convergence holds at all or there is convergence to a certain definite stable law (see [1], §35).

Even more important is A.Ya. Khinchin's theorem, which characterizes a class of possible limit laws for the distributions of sums

$$\zeta_n = \xi_{n1} + \xi_{n2} + \ldots + \xi_{nm_n}$$

of independent "*infinitesimally small*" terms (see [1], §24). By virtue of Khinchin's theorem, this class coincides with the class of infinitely divisible distributions.[4] The theorems of Lévy and Khinchin deal with weak convergence. It is clear that passing to a stronger convergence can only narrow the class of possible limit laws. In reality, when we pass to strong convergence in the sense of the distance ρ_2 or even to convergence in variation, this class of laws remains the same.

There is the perfectly justifiable view that for sums of independent terms Khinchin's theorem indicates the most general class of limiting distributions that are of interest. Soon after this theorem was established, B.V. Gnedenko completed the study of conditions for convergence to any infinitely divisible distribution (see [1], §25–27), which essentially exhausted the question of the limiting behaviour of the sums of an increasing number of independent terms each of which is "negligible in the limit".

For schemes with independent tests similar questions are not yet sufficiently studied. A recent paper by R.L. Dobrushin [2] can serve as an example

---

[4] The fact that we do not include in the expression for ζ_n the non-random term A_n (cf. (1) on p.119 in [1]) does not make any difference: Khinchin's theorem holds in this form as well.

of a completely definitive study for a somewhat particularized scheme. He studied the case of sums

$$\zeta_n = \xi_{n1} + \xi_{n2} + \ldots + \xi_{nn},$$

where each ξ_{nk} takes only two values, 0 and 1, and are connected within each sequence in a simple homogeneous Markov chain with matrix of transition probabilities

$$\begin{vmatrix} p_n & 1-p_n \\ q_n & 1-q_n, \end{vmatrix}$$

depending only on the number n of the sequence, but not on the number k of the test in the sequence. If the ξ_{nk} are independent (that is, when $p_n = q_n$) then only the degenerate distribution ϵ_n, the normal distribution, the Poisson distribution and distributions obtainable from these three by linear transformations[5] can serve as limit distributions for

$$\eta_n = (\zeta_n - A_n)/B_n.$$

For arbitrary p_n and q_n the problem becomes much more difficult and in order to enumerate all possible limit distributions of η_n a number of new ingenious considerations are required.

The study of similar questions for the case of Markov chains with any finite number of states (not necessarily homogeneous) was started by Koopman (see, for instance [14]). It would be very interesting to obtain as complete results as those of R.L. Dobrushin for $s = 2$, at least for the homogeneous case with an arbitrary number of states s.

In estimating the proximity of distributions with the help of some distance $\rho(P_1, P_2)$ between distributions we can raise the question of convergence not to an individual distribution, but to a whole class of distributions. For instance, we can ask if the distribution of the sum

$$\zeta = \xi_1 + \xi_2 + \ldots + \xi_n$$

of independent identically distributed summands ξ_k should necessarily be close to an infinitely divisible distribution in the sense of some distance ρ as $n \to \infty$.

[5] This result by P.A. Kozulyaev is contained in general theorems of §26 and §27 in [1].

More exactly, the question is whether for any $\epsilon > 0$ there exists an N such that for any $n \geq N$ there exists an infinitely divisible distribution S such that for the distribution P of the sum ζ we have

$$\rho(P, S) < \epsilon.$$

So far there are few papers in which problems of this sort are solved in a setting essentially non-reducible to limit theorems of the usual type. Note that in Dobrushin's work mentioned above, the question of approximating the distributions of the sums ζ_n is solved (for the case he considered) also in the sense of this kind of uniform approximation (as $n \to \infty$) with respect to the distance ρ_1.

Dobrushin's result is quite complicated because of case-by-case checking of many particular distributions. The principle underlying his result can be clarified by the example of a particular case of independent variables ξ_{nk} considered earlier by Yu.V. Prokhorov [3]. In this case we speak about uniform approximation in the sense of the distance ρ_1 of the binomial distribution B_{np}, given by the formula

$$\mathsf{P}\{\zeta = z\} = C_n^z p^z (1-p)^{n-z}.$$

Denote by L_{np} the approximation to B_{np} used in the local Laplace theorem, P_{np} the approximation to B_{np} in Poisson's sense (used for p close to zero), and $\widetilde{P}_{np}$ the approximation obtained by Poisson's approximation of $n-\zeta$ (which is natural for p close to 1). It is reasonable to use each of these approximations when the corresponding distance

$$\rho_1(B_{np}, L_{np}), \quad \rho_1(B_{np}, P_{np}), \quad \rho_1(B_{np}, \widetilde{P}_{np})$$

is smaller. Prokhorov's theorem states that

$$\epsilon_{np} = \min\{\rho_1(B_{np}, L_{np}), \rho_1(B_{np}, P_{np}), \rho_1(B_{np}, \widetilde{P}_{np})\}$$

tends to zero uniformly with respect to p as $n \to \infty$. Thus, the family of distributions L_{np}, P_{np} and $\widetilde{P}_{np}$ is *sufficient for uniform approximation* of B_{np} as $n \to \infty$.

In a simpler setting, when the approximating distributions are linear transforms of the normal distribution given by the distribution function

$$\Phi(x) = \frac{1}{\sqrt{2\pi}} \int_{-\infty}^{x} e^{-x^2/2} dx,$$

and ρ_2 is taken for the distance, these theorems on uniform approximation are not essentially new (see §4). However, a deliberate search for families sufficient for uniform approximation in more complicated cases has been started quite recently, and there still remains much to be done in this direction.

§3. Local theorems and convergence in variation

Any distribution $P(A)$ on the straight line can be expressed as

$$P(A) = \int_A p(x)dx + \psi(A),$$

where $p(x)$ is the corresponding probability density and $\psi(A)$ is the singular part of the distribution. If $\psi \equiv 0$, then P is continuous. If at least one of the distributions P_1 and P_2 is continuous, then

$$\tfrac{1}{2}\int_{-\infty}^{\infty} |p_1 - p_2|dx \le \rho_1(P_1, P_2) \le \int_{-\infty}^{\infty} |p_1 - p_2|dx.$$

Therefore for a continuous limit distribution P, convergence in variation

$$\rho_1(P_n, P) \to 0$$

is equivalent to convergence in mean of the densities

$$\int_{-\infty}^{\infty} |p_n - p|dx \to 0.$$

A number of researchers have considered local theorems of convergence of probability densities to the probability density of the limit distribution.[6] They have usually been concerned with conditions of uniform convergence of distributions. As was mentioned above, this is excessive from the probabilistic viewpoint. In Yu.V. Prokhorov's work [4] it is shown that passing to the convergence in variation (that is, in the case of a continuous limit distribution, to the convergence of densities in the mean) we arrive at very simple necessary and sufficient conditions of convergence for the case of identically distributed independent summands. Prokhorov demonstrated that under the conditions of Lévy's theorem given above for the convergence of the distributions of the

[6] See [1], §46 and §47 and the Hungarian translation of this book, in which the formulations of local theorems are improved.

variables $(\zeta_n - A_n)/B_n$ to a stable distribution, two requirements are necessary and sufficient: 1) weak convergence to S; and 2) for some n the distribution of ζ_n must have a non-zero continuous component.

Together with local theorems for densities, local theorems of discrete arithmetic type for "lattice" distributions (see [1], §§48–50) are usually considered. Local theorems of this type are in connection with transferring from a continuous distribution with density $p(x)$ to the "lattice"

$$x_k = kh + a.$$

This operation can be performed, for example, by ascribing the probability

$$P_k = \int_{x_k - h/2}^{x_k + h/2} p(x)dx$$

to each point x_k of this lattice.

If in the setting of Lévy's theorem the ξ_n are lattice-like with step h, then

$$\eta_n = (\zeta_n - A_n)/B_n$$

are lattice-like with step

$$h_n = h/B_n.$$

B.V. Gnedenko [5] has shown that when the step h for ξ_n is maximal and the distributions of η_n converge weakly to a stable distribution S, then the variational distance between the distribution F_n of η_n and the distribution S_n obtained by transferring S to the lattice of possible values of η_n tends to zero as $n \to \infty$:

$$\rho(F_n, S_n) \to 0.$$

The theorems of Prokhorov and Gnedenko are limited to the case of identically distributed summands. But in any case they show that more general and complete results on "the localization of limit theorems" are possible than it seemed before. To conclude this section, note also the papers on arithmetic local theorems [6]–[10]. The first three of them, [6]–[8], deal with multidimensional generalizations of Gnedenko's result mentioned above. The fourth, [9], gives an exhaustive answer to the question of approximating in variation the distribution of the number of distinct states visited in a homogeneous Markov chain with a finite number of states. The fifth, [10], contains quite deep, though less complete, results for inhomogeneous Markov chains.

§4. Precise estimates of remainder terms and improvement of approximations

In [3] Prokhorov considers not only

$$\epsilon_{np} = \min\{\rho_1(B_{np}, L_{np}), \rho_1(B_{np}, P_{np}), \rho_1(B_{np}, \widetilde{P}_{np})\},$$

but also

$$\epsilon_n = \sup_{0<p<1} \epsilon_{np}$$

and shows that

$$\epsilon_n = \lambda_n^{-1/3} + O(n^{-2/3}),$$

where

$$\lambda = \left(\frac{2}{9}\right)^{1/3} \frac{1}{\sqrt{2\pi}}(1 + 4e^{-3/2})^{2/3} e^{-1/6} = 0.42\ldots.$$

Thus, in terms of the variational distance he found an asymptotically precise estimate of the largest deviation (for given n and variable p) of the binomial distribution from the most suitable of the approximations L_{np}, P_{np} and $\widetilde{P}_{np}$.

This somewhat narrow result can serve as an example of the technique for estimating remainder terms that have taken shape in the papers of the last decade. There are no complete results of the same kind for sums of arbitrarily distributed independent summands. Below we consider problems of this kind connected with the so-called Lyapunov's ratio.

For simplicity we assume that the sum

$$\zeta = \xi_1 + \xi_2 + \ldots + \xi_n$$

of independent summands ξ_k is normalized, that is, it has expectation 0 and variance 1. Then "Lyapunov's ratio" takes the form

$$L = \sum_{k=1}^{n} \mathsf{E}|\xi_k - \mathsf{E}\xi_k|^3.$$

It is well known that as $L \to 0$ the following limit relation for the distribution function $F(z)$ of the sum ζ holds:

$$F(z) \to \Phi(z) = \frac{1}{\sqrt{2\pi}} \int_{-\infty}^{z} e^{-z^2/2} dz.$$

We introduce the new variables

$$\rho(l, z) - \sup_{L=l} |F(z) - \Phi(z)|,$$

$$\rho(l) = \sup_{z} \rho(l, z) = \sup_{L=l} \rho_2(F, \Phi).$$

Cramér showed that

$$c_1 l < \rho(l) < c_2 l,$$

where c_1 and c_2 are positive constants. Therefore it is natural to consider $\rho(l)/l$, its least upper bound

$$c = \sup \frac{\rho(l)}{l}$$

and its lim sup,

$$c^* = \limsup_{l \to 0} \frac{\rho(l)}{l}.$$

The estimates

$$\rho_2(F, \Phi) \leq cL, \quad \rho_2(F, \Phi) \leq c^* L + o(L)$$

are best possible as regards the order of L.

Clearly

$$c^* \leq c.$$

After Berry's work [15], where a more particular problem was considered, Esseen [16] proved that

$$1/\sqrt{2\pi} \leq c \leq 15/2.$$

Of great interest is the question whether more precise estimates or computations of c and c^* are possible, especially for the functions

$$c(z) = \sup_{l} \frac{\rho(l, z)}{l} = \sup \frac{|F(z) - \Phi(z)|}{L},$$

$$c^*(z) = \limsup_{l \to 0} \frac{\rho(l, z)}{l} = \limsup_{L \to 0} \frac{|F(z) - \Phi(z)|}{L}.$$

If we impose certain restrictions on the character of the terms ξ_k, then the values of c and c^*, $c(z)$ and $c^*(z)$ may decrease. Assuming symmetry of the distribution of the terms and imposing certain other restrictions, whose necessity is not clear, Yu.V. Linnik [11] proved that

$$c^* = \frac{1}{\sqrt{2\pi}}, \quad c^*(z) = \frac{1}{\sqrt{2\pi}} e^{-z^2/2}.$$

Perhaps these formulas hold under the single restriction of symmetry of the distributions of the ξ. The first of these formulas might also be true in the general case. Even the conjecture

$$c = \frac{1}{\sqrt{2\pi}}$$

has not been disproved.

It is also very important to study from the same viewpoint (of obtaining uniformly precise or asymptotically precise estimates of remainder terms) approximations for the distributions of sums of a large number of independent summands, which make use of the corrections to normal distribution that depend on the highest moments and are expressed via Chebyshev-Hermite polynomials. Nothing has been done in this direction so far. On the other hand, a remarkable paper by S.H. Sirazhdinov [13], in which local theorems with specifications depending on the higher moments and giving remainder terms of order $1/n^{\alpha}$ with arbitrary large exponents α are extended to the distribution of the number of distinct states visited during n steps in a homogeneous Markov chain with a finite number of states.

3 September 1953

References

1. B.V. Gnedenko and A.N. Kolmogorov, *Limit distributions for sums of independent random variables*, Gostekhizdat, Moscow-Leningrad, 1949 (in Russian).
2. R.L. Dobrushin, *Izv. Akad. Nauk SSSR Ser. Mat.* **17** (1953), 285 (in Russian).
3. Yu.V . Prokhorov, *Uspekhi Mat. Nauk* **8**:3 (1953), 135–142 (in Russian).
4. Yu.V. Prokhorov, *Dokl. Akad. Nauk SSSR* **83** (1952), 797–800 (in Russian).
5. B.V. Gnedenko, *Dokl. Akad. Nauk SSSR* **71** (1950), 425–428 (in Russian).
6. D.G. Meĭzler, O.S. Parasyuk and E.L. Rvacheva, *Dokl. Akad. Nauk SSSR* **60** (1948), 1127–1128 (in Russian).
7. D.G. Meĭzler, O.S. Parasyuk and E.N. Rvacheva, *Ukr. Mat. Zh.* (1949), 9–20 (in Russian).
8. E.L. Rvacheva, *Proc. Inst. Math. Mekh. Uzbek. Akad. Nauk* No.10, Part I (1953) 106–121 (in Russian).

9. A.N. Kolmogorov, *Izv. Akad. Nauk SSSR Ser. Math.* **13** (1949), 281–300 (in Russian).
10. Yu.V. Linnik and N.A. Sapogov, *Izv. Akad. Nauk SSSR Ser. Mat.* **13** (1949), 533–566 (in Russian).
11. Yu.V. Linnik, *Izv. Akad. Nauk SSSR Ser. Mat.* **11** (1947), 111–138 (in Russian).
12. Yu.V. Prokhorov, *Izv. Akad. Nauk SSSR Ser. Mat.* **16** (1952), 281–292 (in Russian).
13. S.Kh. Sirazhdinov, *Dokl. Akad. Nauk SSSR* **84** (1952), 1143–1146 (in Russian).
14. B.O. Koopman, *Trans. Amer. Math. Soc.* **70** (1951), 277–290.
15. A.G. Berry, *Trans. Amer. Math. Soc.* **49** (1941), 122–136.
16. C.G. Esseen, *Acta Math.* **77** (1945), 1–125.

42. ON A.V. SKOROKHOD'S CONVERGENCE *

When describing the course of a real process in time with the help of a function $f(t)$ of time t that takes values in an appropriate "phase space" X, it is often natural and correct to assume that f has only discontinuities of the first kind (jumps). To study such processes in detail it is useful to introduce in a set D of functions with discontinuities of the first kind (for the sake of being specific we will consider such functions on the unit interval $0 \leq t \leq 1$) an appropriate topology.

The topology of uniform convergence, natural when studying continuous processes, appears to be too strong when studying processes with discontinuities of the first kind. For example, it is natural to require that the sequence of functions

$$f_n(t) = \begin{cases} x_1 & \text{for } t < t_n, \\ x_2 & \text{for } t > t_n, \end{cases}$$

where $t_n \to t_0$ as $n \to \infty$, converge to the function

$$f(t) = \begin{cases} x_1 & \text{for } t < t_0, \\ x_2 & \text{for } t > t_0, \end{cases}$$

since f_n differs from f for large n only by a small shift in the jump from the state x_1 to the state x_2. As is known, this convergence does not hold in the space of uniform convergence for $x_1 \neq x_2$.

On the other hand, the topology in D should not be too weak, since we wish to retain the most essential properties of $f \in D$ when passing to the limit. For example, we wish that under the assumption that f_n converges to f, the conditions

$$t_n \to t, \quad f_n(t_n + 0) - f_n(t_n - 0) \to c \neq 0$$

imply that

$$f(t + 0) - f(t - 0) = c.$$

Even for these two assumptions we need a new topology in D, adapted especially to the needs of studying processes with discontinuities of the first kind. This kind of topology is suggested in the papers [1], [2] by A.V. Skorokhod. In these papers Skorokhod defines a certain convergence in D (we will call it S-convergence) for the case of real functions $f(t)$ (that is, when X is the

* *Teor. Veroyatnost. i Primenen.* (Probability Theory and its Applications) 1:2 (1956), 239–247 (in Russian).

real line) and notes that this convergence turns the set D into a topological Hausdorff space. Further, in §2, a new definition of S-topology is given which can be used under the assumption that X is an arbitrary metric space and it is shown that the topological space SD arising in this way is a metric space. The distance $s(f,g)$ introduced in §2 for this purpose and the corresponding uniform topology seem to be quite natural and convenient in applications, despite the fact that even for a complete space X the metric space sD is not complete. At the end of §2 a simple necessary and sufficient condition for S-compactness is given.

The results of §3 (as well as Theorem IV of §2) are essentially based on the assumption that the phase space X is complete (for simplicity we will assume this from the very beginning).

In §3 the completion of the metric space sD is given a specific interpretation using D-curves in the space $R = T \times T$. This interpretation is used for proving the possibility of introducing into D another distance $s^*(f,g)$ inducing the same Skorokhod convergence and topology, but turning D into a complete metric space s^*D. When presenting the results of this paper at the seminar on probability theory at Moscow University, I raised the problem of the simplest possible explicit construction of the distance $s^*(f,g)$. This problem was solved by Yu.V. Prokhorov [3].

§1 contains some prerequisites necessary for the considerations of §3. The contents of §2 could be as well presented without introducing the set Θ, using instead a useful technique for representing a process with discontinuities of the first kind, for example, using functions of a real variable t that are continuous from the right:

$$f^+(t) = f(t+0).$$

§1. The class of functions $D = D_X$

In the description of real processes using functions $f(t)$ with discontinuities of the first kind we can usually assume that the real phenomena taking place at the time t of the jump are fully described by the limit values from the left and from the right: $f(t-0)$ and $f(t+0)$. It is only by mathematical convention that the value $f(t)$ is defined either by $f(t) = f(t-0)$ or $f(t) = f(t+0)$, or for real functions, by

$$f(t) = \tfrac{1}{2}[f(t-0) + f(t+0)].$$

Actually, it is more logical to consider functions $f(\theta)$ of the symbols θ of the type $t-0$ or $t+0$, and only in the case $f(t-0) = f(t+0)$ denote their common value as $f(t)$, thus considering $f(t)$ undefined if $f(t-0) \neq f(t+0)$. In the presentation of the theory of D-curves in §3 this approach has also formal advantages: certain additional efforts made in §1 are rewarded by a more concise and graphic presentation in §3.

Let Θ be the set of symbols θ of one of the two types:

$$\theta = t-0,\ 0 < t \leq 1,$$
$$\theta = t+0,\ 0 \leq t < 1.$$

An order relation on Θ is defined by

1) $t \pm 0 < t' \pm 0$, if $t < t'$,
2) $t-0 < t+0$.

We say that

$$\theta_n \to t-0,$$

if for any $\epsilon > 0$ there exists $n(\epsilon)$ such that for $n \geq n(\epsilon)$,

$$t-\epsilon+0 \leq \theta_n \leq t-0,$$

and that

$$\theta_n \to t+0$$

if for any $\epsilon > 0$ there exists an $n(\epsilon)$ such that for $n \geq n(\epsilon)$,

$$t+0 \leq \theta_n \leq t+\epsilon-0.$$

This convergence turns the set Θ into a compact topological space (see [5]). The fact that this space is not metrizable is no hindrance to its use in further constructions.

By definition, the class D_X consists of functions $f(\theta)$ defined on Θ, with values in a metric space X and continuous in the sense that

$$\theta_n \to \theta$$

implies

$$f(\theta_n) \to f(\theta).$$

It can easily be proved that for any $f \in D = D_X$ the function

$$f^+(t) = f(t+0)$$

is defined for all t in the half-interval $0 \le t < 1$, is right continuous and has a left limit,

$$f^+(t-0) = f(t-0),$$

at every point t of the half-interval $0 < t \le 1$. It is clear that the correspondence between the functions $f \in D$ and f^+ with these properties is one-to-one.

For any set $M \subseteq \Theta$ we define the "oscillation" of $f \in D$ on M by setting

$$\omega f(M) = \sup_{x, x' \in M} \rho(x, x').$$

The "intervals" of the space Θ, that is, the sets

$$[\theta', \theta''] = \{\theta; \theta' \le \theta \le \theta''\}$$

are of special interest in what follows. In particular, the interval $[t-0, t+0]$ consists of two points: its end points $t-0$ and $t+0$. For this interval

$$\omega f[t-0, t+0] = |f(t+0) - f(t-0)|.$$

It is easy to prove

Theorem 1. *A function $f(\theta)$ defined on Θ and with values in X belongs to the class D if and only if for any $\epsilon > 0$ there exists a sequence*

$$0 = t_0 < t_1 < \ldots < t_n = 1,$$

such that for all $k = 1, 2, \ldots, n$,

$$\omega f[t_{k-1} + 0,\ t_k - 0] < \epsilon.$$

§2. *S*-convergence and the distance $s(f, g)$

Definition 1. Two functions $f \in D$ and $g \in D$ are called *ϵ-equivalent* if there exist r and sequences

$$0 = t_0 < t_1 < \ldots < t_r = 1, \quad 0 = t'_0 < t'_1 < \ldots < t'_r = 1$$

such that for $k = 1, 2, \dots, r$, the following inequalities hold:

$$|t_k - t'_k| \leq \epsilon,$$

$$\sup\{\rho(f(\theta), g(\theta')) \leq \epsilon : \theta \in [t_{k-1}+0, t_k - 0],\ \theta' \in [t'_{k-1}+0, t'_k - 0]\}.$$

In what follows ϵ-equivalence is denoted by

$$f \overset{\epsilon}{\sim} g.$$

Definition 2. A sequence of functions $f_n \in D$ is called *S-convergent* to $f \in D$ if for any $\epsilon > 0$ there exists $n(\epsilon)$ such that for $n \geq n(\epsilon)$,

$$f_n \overset{\epsilon}{\sim} f.$$

In what follows S-convergence is denoted by

$$f_n \xrightarrow{S} f.$$

This is the convergence Skorokhod introduced for the case when X is the real line.

We now set

$$s(f,g) = \inf_{f \overset{\epsilon}{\sim} g} \epsilon.$$

Theorem 2. *The set D with the distance $s(f,g)$ is a metric space.*

In fact, it can easily be seen that $s(f,g)$ is real and non-negative for any $f \in D$ and $g \in D$. Theorem 1 implies that

$$s(f,f) = 0.$$

It is fairly easy to show that

$$s(f,g) = 0$$

only if $f = g$. The symmetry of $s(f,g)$ is clear from the definition. The inequality

$$s(f,h) \leq s(f,g) + s(g,h)$$

holds because of the following lemma, which is easily proved:

Lemma. *If*

$$f \overset{\epsilon}{\sim} g, \quad g \overset{\epsilon'}{\sim} h,$$

then

$$f \overset{\epsilon+\epsilon'}{\sim} h.$$

The following theorem is obvious.

Theorem 3. *For the convergence*

$$f_n \xrightarrow{S} f$$

it is necessary and sufficient that

$$s(f_n, f) \to 0.$$

If X contains two different points $x_1 \neq x_2$, then the metric space sD is not complete, since for

$$t_n < t_0 < 1, \; t_n \to t_0$$

the sequence of functions

$$f_n(\theta) = \begin{cases} x_1 & \text{for } 0+0 \leq \theta \leq t_n - 0, \\ x_2 & \text{for } t_n + 0 \leq \theta \leq t_0 - 0, \\ x_1 & \text{for } t_0 + 0 \leq \theta \leq 1 - 0 \end{cases}$$

satisfies the Cauchy criterion but does not converge. In this connection the following theorem is interesting.

Theorem 4. *A sequence $f_n \in D$ satisfying the Cauchy criterion*

$$\lim_{n\to\infty} \sup_{p\geq 0} s(f_n, f_{n+p}) = 0$$

converges to $f \in D$ if and only if the following condition holds:

(*) *for any $\epsilon > 0$ there exists $\delta > 0$ such that for any n and $t \in (0;1)$*

$$\omega f_n[0+0, \delta - 0] < \epsilon, \quad \omega f_n[1-\delta+0, 1-0] < \epsilon,$$

$$\omega^* f_n[t+0, t+\delta-0] < \epsilon,$$

where

$$\omega^* f[\theta, \theta'] = \inf_t \max\{\omega f[\theta, t-0], \omega f(t+0, \theta')\}.$$

It should be noted that the operations $\omega^* f$ are similar to the γ_f that were already used by E.B. Dynkin in [4] for studying the functions of the class D.

Theorem 4 is closely connected with

Theorem 5. *A set $M \subseteq D$ is compact in the space SD of Skorokhod convergence if and only if the following two conditions hold:*

1) *there exists a compact set $K \subseteq D$ such that all values $f(\theta)$ of the function $f \in M$ at the points $\theta \in \Theta$ belong to K;*

2) *for any $\epsilon > 0$ there exists $\delta > 0$ such that for $f \in M$ we have*

$$\omega f[0+0, \delta - 0] < \epsilon, \quad \omega f[1-\delta+0, 1-0] < \epsilon,$$

$$\omega^* f[t+0, t+\delta-0] < \epsilon.$$

§3. Graphs of functions of class D and D-curves in the space $R = T \times T$

Let T be the unit interval

$$T = \{t; 0 \leq t \leq 1\}$$

of the real line and denote by R the direct product

$$R = T \times T = \{(t, x);\ t \in T,\ x \in X\}.$$

We will consider this direct product as a metric space with distance

$$\rho((t,x),(t',x')) = \max\{|t'-t|, \rho(x,x')\}.$$

The set Γ_f of points from R of any of the two types

$$(t, f(t-0)) \quad \text{or} \quad (t, f(t+0))$$

will be called the graph of the function $f \in D$. It can easily be seen that the function

$$\phi_f(t-0) = (t, f(t-0)), \quad \phi_f(t+0) = (t, f(t+0))$$

defined on Θ and with values in R is continuous in the sense of §1. It maps Θ onto Γ_f, and the order on Θ (the relations $\theta < \theta'$ defined in §1) induces a certain direction of travel along Γ_f.

Quite apart from the graphs of functions $f \in D = D_X$, we now consider mappings of Θ into the space R that are are continuous in the sense of §1, that is, functions $\phi(\theta)$ of class D_R. Two functions $\phi \in D_R$ and $\psi \in D_R$ are called equivalent if there exist monotone mappings χ_1 and χ_2 (that is, mappings χ satisfying the condition $\chi(\theta') \geq \chi(\theta)$ when $\theta' > \theta$) of Θ onto itself such that

$$\phi(\chi_1(\theta)) = \psi(\chi_2(\theta)).$$

With respect to this equivalence relation D_R is divided into classes $\tilde{\phi}$, which we call *D-curves* in R. It is natural to introduce the distance between D-curves by the formula

$$\rho(\tilde{\phi}, \tilde{\psi}) = \inf_{\substack{\phi \in \tilde{\phi} \\ \psi \in \tilde{\psi}}} \sup_{\theta \subset \Theta} \rho(\phi(\theta), \psi(\theta)).$$

The theory of D-curves is quite similar to the theory of continuous curves. The distance $\rho(\tilde{\phi}, \tilde{\psi})$ turns the set $\tilde{\Phi}$ of all D-curves in R into a complete metric space.

Theorem 6. *A function $f \in D$ is uniquely determined by the corresponding curve*

$$\tilde{f} = \tilde{\phi}_f,$$

and

$$s(f, g) = \rho(\tilde{f}, \tilde{g}).$$

Theorem 6 gives a new interpretation of the distance $s(f, g)$ and S-convergence: S-convergence of a sequence of functions f_n to f is equivalent to convergence in the sense of the distance $\rho(\tilde{\phi}, \tilde{\psi})$ of the curves $\tilde{f}_n$ to the curve $\tilde{f}$.

In what follows, it is useful to introduce the notations $\tau_\phi(\theta)$, $\xi_\phi(\theta)$ for the components of the function

$$\phi(\theta) = (\tau_\phi(\theta), \xi_\phi(\theta)).$$

Denote by Λ the set of D-curves $\tilde{f}$ corresponding to the functions $f \in D$, and by $\bar{\Lambda}$ the closure of Λ in $\tilde{\Phi}$. It can easily be proved that

$$\phi \in \tilde{\phi} \in \tilde{\Lambda}$$

if and only if

(λ_1) $\tau_\phi(\theta)$ maps θ onto the whole set T;

(λ_2) $\theta < \theta'$ implies that $\tau_\phi(\theta) \le \tau_\phi(\theta')$.

By (λ_1), (λ_2) for a function $\phi \in \tilde{\phi} \in \Lambda$ the sets $\tau_\phi^{-1}(t)$ of those θ for which

$$\tau_\phi(\theta) = t$$

are intervals of the space Θ, that is,

$$\tau_\phi^{-1}(t) = [\theta'_\phi(t), \theta''_\phi(t)].$$

When the parameter θ runs over $\tau_\phi^{-1}(t)$, the value $\tau_\phi(t)$ remains constant, while $\xi_\phi(\theta)$ runs over some sequence of points of the phase space X. Therefore the curves $\tilde{\phi} \in \bar{\Lambda}$ may be considered as the graphs of special generalized processes with discontinuities of the first kind whose behaviour at time t can, in general, be more complex than simple transition from the state $f(t-0)$ to the state $f(t+0)$. Such generalized processes can naturally appear as limiting restrictions similar to condition ($*$) of Theorem 4 that prevent the accumulation of several vanishingly small jumps at one point t. Possibly, introducing this kind of generalized process will appear useful in certain special studies on limit properties of random processes. Without developing this idea any further I use the analysis of the construction of $\bar{\Lambda}$ only as auxiliary means for proving the possibility of introducing, in the space SD, a distance $s^*(f,g)$ different from $s(f,g)$ and turning SD into a complete metric space. For this purpose I will prove the following lemma:

Lemma. *The set Λ is a set of type G_δ in the space $\tilde{\Phi}$.*

Proof. It can easily be shown that for given $t \in T$ and $\tilde{\phi} \in \bar{\Lambda}$ the value of

$$\omega^*_\phi[\tau_\phi^{-1}(t)]$$

is the same for all $\phi \in \tilde{\phi}$, that is, it characterizes the properties of the curve $\tilde{\phi}$, not of its parametric representation ϕ; we denote this value by $\omega^*_\phi(t)$. The condition

(μ) $\omega^*_\phi(t) = 0 \quad$ for all $t \in T$

is necessary and sufficient for $\tilde{\phi} \in \bar{\Lambda}$ to belong to Λ. Therefore

$$\Lambda = \bar{\Lambda} \setminus \bigcup_n \Lambda_n, \tag{1}$$

where Λ_n is the set of $\tilde{\phi} \in \bar{\Lambda}$ for which

$$\sup_{t \in T} \omega^*_{\tilde{\phi}}(t) \geq \frac{1}{n}.$$

The set $\bar{\Lambda}$ is closed in $\tilde{\Phi}$ by definition. It can be proved that the sets Λ_n are also closed in $\tilde{\Phi}$. Our lemma now follows immediately from the closedness of the sets $\bar{\Lambda}$ and Λ_n and (1).

By a well-known theorem of P.S. Aleksandrov this lemma implies that we can introduce a new distance $\rho^*(\tilde{f}, \tilde{g})$ in Λ that is topologically equivalent to the distance $\rho(\tilde{f}, \tilde{g})$ but turns it into a complete metric space.

Setting

$$s^*(f, g) = \rho^*(\tilde{f}, g)$$

and keeping Theorem 6 in mind, we can see that the following is true:

Theorem 7. *In the set D a distance $s^*(f, g)$ can be introduced that is topologically equivalent to $s(f, g)$ but turns D into a complete metric space.*

Moscow, 25 July 1956

References

1. A.V. Skorokhod, *Dokl. Akad. Nauk SSSR* **104** (1955), 364–367 (in Russian).
2. A.V. Skorokhod, *Dokl. Akad. Nauk SSSR* **106** (1956), 781–784 (in Russian).
3. Yu.V. Prokhorov, *Teor. Veroyatnost. i Primenen.* (Probability Theory and its Applications) **1**:2 (1956), 175–237 (in Russian).
4. E.B. Dynkin, *Izv. Akad. Nauk SSSR Ser. Mat.* **16** (1952), 563–572 (in Russian).
5. P.S. Urysohn, *Works on topology and other fields of mathematics*, Gostekhizdat, Moscow, 1951, Vol. 2 (in Russian).

43. TW0 UNIFORM LIMIT THEOREMS FOR SUMS OF INDEPENDENT TERMS *

In what follows

$$\Phi(x) = \mathbf{P}\{\xi < x\}, \quad F_k(x) = \mathbf{P}\{\xi_k < x\},$$

where

$$\xi = \xi_1 + \xi_2 + \ldots + \xi_n$$

and the random variables ξ_k are independent. Let $\mathfrak{E}$ be the family of degenerate distributions of the form

$$E(x) = \begin{cases} 0 & \text{for } x \le a, \\ 1 & \text{for } x > a, \end{cases}$$

and let Θ be the family of infinitely divisible distributions. The purpose of the present paper is to prove the following two theorems:

Theorem 1. *There exists a constant C such that if the inequalities*

$$E_k(x-l) - \epsilon \le F_k(x) \le E_k(x+l) + \epsilon,$$

hold for all $\epsilon > 0, L > 2l > 0$, where $E_k \in \mathfrak{E}$, $k = 1, 2, \ldots, n$, then there exists $\Psi \in \Theta$ for which

$$\Psi(x-L) - \delta \le \Phi(x) \le \Psi(x+L) + \delta,$$

where

$$\delta = C \max\Big(\frac{l}{L}\sqrt{\log\frac{L}{l}}, \epsilon^{1/5}\Big).$$

Theorem 2. *There exists a constant C such that for identically distributed summands ξ_k with arbitrary distribution functions $F_k(x) = F(x)$ there exists $\Psi \in \Theta$ such that for any x,*

$$|\Psi(x) - \Phi(x)| \le Cn^{-1/5}.$$

* *Teor. Veroyatnost. i Primenen.* (Probability theory and its applications) 1:4 (1956), 426–436 (in Russian).

The proof of each theorem makes use of a number of ideas borrowed from papers of P. Lévy (see [1], §48), W. Doeblin [2] and Yu.V. Prokhorov [3], and is based on the following lemmas, in which

$$G_{a,\sigma}(x) = \frac{1}{\sqrt{2\pi}\sigma}\int_{-\infty}^{x} e^{-(x-a)^2/2\sigma^2}\,dx, \quad G_\sigma = G_{0,\sigma},$$

$Q_F(l) = \sup_x[F(x+l+0)-F(x)]$ is the "concentration function" for the distribution $F(x)$ introduced by P. Lévy and $C_1, C_2, C_3, C_4, C_5, C_6, C_7, C'$ are absolute constants.

Lemma 1. *If*

$$R \geq r, \quad R^2 \geq r^2 \log s,$$

where[1]

$$s = \sum_k [1 - Q_{F_k}(r)],$$

then

$$Q_\Phi(R) \leq C_1 R / r\sqrt{s}.$$

Lemma 2.

$$|F * G_\sigma - F| \leq C_2 Q_F(\sigma).$$

Lemma 3.

$$|G_{a,s} - G_{a,\sigma}| \leq C_3(s^2 - \sigma^2)/\sigma^2.$$

Lemma 4. *If*

$$\eta = C_4 e^{-l^2/2\sigma^2},$$

then

$$F * G_{0,\sigma}(x - l) - \eta \leq F(x) \leq F * G_{0,\sigma}(x + l) + \eta.$$

Lemma 5. *Let*

$$p_{m_1 \ldots m_n} = \prod_k p_{m_k}^{(k)}, \quad q_{m_1 \ldots m_n} = \prod_k q_{m_k}^{(k)},$$

---

[1] Throughout, the index k in the sums $\sum_k$ and products $\prod_k$ or $\prod_k^*$ runs through all integers $1 \leq k \leq n$.

where

$$p_m^{(k)} = \begin{cases} 1 - p_k & \text{for } m = 0 \\ p_k & \text{for } m = 1, \\ 0 & \text{for } m > 1. \end{cases} \quad q_m^{(k)} = \frac{p_k^m}{m!} e^{-p_k}, \ m = 0, 1, \ldots$$

If $0 \le p_k \le 1$ *then*[2]

$$\sum_{m_1 \ldots m_n} |p_{m_1 \ldots m_n} - q_{m_1 \ldots m_n}| \le C_5 \sum_k p_k^2.$$

Lemma 6. *If* $0 \le p \le 1$, *then*

$$\sum_{m=0}^{n} \left| C_n^m p^m (1-p)^{n-m} - \frac{(np)^m}{m!} e^{-np} \right| \le C_6 p.$$

Lemma 7. *Let*

$$a_k = \mathbf{E}\xi_k, \quad \sigma_k^2 = \mathbf{D}\xi_k,$$

$$a = \sum_k a_k, \quad \sigma^2 = \sum_k \sigma_k^2, \quad \sigma_*^2 = \sigma^2 + \sigma_0^2.$$

Then the inequalities

$$|\xi_k| \le l, \quad k = 1, 2, \ldots, n,$$

imply that for all x

$$|\Phi * G_{\sigma_0}(x) - G_{a,\sigma_*}(x)| \le C_7 l / \sigma_*.$$

The proof of Lemma 1 will be published elsewhere.[3] Lemmas 2–5 are quite elementary and can easily be proved by the reader. Lemma 6 is due to Yu.V. Prokhorov (see [3]). Lemma 7 easily follows from the known estimate

$$|\Phi(x) - G_{a,\sigma}(x)| \le C' \frac{\sum \mathbf{E}|\xi_k - a_k|^3}{\sigma^{3/2}},$$

[2] Here and in what follows, $\sum_{m_1 \ldots m_n}$ denotes summation over all sets of positive integers $m_1, \ldots, m_n$.

[3] See: A. Kolmogorov, 'Sur les propriétés des fonctions de concentrations de M.P. Lévy', *Ann. Inst. H. Poincaré* **16**:1 (1958), 27–34. No. 45 in this book.

if the additional normal summand with variance σ_0^2 is represented as a sum of several normal summands, each of them with sufficiently small variance.

1. *Proof of Theorem* 1, *general part.* It suffices to prove the theorem for the case of continuous and strictly increasing functions $F_k(x)$. Indeed, assume that F_k, ϵ, l and L are fixed so that the conditions of the theorem hold. Choose l' and L' so that

$$L > L' > 2l' > 2l$$

and set $\epsilon' = 2\epsilon$. Using only the qualitative part of Lemma 4, it is easy to establish that for a sufficiently small s the distributions

$$F_k' = F_k * G_s$$

satisfy the conditions of the theorem with ϵ, l, L replaced by ϵ', l', L'. It is easy to prove that the $F_k'(x)$ are continuous strictly increasing functions. If for the corresponding function $\Phi' = \Phi * G_{s\sqrt{n}}$ we prove the inequality

$$\Psi(x - L') - \delta' \leq \Phi'(x) \leq \Psi(x + L') + \delta',$$

where

$$\delta' = C' \max\Bigl(\frac{l'}{L'}\sqrt{\log\frac{L'}{l'}}, \epsilon'^{1/5}\Bigr),$$

then, using Lemma 4 with $\sigma^2 = ns^2$ and $l = L - L'$, we easily find that for sufficiently small s and $C = 2C'$ the original function $\Phi(x)$ satisfies the inequality stated in the theorem.

In accordance with this, we take $y = F_k(x)$ continuous and strictly increasing, which allows us to consider the inverse functions $x = F_k^{-1}(y)$, which are uniquely defined for all y in the interval $0 < y < 1$, continuous, strictly increasing, and running through all real values of x.

We set

$$\lambda_k(P) = F_k^{-1}(1 - P) - F_k^{-1}(P).$$

Under these assumptions the $\lambda_k(P)$ are continuous and strictly decreasing. They run through all positive values λ, $0 < \lambda < \infty$, when P runs through $0 < P < \frac{1}{2}$. The inverse functions

$$\epsilon_k(\lambda) = \lambda_k^{-1}(\lambda)$$

are also continuous and strictly decreasing: when λ runs through $0 < \lambda < \infty$, they run through all values of P in the interval $\frac{1}{2} > P > 0$. Therefore, provided that

$$\sum_k \epsilon_k(2l) > 2\epsilon^{-4/5}$$

(here l and ϵ are the quantities occurring in the conditions of the theorem and we assume without loss of generality that $\epsilon \leq 1$), the equation

$$\sum_k \epsilon_k(\lambda_0) = 2\epsilon^{-4/5}$$

has the unique solution

$$\lambda_0 > 2l.$$

Now define Λ and σ_0 in the following way:

A) *If*

$$\sum_k \epsilon_k(2l) > 2\epsilon^{-4/5}, \tag{A.1}$$

then

$$\Lambda = \lambda_0, \quad \sigma_0 = 2\epsilon^{-1/5}\Lambda. \tag{A.2}$$

B) *If*

$$\sum_k \epsilon_k(2l) \leq 2\epsilon^{-4/5}, \tag{B.1}$$

then

$$\Lambda = 2l, \quad \sigma_0 = \frac{1}{\sqrt{2}} L\left(\log\frac{L}{l}\right)^{-1/2}. \tag{B.2}$$

To determine the required infinitely divisible distribution Ψ, we set

$$\epsilon_k = \epsilon_k(\Lambda)$$

and introduce the random variables

$$\mu_k = \begin{cases} 0 & \text{if } F_k^{-1}(\epsilon_k) < \xi_k < F_k^{-1}(1-\epsilon_k), \\ 1 & \text{otherwise.} \end{cases}$$

We then set

$$a_k = \mathbf{E}\{\xi_k | \mu_k = 0\}, \quad \sigma_k^2 = \mathbf{D}\{\xi_k | \mu_k = 0\},$$

$$a = \sum_k (1 - 2\epsilon_k) a_k, \quad \sigma^2 = \sum_k (1 - 2\epsilon_k)\sigma_k^2,$$

$$F_k^{(0)}(x) = \mathbf{P}\{\xi_k < x | \mu_k = 0\},$$

$$F_k^{(1)}(x) = \mathbf{P}\{\xi_k < x | \mu_k = 1\}.$$

It can easily be seen that

$$\Phi = \prod_k{}^*[2\epsilon_k F_k^{(1)} + (1-2\epsilon_k)F_k^{(0)}] = \mathbf{E}\prod_k{}^*[\mu_k F_k^{(1)} + (1-\mu_k)F_k^{(0)}] =$$

$$= \sum_{m_1\ldots m_n} p_{m_1\ldots m_n} \prod_k{}^*[F_k^{(1)}]^{m_k} * \prod_k{}^*[F_k^{(0)}]^{1-m_k},$$

where the probabilities

$$p_{m_1\ldots m_n} = \mathbf{P}\{\mu_1 = m_1, \ldots, \mu_n = m_n\}$$

are expressed via the formulas from Lemma 5 by setting

$$p_k = 2\epsilon_k.$$

Now, expressing the $q_{m_1\ldots m_n}$ in terms of $p_k = 2\epsilon_k$ according to the formulas of the same Lemma 5, we set

$$\Psi = \left[\exp\sum_k 2\epsilon_k(F_k^{(1)} - E_0\right] * G_{a,\sigma_*} = \sum_{m_1\ldots m_n} q_{m_1\ldots m_n} \prod_k{}^*[F_k^{(1)}]^{m_k} * G_{a,\sigma_*},$$

where

$$\sigma_*^2 = \sigma^2 + \sigma_0^2.$$

In the formulas for Φ and Ψ, E_0 denotes the unit distribution

$$E_0(x) = \begin{cases} 0 & \text{for } x \leq 0, \\ 1 & \text{for } x > 0, \end{cases}$$

and the powers are understood in the sense of convolution

$$F^0 = E_0, \quad F^1 = F, \quad F^2 = F * F \text{ etc.}$$

Clearly, by passing from ξ_k to

$$\xi_k' = \xi_k - a_k$$

the a_k are replaced by $a_k' = 0$ and Φ and Ψ by $\Phi'(x) = \Phi(x+a)$ and $\Psi'(x) = \Psi(x+a)$. Therefore it is clear that we may restrict ourselves to the case

$$a_k = 0, \ k = 1, \ldots, n; \ a = 0.$$

We shall also need the random variable

$$\zeta = \sum_k (1 - \mu_k)\xi_k$$

and its conditional variances

$$\sigma^2_{m_1 \dots m_n} = \mathbf{D}\{\zeta | \mu_1 = m_1, \dots, \mu_n = m_n\}$$

for fixed $\mu_1, \dots, \mu_n$. Clearly,

$$\sigma^2_{m_1 \dots m_n} = \sum_k (1 - m_k)\sigma_k^2.$$

We now set [4]

$$\Phi^{(0)}_{m_1 \dots m_n} = \prod_k {}^*[F_k^{(0)}]^{1-m_k}, \quad \Phi^{(1)}_{m_1 \dots m_n} = \prod_k {}^*[F_k^{(1)}]^{m_k}.$$

Noting that

$$\Phi = \sum_{m_1 \dots m_n} p_{m_1 \dots m_n} \Phi^{(1)}_{m_1 \dots m_n} * \Phi^{(0)}_{m_1 \dots m_n},$$

$$\Psi = \sum_{m_1 \dots m_n} q_{m_1 \dots m_n} \Phi^{(1)}_{m_1 \dots m_n} * G_{\sigma_*},$$

we introduce the distributions

$$\Phi_1 = \Phi * G_{\sigma_0} = \sum_{m_1 \dots m_n} p_{m_1 \dots m_n} \Phi^{(1)}_{m_1 \dots m_n} * \Phi^{(0)}_{m_1 \dots m_n} * G_{\sigma_0},$$

$$\Phi_2 = \sum_{m_1 \dots m_n} p_{m_1 \dots m_n} \Phi^{(1)}_{m_1 \dots m_n} * G_{\sigma_{m_1 \dots m_n}} * G_{\sigma_0},$$

$$\Phi_3 = \sum_{m_1 \dots m_n} p_{m_1 \dots m_n} \Phi^{(1)}_{m_1 \dots m_n} * G_{\sigma_*}.$$

It can easily be seen from the condition of the theorem that

$$\epsilon_k(2l) \le \epsilon$$

and, since $\Lambda \ge 2l$,

$$\epsilon_k = \epsilon_k(\Lambda) \le \epsilon. \tag{3}$$

---

[4] The definition of $\Phi^{(0)}_{m_1 \dots m_n}$ presupposes that $m_k = 0$ or 1. The definition of $\Phi^{(1)}_{m_1 \dots m_n}$ applies to any non-negative m_k.

In both cases A) and B),

$$\sum_k \epsilon_k \leq 2\epsilon^{-4/5} \tag{4}$$

Formulas (3) and (4) imply that

$$\sum_k \epsilon_k^2 \leq 2\epsilon^{1/5}. \tag{5}$$

Using (5) and Lemma 5 we obtain

$$|\Psi - \Phi_3| \leq \sum_{m_1 \ldots m_n} |p_{m_1\ldots m_n} - q_{m_1\ldots m_n}| \leq C_5 \sum_k p_k^2 = 4C_5 \sum_k \epsilon_k^2 \leq 8C_5\epsilon^{1/5}. \tag{6}$$

By Lemma 3 and assuming that

$$|\sigma^2_{m_1\ldots m_n} - \sigma^2| \leq \epsilon^{1/5}\sigma_*^2 \tag{7}$$

we have

$$|G_{\sigma_{m_1\ldots m_n}} * G_{\sigma_0} - G_{\sigma_*}| \leq C_3\epsilon^{1/5}. \tag{8}$$

Therefore

$$|\Phi_2 - \Phi_3| \leq C_3\epsilon^{1/5} + 2\sum{}', \tag{9}$$

where $\sum'$ is the sum of those $p_{m_1\ldots m_n}$ for which (7) does not hold. Clearly,

$$\sum{}' = \mathsf{P}\{|\rho^2 - \sigma^2| > \epsilon^{1/5}\sigma_*^2\}, \tag{10}$$

where

$$\rho^2 = \sum_k (1 - \mu_k)\sigma_k^2.$$

It can easily be seen that

$$\begin{aligned} \mathsf{E}\rho^2 &= \sum_k (1 - 2\epsilon_k)\sigma_k^2 = \sigma^2, \\ \mathsf{D}\rho^2 &= \sum_k 2\epsilon_k(1 - 2\epsilon_k)\sigma_k^4, \end{aligned} \tag{11}$$

and since $\sigma_k^2 \leq \Lambda^2$, $\epsilon_k \leq \epsilon$,

$$\mathsf{D}\rho^2 \leq 2\epsilon \sum_k (1 - 2\epsilon_k)\sigma_k^4 \leq 2\epsilon\Lambda^2 \sum_k (1 - 2\epsilon_k)\sigma_k^2 = 2\epsilon\Lambda^2\sigma^2. \tag{12}$$

Using Chebyshev's inequality, from (10)–(12) we obtain

$$\sum{}' \le \frac{2\epsilon\Lambda^2\sigma^2}{\epsilon^{2/5}\sigma_*^4} \le 2\epsilon^{3/5}\frac{\Lambda^2}{\sigma_0^2}. \tag{13}$$

Since always $\sigma_0 \ge \Lambda$, $\epsilon^{3/5} \le \epsilon^{1/5}$, (9) and (13) imply

$$|\Phi_2 - \Phi_3| \le (C_3 + 4)\epsilon^{1/5}. \tag{14}$$

By Lemma 7 and the fact that $|\xi_k| \le \Lambda$ for $\mu_k = 1$, we have

$$|\Phi^{(0)}_{m_1\dots m_n} * G_{\sigma_0} - G_{\sigma_{m_1\dots m_n}} * G_{\sigma_0}| \le C_7\frac{\Lambda}{\sigma_0}, \tag{15}$$

which implies that

$$|\Phi_1 - \Phi_2| \le C_7\Lambda/\sigma_0. \tag{16}$$

From (6), (14) and (16) we obtain

$$|\Psi - \Phi_1| \le C_*\epsilon^{1/5} + C_7\Lambda/\sigma_0, \tag{17}$$

where

$$C_* = 8C_5 + C_3 + 4.$$

2. *Case* A). The end of the proof of Theorem 1 differs according to cases A) and B). Let us first consider case A). In this case, by (A.2), (17) turns into

$$|\Psi - \Phi_1| \le (C_* + C_7)\epsilon^{1/5}. \tag{18}$$

We now recall that

$$\Lambda = \lambda_0 > 2l, \tag{19}$$

$$\sum_k \epsilon_k = \sum_k \epsilon_k(\lambda_0) = 2\epsilon^{-4/5}. \tag{20}$$

It can easily be seen that

$$1 - 2\epsilon_k \le Q_{F_k}(\Lambda) \le 1 - \epsilon_k \tag{21}$$

and, consequently, the quantity

$$s = \sum_{k=1}^{n}[1 - Q_{F_k}(\Lambda)]$$

is bounded by

$$2e^{-4/5} \le s \le 4\epsilon^{-4/5}. \tag{22}$$

Lemma 1 and formulas (22) and (A.2) imply that [5]

$$Q_{\Phi}(\tau_0) \le \frac{C_1\sigma_0}{\Lambda\sqrt{s}} \le 2C_1\epsilon^{1/5}. \tag{23}$$

According to Lemma 2, (23) implies that

$$|\Phi - \Phi_1| = |\Phi - \Phi * G_{\sigma_0}| \le C_2 Q_{\Phi}(\sigma_0) \le 2C_1C_2\epsilon^{1/5}. \tag{24}$$

Finally, from (18) and (24) we obtain

$$|\Phi - \Psi| \le (C_* + C_7 + 2C_1C_2)\epsilon^{1/5}. \tag{25}$$

Formula (25) now immediately implies (for case A)) the conclusion of the theorem, and the shifts by L are not needed; they are needed only for case B).

3. *Case* B). In this case, by (B.2) inequality (17) takes the form

$$|\Psi - \Phi_1| \le C_*\epsilon^{1/5} + 2^{3/2}C_7\frac{l}{L}\sqrt{\log\frac{L}{l}}. \tag{26}$$

According to Lemma 4, by (B.2) for

$$\eta = C_4 e^{-L^2/2\sigma_0^2} = C_4 e^{\log(l/L)} = C_4\frac{l}{L} \tag{27}$$

we have

$$\Phi_1(x - L) - \eta \le \Phi(x) \le \Phi_1(x + L) + \eta. \tag{28}$$

Now set

$$C = C_* + 2^{3/2}C_7 + 2C_1C_2 + C_4/\log 2.$$

Since $L > 2l$, $\log(L/l) > \log 2$, it follows from (26) and (28) that

$$\Psi(x - L) - \delta \le \Phi(x) \le \Psi(x + L) + \delta, \tag{29}$$

where

$$\delta = C\max\Big(\frac{l}{L}\sqrt{\log\frac{L}{l}}, \epsilon^{1/5}\Big).$$

[5] It is easy to verify that by (A.2), (22) and under the assumption that $\epsilon \le 1$, the additional condition $R^2 \ge r^2 \log s$ holds for $R = \sigma_0$ and $r = \Lambda$.

As we see from (25), inequality (29) proved here for case B), holds also for case A). This completes the proof of Theorem 1.

4. *Proof of Theorem* 2. The proof of Theorem 2 is essentially the same as that for Theorem 1, case A), where

$$\epsilon = c/n.$$

Naturally, ϵ_k and $p_k = 2\epsilon_k$ do not now depend on k. Since

$$\sum_k \epsilon_k = 2\epsilon^{-1/5}$$

the p_k must be equal to

$$p = c' n^{-1/5}.$$

The construction involving transition from Φ to

$$\Phi' = \Phi * G_{s\sqrt{n}}$$

is now superfluous and would even have somewhat complicated the achievement of the final result. Instead of referring to Lemma 5 it is better to refer to the simpler Lemma 6. Since these simplifications are possible, we now give a proof of Theorem 2 that is independent of Nos. 1, 2. We denote by $F(x) = \mathbf{P}\{\xi_k < x\}$ the general distribution function of the terms ξ_k. The conditions

$$F(F^{-1}(y)) \leq y \leq F(F^{-1}(y) + 0)$$

uniquely define the generalized inverse function F^{-1} to within a countable set of values y corresponding to the intervals where F is constant. Without loss of generality we may assume that

$$\xi_k = F^{-1}(\eta_k),$$

where the random variables η_k are independent and uniformly distributed on the interval $0 < y < 1$.

We set $p = \frac{1}{2} n^{-1/5}$,

$$\mu_k = \begin{cases} 0, & \text{if } p/2 < \eta_k < 1 - p/2, \\ 1 & \text{otherwise,} \end{cases}$$

$$a = (1-p)\mathbf{E}\{\xi_k|\mu_k = 0\}, \quad \sigma^2 = (1-p)\mathbf{D}\{\xi_k|\mu_k = 0\},$$
$$F_0(x) = \mathbf{P}\{\xi_k < x|\mu_k = 0\}, \quad F_1(x) = \mathbf{P}\{\xi_k < x|\mu_k = 1\}.$$

Clearly,

$$\Phi = \mathbf{E}\prod_k{}^{*}[\mu_k F_1 + (1-\mu_k)F_0] = \sum_k C_n^k p^k (1-p)^{n-k} F_1^k * F_0^{n-k}.$$

We give the desired infinitely divisible distribution Ψ in the form

$$\Psi = \sum \frac{(np)^k}{k!} e^{-np} F_1^k * G_{na,\sigma_*}$$

where $\sigma_*^2 = n\sigma^2 + \sigma_0^2$, $\sigma_0^2 = 2n^{1/5}\Lambda$, $\Lambda = F^{-1}(1-p/2) - F^{-1}(p/2)$. As in No. 1, it suffices to consider the case $a = 0$. Set

$$\Phi_1 = \Phi * G_{\sigma_0} = \sum_k C_n^k p^k (1-p)^{n-k} F_1^k * F_0^{n-k} * G_{\sigma_0},$$
$$\Phi_2 = \sum_k C_n^k p^k (1-p)^{n-k} F_1^k * G_{\sigma_k} * G_{\sigma_0}, \quad \sigma_k^2 = \frac{n-k}{1-p}\sigma^2,$$
$$\Phi_3 = \sum_k C_n^k p^k (1-p)^{n-k} F_1^k * G_{\sigma_*}.$$

By Lemma 6

$$|\Psi - \Phi_3| \le C_6 p = \tfrac{1}{2} C_6 n^{-1/5}. \tag{30}$$

Lemma 3, under the assumption that

$$\left|\frac{n-k}{1-p} - n\right| \le n^{4/5} \tag{31}$$

implies that

$$|G_{\sigma_k} * G_{\sigma_0} - G_{\sigma_*}| \le C_3 n^{-1/5}, \tag{32}$$

therefore

$$|\Phi_2 - \Phi_3| \le C_3 n^{-1/5} + 2\sum{}', \tag{33}$$

where $\sum'$ is the sum of those $C_n^k p^k (1-p)^{n-k}$ for which (31) fails. Clearly

$$\sum{}' = \mathbf{P}\{|\chi - n(1-p)| > n^{4/5}\},$$

where $\chi = \sum_k \mu_k$. Therefore, using Chebyshev's inequality,

$$\sum{}' \le \frac{D\chi}{n^{8/5}} = \frac{np(1-p)}{n^{8/5}} \le n^{-4/5} \le n^{-1/5}. \tag{34}$$

From (33) and (34) we obtain

$$|\Phi_2 - \Phi_3| \leq (C_3 + 2)n^{-1/5}. \tag{35}$$

Since for $\mu_k = 1$ and $a = 0$ we have $|\xi_k| \leq \Lambda$, by Lemma 7 we have

$$|F_0^{n-k} * G_{\sigma_0} - G_{\sigma_k} * G_{\sigma_0}| \leq C_7\Lambda/\sigma_0 = \tfrac{1}{2}C_7 n^{-1/5},$$

which implies that

$$|\Phi_1 - \Phi_2| \leq \tfrac{1}{2}C_7 n^{-1/5}. \tag{36}$$

Clearly,

$$Q_F(\Lambda/2) \leq 1 - p/2.$$

By Lemma 1 this implies that

$$Q_\Phi(\sigma_0) \leq \frac{2C_1\sigma_0}{\Lambda\sqrt{np/2}} = 4C_1 n^{-1/5}. \tag{37}$$

By Lemma 2, it follows from formula (37) that

$$|\Phi - \Phi_1| \leq C_2 Q_\Phi(\sigma_0) \leq 4C_1C_2 n^{-1/5}. \tag{38}$$

From (30), (35), (36), (38) we obtain

$$|\Phi - \Psi| \leq (C_6/2 + C_3 + 2 + C_7/2 + 4C_1C_2)n^{-1/5},$$

which completes the proof of Theorem 2.

Moscow, 12 November 1956

References

1. P. Lévy, *Théorie de l'addition des variables aléatoires*, Gauthier-Villars, Paris, 1937.
2. W. Doeblin, 'Sur les sommes d'un grand nombre des variables aléatoires indépendantes', *Bull. Sci. Math.* **63** (1939), 23–32; 35–64.
3. Yu.V. Prokhorov, 'On the sums of identically distributed random variables', *Dokl. Akad. Nauk SSSR* **105**:4 (1955), 645–647 (in Russian).

44. RANDOM FUNCTIONS AND LIMIT THEOREMS *[1]

Jointly with Yu.V. Prokhorov

Without claiming to present an exhaustive review of the literature on the question, we intend to give an outline of the first steps of the theory of random functions, possible basic methods for a systematic construction of this theory and basic problems concerning functional methods in limit theorems; in addition, we report certain comparatively new results.

§1. Preliminary remarks on random elements of arbitrary nature

We will proceed from the axiomatics of probability theory given in [1]. A family of $\{\Omega, \Phi, \mathbf{P}\}$ consisting of

1) a set Ω whose elements ω are called "elementary events";
2) a σ-algebra Φ of subsets of Ω;
3) a measure $\mathbf{P}(A)$ defined on Φ and satisfying the additional requirement

$$\mathbf{P}(A) = 1,$$

will be called a probability field (the terminology corresponds to [2]). Sometimes it is also useful to assume that

(ω') The measure $\mathbf{P}(A)$ is complete, that is $\mathbf{P}(A) = 0$, $B \subseteq A$ implies that

$$B \in \Phi.$$

In studying random variables, random vectors and other "random elements" of an arbitrary nature it is sometimes useful to leave the nature of the basic set of elementary events ω undetermined and to consider specific random

---

* 'Zufällige Funktionen und Grenzverteilungssätze', In: *Bericht über die Tagung Wahrscheinlichkeitsrechnung und mathematische Statistik*, Berlin, 1956, pp.113–126.

[1] The papers presented by Kolmogorov and Prokhorov at the conference are close to those of M. Fréchet and R. Fortet. In his talk given on the last day of the conference Kolmogorov presented new results obtained by Prokhorov in between his talk and that of A.N. Kolmogorov. Therefore the authors deemed it appropriate to publish the joint text of their two papers. §4 (without Theorems IV and V) and §6 represent Prokhorov's contribution, while §§1–3, 5 and Theorems IV and V from §4 constituted Kolmogorov's paper. Theorems IV and V from §4 were proved by Prokhorov between his talk and that of Kolmogorov.

elements ξ as functions of ω. For instance, it is customary to call any real function ξ_ω such that the set

$$\{\omega : \xi_\omega < a\}$$

belongs to Φ for any real a, a (real) random variable (see [1]).

An arbitrary function ξ_ω with values in a set X determines a probability field $\{X, \Phi_\xi, P_\xi\}$ where Φ_ξ consists of all $A \subseteq X$ for which

$$\xi^{-1}(A) \in \Phi,$$

and on Φ_ξ

$$P_\xi(A) = \mathbf{P}\{\xi^{-1}(A)\}.$$

Since in any problem in probability theory the basic set Ω is considered to be quite definite while its elements ω are implicit in the formulations of more specific problems, the index ω at ξ_ω is usually omitted, and we speak about a "random element ξ of the set X". Note that under the assumption (ω') the condition

(ξ_1) if $P_\xi(A) = 0$, $B \subseteq A$, then $B \in \Phi_\xi$

holds automatically.

If X is a topological space, then it is natural to require for a random element $\xi \in X$ that certain topologically very simple sets $A \subseteq X$ are contained in Φ_ξ. If X is a metric space (this space is the main one in what follows), then these requirements are undoubtedly quite sensible. Namely, it is reasonable to confine ourselves to random variables satisfying the condition

(ξ_2) any open set $G \subseteq X$ belongs to Φ_ξ.

As is well known, (ξ_2) implies that any Borel set $A \subseteq X$ belongs to Φ_ξ. If X is the real axis, then (ξ_2) is equivalent to the usual requirement in the definition of random variables. We can further simplify the problem if, in addition,

(ξ_3) For any $A \in \Phi_\xi$, $P_\xi(A) = \inf_{G \supseteq A} \mathbf{P}_\xi(G)$,

where the infimum is taken over all open sets G containing A.

For (ξ_3) to hold for any random element ξ in a separable complete metric space X satisfying (ξ_2), it is necessary and sufficient that

(ω'') the measure $\mathbf{P}$ on Ω is perfect (see [3]).

The question whether it is worth giving a systematic construction of probability theory using perfect measures remains debatable until such a systematic construction is actually carried out. We note, however, that this limitation by no means restricts the scope of real applications of probability theory, since we have the following theorem: any complete normed Boolean algebra can be realized as the algebra of metric types of perfect measures (cf. [4]).

Requirements (ξ_2) and (ξ_3) can be generalized in different directions for topological spaces of a more general nature (cf. [2], §§52–54).

§2. Random functions

For the sake of definiteness we will consider complex functions $x(t)$ of an argument $t \in T$, where T is, in general, an arbitrary set. (Mostly studied here are "random processes" where T is the real axis.) Two approaches to introducing the notion of a "random function" of this kind are possible.

(I) A random function $\xi(t)$ defined on T is a system of complex random variables $\xi(t)$ defined at every $t \in T$. If in this case the basic probability field $\{\Omega, \Phi, \mathsf{P}\}$ satisfies (ω'), then the distributions $P_{\xi}(t)$, as well as the n-dimensional distributions $P_{\xi(t_1),\xi(t_2),\dots,\xi(t_n)}$ satisfy (ξ_1) and (ξ_2).

(II) We choose in advance a certain function space $\mathfrak{C}$ consisting of functions $x(t)$ defined on T and a metric is introduced (the distance satisfying the usual axioms of a metric space). A random element of the metric space $\mathfrak{C}$ satisfying (ξ_1) and (ξ_2) is called a random function $\xi(t)$ of type $\mathfrak{C}$.

A systematic study of random functions in a very important particular case was started by N. Wiener in 1923 (see [5]). The first of the concepts given above was most extensively developed by E.E. Slutskii, starting from 1928 (see [6]). The second concept for arbitrary Banach space was developed by V.I. Glivenko from 1928 on (see [7]–[9]). The subsequent abundant literature can be found in modern presentations of the question in [10], [11].

If $\mathfrak{C}$ is such that for any $t_0 \in T$ and any open set G in the complex plane the set

$$\{x(t) : x(t_0) \in G\} \tag{1}$$

of functions $x(t)$ for which $x(t_0)$ belongs to G is a Borel subset of $\mathfrak{C}$, then every random function in the sense of (II) clearly has a corresponding random function in the sense of (I).

For a random function in the sense of (I) the inverse problem can be posed: Is it possible to consider this kind of random function $\xi(t)$ as actually belonging to a certain function space (for example, in the case $T = [a, b]$) as belonging to the space $C_{[a,b]}$ of continuous complex functions with the metric

$$\rho(x(t), y(t)) = \max_{a \le t \le b} |x(t) - y(t)|)\,?$$

This problem can be stated precisely using equivalence of random functions in the sense of (I). Two random functions $\xi_1(t)$ and $\xi_2(t)$ (in the sense of (I)) are called equivalent if for any $t \in T$,

$$\mathbf{P}\{\xi_1(t) \neq \xi_2(t)\} = 0. \tag{2}$$

The question posed above can now be interpreted in the following way: Given a random function $\xi_1(t)$ in the sense of (I) and a space of Borel measurable sets (I), does there exist a random function $\xi_2(t)$ of type $\mathfrak{C}$ for which (2) holds for any $t \in T$? From the point of view of applications, this approach developed by E.E. Slutskii (see [12], [13]) is apparently quite sufficient and allows us to avoid a number of complications of set theory which will be discussed below in §3. From the practical viewpoint the most interesting are the questions: Is it possible to consider a random function as a) continuous, b) having only discontinuities of the first kind ? In the first direction the following result by A.N. Kolmogorov (see [12]) has long been known: in order that a random function $\xi(t)$ in the sense of (I) given on $T = [a, b]$ be equivalent to a random function $\xi^*(t)$ of type $C_{[a,b]}$, it suffices that

$$\mathbf{E}|\xi(t+\tau) - \xi(t)|^a \le K|\tau|^\alpha \tag{3}$$

for some $a > 0$, $\alpha > 1$ and K. In this case $\xi^*(t)$ satisfies with probability 1 a Lipschitz condition

$$|\xi^*(t+\tau) - \xi^*(t)| \le K^*|\tau|^{\frac{\alpha-1}{a} - \delta}, \tag{4}$$

for any $\delta > 0$, where K^* is a random variable. (Note also that for any $\epsilon > 0$ there exists K_0^*, depending only on $\epsilon, \delta, \alpha, a, K$, such that $\mathbf{P}\{K^* > K_0^*\} < \epsilon$. This will be used in §6.)

The conditions for the representability of $\xi(t)$ as a function with discontinuities of the first kind only are especially essential when $\xi(t)$ represents a

Markov process. Among the recent papers on this question we point out those by E.B. Dynkin [14] and Kinney [15].[2]

As is well known, elements of a number of important function spaces are not individual functions, but metric types $\bar{x}(t)$ of these functions with respect to the measure μ introduced in T (a metric type is a class of all functions that differ from a certain fixed function $x(t)$ only on a set of μ-measure 0). We introduce the notion of asymptotic continuity of $x(t)$ at t_0 so that for the metric type $\bar{x}(t)$ the "true values" of $x(t)$ are uniquely determined at almost all points $t_0 \in T$ by asymptotic continuity, and the space $\mathfrak{C}$ is such that the true values of a random function of type $\mathfrak{C}$ at the points where it is defined are with probability 1 random variables. In this setting the same problem on the relation between the two notions of a random function, (I) and (II), may be raised and solved, perhaps in a different form. Now, using the approach (I) we naturally require that random variables $\xi(t)$ be defined only almost everywhere on T and regard two random functions (in the sense (I)) $\xi_1(t)$ and $\xi_2(t)$ as being equivalent if (2) holds for almost all t. Then corresponding to each random function $\xi^*(t)$ in the sense of (II) there is a random function $\xi(t)$ in the sense of (I) which is defined to within equivalence and is equal to the true value of $\xi^*(t)$ at t for almost all t with probability 1. The complete solution of the inverse problem under conditions when a random function in the sense of (I) is equivalent to a measurable function was given in the paper by Ambrose [16] for the case when T is the real axis and μ is Lebesgue measure. A necessary and sufficient condition is for $\xi(t)$ to be asymptotically stochastically continuous almost everywhere (with respect to μ). If this condition is satisfied, we can further consider our problem in the space $\mathfrak{C}$ of measurable functions (more exactly, in the space of metric types) and without any difficulties in principle compute the probability of the summability of $\xi(t)$, its square summability, "essential boundedness", etc.

§3. Finite-dimensional distributions of values of a random function

Let us return to random functions $\xi(t)$ in the sense of (I) defined for all $t \in T$.

---

[2] See also a later work by N.N. Chentsov, *Teor. Veroyatnost. i Primenen.* (Translated as *Theory Probab. Appl.*) 1:1 (1956), 155–161 (in Russian). (Note of Russian editor.)

Each such function ξ automatically defines a probability field $\{\Omega_T, \Phi_\xi, P_\xi\}$ where Ω_T is the set of all complex functions defined on T.

The formula

$$\alpha_k = x(t_k), \quad k = 1, 2, \ldots, n,$$

defines a mapping $\pi_{t_1,t_2,\ldots,t_n}(x)$ from Ω_T into the n-dimensional complex vector space of vectors

$$\alpha = \{\alpha_1, \alpha_2, \ldots, \alpha_n\}.$$

For fixed $t_1, t_2, \ldots, t_n$ this mapping assigns to each random function $\xi(t)$ a random vector

$$\alpha^\xi_{t_1,t_2,\ldots,t_n}.$$

Usually, in specific problems involving random functions the corresponding distributions

$$P^\xi_{t_1,t_2,\ldots,t_n}(B) = \mathbf{P}\{\alpha^\xi_{t_1,t_2,\ldots,t_n} \in B\}$$

play an important role.

Since $P^\xi_{t_1,t_2,\ldots,t_n}$ is defined on the Borel sets B, P_ξ is defined on their pre-images in Ω_T:

$$B' = \pi^{-1}_{t_1,t_2,\ldots,t_n}(B).$$

These are the so-called cylindrical Borel sets in Ω_T.

Let Φ_T be the Borel closure of a system of such sets (the smallest σ-algebra containing these sets). Clearly, for any random function $\xi(t)$ we have

$$\Phi_\xi \supseteq \Phi_T.$$

The values $P_\xi(B)$ for $B \in \Phi_T$ are uniquely defined by the finite-dimensional distributions. In [1] it was proved that, conversely, arbitrary "compatible" finite-dimensional distributions $P_{t_1,t_2,\ldots,t_n}$ given for all finite systems of elements $t_1, t_2, \ldots, t_n$ in T and defined on the Borel sets in an n-dimensional complex vector space give rise to a probability field $\{\Omega_T, \Phi_T, P^*\}$. The following formulation is a slight variation of this result: If the distributions $P_{t_1,t_2,\ldots,t_n}$ are defined on all systems of finite elements of T, are compatible and satisfy (ξ_2) and (ξ_3), then there exist random functions $\xi(t)$ in the sense of (I) with

$$P^\xi_{t_1,t_2,\ldots,t_n} = P_{t_1,t_2,\ldots,t_n}.$$

It is essential, however, that the function P_ξ which, generally speaking, is uniquely defined by the finite-dimensional distributions $P^\xi_{t_1,t_2,\ldots,t_n}$, is not defined beyond Φ_T. At the same time, the class Φ_T is too narrow for applications: for uncountable T it does not include the set of functions $x(t)$ such that $|x(t)| \leq K$ for all $t \in T$, or, if $T = [a, b]$, the set C of continuous functions, etc.

Passing to the concept (II), in all cases of practical importance studied so far we were able to recover P_ξ uniquely from the finite-dimensional distributions $P^\xi_{t_1,t_2,\ldots,t_n}$ (the meaning of this statement for spaces $\mathfrak{C}$ consisting of metric types $\bar{x}(t)$, rather than of individual functions, needs some clarification which we will not go into now) and at the same time, when the space $\mathfrak{C}$ is properly chosen, all the sets needed for specific problems appear to belong to Φ_ξ.

§4. Weak convergence of distributions

In this section we will confine ourselves to the consideration of random elements ξ taking values in a complete separable metric space X and satisfying the conditions (ξ_2) and (ξ_3). In the study of these sequences

$$\xi_1, \xi_2, \ldots, \xi_n, \ldots$$

of random elements from X, there arises the question of the convergence of the corresponding distributions P_{ξ_n} to a distribution P on X. Here convergence naturally means "weak convergence" of distributions in the usual functional analytic sense. Recall that a sequence P_n of distributions on X weakly converges to a distribution P, if for any function $f(x)$, $x \in X$ that is continuous and bounded on X,

$$\int_X f(x)dP_n \to \int_X f(x)dP.$$

If X is the real line, then weak convergence appears to be equivalent to the well-known convergence "in general" of the corresponding distribution functions. Denote by $P_n \Rightarrow P$ the weak convergence of distributions. The meaning of weak convergence is adequately clarified by the following two theorems, of which the first is well known and the second can easily be proved.

I) $P_n \Rightarrow P$ if and only if

$$P_n(A) \to P(A)$$

for any set A whose boundary has P-measure 0 (that is, for any continuity set of the distribution P);

II) $P_n \Rightarrow P$ if and only if for any functional $f(x)$ continuous almost everywhere with respect to P, the sequence of distribution functions

$$F_f^{(n)}(\alpha) = P_n\{x : f(x) \leq \alpha\}$$

converges to the distribution function

$$F_f(\alpha) = P\{x : f(x) < \alpha\}$$

at every continuity point of the latter.

Now let

$$\xi; \xi_1, \xi_2, \ldots, \xi_n, \ldots$$

be a sequence of random elements of X for which

$$P_{\xi_n} \Rightarrow P_\xi. \tag{5}$$

Further, let

$$\eta; \eta_1, \eta_2, \ldots, \eta_n, \ldots$$

be a sequence of random elements of a complete metric space Y with $\eta = f(\xi)$ and $\eta_n = f_n(\xi_n)$, where f and f_n are continuous mappings of X into Y. Then, under rather general assumptions (for example, that $f_n \to f$ uniformly on any compact set $K \subset X$), (5) implies that

$$P_{\eta_n} \Rightarrow P_\eta.$$

This simple remark is useful when proving a number of limit theorems.

Finally, the following theorem may be of use for studying limit theorems for a sequence of random variables (see [17]):

III) For a family $\mathfrak{M}$ of distributions in X to be compact it is sufficient[3] that for any $\epsilon > 0$ there exists a compact set K_ϵ such that $P(K_\epsilon) > 1 - \epsilon$ for all $P \in \mathfrak{M}$.

We consider an example illustrating Theorem III. It is known that a set $K \subset C$, where C is the space of functions $x(t)$, continuous on the interval

[3] Yu.V. Prokhorov proved also the necessity of this condition.

$[0,1]$, $x(0)=0$, is compact if and only if there exists a function $\phi(\tau)$, tending to zero as $\tau \to 0$, for which

$$\sup_{t,x\in K} |x(t+\tau)-x(t)| \leq \phi(\tau).$$

Now using the continuity condition with probability 1 of a random process (see (3) and (4) in §2) we obtain the following sufficiency condition for the compactness of a family $\mathfrak{M}$ of distributions in C:

There exist constants $a>0$, $\alpha>1$, $K>0$ such that for all $P\in\mathfrak{M}$

$$\mathbf{E}_P|\xi(t+\tau)-\xi(t)|^a \leq K|\tau|^\alpha,\ \xi(0)=0,$$

where $\mathbf{E}_P\xi$ denotes

$$\int_C \xi dP.$$

Sometimes it is relatively simple to establish convergence

$$P_n(A)\to P(A),\ A\in S, \tag{6}$$

for some system S of subsets of X. Weak convergence $P_n \Rightarrow P$ can be derived from (6) using, for example, the following theorem:

IV) A system S of subsets X is called a "uniqueness system" if any distribution P defined on S is uniquely defined on X.

In order that $P_n \Rightarrow P$ it is necessary and sufficient that the following two conditions hold:

a) $P_n(A)\to P(A)$ on a "uniqueness system" S;

b) the family $\{P_n\}$ is compact.

This theorem implies, for example, that if a sequence of processes $\{\xi_n(t)\}$ defined on $[a,b]$ is such that

a) the finite-dimensional distributions of the processes $\xi_n(t)$ converge to the finite-dimensional distributions of some process $\xi(t)$,

b) there exist constants a, α, $K (a>0,\ \alpha>1,\ K>0)$ independent of n such that

$$\mathbf{E}|\xi_n(t+\tau)-\xi_n(t)|^a < K|\tau|^\alpha,$$

then for the corresponding distributions in $C_{[a,b]}$ we have

$$P_{\xi_n} \Rightarrow P_\xi.$$

V) Let S be a system of open sets $A \subseteq X$ such that:

a) S is a basis in X;

b) $A_1 \in S$ and $A_2 \in S$ imply $A_1 \cap A_2 \in S$. Then the convergence

$$P_n(A) \to P(A), \ A \in S,$$

implies that

$$P_n \Rightarrow P.$$

Examples illustrating the use of these general results can be found in the last section of the present paper.

§5. Characteristic functionals

If ξ is a random element from a Banach space X,[4] then for the corresponding distribution P_ξ we can define an analogue of the usual characteristic function. For any element f of the space X^* dual to X we define the value of the characteristic functional H by the formula

$$H(f, P_\xi) = \mathbf{E}e^{i(f,\xi)} = \int_X e^{i(f,x)} dP_\xi.$$

This definition, as well as the basic properties of characteristic functionals, was given in the note by A.N. Kolmogorov [18]. Recently interest in distributions in function spaces has grown considerably, and it seems that the apparatus of characteristic functionals will find broader applications in probability theory and in certain fields of mathematical physics (see, for example, Hopf [19]). In view of this we consider it useful to take note of the results of [18] and point out certain unsolved problems in this connection. For example, in [18] it has been established that

1. P_ξ is uniquely determined by $H(f, P_\xi)$.

2. When summing independent random variables with values in X, the corresponding characteristic functionals are multiplied.

3. If

$$R_{n+1} = \int_X \|x\|^{n+1} dP_\xi < +\infty,$$

then

$$H(f, P_\xi) = 1 + \mathbf{E}f + \frac{1}{2!}\mathbf{E}(f, f) + \ldots + \frac{1}{n!}\mathbf{E}(f, \ldots, f) + \frac{1}{(n+1)!} R_{n+1}\|f\|^{n+1},$$

[4] In what follows we shall consider separable spaces only.

where

$$\mathbf{E}(f_1, f_2, \ldots, f_n) = \int_X f_1 f_2 \ldots f_n dP_\xi$$

are multilinear functionals which are natural generalizations of the moments.

Clearly, the functions

$$\mathbf{E}(f_1, f_2, \ldots, f_n)$$

are uniquely determined by $H(f, P_\xi)$.

For the purposes of probability theory it is essential to have necessary and sufficient conditions for weak convergence of distributions in terms of the corresponding characteristic functionals. However, these conditions are not yet known. Certain applications of characteristic functionals for deriving some limit theorems are studied in work by R. Fortet and E. Mourier (see [20]–[23] and Fortet's talk at the present conference). Here we note only one result following from Theorems III and IV of §4.

Let X be a separable Banach space and let $\{P_n\}$ be a sequence of distributions on X. Convergence $P_n \Rightarrow P$ holds if and only if the following two conditions hold simultaneously:

a) $H(f, P_n) \to H(f, P)$ for every $f \in X^*$;

b) for any $\epsilon > 0$ there exists a compact set K_ϵ such that $P_n(K_\epsilon) > 1 - \epsilon$ for all n.

The question of sufficient conditions for a functional $H(f)$ to be characteristic acquires new aspects in the infinite-dimensional case. Here, we only make the following remark. In accordance with [18] it is natural to call a distribution P in X normal if its characteristic functional has the form

$$H(f) = e^{imf - \frac{1}{2}Q(f)},$$

where mf is a linear and $Q(f)$ a quadratic functional of f.

For a Hilbert space this brings us to the formula

$$H(f) = e^{i(m,f) - \frac{1}{2}(f,Sf)},$$

where m is an element of X and S is a symmetric bounded linear operator replacing the matrix of second moments.

However, the condition that

$$(f, Sf) \geq 0$$

be non-negative, which is sufficient for obtaining the characteristic function in the finite-dimensional case is not sufficient here.[5] In particular, the functional $H(f) = e^{-\frac{1}{2}(f,f)}$ which is obtained for $m = 0$, $Sf \equiv f$, is not characteristic. Note also that the necessary and sufficient conditions for a functional to be characteristic given in [23] can hardly be considered final, since they are of an extremely complex character.[6]

§6. Applications to limit theorems

6.1. Passage from sums of random variables to processes. The notion that limit distributions for sums of random variables are exact for properly chosen random processes took shape long ago (see, for example, [25]–[29] and especially A.Ya. Khinchin's monograph [30]). Here we will give the results of more recent works touching on this group of questions. Note that posing and solving the problems on passing from sums of random variables to processes seems impossible without the methods which essentially have to do with functional analysis in their most general form.

Let

$$\xi_{n,1},\ \xi_{n,2}, \ldots,\ \xi_{n,n}$$

be random variables with zero expectations $\mathbf{E}\xi_{n,k}$ and finite variances $\mathbf{D}\xi_{n,k}$ such that

$$\sum_{k=1}^{n} \mathbf{D}\xi_{n,k} = 1.$$

We denote by

$$s_{n,k} = \xi_{n,1} + \xi_{n,2} + \ldots + \xi_{n,k},\quad k = 1, 2, \ldots, n, \tag{7}$$

the "accumulated sums" of the random variables $\xi_{n,k}$, and by $t_{n,k}$ the quantities

$$\mathbf{D}s_{n,k} = \sum_{j=1}^{k} \mathbf{D}\xi_{n,j}.$$

[5] It is sufficient if the trace of S is finite (which is also necessary).

[6] The modern situation is reviewed in: N.N. Vakhaniya, V.I. Tarieladze and S.A. Chobanyan, *Probability distributions in Banach spaces*, Nauka, Moscow, 1985 (in Russian) (Note of Russian editor).

We assign to the sequence of sums $\{s_{n,k}\}$, $k = 1, 2, \ldots, n$, a random element ξ_n of the space C of functions $x(t)$ continuous on the segment $[0,1]$ and vanishing at $t = 0$, by setting

$$\xi_n(t) = s_{n,k} + \frac{t - t_{n,k}}{t_{n,k+1} - t_{n,k}} \xi_{n,k+1}$$

for $t_{n,k} \leq t < t_{n,k+1}$.

It can be shown (see [17]) that if the sums s_{nn} satisfy Lindeberg's condition used in the ordinary central limit theorem, then

$$P_{\xi_n} \Rightarrow P_\xi \tag{8}$$

as $n \to \infty$, where $\xi(t)$ is a Wiener random process, that is, a process continuous with probability 1 and satisfying the conditions:

1) $\xi(0) = 0$ with probability 1;

2) the increments of $\xi(t)$ are independent on non-intersecting intervals;

3) $\xi(t + \Delta) - \xi(t)$ has normal distribution with mean 0 and variance Δ.

Formula (8) and Theorem II of §4 imply a number of limit relations. Thus, for instance, by considering the functional

$$f(x) = \max_{0 \leq t \leq 1} x(t), \quad x(t) \in C,$$

we obtain for $a \geq 0$,

$$\mathsf{P}\{\max_{0 \leq t \leq 1} \xi_n(t) > a\} = \mathsf{P}\{\max_{1 \leq k \leq n} s_{n,k} > a\} \to \mathsf{P}\{\max_{0 \leq t \leq 1} \xi(t) > a\} =$$

$$= 2\Big\{1 - \frac{1}{\sqrt{2\pi}} \int_\alpha^\infty e^{-u^2/2} du\Big\}.$$

Given continuous functions $a(t)$ and $b(t)$ on [0,1] such that $a(0) < 0 < b(0)$ and $a(t) \leq b(t)$, consider the functional $f(x)$ equal to 1 if $a(t) < x(t) < b(t)$ for all t and 0 otherwise. We find that

$$\mathsf{P}\{\text{for all } t,\ a(t) < \xi_n(t) < b(t)\} \to \mathsf{P}\{\text{for all } t,\ a(t) < \xi(t) < b(t)\}$$

and hence

$$\mathsf{P}\{\text{for all } k,\ a(t_{n,k}) \leq s_{n,k} \leq b(t_{n,k})\} \to \mathsf{P}\{\text{for all } t,\ a(t) < \xi(t) < b(t)\}.$$

The latter result was obtained by A.N. Kolmogorov in 1931 under somewhat different assumptions (see [28], [29]). This remark allows us to obtain similar limiting relations for the joint distribution of any finite number of functionals, for example

$$\mathbf{P}\{\min_{1\le k\le n} s_{n,k} < x,\ \max_{1\le k\le n} s_{n,k} < y\} \to \mathbf{P}\{\min_{0\le t\le 1} \xi(t) < x,\ \max_{0\le t\le 1} \xi(t) < y\}.$$

Statement (8) contains theorems of Erdös, Kac and Donsker (see [31]–[33]) as particular cases. According to these theorems, the distributions of functions of "accumulated sums" converge to the distributions of the corresponding functionals of Wiener's process. The method by Erdös and Kac in its final form given by Donsker [33] is first to establish the convergence of

$$P_{\xi_n}(A) \to P_{\xi}(A) \tag{9}$$

for "strips", that is, for the sets A whose elements satisfy $a(t) < x(t) < b(t)$ for all t, where $a(t)$ and $b(t)$ are step functions (see Figure).

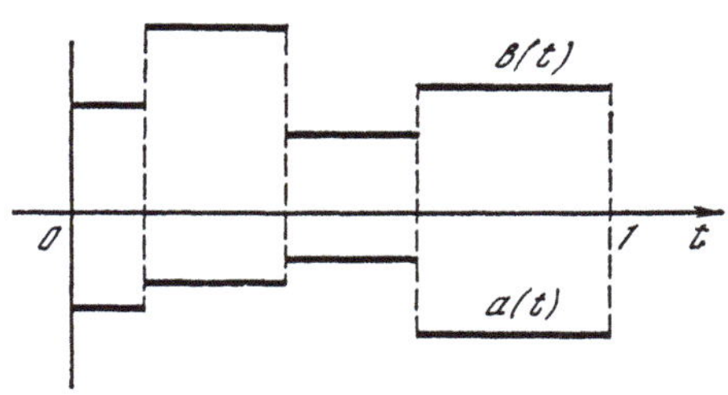

Then by approximating functionals f that are continuous almost everywhere with respect to P_ξ by characteristic functionals (in the set-theoretic sense) of "strips", Donsker established the convergence of the distribution functions $F_f^{(n)}(\alpha) = \mathbf{P}\{f(\xi_n) < \alpha\}$ which, by Theorem II of §4, is equivalent to weak convergence (8). To pass from (8) to (9) we can also use Theorem V of §4, since the system S of all "strips" satisfies the conditions of this theorem. Another method for proving (8) based on Theorem III of §4 is given in [17].

Remark. It is known that by using the method of upper and lower functions (see [30]) for the case when the process $\xi_n(t)$ is constructed by the accumulated sums of terms that are independent or connected in a Markov chain, and if the process $\xi(t)$ is a properly chosen Markov process satisfying the Fokker-Planck differential equations, we can obtain a statement of the following type: for any piecewise smooth functions $a(t)$ and $b(t)$

$$\mathsf{P}\{\text{for all } t,\ a(t) < \xi_n(t) < b(t)\} \to \mathsf{P}\{\text{for all } t,\ a(t) < \xi(t) < b(t)\}. \qquad (10)$$

Clearly, using Theorem III of §4, we can infer from the convergence (10) that $P_{\xi_n} \Rightarrow P_\xi$. Nowadays the method of upper and lower functions is successfully used by I.I. Gikhman [34]–[39], who has applied it to a number of problems concerning sums of random variables and certain problems in mathematical statistics (see below).

6.2. Approximation of the empirical distribution to the theoretical one. Let ξ be a random variable with uniform probability distribution on [0,1], let $F_n(t)$ be the empirical distribution function constructed by n independent observations of ξ and let

$$\xi_n(t) = (F_n(t) - t)\sqrt{n}, \quad 0 \le t \le 1.$$

Denote by $\xi(t)$ the Gaussian process (that is, the process all finite-dimensional distributions of which are Gaussian), continuous with probability 1 and with correlation function

$$\mathsf{E}\{\xi(t), \xi(s)\} = t(1-s),\ 0 \le t \le s \le 1.$$

Using the higher-dimensional Laplace theorem it is easy to show that the finite-dimensional distributions of the process $\xi_n(t)$ converge to the finite-dimensional distributions of the process $\xi(t)$ as $n \to \infty$.

This makes it possible to derive convergence of distributions to the distribution $f(\xi)$ for a broad class of functionals $f(\xi_n)$. This can be done by all the methods described in §6.1. (See, for example, the papers by Donsker [45], Gikhman [38], Fortet and Mourier [23]. The latter makes use of the method of characteristic functionals when ξ_n and ξ are considered as random elements in the space $L^2[0,1]$. The paper by Donsker [45] is close to [33].)

Based on this result we can obtain a number of new non-parametric tests (see, for example, [46]), generalizing the known tests of Kolmogorov [42] and Smirnov [43], [44].

References

1. A.N. Kolmogorov, *Basic notions of probability theory*, ONTI, Moscow-Leningrad, 1936 (in Russian).
2. P. Halmos, *Measure theory*, Van Nostrand, Princeton, N.J.,1950.
3. B.V. Gnedenko and A.N. Kolmogorov, *Limit distributions for sums of independent random variables*, GTTI, Moscow, 1949 (in Russian).
4. A. Kolmogoroff, 'Algèbres de Boole métriques complètes', In: *VI Zjazd Matematykow Polskich Warszawa*, 1948, pp. 22–30.
5. N. Wiener, *Publ. Mass. Inst. Techn.*, **2** (1923), 60.
6. E.E. Slutsky, *C. R. Acad. Sci. Paris*, **187** (1928), 370–372.
7. V.I. Glivenko, *Atti Accad. Naz. Lincei. Rend.* **8** (1928), 480–483.
8. V.I. Glivenko, *Atti Accad. Naz. Lincei. Rend.* **8** (1928), 673–676.
9. V.I. Glivenko, *Atti Accad. Naz. Lincei. Rend.* **9** (1929), 830–833.
10. J.L. Doob, *Stochastic processes*, New York, 1953.
11. A. Blanc-Lapierre and R. Fortet, *Théorie des fonctions aléatoires*, Paris, 1953.
12. E.E. Slutsky, *G. Ist. Ital. Attuar.* **8** (1937), 183–199.
13. E.E. Slutskii, *Trudy Sredneaz. Univ.* **31** (1939), 1–15 (in Russian).
14. E.B. Dynkin, *Izv. Akad. Nauk SSSR Ser. Mat.* **16** (1952), 569–572 (in Russian).
15. J.R. Kinney, *Trans. Amer. Math. Soc.* **74** (1953), 280–302.
16. W. Ambrose, *Trans. Amer. Math. Soc.* **47** (1940), 66–79.
17. Yu.V. Prokhorov, *Uspekhi Mat. Nauk* **8**:3 (1953), 165–166 (in Russian).
18. A.N. Kolmogoroff, *C. R. Acad. Sci. Paris* **200** (1935), 1717–1718.
19. E. Hopf, *J. Rat. Mech. and Analysis* **1** (1952), 87–123.
20. E. Mourier, *C. R. Acad. Sci. Paris* **231** (1950), 28–30.
21. R. Fortet and E. Mourier, *C. R. Acad. Sci. Paris* **238** (1954), 1557–1559.
22. R. Fortet and E. Mourier, *Bull. Sci. Math.* **78** (1954), 14–30.
23. R. Fortet and E. Mourier, *Ann. Ecole Norm. Super.* (1954), 267–285.
24. E. Mourier, 'Eléments aléatoires dans un espace de Banach', Thèse Doct. Paris, 1954.
25. L. Bachelier, *Les méthodes du calcul des probabilités*, Paris, 1939.
26. R. Lüneburg, *Math. Ann.* **104** (1931), 700.
27. I.G. Petrovsky, *Math. Ann.* **109** (1934), 425–444.

28. A.N. Kolmogorov, *Izv. Akad. Nauk SSSR Ser. Fiz.-Mat.* (1931), 959–962 (in Russian).
29. A.N. Kolmogorov, *Izv. Akad. Nauk SSSR Ser. Fiz.-Mat.* (1933), 363–372 (in Russian).
30. A.Ya. Khinchin, *Asymptotic laws of probability theory*, ONTI, Moscow-Leningrad, 1936 (in Russian).
31. P. Erdös and M. Kac, *Bull. Amer. Math. Soc.* **52** (1946), 292–302.
32. P. Erdös and M. Kac, *Bull. Amer. Math. Soc.* **53** (1947), 1011–1020.
33. M. Donsker, *Mem. Amer. Math. Soc.* **6** (1951), 1–12.
34. I.I. Gikhman, *Ukrain. Mat. Zh.* **6**:1 (1954), 28–36. (in Russian).
35. I.I. Gikhman, *Mat. Sb. Kiev. Gos. Univ.* 1953, No. 7, 8–15 (in Russian).
36. I.I. Gikhman, *Dokl. Akad. Nauk SSSR* **82**:6 (1952), 837–840 (in Russian).
37. I.I. Gikhman, *Dokl. Akad. Nauk SSSR* **56** (1947), 961–964 (in Russian).
38. I.I. Gikhman, *Ukrain. Mat. Zh.* **5** (1953), 413–433 (in Russian).
39. I.I. Gikhman, *Dokl. Akad. Nauk SSSR* **91**:4 (1953), 1003–1006 (in Russian).
40. M. Kac, In: *Proc. Second Berkeley Sympos.* 1951, pp. 189–215.
41. G. Maruyama, *Nat. Sci. Rept Ochanomicu Univ.* **1** (1953).
42. A.N. Kolmogoroff, *G. Ist. Ital. Attuar.* **4** (1933), 83–91.
43. N.V. Smirnov, *Rev. Math. Moscow* **6** (1939), 3–26.
44. N.V. Smirnov, *Byull. Moskov. Gos. Univ. Ser. A* **2**:2 (1939), 3–14 (in Russian).
45. M. Donsker, *Ann. Math. Statist.* **23** (1952), 277–281.
46. T. Anderson and D. Darlong, *Ann. Math. Statist* **23** (1952), 193–212.

45. ON THE PROPERTIES OF P. LEVY'S CONCENTRATION FUNCTIONS *[1]

In what follows

$$\xi = \xi_1 + \ldots + \xi_n$$

denotes the sum of independent random variables ξ_k,

$$Q(l) = \sup_x \mathbf{P}\{x \leq \xi \leq x + l\}$$

the concentration function of the sum ξ and Q_k the concentration function of the ξ_k. Theorem 48.2 from the well-known monograph by P. Lévy ' Théorie de l'addition des variables aléatoires', can be formulated in the following way:

For any ϵ, $1 \geq \epsilon > 0$, and $\beta > 0$ there exist two constants $\delta > 0$ and $N > 0$ such that for $n \geq N$ the inequality

$$Q_k(l) \leq 1 - \epsilon$$

implies

$$Q(\delta l \sqrt{n}) \leq \beta. \tag{1}$$

In my recent work I happened to need a generalization of (1). The following theorem [1] suited my requirements:

Theorem. *There exists a constant C such that*[2]

$$L \geq l, \quad L^2 \geq l^2 \log s,$$

where

$$s = \sum_{k=1}^{n} (1 - Q_k(l)),$$

imply that

$$Q(L) \leq CL / l\sqrt{s}. \tag{2}$$

---

* 'Sur les propriétés des fonctions de concentrations de M.P. Lévy', *Ann. Inst. H. Poincaré* **16**:1 (1958), 27–34.

[1] In two issues of the *Proc. Inst. Statistics Paris Univ.* articles in honour of Paul Lévy have been published. This paper was submitted too late to be included in these issues.

[2] Here and in what follows logarithms to the base 2 are considered.

In this paper I give a proof of this theorem. Another refinement of (1) was given with a detailed proof by W. Doeblin in [2]. His result is not clearly formulated in [2], and there is a misprint.[3] It seems evident, however, that in terms of concentration functions the result by W. Doeblin may be formulated in the following way:

For any $\epsilon > 0$ and $\beta > 0$, there exist constants $\delta > 0$ and $N > 0$ such that for

$$\max l_k \geq \frac{L}{N}, \quad L = \sqrt{\sum_{k=1}^{n} l_k^2},$$

the inequality

$$Q_k(l_k) \leq 1 - \epsilon$$

implies

$$Q(\delta L) \leq \beta. \tag{3}$$

It would be interesting to find inequalities that include (2) and (3) as particular cases. In general, I would like to point out that the further development of elementary methods for direct probabilistic computations, so brilliantly developed in France by P. Lévy and W. Doeblin, seems to remain as urgent as the development of classical or functional analytic methods. In any case, I could not prove the results of [1] without these elementary direct methods. It is quite possible that mathematicians that have a better command of the subtle properties of characteristic functions will sooner or later be able to prove and even generalize theorems from [1] using purely analytical methods, as already has happened with results initially proved by direct methods of probability theory. It seems, however, that today we are still in a period when the competition between these two trends is bringing about the best results. While P. Lévy who masters these two techniques equally well, makes use of these two kinds of method in his own work, it is very desirable that, after the premature death of Doeblin, the younger generation of probabilists should not forget the direct methods despite their admiration (admittedly, quite justified) for the power of the methods which make use of distributions in function spaces.

We will use the following notation:

$$F(x) = \mathbf{P}\{\xi < x\}, \quad a = \mathbf{E}\xi, \quad \sigma^2 = \mathbf{D}\xi.$$

[3] It says $\sum_i^2 l$ instead of $\sum_i l_i^2$.

The following lemma is well known.

Lemma 1. *There exists a constant C_1 such that*

$$|\xi_k - a_k| \leq l,$$

where $a_k = \mathbf{E}\xi_k$ $(k = 1, 2, \dots, n)$, implies the inequality

$$|F(x) - G_{a,\sigma}(x)| \leq C_1 l/\sigma,$$

$G_{a,\sigma}$ being the normal distribution function with parameters a, σ.

The following lemma is a generalization of Lemma 48.2 from 'Théorie de l'addition des variables aléatoires'.

Lemma 2. *There exists a constant C_2 such that*

$$\mathbf{P}\{\xi_k = x_k\} = \mathbf{P}\{\xi_k = -x_k\} = \tfrac{1}{2}, \quad x_k \geq l \quad (k = 1, 2, \dots, n),$$

$$L \geq l\sqrt{\log n}(n \geq 2)$$

imply that

$$Q(L) \leq C_2 L/l\sqrt{n}.$$

To prove this lemma, it is necessary to distinguish between the following three different cases:

1°. $L > l\sqrt{n}$.

2°. $L \leq l\sqrt{n}$ and moreover, there exists an $r \geq 1$ such that the number n_r of numbers x_k such that

$$4^{r-1}l \leq x_k < 4^r l \tag{4}$$

satisfies the inequalities

$$n_r \geq n/4^r, \quad n_r \geq n/4\log n;$$

3°. $L \leq l\sqrt{n}$ and there is no r satisfying the properties indicated in 2°.

In the first case

$$Q(L) \leq 1 < L/l\sqrt{n}.$$

In the second case it is natural to consider the sum ξ_* of those ξ_k satisfying (4). Since all the terms of this sum satisfy the inequalities

$$|\xi_k| < l_* = 4^r l, \quad \sigma_* = \sqrt{\mathbf{D}\xi_*} \geq 4^{r-1}l\sqrt{n_r} \geq \frac{l}{2}\sqrt{n},$$

according to Lemma 1 for the corresponding function $F_*(x)$ we have

$$|F_*(x) - G_{a_*\sigma_*}(x)| \leq C_1 l_*/\sigma_* \leq \frac{4C_1}{\sqrt{n_r}} \leq \frac{8C_1\sqrt{\log n}}{\sqrt{n}} \leq \frac{8C_1 L}{l\sqrt{n}}. \tag{5}$$

Since the concentration function $Q_{a,\sigma}(L)$ of the normal distribution function $G_{a,\sigma}$ satisfies the inequality

$$Q_{a,\sigma}(L) \leq \frac{L}{\sqrt{2\pi}\sigma},$$

we have in the second case

$$Q(L) \leq Q_*(L) \leq \frac{L}{\sqrt{2\pi}\sigma_*} + \frac{8C_1 L}{l\sqrt{n}} \leq \left(\sqrt{\frac{2}{\pi}} + 8C_1\right)\frac{L}{l\sqrt{n}}, \tag{6}$$

where $Q_*(L)$ denotes the concentration function of ξ_*.

In the third case, for each $r \geq 1$ one of the following two inequalities is satisfied:

$$\text{either} \quad n_r < n/4^r, \tag{7}$$

$$\text{or} \quad n_r < n/4\log n. \tag{8}$$

The sum of the n' numbers n_r satisfying (7) is not greater than

$$n\sum_{r=1}^{\infty}\frac{1}{4^r} = \frac{n}{3}.$$

Since the sum of all the n_r equals n, the sum of the n'' numbers n_r satisfying (8) is not smaller than $\frac{1}{2}n$. Clearly, among all these n_r there are at least $2\log n$ non-zero numbers. Since the number of r satisfying

$$4^r l < L \leq l\sqrt{n},$$

is smaller than $\log n$, there exists some number, not less than $\log n$, of numbers r for which

$$n_r > 0, \quad 4^r l \geq L.$$

Taking from these r either only even, or only odd numbers in increasing order we obtain the sequence

$$r_1 < r_2 < \ldots < r_s, \; s > \tfrac{1}{2}\log n,$$

for which

$$4^{r_i} l \geq L, \quad r_{i+1} \geq r_{i+2}.$$

We choose the terms $\xi_{k_1}, \xi_{k_2}, \ldots, \xi_{k_s}$ so that

$$4^{r_{i-1}} l \leq x_{k_i} < 4^{r_i} l,$$

and study their sum ξ'. Since

$$|\xi_{k_i}| < 4^{r_i} l = L_i \geq L, \quad 4L_{i-1} \leq |\xi_{k_i}| < L_i \quad (i = 2, 3, \ldots, s)$$

a more or less elementary argument, which is left to the reader, shows that the sum ξ' can belong to an interval of length L only for one single quite specific combination (if $n > 1$) of the signs of the terms ξ_{k_i}. In other words, with probability at most $2^{-s} \leq n^{-\frac{1}{2}}$.

Hence, in the third case, because of the condition $L \geq l$ we have

$$Q(L) \leq Q'(L) \leq 1/\sqrt{n} \leq L/l\sqrt{n}. \tag{9}$$

Comparing (5), (6) and (9) in these three cases we see that Lemma 2 is proved.

Now let us prove the theorem. The relations

$$F_k(u_k(y)) \leq y \leq F_k(u_k(y) + 0) \tag{10}$$

determine $u_k(y)$, $0 < y < 1$. (Since $F_k(x)$ is the distribution function of a random variable ξ_k, it uniquely determines the values of $u_k(y)$ for all y, $0 < y < 1$ except perhaps on a countable set of points y.)[4] It can be considered as the inverse function

$$u_k(y) = F_k^{-1}(y),$$

whose definition must be made suitably precise for all y corresponding to discontinuity points of $F_k(x)$. If we assume that the random variables η_k are uniformly distributed on (0,1) and are independent, then $\xi'_k = u_k(\eta_k)$ are also independent and have distributions $F_k(x)$. Clearly, it may be assumed without

[4] There is no uniqueness for the points y corresponding to the intervals of constancy of $F_k(x)$, but clearly $u_k(y)$ can be well-defined on the whole interval (0,1) so that (10) holds.

loss of generality that the initially given random variables ξ_k are expressed in this way.

It may be assumed that all the $Q_k(l)$ are smaller than 1, since by excluding from our discussion those terms for which $Q_k(l) = 1$ and obtaining (2) for the sum of all other terms we can include the excluded terms again, and this will not enlarge $Q(L)$. We set

$$4\epsilon_k = 1 - Q_k(l),$$
$$x'_k = u_k(\epsilon_k), \quad x''_k = u_k(1 - \epsilon_k).$$

Clearly,

$$x''_k - x'_k > l.$$

Denote by $k_1, k_2, \ldots, k_m$ the indices k for which either $\eta_k < \epsilon_k$ or $\eta_k > 1 - \epsilon_k$.

We fix the quantities

$$Z_r = \begin{cases} \eta_{k_r}, & \text{if } \eta_{k_r} < \epsilon_{k_r}, \\ 1 - \eta_{k_r}, & \text{if } \eta_{k_r} > 1 - \epsilon_{k_r}. \end{cases}$$

It is easy to see that the joint conditional distribution of the random variables ξ_{k_i} is such that they remain independent and each of them has the distribution

$$\mathsf{P}\{\xi_{k_r} = a_r + x_r\} = \mathsf{P}\{\xi_{k_r} = a_r - x_r\} = \tfrac{1}{2},$$

where

$$x_r = \tfrac{1}{2}[u_{k_r}(1 - Z_r) - u_{k_r}(Z_r)],$$
$$a_r = \tfrac{1}{2}[u_{k_r}(1 - Z_r) + u_{k_r}(Z_r)].$$

Applying (for fixed k_r and Z_r) Lemma 2 to the $\xi'_r = \xi_{k_r} - a_r$ for which $|\xi'_r| \geq l' = \frac{1}{2}l$, we obtain for $L \geq l'\sqrt{\log m}$ the inequality

$$Q(L) \leq C_2 L / l'\sqrt{m} = 2C_2 L / l\sqrt{m}. \tag{11}$$

Inequality (11) is proved for the conditional function $Q(L)$ (when m, k_r, a_r, Z_r are fixed and $L \geq l'\sqrt{\log m}$.) This immediately implies the inequality

$$Q(L) \leq p + 4C_2 L / l\sqrt{s}$$

for the absolute function $Q(L)$, where

$$p = 1 - \mathsf{P}\{\tfrac{1}{4}s \leq m \leq s\}.$$

We now merely have to estimate this latter probability. Since

$$\mathsf{E}m = \sum_k 2\epsilon_k = \tfrac{1}{2}s, \quad \mathsf{D}m = \sum_k 2\epsilon_k(1-\epsilon_k) \le \tfrac{1}{2}s,$$

it follows from Chebyshev's inequality for $k = \frac{1}{4}s$ that

$$1 - \mathsf{P}\left\{\tfrac{1}{4}s \le m \le s\right\} \le \mathsf{P}\{|m - \mathsf{E}m| \ge k\} \le \frac{\mathsf{D}m}{k^2} \le \frac{8}{s}.$$

Thus we finally obtain

$$Q(L) \le \frac{8}{s} + \frac{4C_2L}{l\sqrt{s}} \quad \text{for } s \ge 1$$

or, since $Q(L) \le 1$, and by our assumptions $L \ge l$,

$$Q(L) \le CL/l\sqrt{s},$$

where $C = 8 + 4C_2$.

References

1. A.N. Kolmogorov, 'Two uniform limit theorems for sums of independent terms', *Teor. Veroyatnost. i Primenen.* **1**:4 (1956), 426–436 (in Russian). (Translated as *Theory Probab. Appl.*)
2. W. Doeblin, 'Sur les sommes d'un grand nombre des variables aléatoires indépendantes', *Bull. Sci. Math.* **63** (1939), 23–32; 35–64.

46. TRANSITION OF BRANCHING PROCESSES TO DIFFUSION PROCESSES AND RELATED GENETIC PROBLEMS *

Survey Report

The abstract theory of "branching processes" with the "number of particles" taking arbitrary non-negative real values (see [1]) was obviously developed to include the limiting laws of behaviour of ordinary branching processes for a large number of particles. Apparently, the most important particular case of these processes is a diffusion process. In the simplest one-dimensional case the corresponding Fokker-Planck equation is of the form

$$u_\tau = \tfrac{1}{2}b(xu)_{xx} - a(xu)_x. \tag{1}$$

A more complex related problem was considered by R.A. Fisher in his remarkable book [2]. In a population of N individuals a gene A can be present in ν copies, $0 \le \nu \le 2N$. For $\xi = \nu/2N$, $0 \le \xi \le 1$, Fisher obtains the Fokker-Planck equation

$$u_\tau = \tfrac{1}{2}b[x(1-x)u]_{xx} \tag{2}$$

which, even for large N, is only reasonable for x not too close to 0 or 1. On the other hand, near $x = 0$, ν obeys the laws of a branching process, and near $x = 1$ so does $2N - \nu$. Similar problems may appear also in physics and chemistry.

Feller's paper [3] may serve as an introduction to the whole range of problems. Apart from [2], the corresponding genetic problems are discussed in [4], [5] (without mathematical rigour). Basic information on branching processes is given in Sevastyanov's review [6].

It is interesting that, from the viewpoint of the theory of diffusion processes here we have a degenerate case: the diffusion coefficient $bx/2$, or $bx(1-x)/2$, vanishes on natural boundaries. In [3] Feller stresses the difficulties that appear because of this. However, I do not agree with him when (on p.238) he says that certain results obtained by biologists are meaningless: these results are true under proper interpretation. I think that in this field there are many subjects that deserve further development. In what follows I only treat certain very simple results obtained by other methods and give a correct interpretation

* *Teor. Veroyatnost. i Primenen.* 4:2 (1959), 233–236 (in Russian). (Translated as *Theory Probab. Appl.*)

of certain results of Fisher which were tackled by Feller, from the diffusion viewpoint.

Suppose that during one step of a process one particle turns into k particles with probability $p_k(N)$,

$$\sum_k p_k(N) = 1, \quad \sum_k kp_k(N) = 1 + \alpha(N), \quad \sum_k k(k-1)p_k(N) = b(N),$$
$$\sum_k k(k-1)(k-2)p_k(N) = c(N).$$

We let N tend to infinity and suppose that

$$p_k(N) \to p_k, \tag{3}$$

$$N\alpha(N) \to a, \tag{4}$$

$$b(N) \to b, \tag{5}$$

$$c(N) \to c, \tag{6}$$

where a, b and c are finite.

The meaning of assumption (4) is that for the number of particles of order N and α of order $1/N$, the role of the number of particles in the process of systematic increase (or decrease) and the role of random scattering are of the same order. If the order of α is greater, then for a large number of particles the process becomes almost deterministic, whereas if α is of a smaller order, it can be neglected. "Borderline" phenomena for α close to zero are studied in Sevastyanov's work [7]. It is here that the idealized diffusion model can be expected to bring about some results.

If we now set $\nu = N\xi$, then a natural time variable for studying the behaviour of ξ is given by τ which is related to the number of steps t of the process by the relation $t = N\tau$. Under these assumptions the process $\xi(\tau)$ gives a diffusion process with Fokker-Planck equation (1) in the limit as $N \to \infty$.

At $a = 0$ equation (1) has the solution

$$u(\tau, x) = (c/\tau^2)e^{-2x/b\tau} \tag{7}$$

with singularity at $\tau = 0$, $x = 0$. In a certain sense it describes the behaviour of one particle that appeared at time $\tau = 0$, meaning that if for $\tau_0 > 0$ we choose c so that

$$\int_0^\infty u(\tau_0, x)dx = 1,$$

then, provided that $\xi(\tau_0) > 0$, the conditional distribution of $\xi(\tau)$ for $\tau > \tau_0$ converges to the distribution

$$\mathbf{P}\{\xi(\tau) < x\} = 0, \quad x \leq 0,$$
$$\mathbf{P}\{\xi(\tau) < x\} = P(\tau) + \int_0^x u(\tau, x)dx, \quad x > 0,$$
$$P(\tau) = \mathbf{P}\{\xi(\tau) = 0\} = 1 - \int_0^\infty u(\tau, x)dx$$

as $N \to \infty$.

The general solution of the stationary equation at $a = 0$, that is, of the equation

$$(xu)_{xx} = 0$$

has the form

$$u(x) = c_1/x + c_2.$$

This is not a probability distribution. However, the solution

$$u(x) = c_1/x$$

has the following statistical meaning: if at each step a new particle appears independently of time with probability p, and $\mu(x_1, x_2)$ is the number of those initial particles whose progeny at a certain fixed time t satisfies

$$Nx_1 \leq \nu \leq Nx_2,$$

then the expectation $\mathbf{E}\mu(x_1, x_2)$ tends to the limit

$$\frac{2}{b}\int_{x_1}^{x_2} \frac{dx}{x}$$

as $N \to \infty$.

The solution

$$c_1/x + c_2/(1 - x)$$

of the stationary equation

$$(x(1 - x)u)_{xx} = 0$$

in Fisher's theory can be interpreted in a similar way. This interpretation was even verified in an experiment and appeared to be true.

18 November 1958

References

1. M. Irzhina, *Czechoslovak Math. J.* **8** (1958), 83.
2. R.A. Fisher, *The genetical theory of natural selection*, Oxford, 1930.
3. W. Feller, In: *Proc. Second Berkeley Sympos.*, 1951, pp. 227–246.
4. J. Malécot, *Les mathématiques de l'hérédité*, Paris, 1948.
5. S. Wright, *Bull. Amer. Math. Soc.* **48** (1942), 223–246.
6. B.A. Sevastyanov, *Uspekhi Mat. Nauk* **6**:6 (1951), 37–99 (in Russian).
7. B.A. Sevastyanov, *Teor. Veroyatnost. i Primenen.* **4**:2 (1959), 121–135 (in Russian). (Translated as *Theory Probab. Appl.*)

47. ON THE CLASSES $\Phi^{(n)}$ OF FORTET AND BLANC-LAPIERRE*

If for a real process $\xi(t)$ there exist continuous absolute moments of order n such that

$$\mathsf{E}|\xi(t)|^n \leq C(1+|t|^s), \tag{1}$$

then for $k \leq n$ the moments

$$m_\xi^{(k)}(t_1,\dots,t_k) = \mathsf{E}[\xi(t_1)\dots\xi(t_k)]$$

may be represented as

$$m_\xi^{(k)}(t_1,\dots,t_k) = \int\dots\int e^{i(t,\lambda)}\mu_\xi^{(k)}(\lambda_1,\dots,\lambda_k)d\lambda_1\dots d\lambda_k,$$

where the $\mu_\xi^{(k)}$ are generalized functions.

A process belongs to the class $\Phi^{(n)}$ (see [1]), if all the $\mu_\xi^{(k)}$ are of "finite measure type", that is, if

$$m_\xi^{(k)}(t_1,\dots,t_k) = \int\dots\int e^{i(t,\lambda)}M_\xi^{(k)}(d\lambda),$$

where $M_\xi^{(k)}(A)$ is a finite complex measure in k-dimensional space.

It is well known that when $n=2$, for a strictly stationary process $\xi(t)$ the existence of the moments $\mathsf{E}|\xi|$ and $\mathsf{E}|\xi|^2$ implies that it is of the class $\Phi^{(2)}$.

It can be shown by an example that similar statements fail for $n>2$.

References

1. A. Blanc-Lapierre and R. Fortet, *Théorie des fonctions aléatoires*, Paris, 1953.

* *Teor. Veroyatnost. i Primenen.* 5:3 (1960), 373 (in Russian). (Translated as *Theory Probab. Appl.*)

48. ON CONDITIONS OF STRONG MIXING OF A GAUSSIAN STATIONARY PROCESS *

Jointly with Yu.A. Rozanov

As is well known, two σ-algebras of events $\mathfrak{M}'$ and $\mathfrak{M}''$ are said to be independent if for any $A' \in \mathfrak{M}'$, $A'' \in \mathfrak{M}''$, $P(A', A'') = P(A')P(A'')$.

M. Rosenblatt [1] has suggested a natural measure of dependence between two σ-algebras of events:

$$\alpha(\mathfrak{M}', \mathfrak{M}'') = \sup_{A' \in \mathfrak{M}', A'' \in \mathfrak{M}''} |P(A' \cap A'') - P(A')P(A'')|.$$

For a stationary random process $\xi(t)$ the measure $\alpha(\mathfrak{M}^t_{\infty}, \mathfrak{M}^{\infty}_{t+\tau})$ (where $\mathfrak{M}^t_s$ denotes the σ-algebra of events generated by $\xi(u)$, $s \le u \le t$) depends only on τ and will be denoted by $\alpha(\tau)$. If $\alpha(\tau) \to 0$ as $\tau \to \infty$, then it is said that the process $\xi(t)$ has the strong mixing property. In this paper we indicate which properties of the spectral function $F(\lambda)$ of the process guarantee the strong mixing condition for Gaussian processes.

1. For any two systems $\{\xi\} = \mathfrak{A}'$ and $\{\eta\} = \mathfrak{A}''$ with finite second moments we introduce the index

$$\rho(\mathfrak{A}', \mathfrak{A}'') = \sup_{\xi, \eta} \frac{|\mathsf{E}(\xi - \mathsf{E}\xi)(\eta - \mathsf{E}\eta)|}{[\mathsf{E}(\xi - \mathsf{E}\xi)^2 \mathsf{E}(\eta - \mathsf{E}\eta)^2]^{\frac{1}{2}}}.$$

If $\mathfrak{A}'$ and $\mathfrak{A}''$ are, respectively, the families of all variables with finite second moments measurable with respect to the σ-algebras $\mathfrak{M}'$ and $\mathfrak{M}''$, then, by definition (see [2]), $\rho(\mathfrak{M}', \mathfrak{M}'') = \rho(\mathfrak{A}', \mathfrak{A}'')$ is the maximal correlation coefficient between $\mathfrak{M}'$ and $\mathfrak{M}''$.

Clearly,

$$\alpha(\mathfrak{M}', \mathfrak{M}'') \le \rho(\mathfrak{M}', \mathfrak{M}''). \tag{1}$$

Now let $\{\xi\}$ and $\{\eta\}$ be two sets of random variables which have (for any finite set $\xi_1, \dots, \xi_m, \eta_1, \dots, \eta_m$) Gaussian joint distributions, and let $\mathfrak{M}_\xi$ and $\mathfrak{M}_\eta$ be the σ-algebras generated by the events $(\xi \in \Gamma')$ and $(\eta \in \Gamma'')$ respectively, where Γ' and Γ'' are arbitrary Borel sets on the line, and let H_ξ and H_η be the closed (with respect to the mean square) linear spans of $\{\xi\}$ and $\{\eta\}$.

* *Teor. Veroyatnost. i Primenen.* (Probability theory and its applications) **5**:2 (1960), 222–227 (in Russian).

Theorem 1.[1]

$$\rho(\mathfrak{M}_\xi, \mathfrak{M}_\eta) = \rho(H_\xi, H_\eta). \tag{2}$$

Theorem 2. *The maximal correlation coefficient satisfies the following inequalities:*

$$\alpha(\mathfrak{M}_\xi, \mathfrak{M}_\eta) \le \rho(\mathfrak{M}_\xi, \mathfrak{M}_\eta) \le 2\pi\alpha(\mathfrak{M}_\xi, \mathfrak{M}_\eta). \tag{3}$$

Proof of Theorem 1. Clearly, we may restrict ourselves to the case when $\{\xi\}$ and $\{\eta\}$ consist of a finite number of variables.

Further, in H_ξ and H_η we can choose $\xi_1, \dots, \xi_m$ and $\eta_1, \dots, \eta_n$ so that only those ξ_k and η_k with the same indices are dependent, and each $\xi \in \{\xi\}$, $\eta \in \{\eta\}$ is a function of $\xi_1, \dots, \xi_m$ and $\eta_1, \dots, \eta_n$ respectively. We may also assume that $\mathbf{E}\xi_k = 0$, $\mathbf{E}\eta_j = 0$, $\mathbf{D}\xi_k = \mathbf{D}\eta_j = 1$, $k = \overline{1,m}$, $j = \overline{1,n}$.

Then the quantities $f = f(\xi_1, \dots, \xi_m)$ and $g = g(\eta_1, \dots, \eta_n)$ may be represented as

$$f = \sum_1^m f_k(\xi_1, \dots, \xi_k), \quad g = \sum_1^n g_j(\eta_1, \dots, \eta_j),$$

where

$$f_k = \mathbf{E}(f|\xi_1, \dots, \xi_k) - \mathbf{E}(f|\xi_1, \dots, \xi_{k-1}),$$
$$g_j = \mathbf{E}(g|\eta_1, \dots, \eta_j) - \mathbf{E}(g|\eta_1, \dots, \eta_{j-1}).$$

Note that for $k \ge j$,

$$\mathbf{E}[\mathbf{E}(f|\xi_1, \dots, \xi_k)|\eta_1, \dots, \eta_j] = \mathbf{E}[\mathbf{E}(f|\xi_1, \dots, \xi_j)|\eta_1, \dots, \eta_j]$$

and similarly for $j \ge k$

$$\mathbf{E}[\mathbf{E}(g|\eta_1, \dots, \eta_j)|\xi_1, \dots, \xi_k] = \mathbf{E}[\mathbf{E}(g|\eta_1, \dots, \eta_k)|\xi_1, \dots, \xi_k],$$

which implies that $\mathbf{E}f_k g_j = 0$ for $k \ne j$ and, taking $m \le n$ for the sake of definiteness,

$$\mathbf{E}fg = \sum_1^m \mathbf{E}f_k g_k = \sum_1^m \mathbf{E}[\mathbf{E}(f_k g_k|\xi_1, \dots, \xi_{k-1}, \eta_1, \dots, \eta_{k-1})].$$

[1] Theorem 1 is a multidimensional generalization of the result by Sarmanov [2], according to which (2) holds for any two variables ξ and η with Gaussian joint distribution.

The variables ξ_k and η_k with Gaussian distribution do not depend on $\xi_1, \dots, \xi_{k-1}$, $\eta_1, \dots, \eta_{k-1}$, and Sarmanov's result [2] implies that

$$|\mathbf{E}(f_k g_k | \xi_1, \dots, \xi_{k-1}, \eta_1, \dots, \eta_{k-1})| \le \rho a_k b_k,$$

where

$$\rho = \rho(H_\xi, H_\eta), \quad a_k^2 = \mathbf{E}(f_k^2 | \xi_1, \dots, \xi_{k-1}, \eta_1, \dots, \eta_{k-1}),$$

$$b_k^2 = \mathbf{E}(g_k^2 | \xi_1, \dots, \xi_{k-1}, \eta_1, \dots, \eta_{k-1}).$$

Hence,

$$|\mathbf{E} fg| \le \rho \sum_1^m \mathbf{E} a_k b_k \le \rho,$$

which proves our theorem.

Proof of Theorem 2.[2] Take an arbitrary $\epsilon > 0$ and $\xi_\epsilon \in H_\xi$, $\eta_\epsilon \in H_\eta$ with $\mathbf{E}\xi_\epsilon = \mathbf{E}\eta_\epsilon = 0$, $\mathbf{D}\xi_\epsilon = \beta\eta_\epsilon = 1$ such that $r = \mathbf{E}\xi_\epsilon\eta_\epsilon > \rho - \epsilon$. Consider the events

$$A_\epsilon = \{\xi_\epsilon > 0\} \in \mathfrak{M}_\xi \text{ and } B_\epsilon = \{\eta_\epsilon > 0\} \in \mathfrak{M}_\eta.$$

We have ([9], p.321)

$$P(A_\epsilon B_\epsilon) = \frac{1}{4} + \frac{1}{2\pi}\sin^{-1} r, \quad P(A_\epsilon)P(B_\epsilon) = \frac{1}{4}$$

and, clearly ,

$$\frac{1}{2\pi}\sin^{-1} r = P(A_\epsilon B_\epsilon) - P(A_\epsilon)P(B_\epsilon) \le \alpha.$$

Further, if $\alpha > \frac{1}{4}$, then the inequality $\rho \le 2\pi\alpha$ is trivial. If, on the other hand, $\alpha \le \frac{1}{4}$, then

$$\rho - \epsilon \le r \le \sin 2\pi\alpha, \quad \rho \le 2\pi\alpha + \epsilon, \quad \rho \le 2\pi\alpha,$$

which, together with (1) and (2) proves our theorem.

Theorems 1 and 2 imply, in particular, that a Gaussian stationary process $\xi(t)$ satisfies the strong mixing condition if and only if the maximal correlation coefficient $\rho(\mathfrak{M}_{-\infty}^t, \mathfrak{M}_{t+\tau}^\infty) \to 0$ as $\tau \to \infty$ (by Theorem 1 it coincides with $\rho(\tau) = \rho(H_{-\infty}^t, H_{t+\tau}^\infty)$, where $H_{-\infty}^t$ and $H_{t+\tau}^\infty$ are the linear closures with

[2] This simple proof was suggested to the authors by Yu.V. Prokhorov.

respect to the mean square of the quantities $\xi(u)$, $u \leq t$ and $\xi(v)$, $v \geq t$ respectively). Moreover, (cf. [4], Introduction),

$$\alpha(\tau) \leq \rho(\tau) \leq 2\pi\alpha(\tau). \tag{3'}$$

2. Let $\xi(t)$ be a process that is stationary in the wide sense and let

$$\rho(\tau) = \rho(H_{-\infty}^{t}, H_{t+\tau}^{\infty})$$

(stationarity implies that $\rho(\tau)$ does not depend on t).

If the spectral function $F(\lambda)$ of our process is not absolutely continuous, then $\rho(\tau) = 1$ for all τ (see [5], [6]).

Let $F(\lambda)$ be absolutely continuous and $f(\lambda)$ the spectral density of the process $\xi(t)$, $f(\lambda) = F'(\lambda)$.

Theorem 3. *For integer time*

$$\rho(\tau) = \inf_{\phi} \operatorname*{ess\,sup}_{\lambda} \left[|f(\lambda) - e^{i\lambda\tau}\phi(e^{-i\lambda})| \frac{1}{f(\lambda)} \right], \tag{4}$$

where $\inf_\phi$ *is taken over all functions* $\phi(z)$ *that can be analytically continued inside the unit disc; for continuous time*

$$\rho(\tau) = \inf_{\phi} \operatorname*{ess\,sup}_{\lambda} \left[|f(\lambda) - e^{i\lambda\tau}\phi(\lambda)| \frac{1}{f(\lambda)} \right], \tag{4'}$$

where $\inf_\phi$ *is taken over all functions* $\phi(z)$ *that can be analytically continued into the lower half-plane.*

The proof of this theorem is based on a general lemma taken from functional analysis.

Lemma. *Let L be a Banach space and L^* its dual.*

Let H be a subspace of L and H^0 the set of linear functionals that vanish on H. Then

$$\sup_{h \in H, \|h\|=1} h^*(h) = \inf_{h^0 \in H^0} \|h^* - h^0\| \tag{5}$$

for any $h^* \in L^*$.

Proof. Since $(h^* - h^0)(h) = h^*(h)$, it follows that $h^*(h) \leq \|h^* - h^0\|$ for $h \in H$, $\|h\| = 1$, therefore $\sup_{h \in H, \|h\|=1} h^*(h) \leq \inf_{h^0 \in H^0} \|h^* - h^0\|$. Further, according to the Hahn-Banach extension theorem there exists a functional h_1^*, coinciding with h^* on H, with norm $\|h_1^*\|$ not exceeding $\sup\limits_{h\in H,\,\|h\|=1} h^*(h)$. Then $h^* - h_1^* = h_1^0 \in H^0$ and

$$\|h^* - h_1^0\| = \|h_1^*\| = \sup_{h\in H,\,\|h\|=1} h^*(h),$$

which proves (5).

Let us come back to our theorem. Clearly

$$\rho(\tau) = \sup_{p_1,p_3} \int e^{i\lambda\tau} p_1(\lambda)p_2(\lambda)f(\lambda)d\lambda,$$

where

$$p_j(\lambda) = \sum_{t_k^j \geq 0} c_k^j e^{-i\lambda t_k^j}, \qquad \int |p_j|^2 f(\lambda)d\lambda \leq 1$$

and the integration is carried out from $-\pi$ to π for integer time, and from $-\infty$ to ∞ for continuous time. Using certain properties of boundary values of analytic functions it can be shown that in fact

$$\rho(\tau) = \sup_{p} \int e^{i\lambda\tau} p(\lambda)f(\lambda)d\lambda, \tag{6}$$

where

$$\rho(\lambda) = \sum_{t_k \geq 0} c_k e^{-i\lambda t_k}, \qquad \int |p(\lambda)| f(\lambda)d\lambda \leq 1.$$

Let us take for the space L the space of functions $h(\lambda)$ that are integrable with weight $f(\lambda)$, that is, $\|h\| = \int |h(\lambda)|f(\lambda)d\lambda < \infty$, and as subspace H the linear closure of the functions $p(\lambda)$ of the form

$$p(\lambda) = \sum_{t_k \geq 0} c_k e^{-i\lambda t_k}.$$

Every linear functional h^* on L has the form

$$h^*(h) = \int h^*(\lambda)h(\lambda)f(\lambda)d\lambda,$$

where $\|h^*\| = \operatorname*{ess\,sup}\limits_{\lambda} |h^*(\lambda)|$.

The subspace H^0 is clearly the subspace of linear functionals h^0 corresponding to functions $h^0(\lambda)$ such that $\int e^{-it\lambda} h^0(\lambda)f(\lambda)d\lambda = 0$ for all $t \geq 0$. Hence the function $\phi(e^{-i\lambda}) = h^0(\lambda)f(\lambda)$ has an analytic continuation to the unit disc for integer time, and for continuous time $\phi(\lambda) = h(\lambda)f(\lambda)$ may be analytically continued into the lower half-plane.

Taking as h^* the linear functional corresponding to the function $h^*(\lambda) = e^{-i\lambda\tau}$, we obtain (4) and (4′) from (5).

Theorem 4. *If there exists $\phi_0(z)$ that has an analytic continuation inside the unit disc for integer time (to the lower half-plane for the case of continuous time) with boundary value $\phi_\rho(e^{-i\lambda})$ (respectively, $\phi_\rho(\lambda)$), such that the ratio f/ϕ_0 is a uniformly continuous function of λ with $|f/\phi_0| \geq \epsilon > 0$ for almost all λ, then*

$$\rho(\tau) \to 0 \tag{7}$$

as $\tau \to \infty$. If there exists an analytic function $\phi_0(z)$ such that $|f/\phi_0| \geq \epsilon > 0$ and the derivative $(f/\phi_0)^{(k)}$ is uniformly bounded, then

$$\rho(\tau) \leq c\tau^{-k}.$$

Proof. Let $\phi(z)$ be a polynomial of degree at most $[\tau/2]$ for integer time (or an analytic function of exponential type with type at most $\tau/2$ for continuous time). We have

$$\begin{aligned}\inf_{\phi} \operatorname*{ess\,sup}_{\lambda}\left[|f - e^{-i\lambda\tau}\phi|\frac{1}{f(\lambda)}\right] &\leq \inf_{\phi=\phi_0\psi} \operatorname*{ess\,sup}_{\lambda}\left[\left|\frac{f}{\phi_0} - e^{i\lambda\tau}\psi\right|\frac{|\phi_0|}{f}\right] \leq \\ &\leq \frac{1}{\epsilon}\inf_{\psi} \operatorname*{ess\,sup}_{\lambda}\left|\frac{f}{\phi_0} - e^{i\lambda\tau}\psi\right| \to 0\end{aligned} \tag{8}$$

as $\tau \to \infty$ (since f/ϕ_0 can be uniformly approximated by functions $\psi(z)$ of the above type (see [7], p.207; [8]).

For example, (7) always holds if the spectral density $f(\lambda)$ is continuous and does not vanish at any λ, $-\pi \leq \lambda \leq \tau$ (integer time) or if $f(\lambda)$ is uniformly continuous over the whole line, does not vanish and for sufficiently large λ satisfies the inequality

$$m/\lambda^k \leq f(\lambda) \leq M/\lambda^{k-1} \tag{9}$$

for some positive m, M and integer $k > 0$ (for the case of continuous time).

It then follows from [5], [6] that if $\rho(\tau) \to 0$ as $\tau \to \infty$, then the spectral density cannot vanish "too strongly": namely, it must be positive almost everywhere and satisfy the inequality

$$\int \frac{\log f(\lambda)}{1+\lambda^2} d\lambda > -\infty. \tag{10}$$

As follows from Theorem 4, weaker vanishing of the spectral density $f(\lambda)$ compatible with (10) does not by itself contradict the strong mixing condition (equivalent to (7)). If $f(\lambda)$ is rational with respect to $e^{i\lambda}$ (for integer t) or with respect to λ (for continuous t), then $\rho(\tau)$ decreases exponentially fast; for example, if $f(\lambda) = c/(a^2 + \lambda^2)$ ($\xi(t)$ is a Markov Gaussian process), then $\rho(\tau) = e^{-a\tau}$.

Apparently, the strong mixing condition might fail if the spectral density $f(\lambda)$ has a discontinuity (even if it is everywhere greater than a positive constant for integer t).

The authors thank Yu.V. Prokhorov for the remarks on the manuscript, which undoubtedly has improved it.

References

1. M. Rosenblatt, 'A central limit theorem and a strong mixing condition', *Proc. Nat. Acad. Sci.* **42**:1 (1956), 43–47.
2. O.V. Sarmanov, 'Maximal correlation coefficients', *Dokl. Akad. Nauk SSSR* **121**:1 (1958), 52–55 (in Russian).
3. O.V. Sarmanov and V.K. Zakharov, 'Maximal coefficients of multiple correlation', *Dokl. Akad. Nauk SSSR* **130**:2 (1960), 269–271 (in Russian).
4. V.A. Volkonskii and Yu.A. Rozanov, 'Some limit theorems of random functions, I', *Teor. Veroyatnost. i Primenen.* **4**:2 (1959), 186–207 (Russian).
5. A.N. Kolmogorov, 'Stationary sequences in Hilbert space', *Byull. Moskov. Gos. Univ. Ser. Mat.* **2**:6 (1941), 1–40 (in Russian).
6. M.G. Krein, 'On an extrapolation problem of A.N. Kolmogorov', *Dokl. Akad. Nauk SSSR* **46** (1945), 339–342 (in Russian).
7. N.I. Akhiezer, *Theory of approximation*, F. Ungar, 1956 (translated from the Russian).
8. S.N. Bernshtein, 'On best approximation of continuous functions on the real axis', *Dokl. Akad. Nauk SSSR* **51** (1946), 327–330; 485–488 (in Russian).
9. H. Cramér, *Mathematical methods of statistics*, Princeton Univ. Press, 1946.

49. RANDOM FUNCTIONS OF SEVERAL VARIABLES ALMOST ALL REALIZATIONS OF WHICH ARE PERIODIC *

A set

$$R = (l_1, \ldots, l_n)$$

of n linearly independent vectors is called a frame in Euclidean space E^n. A function $k(t)$ of an n-dimensional vector is called R-periodic if there exists a frame

$$R^* = (l_1^*, \ldots, l_n^*)$$

congruent to R such that

$$k = (z^1 l_1^* + \ldots + z^n l_n^*) = f(z^1, \ldots, z^n)$$

and if f is periodic in each variable z^k with period 2π.

We give necessary and sufficient conditions for R-periodicity for almost all realizations of a random function $\xi(t)$ in terms of "spectral moments".

In particular, this implies that a stationary isotropic non-constant random function cannot be Gaussian for $n = 1$. This may be of interest when studying potentials and solutions of Schrödinger's equation in crystals and media with "close order".

---

* *Teor. Veroyatnost. i Primenen.* (Probability theory and its applications) 5:3 (1960), 374 (in Russian).

50. AN ESTIMATE OF THE PARAMETERS OF A COMPLEX STATIONARY GAUSSIAN MARKOV PROCESS *

Jointly with M. Arato and Ya.G. Sinai

§1. Here we consider a two-dimensional stationary random process whose components $\xi(t)$ and $\eta(t)$ satisfy the stochastic differential equations

$$\begin{aligned} d\xi &= -\lambda\xi dt - \omega\eta dt + d\phi, \\ d\eta &= \omega\xi dt - \lambda\eta dt + d\psi, \end{aligned} \tag{1}$$

where $\phi(t)$ and $\psi(t)$ are two independent Wiener processes with

$$\mathbf{E}d\phi = \mathbf{E}d\psi = 0, \quad \mathbf{E}(d\phi)^2 = \mathbf{E}(d\psi)^2 = adt.$$

Setting $\zeta = \xi + i\eta$, $\quad \chi = \phi + i\psi$, $\quad \gamma = \lambda - i\omega$, we can write the system (1) as one equation

$$d\zeta = -\gamma\zeta dt + d\chi. \tag{1a}$$

The complex correlation function of our process is of the form

$$C(\tau) = A(\tau) + iB(\tau) = \mathbf{E}[\zeta(t)\overline{\zeta(t+\tau)}] = \sigma^2 \exp(-\lambda|\tau| - i\omega\tau), \tag{2}$$

where $\sigma^2 = a/\lambda$.

If a process is observed on the interval $[0, T]$, then we can determine the empirical correlation function

$$c(\tau) = a(\tau) + ib(\tau) = \frac{1}{T-\tau}\int_0^{T-\tau} \zeta(t)\overline{\zeta(t+\tau)}dt. \tag{3}$$

With probability 1 the right-hand derivative of the empirical correlation function at 0 is

$$c'(0) = -a - \frac{1}{T}s_1^2 + \frac{1}{T}s_2^2 - ir,$$

where the parameter a introduced above characterizes the intensity of the "white noises" $\phi'(t)$ and $\psi'(t)$, and

$$s_1^2 = \tfrac{1}{2}[|\zeta(0)|^2 + |\zeta(T)|^2],$$

$$s_2^2 = \frac{1}{T}\int_0^T |\zeta(t)|^2 dt, \quad r = \frac{1}{T}\int_0^T |\zeta(t)|^2 d\theta.$$

* *Dokl. Akad. Nauk SSSR* **146**:4 (1962), 747–750 (in Russian).

In the expression for r the integration is over the angle defined from

$$\zeta(t) = |\zeta(t)|e^{i\theta(t)}.$$

The figure shows the empirical correlation function for Chandler variations of the coordinates of the earth's pole.[1]

§2. The parameter a is determined precisely from the realization. Now it remains only to consider the problem of estimating the parameters λ and ω. We denote by P the probability measure in the space of realizations of our process on $[0, T]$. In the same space we introduce the standard measure

$$V = L \times W$$

where L is ordinary Lebesgue measure on the $\zeta(0)$-plane and W is two-dimensional Wiener measure in the space of increments $\zeta(t) - \zeta(0)$ with the same characteristics as those taken for the random process $\chi(t)$. It can be shown that (cf. [2], [3])

$$\frac{dP}{dV} = C\lambda \exp\Big[-\frac{\lambda^2+\omega^2}{2a}Ts_2^2 - \frac{\lambda}{a}s_1^2 + \lambda T + \frac{\omega}{a}Tr\Big], \tag{4}$$

where C is a constant. Formula (4) shows that the system of three statistics s_1^2, s_2^2, r is a sufficient system of statistics of the problem. Differentiating

$$L = \log\frac{dP}{dV} = c' + \log\lambda - \frac{\lambda^2+\omega^2}{2a}Ts_2^2 - \frac{\lambda}{a}s_1^2 + \lambda T + \frac{\omega}{a}Tr$$

[1] The instantaneous axis of the earth's rotation moves with respect to the small axis of the earth's ellipsoid (so called free nutation). These movements have a periodic component with a period of one year. After eliminating this component there remain Chandler's movements with tendency to fluctuate with period of about 14 months, but which are not strictly periodic and have large and mainly smooth variations of the amplitudes (waves of about 10–20 years). The figure shows that the Chandler component of pole movement is in good agreement with the hypothesis at the beginning of this paper.

The figure was obtained by processing the data of Table 6 from the book by A.Ya. Orlov [1]. The component with one year period is singled out from the coordinates $x(t), y(t)$ of Table 6, and the remainder is taken to be $\xi(t)$ and $\eta(t)$. The nodes on the figure indicate the points corresponding to the increments of τ in 0.1 year. The figure shows straight away that the period $2\pi/\omega$ approximately equals 14 months. The regular pattern of the spiral obtained might suggest that the parameter λ can also be estimated very precisely. This, however, is not true, as will be explained at the end of our paper.

with respect to ω and λ we obtain the equations

$$\frac{\partial L}{\partial \omega} = -\frac{\omega}{a} T s_2^2 + \frac{T}{a} r = 0, \tag{5}$$

$$\frac{\partial L}{\partial \lambda} = \frac{1}{\lambda} - \frac{\lambda}{a} T s_2^2 - \frac{s_1^2}{a} + T = 0 \tag{6}$$

for determining the maximum likelihood estimates $\hat{\omega}$ and $\hat{\lambda}$. From (5) we obtain

$$\hat{\omega} = r/s_2^2. \tag{$*$}$$

It can be shown that

$$(\hat{\omega} - \omega)/\sigma(\hat{\omega}), \quad \sigma^2(\hat{\omega}) = a/T s_2^2$$

has a (0,1)-normal distribution (this result is precise, not asymptotic). Equation (6) always has a unique positive solution.

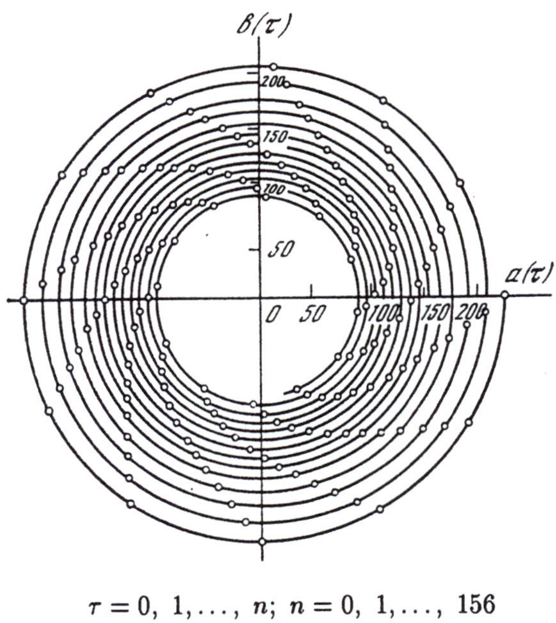

$\tau = 0,\ 1, \ldots,\ n;\ n = 0,\ 1, \ldots,\ 156$

Setting $\lambda T = \kappa$, $\hat{\lambda} T = \hat{\kappa}$, we obtain for $\hat{\kappa}$ the equation

$$h_2 \hat{\kappa}^2 + (h_1 - 1)\hat{\kappa} - 1 = 0,$$

where $h_1 = s_1^2/aT$, $h_2 = s_2^2/aT$.

The distributions of the statistics h_1 and h_2 and consequently $\hat{\kappa}$, depend only on the parameter κ. Since the distribution of $\hat{\kappa}$ is continuous, for any α, $0 < \alpha < 1$, and κ, $0 < \kappa < \infty$, we can find k such that

$$\mathbf{P}\{\hat{\kappa} > k|\kappa\} = \alpha. \tag{7}$$

Taking the inverse of the dependence $k = k_\alpha(\kappa)$ we obtain $\kappa = \kappa_\alpha(k)$ (it has been established that $k_\alpha(\kappa)$ is monotone increasing from 0 to ∞ as κ increases from 0 to ∞, so that taking the inverse is possible and unique). Clearly,

$$\mathbf{P}\{\kappa < \kappa_\alpha(\hat{\kappa})|\kappa\} \equiv \alpha. \tag{8}$$

We have initiated computations of $\kappa_\alpha(\hat{\kappa})$ for $\alpha = 0.1$; 0.05; 0.025; 0.01; 0.005; 0.001; 0.9; 0.95; 0.975; 0.99; 0.995; 0.999. The results thus obtained will be published when the computations are completed.

For small $\hat{\kappa}$, (8) is equivalent to the relation

$$\mathbf{P}\{\hat{\kappa} < u\kappa\} = \exp(-1/u), \tag{9}$$

that is, $\hat{\kappa}/\kappa$ has a χ^2 distribution with two degrees of freedom. For large $\hat{\kappa}$, (8) is equivalent to

$$\mathbf{P}\{\kappa < \hat{\kappa} + u\sqrt{\hat{\kappa}}\} \approx \frac{1}{\sqrt{2\pi}} \int_{-\infty}^{u} e^{-t^2/2} dt, \tag{10}$$

that is, the estimate $\hat{\kappa}$ is asymptotically normal with variance

$$\sigma^2(\hat{\kappa}) \sim \hat{\kappa}. \tag{11}$$

§3. For the case mentioned in the beginning of the paper, that is, the movement of the earth's pole, we have [2]

$$\hat{\omega} = 5.274, \quad \hat{\kappa} = 3.6, \quad 2\pi : \hat{\omega} = 1.191, \quad \sigma(2\pi : \hat{\omega}) = 0.006$$

[2] The introduction of Wiener processes ϕ and ψ, that is, perturbations of "white noise" type in (1) is, of course, a gross generalization in the case of the movement of the earth's poles. It would be more correct to write

$$\xi' = -\lambda\xi - \omega\eta + f, \quad \eta' = \omega\xi - \lambda\eta + g.$$

However, data from [1] show that the values $f(t)$ and $g(t)$ at times t separated by several years are actually independent, so the replacement of f and g by "equivalent white noise" is possible. Apparently, the error in determining the intensity based on observations over $T = 60$ years.

The asymptotic formula (11) gives

$$\sigma^2(\hat{\kappa}) = 3.6.$$

Since κ is clearly positive and (10) gives a negative estimate for k_α when $\alpha < 0.03$, it is evident that the asymptotic formula (10) is still unsuitable.

Our computation gives the estimates

$$\kappa_{0.90} = 5.5; \qquad \kappa_{0.95} = 6.2; \qquad \kappa_{0.975} = 7.8;$$

$$\kappa_{0.10} = 1.27; \quad \kappa_{0.05} = 0.82; \quad \kappa_{0.025} = 0.46,$$

which, as regards λ, correspond to the estimates

$$\kappa_{0.90} = 0.09; \qquad \kappa_{0.95} = 0.10; \qquad \kappa_{0.975} = 0.13;$$

$$\kappa_{0.10} = 0.021; \quad \kappa_{0.05} = 0.014; \quad \kappa_{0.025} = 0.008,$$

Moscow State University, 20 February 1962

References

1. A.Ya. Orlov, *Service of latitude*, Akad. Nauk SSSR, Moscow, 1958 (in Russian).
2. Ch. Striebel, *Ann. Math. Statist.* **30**:2 (1959), 559.
3. M. Arato, *Dokl. Akad. Nauk SSSR* **145**:1 (1962) (in Russian).
4. E.P. Fedorov, *Proc. Poltava Gravimetric Observatory* **2**:3 (1948) (in Russian).
5. N.I. Panchenko, In: Proc. 14th Astronomical Conference of the USSR, Akad. Nauk SSSR, Moscow, (in Russian).
6. W.H. Munk and G.J.F. Macdonal, *The rotation of the earth*, Cambridge, 1960. (Monographs on mechanics and applied mathematics.)
7. H. Jeffreys, *Monthly Notices Roy. Astronom. Soc. Geophys. Suppl.* **100** (1942), 139.
8. A.M Walker and A. Young, *Monthly Notices Roy. Astronom. Soc. Geophys. Suppl.* **115** (1955), 443; (1957) **117**, 119.

---

a of this equivalent white noise is small enough not to affect the estimates of λ substantially. The value of $\hat{\omega}$ is computed via the discrete analogue of formula (*) obtained using the maximum likelihood method for the "discrete-time scheme".

Concerning the estimation of the parameters λ and ω for the earth's movement, see also [6]. The results of [6] are close to ours: $\lambda = 1/15$; $2\pi : \omega = 1.193$. Close values were given by Jeffreys [7], but [5], [8] give sharply different values: $\lambda = 0.3$ and $\lambda = 0.01$.

51. ON THE APPROXIMATION OF DISTRIBUTIONS OF SUMS OF INDEPENDENT TERMS BY INFINITELY DIVISIBLE DISTRIBUTIONS *

Introduction

In what follows $\xi = \xi_1 + \ldots + \xi_n$ is the sum of n independent real terms

$$F_k(x) = \mathbf{P}\{\xi_k < x\}, \quad H(x) = \mathbf{P}\{\xi < x\},$$

$$G_{\sigma^2}(x) = \frac{1}{\sqrt{2\pi}\sigma} \int_{-\infty}^{x} e^{-x^2/2\sigma^2} dx, \ \sigma > 0,$$

$$E(x) = G_0(x) = \begin{cases} 0 & \text{for } x \le 0, \\ 1 & \text{for } x > 0, \end{cases}$$

$\mathfrak{D} = \{D\}$ is the set of infinitely divisible distribution functions $D(x)$, and $C_1, C_2, \ldots$ are positive constants.

We will prove the following two theorems, which are strengthened versions of those given in my paper [1].

Theorem 1. *There exists C_1 such that for identically distributed ξ_k, there exists for any $F(x) = F_k(x)$, $k = 1, 2, \ldots, n$, a distribution $D \in \mathfrak{D}$ such that*

$$|D(x) - H(x)| \le C_1 n^{-1/3} \tag{0.1}$$

for all x.

Theorem 2. *There exists C_2 such that for any $\epsilon > 0$, $L \ge 2l > 0$, the inequalities*

$$E(x-l) - \epsilon \le F_k(x) \le E(x+l) + \epsilon \tag{0.2}$$

for all x and $k = 1, \ldots, n$ imply the existence of $D \in \mathfrak{D}$ for which

$$D(x-L) - \delta \le H(x) \le D(x+L) + \delta \tag{0.3}$$

for all x, where

$$\delta = C_2 \max\left(\frac{l}{L}\left(\log\frac{L}{l}\right)^{1/2}, \epsilon^{1/3}\right). \tag{0.4}$$

The history of the problem is as follows.

---

* *Trudy Moskov. Mat. Obshch.* (Proc. Moscow Math. Soc.) **12** (1963), 437–451 (in Russian).

1. It follows from the closedness (with respect to weak convergence) of the class of infinitely divisible distributions introduced by Bruno de Finetti [2] that if the distributions of sums

$$\xi^{(k)} = \xi_{k1} + \xi_{k2} + \ldots + \xi_{kn_k}, \quad \lim_{k\to\infty} n_k = \infty, \tag{0.5}$$

whose terms are independent and identically distributed within each series, converge weakly, then the limit distribution is infinitely divisible.

It is tempting to interpret this result as follows: *the distribution of the sum of a large number of identically distributed independent terms is close to an infinitely divisible distribution*. However, prior to my work [1] this interpretation was not quite convincing.

Even for a sequence

$$\xi_1, \xi_2, \ldots, \xi_n, \ldots$$

of independent identically distributed terms the "totally divergent" case is possible (according to Doeblin [3]) when for no normalization

$$\xi^{(n)} = A_n(\xi_1 + \ldots + \xi_n) - B_n$$

and for no sequence

$$n_1 < n_2 < \ldots < n_k < \ldots$$

can the distributions of the sums $\xi^{(n_k)}$ converge to anything other than degenerate distributions $E(x-a)$. This, of course, can be achieved by choosing sufficiently small factors A_n.

Only in 1955 Yu.V. Prokhorov [4] proved that for the case of a sequence of identically distributed independent variables ξ_n there always exists a sequence of infinitely divisible distribution functions

$$D_1(x), D_2(x), \ldots, D_n(x), \ldots,$$

which approximate the distributions $H_n(x)$ of the sums

$$\xi^{(n)} = \xi_1 + \xi_2 + \ldots + \xi_n$$

in the sense that

$$\sup_x |H_n(x) - D_n(x)| \to 0 \tag{0.6}$$

as $n \to \infty$. Still, the work left open the question of whether the convergence in (0.6) is uniform with respect to the distribution function $F(x)$ of the variables ξ_n.

In terms of the supremum metric

$$\rho(F', F'') = \sup_x |F'(x) - F''(x)|$$

the question is whether or not the function [1]

$$\psi(n) = \sup_F \rho(H_n, \mathfrak{D})$$

tends to 0 as $n \to \infty$. The answer to this question was given in my work [1]: it was proved that

$$\psi(n) = O(n^{-1/5}). \tag{0.7}$$

In 1960 Prokhorov [5] strengthened this result by showing that

$$\psi(n) = O(n^{-1/3} \log^2 n) \tag{0.8}$$

Our Theorem 1 states that [2]

$$\psi(n) = O(n^{-1/3}). \tag{0.9}$$

It was natural to try to estimate $\psi(n)$ from below. These estimates were made by the student of Prokhorov, I.P. Tsaregradskii, Prokhorov himself and L.D. Meshalkin. The latest result of Meshalkin [6] is

$$\psi(n) \geq C_3 n^{-2/3} (\log n)^{-4}. \tag{0.10}$$

2. For sums

$$\xi^{(k)} = \xi_{k1} + \xi_{k2} + \ldots + \xi_{kn_k}$$

---

[1] The supremum is taken over all distribution functions F.

[2] After this paper was finished I found out that in 1961 F.M. Kagan had obtained a result that is somewhere between (0.8) and (0.9):

$$\psi(n) = O(n^{-1/3} \log n).$$

Later this result was reported by F.M. Kagan at the Meeting on Probability Theory and Mathematical Statistics in Fergana (September 1962).

all terms of which are independent within each series but have different distributions, A.Ya. Khinchin in 1937 [7] established a sufficient condition for the limit distribution of $\xi^{(k)}$ to be infinitely divisible. This is the condition of infinitesimal terms: *there exist*

$$\epsilon_k \to 0, \; l_k \to 0$$

for which the distributions F_{ki} *of the* ξ_{ki} *satisfy the condition*

$$E(x - l_k) - \epsilon_k \leq F_{ki}(x) \leq E(x + l_k) + \epsilon_k.$$

Our Theorem 2 is an attempt to give a uniform character to this result of Khinchin. The essence of Theorem 2 may be clarified by using the "Lévy distance"

$$\rho_L(F', F'') = \inf \epsilon,$$

where the infimum is taken over all ϵ satisfying the condition

$$F'(x - \epsilon) - \epsilon \leq F''(x) \leq F'(x + \epsilon) + \epsilon.$$

It is easy to see that Theorem 2 implies

Corollary. *If*

$$\sup_i \rho_L(F_i, E) \leq \eta,$$

then

$$\rho_L(H, \mathfrak{D}) \leq C_4 \eta^{1/3}.$$

3. As is well known, the most powerful means of proving limit theorems on the distribution of sums of a large number of independent terms is the *apparatus of characteristic functions.* Now "direct" probabilistic methods *in this area* can very seldom compete with the possibilities of the analytic apparatus of characteristic functions.

Our Theorems 1 and 2 give an interesting example of another state of affairs. An essential element for proving these theorems is Lemma 1, which refers to the "concentration functions" introduced by P. Lévy. I strengthened theorems of Lévy and Doeblin on the properties of concentration functions ([8]) specifically to prove the new versions of Theorems 1 and 2 given in [1].

Further progress in deriving estimates for concentration functions is due to B.A. Rogozin (see [9], [10]). He also makes use of elementary direct probabilistic and set-theoretic methods (Erdös' theorem on the subsets of a finite set).[3]

Mathematicians whose attention I drew to this problem have so far failed to prove theorems of types 1 and 2 without turning to these particular methods.

Throughout this paper, as in [1], I make use of the methods of reasoning proposed by Doeblin (see, for example, [3]). As can be seen from the above, the passage from 1/5 to 1/3 in Theorem 1 was carried out by Prokhorov. In order to eliminate the factor $\log^2 n$ from Prokhorov's estimate (0.8) it was required a) to use the more accurate estimates of concentration functions obtained by Rogozin; b) to make some changes in Prokhorov's proof, namely to introduce Lemmas 5 and 6.[4]

The passage from 1/5 given in [1] to 1/3 in the proof of Theorem 2 was made by similar techniques borrowed from Prokhorov's work [5], which also makes use of Lemmas 5 and 6.

4. Apart from the distance

$$\rho(F', F'') = \sup_x |F'(x) - F''(x)|,$$

it is natural to consider the "variational distance"

$$\rho_v(F', F'') = \tfrac{1}{2}\mathrm{Var}[F'(x) - F''(x)] = \sup_A [F'(A) - F''(A)],$$

where A is an arbitrary Borel-measurable set on the straight line.

As is well known,

$$\rho(F', F'') \le \rho_v(F', F'').$$

Therefore, for the function

$$\psi_v(n) = \sup_F \rho_v(F^n, \mathfrak{D})$$

we have

$$\psi_v(n) \ge \psi(n). \tag{0.11}$$

[3] This result of Rogozin (now know as the Kolmogorov-Rogozin inequality) was proved by Esseen in 1966 by the method of characteristic functions (Editor's note).

[4] The first of these steps was taken by F.M. Kagan somewhat before this paper was written (see footnote 2).

We do not know whether $\psi_v(n) \to 0$ as $n \to \infty$.

§1. Eight Lemmas

After P. Lévy, we introduce for any distribution function $F(x)$ its "concentration function"

$$Q_F(l) = \sup_x [F(x+l+0) - F(x)].$$

Lemma 1. *If* $x_k^+ - x_k^- > l$ *and*

$$\mathsf{P}\{\xi_k \le x_k^-\} = \mathsf{P}\{\xi_k \ge x_k^+\} = \tfrac{1}{2},$$

for $k = 1, 2, \ldots, n$, $L \ge l > 0$ *then*

$$Q_H(L) \le C_5(L/l)n^{-1/2}.$$

Lemma 2. *If* $\sigma > 0$, $l > 0$,

$$\eta = C_6 e^{-l^2/2\sigma^2},$$

then for any distribution function $F(x)$

$$F * G_{\sigma^2}(x-l) - \eta \le F(x) \le F * G_{\sigma^2}(x+l) + \eta.$$

Lemma 3. *If* $\sigma > 0$, $\sigma_1 > 0$, *then*

$$|G_{\sigma_1^2}(x) - G_{\sigma^2}(x)| \le C_7 |\sigma_1^2/\sigma^2 - 1|.$$

Lemma 4. *If*

$$\int x F(dx) = 0, \quad \int x^2 F(dx) = \sigma^2, \quad h \ge \sigma > 0,$$

then

$$\sum_{r=-\infty}^{\infty} \sup_{rh \le x \le (r+1)h} |F(x) - E(x)| \le C_8.$$

Lemma 5. *If* $\mathbf{E}\xi_k = 0,\ |\xi_k| \le l,\ \mathbf{D}\xi = \sigma^2,\ h \ge \sigma > 0$, *then*

$$\sum_{r=-\infty}^{\infty} \sup_{rh \le x \le (r+1)h} |H(x) - G_{\sigma^2}(x)| \le C_9 \frac{l}{\sigma}.$$

Lemma 6. *If* $\mathbf{E}\xi_k = 0,\ |\xi_k| \le l,\ \mathbf{D}\xi = \sigma^2,\ \sigma_1^2 = \sigma^2 + \sigma_0^2$, *then*

$$|H * G_{\sigma_0^2}(x) - G_{\sigma_1^2}(x)| \le C_{10} l/\sigma_1.$$

Lemma 7. *For any positive integer* n *and* $0 \le p \le 1$,

$$\sum_{m=0}^{\infty} |C_n^m p^m (1-p)^{n-m} - \frac{(np)^m}{m!} e^{-np}| \le C_{11} p.$$

Lemma 8. *Let*[5] $0 \le p_k \le 1$,

$$p_k(m) = \begin{cases} 1 - p_k & \text{for } m = 0, \\ p_k & \text{for } m = 1, \\ 0 & \text{for } m > 1, \end{cases} \quad q_k(m) = \frac{p_k^m}{m!} e^{-p_k},$$

$$p(\mathbf{m}) = \prod_k p_k(m_k), \quad q(\mathbf{m}) = \prod_k q_k(m_k).$$

Then

$$\sum_{\mathbf{m}} |p(\mathbf{m}) - q(\mathbf{m})| \le C_{12} \sum_k p_k^2.$$

Lemmas 2 and 3 are proved by direct computation. Lemma 1 is a direct corollary of Theorem 1 from Rogozin's paper [9]. Lemma 7 is proved in Prokhorov's work [11], Lemma 4 follows from the estimate (F is the distribution of ξ):

$$F(x) - E(x) \le \mathbf{P}\{|\xi| \ge |x|\} \le \sigma^2/x^2$$

(Chebyshev's inequality).

Lemmas 5 and 6 are close to the known estimate

$$|H - G_{\sigma^2}| \le C_{13} l/\sigma, \tag{1.1}$$

[5] Here and in what follows, $\mathbf{m} = (m_1, \dots, m_n)$ is an n-dimensional vector, $\sum_{\mathbf{m}}$ denotes summation over the vectors $\mathbf{m}$ with non-negative integer components.

which follows from Lyapunov's theorem in Esseen's formulation ([12], p.216):

$$|H - G_{\sigma^2}| \leq \frac{C_{14}}{\sigma^3} \sum_k \mathbf{E}|\xi_k|^3. \tag{1.2}$$

Lemma 6 may be derived from (1.2) if the additional normal term with variance σ_0^2 is represented as a sum of a large number of terms with sufficiently small variances.

The proof[6] of Lemma 5 is somewhat more difficult.

§2. Proof of Theorem 1.

1. In what follows we consider $n > 1$. It can easily be seen that this requirement is inessential.

2. We suppose further that the $\xi_k = F_k^{-1}(\eta_k)$ are non-decreasing functions of the independent variables η_k with uniform distribution on [0,1]:

$$\mathbf{P}\{\eta_k \leq y\} = y \text{ for } 0 \leq y \leq 1.$$

It can easily be shown that in an appropriate extension of the basic probability field $(\Omega, \mathfrak{M}, \mathbf{P})$ such η_k do exist.[7]

3. We set

$$p = n^{-1/3},$$

$$\mu_k = \begin{cases} 0 & \text{for } p/2 < \eta_k < 1 - p/2, \\ 1 & \text{otherwise.} \end{cases}$$

$$a = \mathbf{E}\{\xi_k | \mu_k = 0\}, \quad \sigma^2 = \mathbf{D}\{\xi_k | \mu_k = 0\},$$

$$A(x) = \mathbf{P}\{\xi_k < x | \mu_k = 0\}, \quad B(x) = \mathbf{P}\{\xi_k < x | \mu_k = 1\}.$$

All these constructions hold when passing from ξ_k to

$$\xi_k' = \xi_k - a,$$

---

[6] Lemma 5 can easily be derived from the following estimate due to A. Bikyalis:

$$|H(x) - G_{\sigma^2}(x)| \leq \frac{C}{\sigma^3(1 + |x|/\sigma)^3} \sum_k \mathbf{E}|\xi_k|^3,$$

where C is an absolute constant (see *Litov. Mat. Sb.* **6**:3 (1966), 323–346) (Editor's note).

[7] The functions F_k^{-1} must be suitably defined. This is left to the reader.

only $a' = 0$ appears instead of a, and the functions $A(x)$ and $B(x)$ are replaced by

$$A'(x) = A(x+a), \quad B'(x) = B(x+a).$$

Therefore we need only consider the case $a = 0$, to which we shall restrict ourselves in what follows.

4. In the decomposition

$$F(x) = pB(x) + (1-p)A(x)$$

the support of the distribution A lies in the interval $[x^-, x^+]$,

$$x^- = F^{-1}(p/2), \quad x^+ = F^{-1}(1-p/2).$$

of length

$$\lambda = x^+ - x^-,$$

while the support of the distribution B lies outside this interval, with each of the rays $(-\infty, x^-]$ and $[x^+, \infty)$ having probability $\frac{1}{2}$ in the distribution B. Lemma 1 may be applied to the distributions B^m (throughout, the powers of a distribution are understood in the sense of convolution), which gives the estimate

$$Q_{B^m}(\lambda) \leq C_5 m^{-1/2}. \tag{2.1}$$

5. We approximate the distribution

$$H = [pB + (1-p)A]^n = \sum_m C_n^m p^m (1-p)^{n-m} B^m * A^{n-m}$$

by an infinitely divisible distribution D differently according to the two cases:

A) $\lambda \geq \sqrt{n}\sigma$. B) $\lambda < \sqrt{n}\sigma$.

Case A. We set

$$D = e^{np(B-E)} = \sum_m \frac{(np)^m}{m!} e^{-np} B^m,$$

$$H_1 = \sum_m C_n^m p^m (1-p)^{n-m} B^m.$$

According to Lemma 7,

$$|D - H_1| \le C_{11} p = C_{11} n^{-1/3}. \tag{2.2}$$

By Lemma 4 and estimate (2.1), for $h = \lambda$ we have[8]

$$|B^m * A^{n-m} - B^m| \le \int |A^{n-m}(x-z) - E(x-z)| B^m dz \le$$

$$\le Q_{B^m}(\lambda) \sum_r \sup_{r\lambda \le y \le (r+1)\lambda} |A^{n-m}(y) - E(y)| \le C_5 C_8 m^{-1/2}. \tag{2.3}$$

Therefore

$$|H - H_1| \le \sum_m C_n^m p^m (1-p)^{n-m} |B^m * A^{n-m} - B^m| \le C_5 C_8 n^{-1/3} + 2 \sum{}',$$

where

$$\sum{}' = \sum_{m < \frac{1}{2} n^{2/3}} C_n^m p^m (1-p)^{n-m} = \mathbf{P}\{\mu < \tfrac{1}{2} n^{2/3}\}. \tag{2.4}$$

Noting that

$$\mathbf{E}\mu = np = n^{2/3}, \tag{2.5}$$

$$\mathbf{D}\mu = np(1-p) \le n^{2/3}, \tag{2.6}$$

we obtain by Chebyshev's inequality

$$\sum{}' \le \mathbf{P}\{|\mu - n^{2/3}| > \tfrac{1}{2} n^{2/3}\} \le 4 n^{-2/3},$$

which, together with (2.4), gives

$$|H - H_1| \le (2 C_5 C_8 + 8) n^{-1/3}. \tag{2.7}$$

From (2.6) and (2.2) we obtain (0.1)

Case B. We set

$$D = e^{np(B-E)} * G_{n(1-p)\sigma^2} = \sum_m \frac{(np)^m}{m!} e^{-np} B^m * G_{n(1-p)\sigma^2},$$

$$H_1 = \sum_m C_n^m p^m (1-p)^{n-m} B^m * G_{(n-m)\sigma^2},$$

$$H_2 = \sum_m C_n^m p^m (1-p)^{n-m} B^m * G_{n(1-p)\sigma^2}.$$

---

[8] The variance of the distribution A^{n-m} is $(n-m)\sigma^2$, so that the conditions of Lemma 4 are satisfied for $h = \lambda \ge \sqrt{n}\sigma$.

According to Lemma 7 we have

$$|D - H_2| \le C_{11} n^{-1/3}. \tag{2.8}$$

The difference $H - H_1$ is estimated using Lemma 5, where we now assume that $h = \sqrt{n}\sigma$:

$$|B^m * A^{n-m} - B^m * G_{(n-m)\sigma^2}| \le \int |A^{n-m}(x-z) - G_{(n-m)\sigma^2}(x-z)| \times$$

$$\times B^m(dz) \le Q_{B^m}(\sqrt{n}\sigma) \sum_r \sup_{r\sqrt{n}\sigma \le y \le (r+1)\sqrt{n}\sigma} |A^{n-m}(y) - G_{(n-m)\sigma^2}(y)| \le$$

$$\le C_5 \frac{\sqrt{n}\sigma}{\lambda} m^{-1/2} C_9 \frac{\lambda}{\sqrt{n}\sigma} = C_5 C_9 m^{-1/2}.$$

This estimate is quite similar to estimate (2.3) for case A. As in case A we obtain

$$|H - H_1| \le C_5 C_9 n^{-1/3} + 2 \sum{}' \le C_{15} n^{-1/3}. \tag{2.9}$$

The difference $H_1 - H_2$ is estimated using Lemma 3:

$$|H_1 - H_2| \le C_{16} n^{-1/3} + \sum{}'',$$

where

$$\sum{}'' = \sum_{\left|\frac{n-m}{n(1-p)} - 1\right| > \frac{C_{16}}{C_7}} C_n^m p^m (1-p)^{n-m}.$$

Using Chebyshev's inequality we obtain from (2.5) and (2.6) for $n > 1$ and an appropriate choice of C_{16} the estimate

$$\sum{}'' \le C_{17} n^{-1/3},$$

which leads to

$$|H_1 - H_2| \le (C_{16} + C_{17}) n^{-1/3}. \tag{2.10}$$

From the inequalities (2.8)–(2.10) we obtain (0.1), which completes the proof of Theorem 1.

§3. Proof of Theorem 2

1. We suppose without loss of generality that $\epsilon < 1$.

2. We show that it suffices to consider the case of continuous and strictly increasing functions $F_k(x)$.

Assume that Theorem 2 is proved for continuous and strictly increasing functions with constant C_2' and consider the sum

$$\xi = \xi_1 + \ldots + \xi_n$$

with arbitrary $F_k(x)$ satisfying (0.2). Let $L > 2l$. Choose l' and L' such that

$$L > L' > 2l' > 2l, \quad l'/L' \geq \tfrac{1}{2} l/L. \tag{3.1}$$

By Lemma 2 we can choose σ_0 to be so small that for any distribution function $F(x)$ we have

$$F * G_{\sigma_0^2}(x - \lambda) - \epsilon \leq F(x) \leq F * G_{\sigma_0^2}(x + \lambda) + \epsilon, \tag{3.2}$$

$$F * G_{n\sigma_0^2}(x - \Lambda) - \delta' \leq F(x) \leq F * G_{n\sigma_0^2}(x + \Lambda) + \delta', \tag{3.3}$$

where

$$\lambda = l' - l, \quad \Lambda = L - L',$$

$$\delta' = C_2' \max\left[\frac{l'}{L'}\left(\log\frac{L'}{l'}\right)^{1/2},\ (2\epsilon)^{1/3}\right].$$

Set

$$F_k' = F_k * G_{\sigma_0^2}.$$

According to (0.2) and (3.2) we obtain

$$E(x - l') - 2\epsilon \leq F_k'(x) \leq E(x + l') + 2\epsilon.$$

Since the functions F_k' are continuous and strictly increasing, there exists an infinitely divisible distribution D' for which

$$D'(x - L') - \delta' \leq H(x) \leq D(x + L') + \delta'. \tag{3.4}$$

Noting that

$$H' = H * G_{n\sigma_0^2},$$

(3.3), applied to $F(x) = H(x)$, and (3.4) imply that

$$D'(x-L) - 2\delta' \le H(x) \le D'(x+L) + 2\delta'. \tag{3.4'}$$

Since under a suitable choice of C_2 we have, by virtue of (3.1),

$$2\delta' \le 2C_2' \max\left[\frac{l'}{L'}\left(\log\frac{L'}{l'}\right)^{1/2}, (2\epsilon)^{1/3}\right] \le C_2 \max\left[\frac{l}{L}\left(\log\frac{L}{l}\right)^{1/2}, \epsilon^{1/3}\right],$$

formula (0.3) follows from (3.4′).

3. In accordance with No. 2 we take $F_k(x)$ to be continuous and strictly increasing. Then

$$\lambda_k(p) = F_k^{-1}(1-p/2) - F_k^{-1}(p/2)$$

is defined for all p, $0 < p < 1$, continuous and strictly decreasing. It takes all positive values. Therefore, the inverse function $p_k(\lambda)$ is defined for $0 < \lambda < \infty$, is continuous, strictly decreasing and takes all values in the interval $1 > p > 0$.

The function

$$s(\lambda) = \sum_k p_k(\lambda)$$

is also continuous and strictly decreasing. It takes all values in the interval $n > s > 0$. Therefore for $0 < \epsilon < 1$ there exists a unique solution λ_0 of the equation

$$s(\lambda) = \epsilon^{-2/3}.$$

4. We set

$$\Lambda = \begin{cases} \lambda_0, & \text{if } \lambda_0 \ge l, \\ l, & \text{if } \lambda_0 < l, \end{cases}$$

$$p_k = p_k(\Lambda), \quad s = s(\Lambda) = \sum_k p_k,$$

$$x_k^- = F_k^{-1}(p_k/2), \quad x_k^+ = F_k^{-1}(1-p_k/2),$$

$$\mu_k = \begin{cases} 0 & \text{for } x_k^- < \epsilon_k < x_k^+, \\ 1 & \text{otherwise,} \end{cases}$$

$$a_k = \mathbf{E}\{\xi_k | \mu_k = 0\},$$

$$\sigma_k^2 = \mathbf{D}\{\xi_k | \mu_k = 0\},$$

$$\sigma^2 = \sum_k (1-p_k)\sigma_k^2.$$

Setting

$$A_k(x) = \mathbf{P}\{\xi_k < x | \mu_k = 0\}, \quad B_k(x) = \mathbf{P}\{\xi_k < x | \mu_k = 1\},$$

we represent $F_k(x)$ as

$$F_k(x) = p_k B_k(x) + (1 - p_k) A_k(x),$$

where the support of the distribution A_k lies in the interval $[x_k^-, x_k^+]$, while that of the distribution B_k lies outside this interval; here the probability of the rays $(-\infty, x_k^-]$ and $[x_k^+, \infty)$ is $\frac{1}{2}$.

Using the notation of Lemma 8 and setting[9]

$$B(\mathbf{m}) = \prod_k^* B_k^{m_k}, \quad A(\mathbf{m}) = \prod_k^* A_k^{1-m_k},$$

we obtain

$$H = \prod_k^* [p_k B_k + (1 - p_k) A_k] = \sum_{\mathbf{m}} p(\mathbf{m}) B(\mathbf{m}) * A(\mathbf{m}).$$

The construction of the approximating infinitely divisible distribution differs according to the three cases A, B and C:

A	B	C
$\lambda_0 \geq l$	$\lambda_0 \geq l$	$\lambda_0 < l$
$\lambda_0 \geq \sigma$	$\lambda_0 < \sigma$	
$\lambda = \lambda_0$	$\lambda = \lambda_0$	$\lambda = l$
$s = \epsilon^{-2/3}$	$s = \epsilon^{-2/3}$	$s = \epsilon^{-2/3}$

5. Since always $\Lambda \geq l$, it follows that

$$p_k = p_k(\Lambda) \leq p_k(l) \leq \epsilon. \tag{3.5}$$

[9] The $B(\mathbf{m})$ are defined for any negative m_k, while the $A(\mathbf{m})$ are defined only when the m_k take the values 0 and 1.

This is the only instance in our proof when we use condition (0.2) of the theorem. Since the definition of $p_k(\Lambda)$ and all the other variables essential for our constructions is invariant with respect to the shifts

$$\xi_k' = \xi_k - c_k,$$

we need only consider the case

$$a_k = 0.$$

6. We now make certain calculations that we shall need in the sequel:

$$\tau = \sum_k \mu_k, \quad t(\mathbf{m}) = \sum_k m_k;$$

τ is equal to the *number* of variables ξ_k such that $x_k^- < \xi_k < x_k^+$. It is easily checked that

$$\mathsf{E}\tau = s, \quad \mathsf{D}\tau = \sum_k p_k(1-p_k) < s.$$

Therefore, according to Chebyshev's inequality,

$$\mathsf{P}\{|\tau - s| \geq C\} = \sum_{|t(\mathbf{m})-s|\geq C} p(\mathbf{m}) < \frac{s}{C^2}. \tag{3.6}$$

7. Suppose further that

$$\zeta = \sum_k (1-\mu_k)\xi_k.$$

This is the sum of those ξ_k for which $x_k^- < \xi_k < x_k^+$. In view of the assumption that the $a_k = 0$, for any $\mathbf{m}$

$$\mathsf{E}\zeta = 0, \quad \mathsf{E}(\zeta|\mu = \mathbf{m}) = 0.$$

In what follows we shall be interested in the conditional variance

$$\sigma^2(\mathbf{m}) = \mathsf{D}(\zeta|\mu = \mathbf{m}).$$

For the random variable $\rho^2 = \sigma^2(\mu)$ it is easy to calculate

$$\mathsf{E}\rho^2 = \sigma^2,$$

$$\mathsf{D}\rho^2 = \sum_k p_k(1-p_k)\sigma_k^4.$$

Since

$$\sum_k (1-p_k)\sigma_k^2 = \sigma^2, \quad \sigma_k^2 \le \frac{\lambda^2}{4},$$
$$p_k = p_k(\Lambda) \le p_k(l) \le \epsilon,$$

we have the inequality

$$\mathbf{D}\rho^2 \le \tfrac{1}{4}\sigma^2\lambda^2\epsilon.$$

Therefore

$$\mathbf{P}\{|\rho^2-\sigma^2| \ge C\} = \sum_{|\sigma^2(\mathbf{m})-\sigma^2|\ge C} p(\mathbf{m}) \le \frac{\sigma^2\lambda^2\epsilon}{4C^2}. \tag{3.7}$$

8. Finally, we note that by Lemma 1,

$$Q_{B(\mathbf{m})}(L) \le C_5(L/\lambda_0)t(\mathbf{m}). \tag{3.8}$$

We now proceed to the proof of Theorem 2 for the cases A, B and C.

Case A. In this case

$$H = \prod\nolimits_k^* [p_k B_k + (1-p_k)A_k] = \sum_{\mathbf{m}} q(\mathbf{m})B(\mathbf{m}) * A(\mathbf{m})$$

is approximated by

$$D = \exp \sum_k p_k(B_k - E) = \sum_{\mathbf{m}} q(\mathbf{m})B(\mathbf{m}).$$

To proceed from H to D we also consider

$$H_1 = \sum_{\mathbf{m}} p(\mathbf{m})B(\mathbf{m}).$$

According to Lemma 8 and (3.5),

$$|D - H_1| \le \sum_{\mathbf{m}} |p(\mathbf{m}) - q(\mathbf{m})| \le C_{12}\sum_k p_k^2 \le C_{12}\epsilon \sum_k p_k \le C_{12}\epsilon^{1/3}. \tag{3.9}$$

On the other hand, according to Lemma 4 with $h = \Lambda = \lambda_0 \ge \sigma$ and (3.8),

$$|B(\mathbf{m}) * A(\mathbf{m}) - B(\mathbf{m})| \le \int |A(\mathbf{m})(x-z) - E(x-z)|B(\mathbf{m})(dz) \le$$
$$\le Q_{B(\mathbf{m})}(\lambda_0)\sum_r \sup_{r\lambda_0 \le y \le (r+1)\lambda_0} |A(\mathbf{m})(y) - E(y)| \le C_5C_8[t(\mathbf{m})]^{-1/2}. \tag{3.10}$$

Therefore

$$|H - H_1| \le \sum_{\mathbf{m}} p(\mathbf{m})|B(\mathbf{m}) * A(\mathbf{m}) - B(\mathbf{m})| \le \sqrt{2}C_5C_8\epsilon^{-1/3} + 2\sum{}', \quad (3.11)$$

where

$$\sum{}' = \sum_{t(\mathbf{m})<\frac{1}{2}\epsilon^{2/3}} p(\mathbf{m}).$$

By (3.6) and noting that in our case $s = \epsilon^{2/3}$, we have

$$\sum{}' < 8\epsilon^{-2/3} < 4\epsilon^{1/3},$$

that is, (3.11) implies that

$$|H - H_1| \le (\sqrt{2}C_5C_8 + 8)\epsilon^{1/3}. \quad (3.12)$$

From (3.9) and (3.12) we obtain

$$|H - D| \le C_{18}\epsilon^{1/3}. \quad (3.13)$$

Case B. In this case we set

$$D = \exp\sum_k p_k(B_k - E) * G_{\sigma^2} = \sum_{\mathbf{m}} q(\mathbf{m})B(\mathbf{m}) * G_{\sigma^2},$$

$$H_1 = \sum_{\mathbf{m}} p(\mathbf{m})B(\mathbf{m}) * G_{\sigma^2(\mathbf{m})}, \quad H_2 = \sum_{\mathbf{m}} p(\mathbf{m})B(\mathbf{m}) * G_{\sigma^2}.$$

The inequality

$$|D - H_2| \le C_{12}\epsilon^{1/3} \quad (3.14)$$

is derived in the same way as (3.9) in case A.

By Lemma 5 for $h = \sigma > \lambda = \lambda_0$ and (3.8) we obtain

$$|B(\mathbf{m}) * A(\mathbf{m}) - B(\mathbf{m}) * G_{\sigma^2(\mathbf{m})}| \le$$

$$\le \int |A(\mathbf{m})(x - z) - G_{\sigma^2(\mathbf{m})}(x - z)|B(\mathbf{m})(dz) \le$$

$$\le Q_{B(\mathbf{m})}(\sigma)\sum_r \sup_{r\sigma\le y\le(r+1)\sigma} |A(\mathbf{m}) - G_{\sigma^2(\mathbf{m})}| \le$$

$$\le \frac{C_5\sigma}{\lambda_0}[t(\mathbf{m})]^{-1/2}C_9\frac{\lambda_0}{\sigma} = C_5C_9[t(\mathbf{m})]^{-1/2}. \quad (3.15)$$

Exactly as we obtained (3.12) from (3.10) in case A we now obtain from (3.15)

$$|H - H_1| \leq (C_5C_9 + 8)\epsilon^{1/3}. \tag{3.16}$$

It now remains to estimate the difference $H_1 - H_2$. By Lemma 3,

$$|G_{\sigma^2} - G_{\sigma^2(\mathbf{m})}| \leq C_{19}\epsilon^{1/3}$$

if

$$|\sigma^2(\mathbf{m})/\sigma_2 - 1| \leq C_{19}/C_7.$$

Using (3.7) and taking into account the fact that now $\Lambda < \sigma$, we obtain the estimate

$$\sum\nolimits'' = \sum_{\left|\frac{\sigma^2(\mathbf{m})}{\sigma^2} - 1\right| \leq \frac{C_{19}}{C_7}} p(\mathbf{m}) \leq C_{20}\epsilon^{1/3}. \tag{3.17}$$

Therefore in the same way that we derived (2.10) in the proof of Theorem 1 we obtain

$$|H_1 - H_2| \leq (C_{19} + C_{20})\epsilon^{1/3}.$$

It follows from (3.14), (3.16) and (3.17) that

$$|H - D| \leq C_{21}\epsilon^{1/3}. \tag{3.18}$$

Inequalities (3.12) and (3.18) show that in cases A and B the estimate (0.3), which is the essence of Theorem 2, may be replaced by the following stronger one:

$$D(x) - C_{22}\epsilon^{1/3} \leq H(x) \leq D(x) + C_{22}\epsilon^{1/3}. \tag{3.19}$$

Case C. In this case we have not managed to obtain an estimate of the type (3.19). We set

$$\sigma_0 = \frac{1}{\sqrt{2}}L\left(\log\frac{L}{l}\right)^{-1/2} \tag{3.20}$$

and introduce the auxiliary distribution

$$H' = H * G_{\sigma_0}.$$

By Lemma 2, for

$$\eta = C_6 e^{-L^2/2\sigma_0^2} = l/L \tag{3.21}$$

we have

$$H'(x-L)-\eta \le H(x) \le H'(x+L)+\eta. \tag{3.22}$$

We now show that the infinitely divisible distribution

$$D=\sum_{\mathbf{m}} q(\mathbf{m})B(\mathbf{m})*G_{\sigma_1^2}, \quad \sigma_1^2=\sigma^2+\sigma_0^2,$$

satisfies the inequality

$$|D-H'| \le C_{23}\Big[\epsilon^{1/3}+\frac{l}{L}\Big(\log\frac{L}{l}\Big)^{1/2}\Big]. \tag{3.23}$$

Using (3.23) together with (3.21) and (3.22) we obtain (0.3).

It now only remains to prove (3.23). For this we introduce the distributions

$$H_1'=\sum_{\mathbf{m}} p(\mathbf{m})B(\mathbf{m})*G_{\sigma_1^2(\mathbf{m})}, \quad \sigma_1^2(\mathbf{m})=\sigma^2(\mathbf{m})+\sigma_0^2,$$
$$H_2'=\sum_{\mathbf{m}} p(\mathbf{m})B(\mathbf{m})*G_{\sigma_1^2}.$$

The inequality

$$|D-H_2'| \le C_{12}\epsilon^{1/3} \tag{3.24}$$

is proved in the same way as (3.9) and (3.14) in cases A and B, only now

$$s=\sum_k p_k < \epsilon^{-2/3},$$

which leads to the estimate

$$\sum_k p_k^2 \le \epsilon \sum_k p_k < \epsilon^{1/3}.$$

By Lemma 3,

$$|H_1'-H_2'| \le C_{24}\epsilon^{1/3}+\sum{}''' p(\mathbf{m}), \tag{3.25}$$

where $\sum'''$ runs over those $p(\mathbf{m})$ for which

$$|\sigma_1^2(\mathbf{m})/\sigma_1^2-1| > C_{24}/C_7. \tag{3.26}$$

For suitable C_{24}, inequality (3.25) implies that

$$|\sigma_1^2(\mathbf{m})-\sigma_1^2|=|\sigma^2(\mathbf{m})-\sigma^2|>\sigma_1^2\epsilon^{1/3}.$$

Therefore (3.7) and the inequalities[10] $\sigma^2 < \sigma_1^2$, $\Lambda = l \le \sigma_0$ imply that

$$\sum{}''' \le \frac{\sigma^2\Lambda^2\epsilon}{4\epsilon^{2/3}\sigma_1^4} \le \frac{1}{4}\epsilon^{1/3}.$$

Thus from (3.25) we obtain

$$|H_1' - H_2'| \le (C_{24}+1)\epsilon^{1/3}. \tag{3.27}$$

Finally, according to Lemma 6 we obtain from (3.20)

$$|A(\mathbf{m}) * G_{\sigma_0^2} - G_{\sigma_1^2(\mathbf{m})}| \le C_{10}\frac{l}{\sigma_0} \le \sqrt{2}C_{10}\frac{l}{L}\Big(\log\frac{L}{l}\Big)^{1/2},$$

whence we have

$$|H' - H_1'| = \sqrt{2}C_{10}\frac{l}{2}\Big(\log\frac{L}{l}\Big)^{1/2}. \tag{3.28}$$

Formulas (3.24), (3.27) and (3.28) immediately imply (3.23). This completes the proof of Theorem 2.

Steamboat "Sergei Kirov"
Red Sea–Persian Gulf
13–24 March 1962

References

1. A.N. Kolmogorov, 'Two uniform limit theorems for sums of independent terms', *Teor. Veroyatnost. i Primenen.* (Probability theory and its applications) **1**:4 (1956), 426–436 (in Russian). (Translated as *Theory Probab. Appl.*)
2. B. de Finetti, 'Le funzioni caratteristiche di legge instantanea', *Atti Accad. Naz. Lincei. Rend. Ser.* 6 **12**:7/8 (1930), 278–282.
3. W. Doeblin, 'Sur les sommes d'un grand nombre des variables aléatoires indépendantes', *Bull. Sci. Math.* **63** (1939), 23–32; 35–64.
4. Yu.V. Prokhorov, 'On sums of identically distributed random variables', *Dokl. Akad. Nauk SSSR* **105**:4 (1955), 645–647 (in Russian).

[10] The second inequality is obtained from (3.20):

$$l = \sqrt{2}\frac{l}{L}\Big(\log\frac{L}{l}\Big)^{1/2}\sigma_0 \le \sigma_0.$$

5. Yu.V. Prokhorov, 'On a non-uniform limit theorem of A.N. Kolmogorov', *Teor. Veroyatnost. i Primenen* **5**:1 (1960), 103–113 (in Russian).
6. L.D. Meshalkin, 'On approximating distributions of sums by infinitely divisible laws', *Teor. Veroyatnost. i Primenen.* **6**:3 (1961), 257–275 (in Russian). (Translated as *Theory Probab. Appl.*)
7. A. Khinchin, 'Zur Theorie der unbeschränktteilbaren Verteilungsgesetze', *Mat. Sb.* **2**:1 (1937), 79–119.
8. A.N. Kolmogorov, 'Sur les propriétés des fonctions de concentrations de P. Lévy', *Ann. Inst. H. Poincaré* **16**:1 (1958), 27–34.
9. B.A. Rogozin, 'On an estimate of a concentration function', *Teor. Veroyatnost. i Primenen* **6**:1 (1961), 103–105 (in Russian). (Translated as *Theory Probab. Appl.*)
10. B.A. Rogozin, 'On increasing the variance of sums of independent random variables', *Teor. Veroyatnost. i Primenen.* **6**:1 (1961), 106–108 (in Russian). (Translated as *Theory Probab. Appl.*)
11. Yu.V. Prokhorov, 'Asymptotic behaviour of the binomial distribution', *Uspekhi Mat. Nauk. SSSR* **8**:3 (1953), 135–142 (in Russian).
12. B.V. Gnedenko and A.N. Kolmogorov, *Limit distributions for sums of independent random variables*, Addison-Wesley, 1954 (translated from the Russian).

52. ESTIMATORS OF SPECTRAL FUNCTIONS OF RANDOM PROCESSES *

Jointly with I.G. Zhurbenko

The paper deals with the asymptotic behaviour of efficient statistics of a spectral density as the sample size increases. A certain new statistic introduced by A.N. Kolmogorov is compared with other well-known statistics of a general density, as well as with asymptotically optimal statistics with respect to the mean square deviation. The influence on various statistics of high peaks of the spectral density at neighbouring frequencies is discussed. A new class of statistics is obtained by applying the shift operator to a periodogram calculated from smoothed data.

Let $X(t)$, $t = 0, \pm 1, \ldots$, be a stationary random process with expectation $\mathbf{E}X(t) = 0$, covariance function $C(t)$ and spectral density $f(\lambda)$, where $f(\lambda)$, $-\infty < \lambda < \infty$, is a function with period 2π. For dimensional reasons it is natural to consider all quadratic forms

$$\frac{1}{2\pi N} \sum_{s,t=1}^{N} b_{s,t}^{(N)} X(s) X(t) \tag{1}$$

with arbitrary coefficients $b_{s,t}^{(N)}$ as the admissible class of statistics. As was shown by Grenander and Rosenblatt [1], the asymptotic behaviour as $N \to \infty$ of the first two moments will not be worse if instead of the class of statistics (1) we consider the narrower class of statistics of the form

$$\hat{f}_N(\lambda) = \frac{1}{2\pi N} \sum_{s,t=1}^{N} b^{(N)}(t-s) e^{i(t-s)\lambda} X(t) X(s), \tag{2}$$

which can be represented as

$$\hat{f}_N(\lambda) = \frac{1}{2\pi} \sum_{t=-N+1}^{N-1} e^{i\lambda t} b_N(t) B_N(t), \tag{3}$$

where

$$B_N(t) = \frac{1}{N} \sum_{s=1}^{N-|t|} X(s) X(s+|t|).$$

* Proc. 2nd European Conference on Statistics, Oslo, August 14–18, 1978. First publication.

The statistics $\hat{f}_N(\lambda)$ can also be written in the from

$$\hat{f}_N(\lambda) = \int_{-\pi}^{\pi} \Phi_N(x) I_N(x+\lambda)dx, \tag{4}$$

where

$$I_N(x) = \frac{1}{2\pi N}\Big|\sum_{t=1}^{N} e^{itx}X(t)\Big|^2, \quad b_N(t) = \int_{-\pi}^{\pi} \Phi_N(x)e^{itx}dx.$$

As $N \to \infty$, the asymptotic minimum of the shift $\Delta\hat{f}_N(\lambda) = \mathbf{E}\hat{f}_N(\lambda) - f(\lambda)$ and the variance and consequently also the mean square deviation $\nabla\hat{f}_N(\lambda) = \mathbf{E}(\hat{f}_N(\lambda) - f(\lambda))^2$ of the statistics of the classes (2) and (3) are, according to [1], not greater than those of the statistics of class (1). The function $\Phi_N(x)$ is called the "spectral window" of the statistic $\hat{f}_N(\lambda)$ and will be discussed in detail below.

The asymptotic properties of the estimators $\hat{f}_N(\lambda)$ depend on the smoothness of the spectral density $f(\lambda)$.

We say that

$$X(t) \in \kappa(\lambda, \alpha, C, C_1) = \kappa,$$

if for any μ and certain given λ, $0 < \alpha \leq 2$, $C \geq 0$, $C_1 \geq 0$, one of the inequalities

$$|f(\lambda+\mu) - f(\lambda)| \leq C|\mu|^{\alpha}, \; 0 < \alpha \leq 1,$$

holds, or if $f'(\lambda)$ exists and

$$|f'(\lambda+\mu) - f'(\lambda)| \leq C|\mu|^{\alpha-1}, \; 1 < \alpha \leq 2,$$

and the 4th order spectral density $f_4(x_1, x_2, x_3, x_4)$ of the process $X(t)$ is bounded:

$$|f_4(x_1, x_2, x_3, x_4)| \leq C_1.$$

We say that a sequence of functions $\Phi_N(x)$ is of class $\mathcal{F}$, $\Phi_N(x) \in \mathcal{F}$, if for any integer $N > 0$ $\Phi_N(x)$ is a continuous even periodic function with period 2π such that $\Phi_N(x) \to 0$ uniformly in the region $\epsilon \leq |x| < \pi$ for $\epsilon > 0$ and if the following relations hold:

$$\int_{-\pi}^{\pi} \Phi_N(x)dx = 1, \quad \sup_N \int_{-\pi}^{\pi} |\Phi_N(x)|dx < \infty.$$

Moreover, the following smoothness conditions hold uniformly in $\mathcal{F}$:

1. For an integrable $f(x)$ such that $|f(x)| \le |x|^\alpha$, $0 < \alpha \le 2$, a relation of the form

$$\iint_{-\pi}^{\pi} f(x)\phi_N^2(z)(\Phi_N(x+z) - \Phi_N(x))dx\, dz = o\Big(\int_{-\pi}^{\pi} |x|^\alpha |\Phi_N(x)|dx\Big)$$

holds, where Feller's kernel $\phi_N^2(z)$ is defined as follows:

$$\phi_N^2(z) = \frac{1}{2\pi N}\frac{\sin^2(Nz/2)}{\sin^2(z/2)}. \tag{4a}$$

2. There exists a sequence of numbers $A = A(N) > 0$, $A(N) \to \infty$ as $N \to \infty$, such that

$$\int_{-\pi}^{\pi} \Phi_N(x)\Phi_N(x+y+z)dx =$$
$$= \int_{-\pi}^{\pi} \Phi_N(x)\Phi_N(x+y)dx + o\Big(\int_{-\pi}^{\pi} \Phi_N^2(x)dx\Big)$$

uniformly with respect to y for all $|z| < A/N$.

We define the class $\mathcal{F}(G) \subset \mathcal{F}$ of sequences of functions $\Phi_N(x)$ of the form

$$\Phi_N(x) = A_N G(A_N x)\Big(\int_{-\pi A_N}^{\pi A_N} G(x)dx\Big)^{-1}, \ -\pi \le x \le \pi, \tag{5}$$

for some $A_N > 1$ such that $A_N \to \infty$ and $A_N/N \to 0$ as $N \to \infty$. Let $G(x)$, $-\infty < x < \infty$ be even, piecewise differentiable and satisy the conditions

$$\int_{-\infty}^{\infty} G(x)dx = 1, \quad \int_{-\infty}^{\infty} |x|^\alpha |G(x)|dx < \infty \tag{6}$$

for some $\alpha > 0$.

The problem of finding the optimal statistics in the sense of the mean square deviation is solved under the above natural restrictions on the smoothness of $\Phi_N(x)$. In this case we find not only the type of asymptotics of the mean square deviation, as it was done by E. Parsen [2], but also the values of the coefficients in these asymptotics, which allow us to make a numerical comparison of the most often used statistics of the spectral density proposed by various authors. A considerable weakening of the conditions used in this case is the requirement of smoothness of the spectral density $f(\lambda)$ at one single point as compared to the uniform smoothness

$$\sum_{t=-\infty}^{\infty} |t|^{\alpha} B(t) < \infty,$$

where $B(t)$ is the correlation function of the process $X(t)$.

Note that the smallest possible order of the mean square deviation of the estimator $\hat{f}_N(\lambda)$ is $N^{-2\alpha_0/(1+2\alpha_0)}$, where α_0 is the maximal α for which the latter condition is fulfilled. Unfortunately, however, α_0 is the characteristic of smoothness of $f(\lambda)$ simultaneously with respect to all frequencies [3], [4], [7].

Theorem 1. *If $X(t) \in \kappa(\lambda, \alpha, C, C_1)$, $\Phi_N(x) \in \mathcal{F}$, then for the statistic $\hat{f}_N(\lambda)$ we have, as $N \to \infty$*

$$\inf_{\Phi_N(x)\in\mathcal{F}} \sup_{X(t)\in\kappa(\lambda,\alpha,C,C_1)} \nabla \hat{f}_N(\lambda) \sim O(\alpha) f^2(\lambda) \times$$
$$\times \left(\frac{C^2}{f^2(\lambda)}\right)^{\frac{1}{1+2\alpha}} N^{-\frac{2\alpha}{1+2\alpha}} (1+\eta(\lambda))^{\frac{2\alpha}{1+2\alpha}}, \tag{7}$$

where

$$O(\alpha) = (1+2\alpha)^{-\frac{1}{1+2\alpha}} \left(\frac{\pi(\alpha+1)}{\alpha}\right)^{\frac{2\alpha}{1+2\alpha}}, \quad \eta(\lambda) = \begin{cases} 1, \lambda \equiv 0 \,(\mathrm{mod}\, \pi), \\ 0, \lambda \not\equiv 0 \,(\mathrm{mod}\, \pi). \end{cases} \tag{8}$$

The minimum in (7) *is attained at*

$$\Phi_N(x) = A_N \Phi(A_N x), \tag{9}$$

where

$$\Phi(x) = \begin{cases} -\frac{\alpha+1}{2\alpha}|x|^{\alpha} + \frac{\alpha+1}{2\alpha}, & |x| \le 1, \\ 0, & |x| > 1, \end{cases}$$

and

$$A_N = \left(\frac{\alpha N C^2}{\pi f^2(\lambda)(1+\alpha)(1+2\alpha)}\right)^{1/(1+2\alpha)}. \tag{10}$$

Proof. Isolating the main term of the mean square deviation $\nabla \hat{f}_N(\lambda)$ we obtain, in accordance with Lemmas 1 and 2 in [3], the variational problem of finding the minimum of the functional

$$\left(\int_{-\pi}^{\pi} |x|^{\alpha} \Phi_N(x) x\right)^2 + \frac{2\pi f^2(\lambda)}{NC^2} \int_{-\pi}^{\pi} \Phi_N^2(x) dx$$

over all positive $\Phi_N(x) \in \mathcal{F}$.

In solving the problem of finding the conditional extremum we see that the solution $\Phi_N(x)$ must satisfy the equation

$$|x|^\alpha \int_{-\pi}^{\pi} |u|^\alpha \Phi_N(x)dx + \frac{2\pi f^2(\lambda)}{NC^2}(\Phi_N(x) - C) = 0, \ C = \text{const},$$

which implies that one should look for a solution of the form (9). Taking into account the initial conditions we obtain (10) and with it (7) and (8).

Theorem 1 gives an explicit form of the optimal window in the sense of the mean square deviation of the estimator $\hat{f}_N(\lambda)$ for a given smoothness of the spectral density $f(\lambda)$ under study at the point λ.

In order to compare asymptotic properties of various periodogram estimators of the form (2) and (4) we use the following

Theorem 2. *Let $\Phi_N(x) \in \mathcal{F}$ be defined by* (5) *and suppose that $G(x)$ satisfies* (6) *for some α such that $0 < \alpha \leq 2$. Then for $X(t) \in \kappa$, we have*

$$\inf_{A_N} \sup_{X(t)\in\kappa} \nabla \hat{f}_N(\lambda) \sim f^2(\lambda)\Big(\frac{C^2}{f^2(\lambda)}\Big)^{\frac{1}{1+2\alpha}} N^{-\frac{2\alpha}{1+2\alpha}} g(\alpha) \tag{11}$$

as $N \to \infty$, where

$$\begin{gathered} g(\alpha) = \Big(\frac{1+2\alpha}{2\alpha}\Big)(2\alpha V_1)^{\frac{1}{1+2\alpha}}(2\pi V_2)^{\frac{2\alpha}{1+2\alpha}}, \\ V_1 = \int_{-\infty}^{\infty} |x|^\alpha G(x)dx, \quad V_2 = \int_{-\infty}^{\infty} G^2(x)dx, \end{gathered} \tag{12}$$

with asymptotic maximum attained at

$$A_N = \Big(\frac{\alpha V_1 N C^2}{\pi V_2 f^2(\lambda)}\Big)^{1/(1+2\alpha)}. \tag{12a}$$

The proof of this theorem follows from Lemmas 1 and 2 of [3].

The function $g(\alpha)$ can be calculated by (12) for various estimators of the spectral density, namely, those of Tukey, Parsen, Bartlett et al. (see, for example, [3]), as well as for the optimal statistics given by (10). The results of the calculations are given in [3], [7]. A common drawback of periodogram estimators of the spectral density $\hat{f}_N(\lambda)$ is the weak decrease (of order $O(N^{-1})$) of the dependency on high frequencies and non-stationary phenomena. Earlier, one of the authors (A.N. Kolmogorov) suggested that these drawbacks could be eliminated by using a "time window" followed by averaging periodograms over various time intervals.

Let $a_M(t)$, $t = 0, \pm 1, \ldots$, be a non-negative function vanishing outside $[0, M]$. From a sample $\{X(Q), \ldots, X(Q+M)\}$ we construct the function

$$W_M^Q(\lambda) = \sum_{t=-\infty}^{\infty} a_M(t-Q) X(t) e^{i\lambda t}.$$

We determine the statistic $\bar{f}_N(\lambda)$ of the spectral density $f(\lambda)$ of the random process $X(t)$ in the following way:

$$\bar{f}_N(\lambda) = \frac{1}{T} \sum_{k=0}^{T-1} |W_M^{Lk}(\lambda)|^2.$$

It makes use of $N = T(L-1) + M + 1$ sample values of the process $X(t)$.

Let us estimate the spectral density. We determine the "data window" as follows:

$$a_M(t) = a_{K,P}(t) = \mu(K,P)\Big(\frac{K(P^2-1)}{12\pi}\Big)^{1/4} P^{-K} C_{K,P}(t), \tag{13}$$

where $M = K(P-1)$ and the coefficients $C_{K,P}$ are determined from the relation

$$\sum_{t=0}^{K(P-1)} z^t C_{K,P}(t) = (1 + z + \ldots + z^{P-1})^K = \Big(\frac{1-z^P}{1-z}\Big)^K.$$

Since

$$\phi_{M,P}(x) = \phi_M(x) = \sum_{t=-\infty}^{\infty} a_M(t) e^{itx},$$

it follows from (13) that the "spectral window" of the estimator in question is given by

$$|\phi_{K,P}(x)|^2 = \mu^2(K,P)\Big(\frac{K(P^2-1)}{12\pi}\Big)^{1/2}\Big(\frac{\sin^2(Px/2)}{P^2\sin^2(x/2)}\Big)^K, \tag{13a}$$

where $\mu(K,P) \to 1$ as $K \to \infty$, $P \to \infty$ (see [3]). From the definition of spectral densities of the 2nd and 4th order we obtain the following equalities for the estimator $\bar{f}_N(\lambda)$:

$$\Delta \bar{f}_N(\lambda) = \int_{-\pi}^{\pi} |\phi_M(x-\lambda)|^2 (f(x) - f(\lambda))dx, \tag{14}$$

$$\begin{aligned}\mathbf{D}\bar{f}_N(\lambda) = {} & \frac{1}{T^2}\int_{\Pi^4} \delta^*(x_1 + \ldots + x_4) f_4(x_1, \ldots, x_4)\phi_M(x_1+\lambda)\times \\ & \times \phi_M(x_2-\lambda)\phi_M(x_3-\lambda)\phi_M(x_4-\lambda)\frac{\sin(TL(x_1+x_2)/2}{\sin(L(x_1+x_2)/2}\times \\ & \times \frac{\sin(TL(x_3+x_4)/2)}{\sin(L(x_3+x_4)/2)}dx_1dx_2dx_3dx_4 + \frac{1}{T^2}\int_{\Pi^2} f(x-\lambda)\times \\ & \times f(y+\lambda)|\phi_M(x)|^2|\phi_M(y)|^2\frac{\sin^2(TL(x+y)/2)}{\sin^2(L(x+y)/2)}dxdy+ \\ & + \frac{1}{T^2}\int_{\Pi^2} f(x)f(y)\phi_M(x+\lambda)\phi_M(-x+\lambda)\phi_M(y-\lambda)\phi_M(-y-\lambda)\times \\ & \times \frac{\sin^2(TL(x+y)/2)}{\sin^2(L(x+y)/2)}dxdy,\end{aligned} \tag{15}$$

where $\Pi = [-\pi, \pi]$,

$$\delta^*(x) = \sum_{k=-\infty}^{\infty} \delta(x + 2k\pi), \tag{15a}$$

$\delta(x)$ being the Dirac delta-function.

Theorem 3. *If $X(t) \in \kappa$, then for the statistic $\bar{f}_N(\lambda)$ for $\lambda \not\equiv 0 \pmod{\pi}$ with coefficients defined by* (13) *we have*

$$\inf_{T,L,K,P}\sup_{X(t)\in\kappa} \nabla \bar{f}_N(\lambda) \sim \mu^4(K,P)f^2(\lambda)\Big(\frac{C^2}{f^2(\lambda)}\Big)^{\frac{1}{1+2\alpha}} K(\alpha)N^{-\frac{2\alpha}{1+2\alpha}} \tag{16}$$

as the sample size $N = (L(T-1) + K(P-1) + 1) \to \infty$, where

$$K(\alpha) = \frac{1+2\alpha}{2\alpha}\sqrt{\frac{\pi}{6}}\Big(\frac{\alpha\sqrt{24}}{\sqrt{\pi^3}}\Gamma^2\Big(\frac{\alpha+1}{2}\Big)12^\alpha\Big)^{1/(1+2\alpha)},$$

the asymptotic equality (14) *being attained when*

$$\frac{L}{N^{1/(1+2\alpha)}} \to O(1), \quad \frac{N^{(1-2\alpha)/(1+2\alpha)}}{P} \to O(1), \quad \frac{P}{N^{1/(1+2\alpha)}} \to O(1),$$

$$KP^2\Big(\frac{N\alpha\sqrt{24}}{\sqrt{\pi^3}f^2(\lambda)}\Gamma^2\Big(\frac{\alpha+1}{2}\Big)12^\alpha C^2\Big)^{-2/(1+2\alpha)} = O(1). \tag{17}$$

Proof. A direct application of Laplace's method for estimating the integral on the right-hand side of (14) with kernel $\phi_{K,P}(x)$ calculated via (13a) gives

$$\sup_{X(t)\in\mathcal{K}} \Delta^2 \bar{f}_N(\lambda = \mu^4(K,P)\Big(\frac{K(P^2-1)}{12}\Big)^{-\alpha} C^2\Gamma^2\Big(\frac{\alpha+1}{2}\Big)\frac{1+O(K^{-1})}{\pi}.$$

The main term of $\mathbf{D}\bar{f}_N(\lambda)$ gives an expression which, using the method of steepest descent, can be reduced to the following form:

$$I = \mu^4(K,P)f^2(\lambda)\frac{\sqrt{\pi^2K(P^2-1)}}{TL\sqrt{6}}\Big(1+O\Big(\frac{L}{\sqrt{K(P^2-1)}}\Big)+O\Big(\frac{1}{K}\Big)\Big). \quad (18)$$

The difference between (18) and the second term of (15) is estimated by

$$|I-I_2| \le C_2\frac{P\sqrt{K}}{(TL)^{1+\alpha}}\Big(1+O\Big(\exp\Big\{-\frac{K(P^2-1)}{L^2}\Big\}\Big)+O\Big(\frac{1}{K}\Big)\Big), \quad \alpha<1.$$

If $\alpha > 1$, then $|I - I_2|$ is of order $P\sqrt{K}(TL)^{-2}$. In a similar way we can show that the third term in (15) has smaller order. The first term in (15) under the conditions of the theorem is estimated in the following way:

$$|I_1| \le \frac{2\pi C_1}{TL}\Big(1+O\Big(\frac{L^2}{K(P^2-1)}\Big)+O\Big(\frac{1}{K}\Big)\Big).$$

Setting $K(P^2-1) = \nu N^\nu$ and finding the minimum of $\nabla\bar{f}_N(\lambda)$, we obtain the asymptotic equality (16). This proves our theorem.

As has been shown by calculations, (see [3], [7]), the mean square deviation appears to be closest to the optimal statistic in the case of Kolmogorov's statistic $\bar{f}_N(\lambda)$ as compared to the estimators of Tukey, Parsen, Bartlett, Abel and others. As follows from Theorem 3, the statistic $\bar{f}_N(\lambda)$ by comparison with other statistics, is insensitive to changes in the parameters, which can be chosen in broad ranges. The degree of dependence on high frequencies of this statistic is of order $N^{-2K(1+2\alpha)}$, where K can be arbitrarily large as $N \to \infty$. This enables us to carry out a spectral analysis in the required frequency strip in the presence of strong noises and non-stationary processes concentrated at other frequencies, for example, in the presence of a trend. At the same time it is easy to check stationarity in the frequency strip under study.

We now consider the effect of strong peaks in the spectral density $f(x)$ at a frequency $\lambda + \Delta$ close to λ. Assume that

$$f(x) = f_\alpha(x) + f_\delta(x),$$

where $f_\alpha(x)$ is the spectral density of the process $X_\alpha(t) \in \kappa$ and $f_\delta(x)$ is defined by an equation of the form

$$f\delta(x) = \delta \cdot \delta^*(x - \lambda - \Delta) + \delta\delta^*(x + \lambda + \Delta),\ \Delta \not\equiv 0\ (\mathrm{mod}\, 2\pi), \tag{19}$$

where $\delta > 0$ is a real number and the function $\delta^*(x)$ is given by (15a). We say that $X(t) \in \kappa\ (\lambda, \Delta, \delta)$ if the spectral density of $X(t)$ is determined by (19), the semi-invariant spectral density of the 4th order is bounded and $\mathbf{E}X(t) = 0$.

The asymptotics of the mean square deviation of the statistic with optimal choice of parameters in the presence of a chosen δ at frequency $\lambda+\Delta$ is described by the following theorem (see [5]–[7]):

Theorem 4. *Let $X(t) \in \kappa\ (\lambda, \Delta, \delta)$, let the kernels $\Phi_N(x) \in \mathcal{F}$ be determined by* (5), *let $G(x)$ satisfy* (6) *for $\alpha \leq 2$ and let the parameter A_N of the statistic $\hat{f}_N(\lambda)$ be chosen in accordance with* (12a). *Then as $N \to \infty$*

$$\sup_{X(t))\in\kappa(\lambda,\Delta,\delta)} \nabla \hat{f}_N(\lambda) = f^2(\lambda)\Big(\frac{C^2}{f^2(\lambda)}\Big)^{1/(1+2\alpha)} N^{-2\alpha/(1+2\alpha)} \times$$
$$\times (1 + \eta(\lambda))^{2\alpha/(1+2\alpha)} g(\alpha)(1 + o(1)) +$$
$$+2\delta^2 \Big(\int_{-\pi}^{\pi} \Phi_N(x)\phi_N^2(x + \Delta)dx + \int_{-\pi}^{\pi} \Phi_N(x)\phi_N^2(x + 2\lambda + \Delta)dx\Big)^2 +$$
$$+O\Big(N \int_{-\pi}^{\pi} \Phi_N(x)\phi_N^2(x + \Delta)dx, \tag{20}$$

where

$$\eta(\lambda) = \begin{cases} 1, & \lambda \equiv 0\,(\mathrm{mod}\,\pi), \\ 0, & \lambda \not\equiv 0\,(\mathrm{mod}\,\pi) \end{cases} \tag{21}$$

and the functions $\phi_N^2(x)$ and $g(\alpha)$ are defined by (4a) *and* (12a) *respectively.*

Theorem 5. *Let $X(t) \in \kappa\ (\lambda, \Delta, \delta)$ and let the statistic $\bar{f}_N(\lambda)$ be determined by the coefficients $a_M(t)$ found via* (13) *under the conditions* (16) *as $N = L(T - 1) + M(P - 1) + 1 \to \infty$. Then*

$$\sup_{X(t)\in\kappa(\lambda,\Delta,\delta)} \nabla \bar{f}_N(\lambda) = K(\alpha) f^2(\lambda)\Big(\frac{C}{f^2(\lambda)}\Big)^{1/(1+2\alpha)} \times$$
$$\times N^{-2\alpha/(1+2\alpha)}(1 + \eta(\lambda))^{2\alpha/(1+2\alpha)}(1 + o(1)) + 2\delta^2(|\phi_M(\Delta)|^2 +$$
$$+|\phi_M(\Delta + 2\lambda)|^2)^2 + O(N^{-\alpha/(1+2\alpha)}|\phi_M(\Delta)|^2), \tag{22}$$

where $|\phi_n(x)|^2$, $K(\alpha)$ *and* $\eta(\lambda)$ *are defined by* (13a), (16a) *and* (21) *respectively.*

The proofs of Theorems 4 and 5 are given in [5]–[7].

Note that the order of the remainder terms in (20) for the statistic $\hat{f}_N(\lambda)$ is not less than

$$|\phi_N(\Delta)|^4 = O(N^{-2}) \tag{23}$$

if $\Delta \gg N^{-\epsilon/(1+2\alpha)(1+\epsilon)}$. For $\hat{f}_N(\lambda)$ the order of the remainder terms in (22) is not less than

$$|\phi_M^2(\Delta)| \sim N^{-(2k+1)/(1+2\alpha)} K^k,$$

which can be taken arbitrarily small with respect to (23) for comparatively small k as $N \to \infty$. Under these conditions

$$\Delta \gg P^{-1} \sim C\sqrt{K} N^{-1/(1+2\alpha)}.$$

According to (23) the statistics $\hat{f}_N(\lambda)$ have fixed order of dependence on the choice of δ, which cannot be diminished by choosing a suitable spectral window $\Phi_N(x)$.

References

1. U. Grenander and M. Rosenblatt, *Statistical analysis of stationary time series*, Wiley, New York, 1957.
2. E. Parsen, 'On asymptotically efficient consistent estimates of the spectral density function of a stationary time series', *J. Roy. Statist. Soc. B*, **20** (1958), 303–322.
3. I.G. Zhurbenko, 'On the efficiency of estimators of the spectral density of a stationary process', *Teor. Veroyatnost. i Primenen.* **25**:3 (1980), 476–489 (in Russian). (Translated as *Theory Probab. Appl.*)
4. I.G. Zhurbenko, 'The study of statistics of the spectral density of a stationary random process', *Sibirsk Mat. Zh.* **22**:5 (1981), 40–65 (in Russian).
5. I.G. Zhurbenko, 'On the efficiency of estimators of the spectral density of a stationary random process, II', *Teor. Veroyatnost. i Primenen.* **28**:2 (1983), 388–396 (in Russian). (Translated as *Theory Probab. Appl.*)
6. I.G. Zhurbenko, 'On a limit theorem for statistics of the spectral density with time shift', *Ukrain. Mat. Zh.* **32**:4 (1980), 463–476 (in Russian).
7. I.G. Zhurbenko, *The spectral density of time series*, Moskov. Gos. Univ., 1982, (in Russian).

53. ON THE LOGICAL FOUNDATIONS OF PROBABILITY THEORY *

When we talk about random events in the everyday sense of the word, we mean phenomena in which we do not find any regularities that would allow us to predict their behaviour. Generally speaking, there are no reasons to suggest that events random in this sense obey any probability laws. Consequently we should distinguish between randomness in the wider sense (the absence of any regularity) and stochastic random events (which are the subject of probability theory).

The problem is to describe the reasons why mathematical probability theory can be applied at all to phenomena of the real world. My first attempt to answer questions of this kind was made in [1] (published in an edition of methodological character).

Since a random event is defined as the absence of regularity, we should first define the notion of regularity. The natural means for this is given by the theory of algorithms and recursive functions. The first attempt to apply it to probability theory was made by Church [2].

The purpose of my report is to familiarize the readers with this field, at least in a first approximation.

Paying tribute to tradition, I begin with the classical definition of probability as the ratio

$$P = m/n$$

of the number of favourable outcomes m to the total number of outcomes n. This definition reduces the problem of calculating the probability to combinatorial problems.

However, this definition cannot be applied in many practical situations. This brought to life the so-called statistical definition of probability:

$$P \approx \mu/N, \tag{1}$$

where N is the total number of trials, which is assumed to be sufficiently large, and μ is the number of successes. Strictly speaking, this definition is not mathematical in its initial form. This is why we write the sign of approximate equality in (1).

---

* In: *Lecture Notes in Mathematics*, **1021** (1983), pp.1–5.

The first attempt to make definition (1) more exact was made by R. von Mises. But before we describe his approach, let us discuss (from the viewpoint of the classical definition of probability) why stability of frequencies is so often observed in natural phenomena.

Consider all 0–1 sequences of length n containing exactly m ones and assume that all such sequences are equally probable. Suppose that some method of dividing any sequence of length n into two subsequences has been chosen. Then for each sequence it is important to compare the frequencies of ones in both subsequences by calculating the difference

$$|\mu_1/n_1 - \mu_2/n_2|,$$

where n_1 and n_2 are the lengths of the subsequences and μ_1 and μ_2 the number of ones in them, so that $n_1 + n_2 = n$, $\mu_1 + \mu_2 = m$. We would like to think that this difference would "almost always be small" in the sense that for any $\epsilon > 0$

$$P_{\text{class}}\Big\{\Big|\frac{\mu_1}{n_1} - \frac{\mu_2}{n_2}\Big| < \epsilon\Big\} \to 1 \text{ as } n_1, n_2 \to \infty.$$

Certainly, to make this assertion a theorem, we should narrow the class of possible rules for choosing subsequences (for example, forbid the rule according to which all zeros are placed into one subsequence and all ones into the other). All necessary clarifications of the notion of an admissible rule of choosing subsequences based on von Mises' ideas are given in [3]. The notion of an acceptable selection rule is crucial in von Mises' frequency approach to the notion of probability. According to von Mises, an infinite sequence $X_1, X_2, \ldots$ of zeros and ones is called a *Bernoulli sequence* if:

1) the limit

$$P = \lim_{n\to\infty} \frac{1}{n} \sum_{i\le n} X_i$$

exists;

2) this limit is preserved under transition to a subsequence chosen with the help of an admissible rule:

$$\lim_{m\to\infty} \frac{1}{m} \sum_{j\le m} X_{n_j} = P.$$

Von Mises gave only general characteristics and several examples of admissible rules. His instructions basically mean that the choice of every next term of the subsequence should not depend on the value of the term but only on the values of the previous terms. Of course, this definition is not exact, but no rigorous definition should be expected since the very notion of "rule" did not have a mathematical definition at that time. The situation changed essentially when the notions of algorithm and recursive function appeared, used by Church [2] to clarify von Mises' definition. In [3] I proposed a broader class of admissible selection rules than that given by Church. According to [3], a selection rule is given by an algorithm (or, if you like, a Turing machine). The next term of the sequence is chosen as follows: the input information consists of a finite sequence of numbers $n_1, n_2, \dots, n_k$ and values $X_{n_1}, X_{n_2}, \dots, X_{n_k}$ of the corresponding terms of the original sequence. The output consists, first, of the number n_{k+1} of the next considered element of the sequence (it should not coincide with any of the $n_1, n_2, \dots, n_k$, on whose order no restrictions are imposed), and secondly, indications as to whether the element $X_{n_{k+1}}$ should be chosen only for inspection or is to be included in the chosen subsequence. At the next step the input of the algorithm consists of the longer sequence $n_1, n_2, \dots, n_{k+1}$; naturally, the algorithm starts working from the empty input.

In comparison with [2] the class of admissible selection rules is wider, since the order of terms in the subsequence need not coincide with their order in the initial sequence. Another, even more important, difference is in the strictly finitary character of the entire concept mentioned above and in the quantitative estimate of stability of the frequencies.

Passing to finite sequences inevitably leads to restrictions on the complexity of the selection algorithm. An exact definition of the complexity of a finite object and examples of its use in probability theory were proposed in [3], [6]. The results of frequency and complexity approaches are compared in [4].

Now let us return to the original idea according to which "randomness" consists in the absence of "regularity" and see how the notion of complexity of a finite object enables us to make this idea precise. The notion of complexity is dealt with in a number of papers that may be divided into two groups: on the complexity of calculations and on the complexity of definitions. We will deal with the second group.

We take the definition of complexity from [6]. We define the conditional complexity of a constructible object with respect to a certain algorithm A under the condition that the constructible algorithm Y is known. More precisely, define the conditional complexity $K_A(X|Y)$ of the object X for a known Y as the length of the shortest programme by means of which an algorithm A can derive X from Y:

$$K_A(X|Y) = \min\{l(p)|A(p,Y) = X\}.$$

Here $l(p)$ is the length of the sequence of zeros and ones encoding the programme. There exists an "optimal" algorithm A, that is, one such that for any algorithm A_1 there exists a constant C such that for all X and Y

$$K_A(X|Y) \leq K_{A_1}(X|Y) + C.$$

If A_1 and A_2 are optimal algorithms, then the corresponding complexity functions differ at most by an additive constant (independent of X and Y).

We can now define a random, or more precisely, a Δ-random object in a given finite set M (here Δ is a number). Namely, we say that $X \in M$ is a *Δ-random element* of M if

$$K_A(X|Y) \geq \log_2 |M| - \Delta,$$

where $|M|$ is the number of elements of M. The objects from M that are Δ-random for relatively small Δ's will be called *random* in M. We obtain a definition of a random finite object which can be regarded as definitive.

Taking as M the set D_n of all 0–1 sequences of length n we come to the condition

$$K_A(X|D_n) \geq n - \Delta.$$

It can be proved that the sequence satisfying this condition for sufficiently small Δ possesses the property of stability of frequencies when passing to subsequences. Hence, von Mises' requirements on random sequences can be considered as a particular case of our requirements.

Further results in this direction may be found in [5], [7]–[12].

References

1. A.N. Kolmogorov, 'Probability theory', In: *Mathematics, its contents, methods and significance*, Akad. Nauk. SSSR, Moscow, 1956 (in Russian).
2. A. Church, 'On the concept of a random sequence', *Bull. Amer. Math. Soc.* **46**:2 (1940), 130–135.

3. A.N. Kolmogorov, 'On tables of random numbers', *Sankhya, Ser.* A **25**:4, 369–376.
4. A. Shen, 'A frequency approach to the definition of the notion of random sequence', In: *Semiotics and informatics*, VINITI, No.18, Moscow, 1982, pp. 14–42 (in Russian).
5. D. Loveland, 'A new interpretation of the von Mises concept of random sequence', *Z. Math. Logik und Grundlag. Math.* **12**:4 (1966), 279–294.
6. A.N. Kolmogorov, 'The logical foundations of information theory and probability theory', *Problemy Peredachi Informatsii* **5**:3 (1969), 3–7. (Translated as *Problems Inform. Transmission*)
7. R.J. Solomonoff, 'A formal theory of inductive inference', *Inform. and Control* **7**:1 (1964), 1–22.
8. P. Martin-Löf, *Algorithms and random sequences*, Univ. Press, Erlangen, 1966.
9. P. Martin-Löf, 'The definition of random sequences', *Inform. and Control* **9**:6 (1966), 602–619.
10. V.V. Vyugin, 'Algorithmic entropy (complexity) of finite objects and its applications to the definition of randomness and quantity of information', In: *Semiotics and informatics*, VINITI, No.16, Moscow, 1981, pp.14–43 (in Russian).
11. G. Chaitin, 'On the length of the programs for computing finite binary sequences', *J. ACM* **13**:4 (1966), 547–569.
12. A.K. Zvonkin and L.A. Levin, 'Complexity of finite objects and substantiation of the notions of information and randomness by means of the theory of algorithms', *Uspekhi Mat. Nauk* **25**:6 (1970), 85–127 (in Russian).

Comments

ON THE PAPERS ON PROBABILITY THEORY AND MATHEMATICAL STATISTICS

A.N. Kolmogorov

Here is a classification of my works on probability theory in terms of their contents and the time when they were written.

I. Limit theorems for independent and weakly dependent random variables derived by the methods of metric function theory. Papers Nos. 1–6, 8.

In terms of the methods used, these works, dealing with various forms of the law of large numbers and the law of the iterated logarithm, are adjacent to the work by me and D.E. Men'shov on metrical function theory. They were done in close contact with A.Ya. Khinchin. Of the relevant results obtained by other researchers the first to be mentioned are the necessary and sufficient conditions of applicability of the strong law of large numbers for sums of independent random variables obtained by Yu.V. Prokhorov.

II. Axiomatics and logical foundations of probability theory. Paper No. 7.

Paper No. 7, published in 1929, was the first result of my reflections on the logical structure of probability theory. Here probability theory is presented as one of the fields of application of general measure theory. But the concept developed had not yet revealed the set-theoretic implications of conditional probability, a notion that is fundamental in probability theory. Only after this difficulty was overcome and the theory of distributions in infinite products was constructed did it become possible to speak of a set-theoretic justification of probability theory as a whole which was given in my monograph "Fundamental notions of probability theory" that came out in 1933 in German and in 1936 in Russian. A modernized presentation of these concepts, developed by A.N. Shiryaev and myself, is given in the second edition of this monograph (Nauka, Moscow, 1974).

III. Markov chains (Markov processes with discrete time). Papers Nos. 21, 23, 36.

The fundamental significance of Markov processes was understood in 1906–1907 by A.A. Markov. From a strictly finite viewpoint the essence of the matter is completely characterized already by the example of finite discrete Markov chains. For homogeneous chains with a finite number of states, Markov established a fundamental theorem on the existence of the limit of transition probabilities. A whole range of problems on this was taken up by V.I. Romanovskii (Tashkent) and later by many researchers of the Moscow probability school. For a homogeneous chain with a finite number of states the most exhaustive results were obtained by S. Kh. Sirazhdinov, a student of Romanovskii and Kolmogorov, who determined the asymptotics of multidimensional distributions of the number of distinct states visited in a Markov chain.

IV. Markov processes. Papers Nos. 9, 10, 13, 14, 17, 19, 24, 39.

In 1929 I focused attention on the theory of Markov processes with continuous time. In No. 9 (one-dimensional case) and in No. 17 (multidimensional case) this theory was developed in classical terms without explicitly using trajectory spaces. The first versions of the modern concept of a Markov process with continuous time were developed by J. Doob and E.B. Dynkin (see the comments by A.D. Ventsel). Papers Nos. 14, 19, 24 (see the comments by A.M. Yaglom) deal with various applications of the diffusion-type theory of Markov processes. Paper No. 39 (see the comments by A.A. Yushkevich) discusses the case of a discrete set of states.

V. Limit theorems on the convergence of Markov chains to Markov processes with continuous time. Papers Nos. 9, 12, 16, 41.

The first conception of this sort of theorem was presented in 1931 in §12 of article No. 9. Two examples of these theorems are given in papers Nos. 12, 16. Further work in this direction was done by other researchers (invariance principle), see the review article No. 41.

VI. Stationary processes. Papers Nos. 27, 28, 47, 48, 50, 52.

I became interested in the spectral theory of stationary random processes after the appearance of the work of A.Ya. Khinchin and E.E. Slutskii. A first idea on the range of problems involved can be obtained from Paper No. 34. The hypothesis on representing oscillatory processes by Stieltjes integrals (see (4) in

No. 34) was completely confirmed during the following 40 years. Apparently, even in teaching (in particular, to engineers) it should be more explicitly stated that the spectral decomposition of a process does not allow, in general, a more concrete explanation.

VII. Branching processes. Papers Nos. 25, 32, 33, 46.

A number of researchers have studied particular cases of "branching processes". A review of the early literature can be found in the paper by J. Steffenson referred to in my paper No. 25. In his well-known monograph, R.A. Fisher makes extensive use of the fact that the evolution of the number of genes in a population obeys the scheme of a branching process when the gene concentration is low. Fisher proposed to treat this problem using generating functions. On further development of the theory of branching processes see B.A. Sevastyanov's monograph "Branching processes" (Nauka, Moscow, 1971).

VIII. Various applications. Papers Nos. 14, 18, 22, 26, 29, 37.

Paper No. 14 solves a problem posed by S.I. Vavilov. In this paper §§1 and 2 are written by me, and §3 by M.A. Leontovich. As for paper No. 18, the considered "blow-up" of the empirical correlation coefficients under a small number of observations is quite typical in many applied works (see also the comments by A.M. Yaglom).

IX. Mathematical statistics. Papers Nos. 11, 15, 30, 31, 38, 50.

See the comments by E.V. Khmaladze, E.V. Malyutov and A.N. Shiryaev.

ANALYTICAL METHODS IN PROBABILITY THEORY (No. 9)

(A.D. Ventsel)

The title to this article, which in modern language would rather read "Analytical methods in the theory of Markov processes" may serve as the title for an entire branch of this theory. The essence of this branch is that Markov processes come into consideration only for the purposes of translating the problem in question into the language of transition probabilities $P(s, x, t, E)$ of a Markov process or other related analytical objects; after that the problem is solved as a purely analytical one. In paper No. 9 of this volume, random processes as ensembles of realizations (trajectories) or as objects described by a system of finite-dimensional distributions are not considered explicitly but only as motivations of certain definitions and assumptions. Thus, having introduced the decompositions (111), (112), Kolmogorov writes: "Strictly speaking, only in this case is the process continuous in time". In terms of the modern theory of Markov processes this could be reformulated as follows: (111) (112) are close to the conditions imposed on the transition function necessary for the continuity of the trajectories of the corresponding Markov process (or, more exactly, for the existence of modifications of it with continuous trajectories).

The main themes of the paper are: the general notion of a stochastically determined (Markov) process, differential characteristics of Markov processes, and differential equations related to these processes, that is, ergodic properties.

As for the notion of a Markov process, we can say that as there appeared more and more methods for solving specific problems dealing with the behaviour of Markov processes over a time interval, trajectories became more and more explicitly involved in the schemes under consideration. Thus, in 1933 there appeared papers by Kolmogorov and Leontovich [1] and Andronov, Vitt and Pontryagin [2], which derived differential equations for the functions connected with a diffusion process, under the condition that the process will not leave a certain domain during a given time. General theorems including not only only transition functions, but also trajectories as explicit objects for consideration were presented in the books by Doob [3] and Dynkin [4]. Both theories were based on the notion of conditional probability and conditional expectation with respect to a σ-algebra; but in Doob's theory the main object is a random process on a certain probability space satisfying the independence condition of the past and the future for a fixed present, whereas in Dynkin's case, it is a whole family of such processes that start at an arbitrary time in an arbitrary point of space (which is expressed by considering a family of probability measures $P_{s,x}$ and probability spaces $(\Omega, M_s, P_{s,x})$).

At first the analytic trend in the theory of Markov processes was the only dominating one; then, in the 1950's, direct probabilistic methods started to be used that dealt with constructing realizations of random processes from realizations of other simpler processes, as well as with stochastic equations and random changes of time. Analytic and direct methods complemented each other. There appeared the opposite trend, in which the theory of Markov processes is applied with the purpose of obtaining purely analytical results. Starting from the 1970's we cannot speak about a pure analytical trend in the theory of Markov processes, but only of strong prevalence of analytical methods in certain works.

The central problem in the analytical trend in the theory of Markov processes with continuous time is to obtain differential characteristics of processes and the mechanisms, realizing the connection between them and the processes themselves. Let us review the development of methods of answering this question and some achievements in the analytical trend of the theory of Markov processes.

In No. 9 differential characteristics are introduced by formulas (50) for processes with a discrete state space and by formulas (114), (124), (115), (122) for continuous (diffusion) processes on the real line. The existence of the limits (50), (122), (124) (which are in fact differential characteristics for $s = t$) is established under the assumption of differentiability of the transition probabilities $P_{ij}(s,t)$, (or the transition probability densities $f(s,x,t,y)$) at $s < t$ (additionally, provided that the determinant (119) does not vanish). In both cases differential equations are derived: inverse ones by differentiating with respect to the first time argument s (formulas (57), (125)) and direct ones: by differentiating with respect to the second argument t (formulas (52), (133)). The existence and uniqueness problem is raised for a solution satisfying natural conditions at $s = t$ and the conditions (1), (3) (particular cases are (40), (41) and (85), (86)) required for the existence of the corresponding process. In the continuous case it is solved only when the equation reduces to the classical heat equation. In §19 differential characteristics for pure jump Markov processes are given and a direct differential equation is suggested (175) (as well as (176) for a process with diffusion between the jumps).

The problem of finding transition probability densities for a diffusion process in its analytical formulation is the problem of finding the fundamental solution of a parabolic differential equation. Existence and uniqueness theorems were obtained under general conditions already in the 1930's (Feller [5]); they were also obtained for pure jump processes and processes with diffusion between jumps.

The differential characteristics of Markov processes in No. 9 are introduced separately for different particular classes of processes; these characteristics are functions of the corresponding arguments. The next step was to consider a linear (sometimes unbounded) operator, instead of a set of functions, as a differential characteristic of a Markov process. This step was taken when the theory of semigroups of linear operators was applied to Markov processes. The infinitesimal operator of the semigroup of operators related to the process was introduced as a characteristic of a Markov process (more precisely and more narrowly, of its transition function). The theory of semigroups enabled us to establish, under very weak restrictions, the uniqueness of the transition function corresponding to a given infinitesimal operator and allowed us to find necessary and sufficient conditions for the existence of such an operator. The existence problem for differential characteristics was solved: according to the general theory an infinitesimal operator has a dense domain.

The theory of semigroups is applied directly only to Markov processes homogeneous in time, that is, processes with transition function that depends only on the difference between times:

$$P(s,x,t,E) = P_0(t-s,x,E).$$

In order to be able to apply this theory to non-homogeneous processes, these processes have to be reduced to homogeneous ones by introducing a wider state space.

The first applications of semigroup theory related to semigroups of operators connected with transition functions. These operators mapped initial distributions into distributions at subsequent times (see No. 9 (5)), which is connected with an analogue of the direct Kolmogorov equations (see (52), (133)). It turned out to be more fruitful, however, to use semigroups of operators that act on the functions

$$T_t f(x) = \int_{\mathfrak{X}} P_0(t,x,dy) f(y),$$

rather than on the initial distributions. This is connected with an analogue of the inverse equations (57), (125).

The possibility of considering the values of a process at random times, such as the moment of the first exit from a set, appeared to be essential for the connection between Markov processes and their infinitesimal operators, in particular for deriving the equations for the expectations related to the process. This was provided by introducing the strong Markov property. Using the strong Markov property, Dynkin introduced a new type of differential characteristic for a Markov process, a characteristic operator (see [4]), that is, not a new type of characteristic (because under natural broad assumptions the characteristic operator coincides with the infinitesimal one), but rather a new aspect of the connection between this operator and a Markov process.

New methods made it possible to solve the problem of giving a full description of all one-dimensional diffusion processes homogeneous in time (that is, strong Markov processes with continuous trajectories). At the same time a fruitful setting of new problems was found, in particular, on the behaviour of a Markov process given inside a domain, after going out onto its boundary (analytically it reduces to finding boundary values that restrict the given linear operator to an infinitesimal operator of the semigroup of contracting operators preserving positivity).

A further development of analytical methods, following the work of Hunt [7], was based on considering, on the one hand, general non-negative additive functionals of Markov processes, and on the other hand, from the analytical viewpoint, excessive functions (non-negative functions superharmonic with respect to a given semigroup of operators). By considering the extremal points of the set of these functions, ways were found of constructing an ideal boundary of the domain corresponding to a given Markov process (the Martin boundary).

A new approach to the question of obtaining differential characteristics of a Markov process was started by Fukusima (see [8]). Instead of an infinitesimal operator, he considers the corresponding bilinear Dirichlet form as such a characteristic. However, a complete and closed theory is only obtained for the case when the semigroup consists of operators symmetric with respect to a certain measure (in terms of probability theory, for Markov processes invertible in time). Important results were obtained dealing with possible extensions concerning the exit of a Markov process given inside a domain (in particular, Brownian motion with "reflection in the normal" at an arbitrary, non-smooth, boundary has been considered).

In the 1960–1970's, especially after the works by Stroock and Varadhan [9], the approach to the connection between linear operators and Markov processes based on the notion of martingale and the "martingale problem" became quite popular. Instead of the strong Markov property, it uses the preservation of the martingale property for Markov random moments (the strong Markov property appears to be an automatic consequence of the uniqueness of the solution to the martingale problem). In papers by Krylov [10] and Stroock and Varadhan [9] the existence and uniqueness problem for a diffusion process corresponding to given diffusion and transition coefficients posed in No. 9 was solved practically without any limitations.

Perhaps the analytical trend in the theory of Markov processes will now be developed on a new, higher level based on considering, as main analytical objects, distributions in function spaces, rather than transition probabilities and related operators. This was started in papers by Stroock and Varadhan, and others. Analytical methods give especially wide possibilities for establishing various limit theorems for random processes.

References

1. A.N. Kolmogoroff and M.A. Leontovich, 'Zur Berechnung der mittleren Brounschen Fläche', *Phys. Z. Sow.* **4** (1933), 1–13.
2. A. Andronov, A. Witt and L. Pontryagin, 'On statistical considerations of dynamical systems', *ZhETF* **3**:3 (1933), 165–180 (in Russian).
3. J.L. Doob, *Stochastic processes*, Chapman and Hall, 1953.
4. E.B. Dynkin, *Markov processes*, Springer-Verlag, 1965 (translated from the Russian).
5. W. Feller, 'On the theory of stochastic processes (Existence and uniqueness theorems), *Uspekhi Mat. Nauk* **5** (1938), 57–96.
6. W. Feller, 'Parabolic differential equations and corresponding transformation semigroups', *Matematika* **1**:4 (1957), 105–153.
7. G.A. Hunt, *Markoff processes and potentials* 1, 2, 3, Ill. J. of Math., **1** (1957), 44–93; 316–368; *ibid.* **2** (1958), 151–213.
8. M. Fukusima, *Dirichlet forms and Markov processes*, North-Holland, Amsterdam, 1980.
9. D. Stroock and S. Varadhan, *Multidimensional diffusion processes*, Springer-Verlag, Berlin, 1979.
10. N.V. Krylov, 'On stochastic integral equations', *Teor. Veroyatnost. i Primenen.* **14**:2 (1969), 340–348 (in Russian).

MARKOV PROCESSES WITH A COUNTABLE NUMBER OF STATES (No. 10)

(B.A. Sevastyanov)

An example is given of the application of the equations for Markov processes with a countable number of states to one of the original problems in queuing theory. Namely, the problem of a queue in a service system with n lines under Poisson flow of demands with exponential service time is studied. This result has later become classical and is included in textbooks. (See, for example, W. Feller, *An introduction to probability theory and its applications*, Vols. I, II, Wiley, 1950.)

HOMOGENEOUS RANDOM PROCESSES (No. 13)

(V.M. Zolotarev)

The paper is divided into two parts (the first ends at formula (7)) because of a formal restriction on the length of a paper in a volume of the journal *Atti della Reale Accademia ....*

A brief history preceding this paper is as follows. In 1929–1930 several papers of the well-known Italian mathematician B. de Finetti came out, in which he initiated the study of properties of random processes $X(\lambda)$ that were later called homogeneous processes with independent increments. It turns out that

$$\psi(t,\lambda) = \mathbf{E}\exp\{itX(\lambda)\} = [\psi(t,1)]^{\lambda}.$$

for any $\lambda > 0$.

In this connection, in his last paper (referred to in the paper by A.N. Kolmogorov) de Finetti poses the problem of describing all distributions whose characteristic functions $\psi(t)$ are such that for any $\lambda > 0$, $[\psi(t)]^{\lambda}$ is the characteristic function of a distribution. The class $\mathfrak{G}$ of these distributions is called the class of infinitely divisible laws. In the same paper it is shown that, if properly understood, $\mathfrak{G}$ includes all distributions whose characteristic functions are of the form

$$\psi(t) = \exp\{it\gamma - \tfrac{1}{2}\sigma^2 t^2 + c^2 \int (e^{itx} - 1)dF(x)\}, \qquad (*)$$

where γ, σ, c are real constants and $F(x)$ is a distribution function.

A.N. Kolmogorov, who was aware of de Finetti's work and maintained a correspondence with him, became interested in the problem of describing the class $\mathfrak{G}$. In No. 13 he described only the part of $\mathfrak{G}$ corresponding to laws with finite variance. De Finetti's problem was solved completely in 1934 by the French mathematician P. Lévy [1].

Despite its seemingly intermediate character, the result of Kolmogorov played a fundamental role in the search for a complete description of the class $\mathfrak{G}$.

First of all the representation (15) obtained by Kolmogorov convinced de Finetti that $\mathfrak{G}$ could be described in terms of characteristic functions. Equally important was the fact that the method of proving (15) used by Kolmogorov appeared to be very promising. In 1937 the Soviet mathematician Khinchin used this method in a somewhat elaborated form to reproduce the results by Lévy [2]. The advantages of the Kolmogorov-Khinchin method were so obvious that in modern probability theory courses the canonical Lévy-Khinchin representation (together with its modification, the Lévy representation) is proved by this very method [3]. It is also effective when describing the analogues of infinitely divisible distributions in Banach spaces and locally compact groups [4].

It should also be mentioned that A.N. Kolmogorov was aware of both forms of the canonical representation of characteristic functions of infinitely divisible laws that are currently in use. Indeed, the form of the representation (15) corresponds exactly to the Lévy-Khinchin canonical representation and Lévy's canonical representation can be derived from it if we use (13), (14) for $\Omega(x)$ involving $P_1(x)$ and $P_2(x)$. The latter two are, in fact, the spectral functions in the Lévy representation.

The paper by Kolmogorov contains another important result that the author did not care to emphasize. This is a probabilistic interpretation of the spectral functions P_1 and P_2. Here we should clarify the meaning of the statement that "$P_2(x)d\lambda$ is the probability that over time $d\lambda$ a positive jump greater than x took place". It is understood that $d\lambda$ is small and that $P_2(x)d\lambda$ represents this probability with accuracy $o(d\lambda)$.

References

1. P. Lévy, 'Sur les intégrales dont les éléments sont des variables aléatoires

indépendantes', *Ann. Scuola Norm. Super. Pisa, Ser.* 2 **3** (1934), 337–366.
2. A.Ya. Khinchin, 'A new derivation of a formula of P. Lévy', *Byull. Moskov. Gos. Univ.* **1**:1 (1937) (in Russian).
3. M. Loève, *Probability theory*, Princeton Univ. Press, 1963.
4. H. Heyer, *Probability measures on locally compact groups*, Mir, Moscow, 1981.

HOMOGENEOUS MARKOV PROCESSES (No. 39)

(A.A. Yushkevich)

This paper by A.N. Kolmogorov, together with those by Doob and Lévy quoted below, laid the foundations of the theory of Markov processes dealing with homogeneous processes with a countable number of states. The peculiar effects taking place in these processes made them a separate branch of the general theory. At the same time, studies of the countable case helped to work out important concepts having a wider scope of application, such as the strong Markov property and boundary conditions for Markov processes.

The densities a_α^β of transition from α to β and the densities $a_\alpha = -a_\alpha^\alpha$ of leaving α, as well as the direct and inverse systems of differential equations $P'(t) = P(t)A$ and $P'(t) = AP(t)$, for the transition probabilities $p_\alpha^\beta(t)$ of a homogeneous Markov process with a finite or countable set of states E were introduced by Kolmogorov in his fundamental work [1] (No. 9 of the present volume), in which he assumed differentiability of the $p_\alpha^\beta(t)$ and, when E is infinite, certain conditions of uniform convergence (A and $P(t)$ denote the matrices with entries a_α^β and $p_\alpha^\beta(t)$, $t \geq 0$). If E is finite, it is easy to show that $a_\alpha = \sum_{\beta\neq\alpha} a_\alpha^\beta$, and for the initial value $P(0) = I$ (I is the identity matrix) each of the systems of Kolmogorov differential equations has the unique solution $P(t) = e^{At}$. After Doeblin [2] established that for finite E differentiability of the transition probabilities follows from stochastic continuity of the process, that is, from the condition $P(+0) = I$, the case of a finite set of states was essentially exhausted.

Processes with a countable state space proved to be more difficult. The analysis of their infinitesimal characteristics, smoothness of transition functions, behaviour of trajectories, possible "pathologies", etc., accounted for a whole trend notable in its time.

In the papers [3], [4], which came out earlier and were related to his notion of separable process, Doob studied the transition matrix $P(t)$ based on an examination of the jumps in the trajectories. Assuming that $P(+0) = I$ (this condition is considered to hold throughout) Doob took the following steps: 1) he proved the existence of finite or infinite derivatives a_α^β of the functions $p_\alpha^\beta(t)$ at the origin; 2) he showed that the inverse system holds when there exists a first jump after 0, that is, when $a_\alpha = \sum_{\beta\neq\alpha} a_\alpha^\beta < \infty$ (we then say that the matrix A is conservative) and found a similar connection between the direct system and the last jump before t; and 3) by continuing the process in different ways after accumulating the jumps, he obtained examples of non-uniqueness of a process with a given matrix A.

In No. 39 of [5], by making use of remarkably simple and concise purely algebraic or analytical techniques (in terms of which probabilistic considerations can be reflected), Kolmogorov 1) proved the existence of finite densities a_α^β for $\beta \not\equiv \alpha$; 2) proved the existence of the densities $a_\alpha \leq \infty$; 3) constructed an example of a process where $a_1 = \infty$; and 4) constructed an example of a process where $\sum_{\beta\neq\alpha} a_\alpha^\beta < a_\alpha < \infty$ (we will denote these examples by K1 and K2). Further, Kolmogorov conjectured [5] that the $p_\alpha^\beta(t)$ are always differentiable at $t > 0$.

In a paper that came out at the same time, Lévy [6], who was interested primarily in the asymptotics of $p_\alpha^\beta(t)$ as $t \to \infty$, classified the processes considered from the viewpoint of the behaviour of trajectories, treating them on a less formal level than Doob. In particular, Lévy established the effects observed in the examples K1 and K2. Lévy suggested calling the states α with $a_\alpha < \infty$ stable, and those with $a_\alpha = \infty$ instantaneous ones. The theorem [6] stating that not all states are instantaneous turned out to be erroneous. Lévy continued his analysis in [7], [8].

Further studies of countable homogeneous Markov processes were considerably influenced by Kolmogorov's works and ideas. At Moscow University this subject was discussed at the seminars of Dynkin and Kolmogorov. Kolmogorov suggested the study of the differentiability of $p_\alpha^\beta(t)$ as the subject for diploma work to Yushkevich, who managed with the help of trajectory analysis to prove the continuous differentiability of $p_\alpha^\beta(t)$ under the assumption that at least one of the states α or β is stable, and to construct an example of a process with infinite second derivative of $p_\alpha^\beta(t)$ at certain $t > 0$ (the idea of the example was suggested by A.N. Kolmogorov). This paper was published six years after [9]. Yushkevich also noticed the gap in the proof of the above-mentioned theorem of Lévy on instantaneous states, which was connected with the absence of a definite value of the trajectory at the moment of a discontinuity of the second kind (see [28]).

These questions were discussed at the Mathematical Congress in Amsterdam, where Kendall and Reuter [10] gave a detailed analysis of examples K1 and K2 from the viewpoint of semigroups. Kolmogorov was present at the congress. From conversations, some participants got the wrong impression that in [9] an example is constructed that refuted the hypothesis of differentiability of $p_\alpha^\beta(t)$ [12]. Soon Austin [11], [12], published a purely analytical proof of the continuous differentiability of $p_\alpha^\beta(t)$ for $t > 0$ in the case of a stable state α. Then Chung [13] proved the same for $t \geq 0$ using a probabilistic method close to [9]. Later Austin generalized his proof to the case of a stable state β [14]. Improvements of this proof were proposed by Reuter [15], who further developed the semigroup approach, as well as by Jurkat [16] and Chung [17], [18]. Finally a complete purely analytical proof of Kolmogorov's conjecture was found by Ornstein [19]. Smith [20] gave a negative answer to Chung's question on whether or not the derivative of $p_\alpha^\alpha(t)$ tends as $t \to 0$ to its value $-\infty$ at $t = 0$ for an instantaneous state α. On the other hand, Orey [21] showed that the total probability $p^\alpha(t)$ of the state α at time t need not be differentiable for a suitable initial distribution and time $t > 0$. Subtle questions on the structure of the functions $p_\alpha^\beta(t)$ as $t \to 0$ were also dealt with by Kendall [22], Blackwell and Freedman [23] and Reuter [24]. Studies on the class of possible functions $f(t) = p_\alpha^\alpha(t)$ were one of the factors that stimulated Kingman to develop the theory of regenerative events [25].

Soon after No. 39 came out, Kendall partially expanded Kolmogorov's results on the existence of densities a_α and a_α^β to Markov jump processes with arbitrary set of states [26] and, by combining the ideas of examples K1 and K2, constructed an example of a process with $a_1 = \infty$ and $a_1^\beta = 0$ for $\beta \neq 1$ [27]. A positive answer to the intriguing question whether there exists a process whose states are all instantaneous, was given independently by Dobrushin [28] and Feller and McKean [29]; other examples were soon suggested by Kendall [30] and Blackwell [31]. Later Reuter returned to the studies of K1 and showed that in this example the infinitesimal matrix A uniquely determines the transition matrix $P(t)$. All matrices A corresponding to processes with only instantaneous states were described in subsequent works by Williams [32], [33].

The results on differentiability of transition probabilities and trajectory structures of homogeneous Markov processes with a countable number of states are given in detail in the book by Chung [17] and Freedman [34], in which also a detailed description of examples of various "pathologies" can be found. Another book by Freedman [35] is devoted exclusively to the construction of examples with instantaneous states by passing to the limit from processes with a finite number of states.

The range of problems originating in paper No. 39 promoted the development of the notions of Markov moment and the strong Markov property. This property was formulated and proved by Doob [4], Yushkevich [9] and Chung [36], [13] for various classes of Markov moments that are encountered in homogeneous Markov processes with a countable number of states. On the other hand, in the study of Markov processes in an arbitrary metric space by means of characteristic operators, the strong Markov property was first required by Dynkin [37], [38] and was introduced by him as a quite plausible assumption [37]. In private conversations with Yushkevich, Kolmogorov accepted the possibility of Markov processes that are not strongly Markov, and Yushkevich constructed appropriate examples mentioned in [37] and included in [39]. Then Dynkin, comparing his results with those of Feller, which were obtained in a purely analytical way, conjectured that any Feller process without discontinuities of the second kind is a strong Markov process [37]. Yushkevich proved this by replacing the second condition by the assumption of right continuity of the trajectory [39]. In parallel, the strong Markov property for particular cases was established by Hunt [40] and Ray [41] and was introduced in its general form by Blumenthal [42]. In order that a homogeneous Markov process with a countable set of states E have the strong Markov property, the space E must, in general, be compactified in an appropriate way by introducing fictitious states. This was done much later by Doob [43].

Kolmogorov's paper No. 39 also became one of the starting points in the study of boundaries and boundary conditions for Markov processes. If a matrix A is conservative, then the inverse system of differential equations $P'(t) = AP(t)$ holds but, in general, its solution is not unique for the initial value $P(0) = I$. General results by Feller imply that the probabilities of transition during a finite number of jumps form the minimal solution for this system. The problem arises of constructing possible continuations of the minimal process beyond an accumulation point of the jumps, preserving the Markov property and the infinitesimal matrix A, so that the process does not terminate. After Feller [44], the preliminary question as to when accumulation of jumps (in other words, going away to infinity) can take place over finite time, was studied by Dobrushin for the countable case [45]. The further construction requires constructing exit boundaries, entrance boundaries (if any) and "glue together" the exits to the entrances, or to the initial states by means of boundary conditions. Without mentioning the vast range of corresponding research for diffusion-type processes and their generalizations, we merely note that for a countable state space such a programme was sketched by Dynkin in his talk at the All-Union Meeting on Probability Theory in Leningrad in 1955. Apparently, it could only be realized for the case with a finite number of exits. Without attempting to give a complete review, we mention the relevant papers by Dobrushin [46], Feller [47], Reuter [15], [48], Neveu [49], Chung [50], Williams [51], [52] and Dynkin [53].

References

1. A.N. Kolmogoroff, 'Über die analytischen Methoden in der Wahrscheinlichkeitsrechnung', *Math. Ann.* **104** (1931), 415–457.
2. W. Doeblin, 'Sur l'équation matricielle $A^{(t+s)} = A^{(t)}A^{(s)}$ et ses applications aux probabilités en chaîne', *Bull. Sci. Math.* **62** (1938), 21–32; **64** (1940), 35–37.
3. J.L. Doob, 'Topics in the theory of Markoff chains', *Trans. Amer. Math. Soc.* **52** (1942), 37–64.
4. J.L. Doob, 'Markoff chains — denumerable case', *Trans. Amer. Math. Soc.* **58** (1945), 455–473.
5. A.N. Kolmogorov, 'On differentiability of transition probabilities in time-homogeneous Markov processes with a countable number of states', *Uch. Zap. Moskov. Gos. Univ. Mat.* **148**:4 (1951), 53–59 (in Russian).
6. P. Lévy, 'Systèmes markoviens et stationnaires. Cas dénombrable', *Ann. Sci. École Norm. Supér.* **68** (1951), 327–381.
7. P. Lévy, 'Processus markoviens et stationnaires du cinquième type (infinité dénombrable des états possibles, paramètre continu), *C. R. Acad Sci. Paris* **236** (1953), 1630–1632.

8. P. Lévy, 'Remarques sur les états instantanés des processus markoviens et stationnaires, à une infinité dénombrable d'états possibles', *C. R. Acad. Sci. Paris* **264** (1967), 844–848.
9. A.A. Yushkevich, 'On differentiability of transition probabilities of homogeneous Markov processes with a countable number of states', Diploma work, Moskov. Gos. Univ., *Uch. Zap. Moskov. Gos. Univ. Mat.*,**186**:9 (1959), 141–159 (in Russian).
10. D.G. Kendall and G.E.H. Reuter, 'Some pathological Markov processes with a denumerable infinity of states and the associated semigroups of operators on *l*', In: *Proc. Intern. Congr. Math. Amsterdam* **3** (1954), 377–415.
11. D.G. Austin, 'On the existence of the derivative of Markoff transition probability functions', *Proc. Nat. Acad. Sci. USA* **41** (1955), 224–226.
12. D.G. Austin, 'Some differentiation properties of Markoff transition probability functions', *Proc. Amer. Math. Soc.* **7** (1956), 756–761.
13. K.L. Chung, 'Some new developments in the theory of Markov chains', *Trans. Amer. Math. Soc.* **81** (1956), 195–210.
14. D.G. Austin, 'Note on differentiating Markoff transition functions with stable terminal states', *Duke Math. J.* **25** (1958), 625–629.
15. G.E.H. Reuter, 'Denumerable Markov processes and the associated contraction semigroups on *l*', *Acta Math.* **97** (1957), 1–46.
16. W.B. Jurkat, 'On semigroups of positive matrics. I, II', *Scripta Math.* **24** (1959), 123–131; 207–218.
17. K.L. Chung, *Markov chains with stationary transition probabilities*, Springer-Verlag, Berlin, 1960.
18. K.L. Chung, 'Probabilistic methods in Markov chains', In: *Proc. Fourth Berkeley Sympos. Math. Statist. Probab.* Univ. Calif. Press, Vol. 2, 1961, pp.35–56.
19. D. Ornstein, 'The differentiability of transition functions', *Bull. Amer. Math. Soc.* **66** (1960), 36–39.
20. G. Smith, 'Instantaneous states of Markov processes', *Trans. Amer. Math. Soc.* **110** (1964), 185–195.
21. S. Orey, 'Non-differentiability of absolute probabilities of Markov chains', *Quart. J. Math.* **13** (1962), 252–254.
22. D.G. Kendall, 'On the behaviour of a standard Markov transition function

near $t = 0$', *Z. Wahrscheinlichkeitstheorie* **3** (1965), 276–278.

23. D. Blackwell and D. Freedman, 'On the local behaviour of Markov transition probabilities', *Ann. Math. Statist.* **39** (1968), 2123–2127.
24. G.E.H. Reuter, 'Remarks on a Markov chain example of Kolmogorov', *Z. Wahrscheinlichkeitstheorie* **13** (1969), 315–320.
25. J.F.C. Kingman, 'The stochastic theory of regenerative events', *Z. Wahrscheinlichkeitstheorie* **2** (1964), 180–224.
26. D.G. Kendall, 'Some analytical properties of continuous stationary Markov transition functions', *Trans. Amer. Math. Soc.* **78** (1955), 529–540.
27. D.G. Kendall, 'Some further pathological examples in the theory of denumerable Markov processes', *Quart. J. Math.* **7** (1956), 39–56.
28. R.L. Dobrushin, 'An example of a countable homogeneous Markov process in which all states are instantaneous', *Teor. Veroyatnost. i Primenen.* **1** (1956), 481–485 (in Russian). (Translated as *Theory Probab. Appl.*)
29. W. Feller, and H.P. McKean, Jr., 'A diffusion equivalent to a countable Markov chain', *Proc. Nat. Acad. Sci. USA* **42** (1956), 351–354.
30. D.G. Kendall, 'A totally unstable denumerable Markov process', *Quart. J. Math.* **9** (1958), 149–160.
31. D. Blackwell, 'Another countable Markov process with only instantaneous states', *Ann. Math. Statist.* **29** (1958), 313–316.
32. D. Williams, 'The Q-matrix problem for Markov chains', *Bull. Amer. Math. Soc.* **81** (1975), 1115–1118.
33. D. Williams, 'The Q-matrix problem. 2: Kolmogorov backward equations', *Lecture Notes in Math.* **511** (1976), 216–234; 505–520.
34. D. Freedman, *Markov chains*, Holden-Day, San Francisco, 1971.
35. D. Freedman, *Approximating countable Markov chains*, San Francisco, 1972.
36. K.L. Chung, 'Foundations of the theory of continuous parameter Markov chains', In: *Proc. Third Berkeley Sympos. Math. Statist. Probab.* Univ. Calif. Press, Vol 2, 1956, pp. 29–40.
37. E.B. Dynkin, 'Infinitesimal operators of Markov random processes, *Dokl. Akad. Nauk SSSR* **105** (1955), 206–209 (in Russian).
38. E.B. Dynkin, 'Infinitesimal operators of Markov processes', *Teor. Veroyatnost. i Primenen.* **1** (1956), 38–60 (in Russian). (Translated as *Theory Probab. Appl.*)

39. E.B. Dynkin and A.A. Yushkevich, 'Strict Markov processes', *Teor. Veroyatnost. i Primenen.* **1** (1956), 149–155 (in Russian). (Translated as *Theory Probab. Appl.*)
40. G.A. Hunt, 'Some theorems concerning Brownian motion', *Trans. Amer. Math. Soc.* 81 (1956), 294–319.
41. D. Ray, 'Stationary Markov processes with continuous paths', *Trans. Amer. Math. Soc.* **82** (1956), 452–493.
42. R.M. Blumenthal, 'An extended Markov property', *Trans. Amer. Math. Soc.* **85** (1957), 52–72.
43. J.L. Doob, 'Compactification of the discrete state space of a Markov process', *Z. Wahrscheinlichkeitstheorie* **10** (1968), 236–251.
44. W. Feller, 'On the integro-differential equations of purely discontinuous Markoff processes', *Trans. Amer. Math. Soc.* **48** (1940), 488–515; Errata — Ibid. **58** (1945), 474.
45. R.L. Dobrushin, 'On the regularity conditions for time-homogeneous Markov processes with a countable number of possible states', *Uspekhi Mat. Nauk* **7**:6 (1952), 185–191 (in Russian).
46. R.L. Dobrushin, 'Some classes of homogeneous countable Markov processes', *Teor. Veroyatnost. i Primenen.* **2** (1957), 377–380 (in Russian). (Translated as *Theory Probab. Appl.*)
47. W. Feller, 'On boundaries and lateral conditions for the Kolmogoroff differential equations', *Ann. of Math.* **65** (1957), 527–570.
48. G.E.H. Reuter, 'Denumerable Markov processes. II, III', *J. London Math. Soc.* **34** (1959), 81–91.
49. J. Neveu, 'Sur les états d'entrée et les états fictifs d'un processus de Markoff', *Ann. Inst. H. Poincaré* **17** (1962), 324–337.
50. K.L. Chung, 'On the boundary theory for Markov chains', *Acta Math.* **110** (1963), 19–77; **115** (1966), 111–163.
51. D. Williams, 'The process extended to the boundary', *Z. Wahrscheinlichkeitstheorie* **2** (1962), 332–339.
52. D. Williams, 'On the construction problem for Markov chains', *Z. Wahrscheinlichkeitstheorie* **3** (1964), 227–246.
53. E.B. Dynkin, 'General boundary conditions for Markov processes with a countable set of states', *Teor. Veroyatnost. i Primenen.* **12** (1967), 222–257 (in Russian). (Translated as *Theory Probab. Appl.*)

BRANCHING PROCESSES (Nos. 25, 32, 33, 46)
(B.A. Sevast'yanov)

The general notion of and the term "branching random process" itself that immediately became commonly accepted, were first explicitly coined by A.N. Kolmogorov at his seminar at Moscow University in 1946–1947. Various problems relating to simple models of branching processes had been considered previously. In particular, one of these problems was solved by A.N. Kolmogorov in his earlier work, No. 25. However, it was the publication of papers Nos. 31, 32, that triggered an intensive development of the theory of branching processes. By now there are several monographs that deal with branching processes (books by Harris [1], Sevast'yanov [2], Athreya and Ney [3], and others). The model of branching processes described in papers Nos. 31, 32 are Markov branching processes with several types of particle and with continuous and discrete time. This model appeared to be very efficient, both in terms of the number of results obtained and in its possible applications to biology, chemistry, physics and technology. A special case of this model is a branching process with immigration, when, in addition to multiplying particles, there also appear immigrating particles (see, for example, [2]).

Later on, more complex models of branching processes were developed which took into account the following parameters: dependence of multiplication on the age of particles (the Bellman-Harris process), a particle's position in a certain space, a particle's dependence on energy, random environment, etc.

The asymptotic formulas of the probability of continuing the process obtained in No. 25, as well as related limit theorems proved by Yaglom [4], were later generalized by other authors for other models, including non-Markov ones. Papers [5]–[7] deal with the convergence of branching processes to diffusion processes, which is discussed in the review paper No. 46.

References

1. T.E. Harris, *The theory of branching processes*, Springer-Verlag, Berlin, 1963.
2. B.A. Sevast'yanov, *Branching processes*, Nauka, Moscow, 1971 (in Russian).
3. K.B. Athreya and P.E. Ney, *Branching processes*, Springer-Verlag, Berlin-Heidelberg-New York, 1972.

4. A.M. Yaglom, 'Some limit theorems of the theory of random branching processes', *Dokl. Akad. Nauk SSSR* **56**:8 (1947), 795–798 (in Russian).
5. M. Jiřina, 'On Feller's branching diffusion processes', *Časop. Pestov. Mat.* **94**:1 (1969), 84–90.
6. J. Lamperty, 'The limit of a sequence of branching processes', *Z. Wahrscheinlichkeitstheorie* **7**:4 (1967), 271–288.
7. V.P. Chistyakov, 'On the convergence of branching processes to diffusion processes', *Teor. Veroyatnost. i Primenen.* **15**:4 (1970), 727–730 (in Russian). (Translated as *Theory Probab. Appl.*)

STATIONARY SEQUENCES (No. 27)
(Yu.A. Rozanov)

Paper No. 27 has by now become classical and opened the main road in the theory of stationary (and related) random processes, giving a proper setting[1] for prediction problems and solving them for stationary processes with discrete time (stationary sequences). This work is remarkable because of its deep connection with various questions of approximation theory, spectral theory of operators in Hilbert space and the theory of analytic functions. The two central notions in this theory are the regularity of a random process and subordination of one process to another.

A stationary sequence can be regarded as a sequence

$$x(t) = e^{i\lambda t}, \quad t = 0, \pm 1, \dots,$$

in the Hilbert space L_2 with measure $F(d\lambda)$ on $-\pi \leq \lambda \leq \pi$. Let $H_x(n)$ be the closed linear span of all $x(s)$, $s \leq n$, and let $\hat{x}_n(t)$ be the best approximation for $x(t)$ by functions from $H_x(n)$; regularity means that $\hat{x}_n(t) \to 0$ as $n \to -\infty$. A well-known result of No. 27 consists in establishing the following regularity criterion: $F(d\lambda)$ is absolutely continuous and the density $f(\lambda) = F(d\lambda)/d\lambda$ satisfies the condition

$$\int_{-\pi}^{\pi} \log f(\lambda) d\lambda > -\infty. \tag{*}$$

[1] Prediction problems themselves are considered in another work by A.N. Kolmogorov: 'Interpolation and extrapolation of stationary random sequences', *Izv. Akad. Nauk SSSR Ser. Mat.* **5** (1941), 3–14 (see No. 28 of this volume).

The general model of a stationary sequence is:

$$x(t) = U_t x(0), \quad t = 0, \pm 1, \dots,$$

where $U_t = \int_{-\pi}^{\pi} e^{i\lambda t} E(d\lambda)$ is a group of unitary operators in Hilbert space. The fact that $y(t)$, $t = 0, \pm 1, \dots$ is subordinate to this $x(t)$ means that y is obtained from x by a linear transformation

$$y(t) = Ax(t)$$

with the operator

$$A = \int \phi(\lambda) E(d\lambda)$$

in $H_x = H_x(\infty)$. What is surprising here is that this linear dependence of y and x may be expressed in terms of their correlation dependence (the only previously known fact of this sort was the trivial fact of linear dependence of random variables x and y with correlation coefficient 1). A description of the spectral structure of subordinate sequences gives, in particular, a spectral description of cyclic subspaces for the unitary group U_t in the space H_x.

Of particular importance here are subordinate sequences y with the non-anticipating condition

$$H_y(t) \subseteq H_x(t)$$

for which conditions under which

$$H_y(0) = H_x(0)$$

have been established. For example, for

$$x(t) = e^{i\lambda t}$$

in L_2 with $F(d\lambda) = d\lambda$ the answer is the following: $H_y(0) = H_x(0)$ holds if and only if

$$y(t) = e^{i\lambda t} \phi(\lambda)$$

where $\phi(\lambda)$ is a so-called outer function of the well-known analytic class $H^2 = H_x(0)$. (In No. 27

$$\phi(\lambda) = \sum_{-\infty}^{0} c_k e^{i\lambda k} = \Gamma(e^{-i\lambda})$$

is determined by the condition $\Gamma(z) \neq 0$, $|z| < 1$.) It should be mentioned that many questions for the class H^2 are easily solved within the scheme of subordinate sequences (and processes) proposed in No. 27. This is the case with the well-known question of invariant subspaces in H^2, namely: Every subspace invariant with respect to multiplication by $e^{-i\lambda}$,

$$H \subseteq H^2 = H_x(0),$$

is given by $H = H_y(0)$ for an appropriate regular subordinate sequence $y(t) = e^{i\lambda t}\,\phi(\lambda)$, $\phi \in H^2$ and

$$H = \psi H^2,$$

where the so-called inner function $\psi \in H^2$ gives a fundamental sequence for y,

$$u(t) = e^{i\lambda t}\psi(\lambda),$$

with spectral density

$$1/2\pi = |\psi(\lambda)|^2.$$

The studies related to No. 27 are reviewed in detail, for example, in the monographs [1]–[5]. We will mention some of them.

For some time paper No. 27 remained unknown to many researchers who in the 1940's dealt with closely related questions, among them Wiener (see his paper on filtering stationary processes). Here we should also mention Beurling's problem concerning "translations" in the analytic class H^2, etc.

A direct generalization of regularity condition (*) for stationary processes with continuous time was made in 1949 by Krein in the form

$$\int_{-\infty}^{\infty} \frac{\log f(\lambda)}{1+\lambda^2} d\lambda > -\infty.$$

Later Krein found the connection between the prediction problem and the inverse spectral problem for the string equation. A direct generalization of the regularity criterion for the non-degenerate multidimensional case was given in 1941 by Zasukhin in the form:

$$\int_{-\pi}^{\pi} \log\det f(\lambda) d\lambda > -\infty.$$

The problems of predicting processes with rational spectrum, important in applications, were considered by Yaglom. In the general case, the regularity condition is equivalent to factorizability of $f(\lambda)$:

$$f(\lambda) = (1/2\pi)\phi(\lambda)\phi(\lambda),$$

where $\phi(\lambda)$ is an operator-valued function from the corresponding analytic class H^2. The factorization problem of matrix-valued (operator-valued) functions $f(\lambda)$ was solved by Rozanov, Wiener and Mazani, Matveyev and others. In particular, Lax showed that in the infinite-dimensional case even a very strong generalized version of (*) in the form

$$\int_{-\pi}^{\pi} \log f(\lambda) d\lambda > -cI$$

(where $c > 0$ is a constant and I the identity operator) does not guarantee the required regularity of $f(\lambda)$. A general criterion for regularity was proposed by Rozanov.

References

1. Yu.A. Rozanov, *Stationary random processes*, Fizmatgiz, Moscow, 1963 (in Russian).
2. C. Goffman, *Banach spaces of analytic functions*, Prentice-Hall, 1962.
3. B. Szökefalvi-Nagy and Ch. Foias, *Harmonic analysis of operators on Hilbert space*, North-Holland, 1970 (translation from the French).
4. Yu.A. Rozanov, *Theory of renewal processes*, Nauka, Moscow, 1974 (in Russian).
5. H. Dym and H.P. McKean, *Gaussian processes, function theory and inverse spectral problems*, Academic Press, New York, 1976.

STATIONARY PROCESSES (No. 48)

(V.A. Statulyavichus)

The authors solve the problem for a Gaussian stationary process $X(t)$ under the strong mixing (SM) condition introduced by Rosenblatt [1]. It is proved that in this case the SM function $\alpha(\tau)$ is equivalent to the maximal correlation coefficient $\rho(\tau)$ ($\alpha(\tau) \leq \rho(\tau) \leq 2\pi\alpha(\tau)$) and the representation (4) for $\rho(\tau)$ is found in terms of the spectral density of the process $f(\lambda)$, which exists in view of the SM condition. Thus, the problem is solved completely. For example, in the case of integer time the process X_t possesses the SM property if and only if

$$\rho(\tau) = \sup_p \left\| (f(\lambda) - e^{-i\tau\lambda} p(e^{i\lambda})) \frac{1}{f(\lambda)} \right\|_\infty \to 0 \text{ as } \tau \to \infty,$$

where the supremum is taken over all one-sided trigonometric polynomials $p(e^{i\lambda}) = \sum_{t=0}^m c_t e^{it\lambda}$ $(m \geq 0)$ and $\|\cdot\|_\infty$ is the essential supremum norm. This implies that if $f(\lambda)$ is continuous, $f(-\pi) = f(\pi)$, and bounded everywhere, except perhaps at zero, then X_t has the SM property. Ibragimov [2] obtained further results on the form of $f(\lambda)$ when X_t possesses the SM property. Helson and Sarason [3] reconsidered this problem as a question in harmonic analysis. Generalizations were also obtained for functions X_t with values in a space of matrices [6]. In the case of a random walk on a compact abelian group Rosenblatt [4] showed that a stationary process has the SM property if and only if

$$\sup_{f \perp 1} \frac{\|T^n f\|_2}{\|f\|_2} \to 0 \ (n \to \infty),$$

where T is the transition operator of the random walk. A.N. Kolmogorov also posed the problem of finding an effective criterion for complete regularity of a Gaussian stationary process; it was solved by Volkonskii and Rozanov [7]. A student of A.N. Kolmogorov, Leonov [8], studied various SM conditions of random processes using higher semi-invariants.

References

1. M. Rosenblatt, 'A central limit theorem and a strong mixing condition', *Proc. Nat. Acad. Sci. USA* **42** (1956), 43–47.
2. I.A. Ibragimov, 'On spectral functions of certain classes of stationary Gaussian processes', *Dokl. Akad. Nauk SSSR* **137** (1956), 1046–1048 (in Russian).
3. H. Helson and D. Sarason, 'Past and future', *Math. Scand.* **21** (1967), 5–16.
4. M. Rosenblatt, *Markov processes. Structure and asymptotic behaviour*, Springer-Verlag, Berlin-Heidelberg-New York, 1971.
5. Yu.A. Rozanov, *Stationary random processes*, Fizmatgiz, Moscow, 1963 (in Russian).

6. I.H. More, and L.B. Page, 'The class w of operator valued weight functions', *J. Math. Mech.* **19** (1970), 1011–1017.
7. V.N. Volkonskii and Yu.A. Rozanov, 'Some limit theorems for random functions. I, II', *Teor. Veroyatnost. i Primenen.* **4**:2 (1959), 186–207; **6**:2 (1961), 202–214 (in Russian). (Translated as *Theory Probab. Appl.*)
8. V.P. Leonov, *Some applications of higher semi-invariants to the theory of stationary random processes*, Nauka, Moscow, 1964 (in Russian).

STATISTICS OF PROCESSES (No. 50)

(A.N. Shiryaev)

In this work the estimators $\hat{\lambda}_T$ and $\hat{\omega}_T$ for the parameter λ (damping coefficient) and ω (frequency) in a complex stationary Gaussian Markov process are constructed using the maximum likelihood method. It turned out that the normalized ratio

$$\frac{\hat{\omega}_T - \omega}{\sqrt{\int_0^T (\xi_s^2 + \eta_s^2)ds}}$$

does not depend on ω and λ and is exactly normally distributed with parameters 0 and 1 (when $a = 1$). For a proof of this interesting fact (not mentioned in the paper) see [1] and the monograph [2], Chap. 17, §4.

The paper under review is also interesting because it shows how Kolmogorov interprets a physically posed problem (in this case the problem of studying the movements of the earth's rotation axis) in strict terms of statistics of random processes and, based on results for the densities of certain measures with respect to others, finds estimators of the parameters considered.

Nowadays the statistics of diffusion type processes is a big chapter in the statistics of random processes. A significant part of the monograph [2] deals with the statistics of these processes. In particular, it considers the problem of the paper No. 50, as well as a number of related problems.

References

1. A.A. Novikov, 'Sequential estimators of the parameters of a diffusion process', *Teor. Veroyatnost. i Primenen.* **16**:2 (1971), 394–396 (in Russian). (Translated as *Theory Probab. Appl.*)
2. R. Sh. Liptser and A.N. Shiryaev, *Statistics of random processes*, Nauka, Moscow, 1974 (Translation: Reidel, Dordrecht, to appear).

SPECTRAL THEORY OF STATIONARY PROCESSES (No. 34)
(A.M. Yaglom)

Paper No. 34 by A.N. Kolmogorov published in 1947 is the first popular review of the spectral theory of stationary random processes, one of the most important sections of the mathematical theory of random functions which had been developed shortly before (with an active contribution by Kolmogorov himself), but at that time hardly known to anyone outside the narrow circle of experts. The first similar review of this theory outside the Soviet Union appeared two years later (see [1]). The editions in Russian dealing with this theory include a scientific monograph [2], the rather elementary book [3] and large sections in many textbooks (see, for example, [4]–[7]).

Kolmogorov starts his review by giving classical results by Khinchin *[1]*[1] who was the first to define a stationary random process $\xi(t)$ and proved that its correlation function $B(\tau) = \mathbf{E}\{\xi(t+\tau)\xi(t)\}$ can always be represented as a Fourier-Stieltjes integral (see formula *(6)* which relates to the somewhat more general case of a multidimensional process $\xi(t) = \{\xi_1(t), \xi_2(t), \dots, \xi_s(t)\}$). However he paid main attention to the question of substantiating the possibility of representing the stationary process $\xi(t)$ as a Fourier-Stieltjes integral of the form *(4)* and on the physical meaning of such a representation. He remarks that the spectral representation *(4)* can be easily derived from the well-known theorem in functional analysis on the spectral decomposition of one-parameter groups of unitary operators in Hilbert space, and the first deduction of this type was given in *[6]*–*[7]* in 1940. However, for simplicity he does not stress the point that these works solved a more general (and more complex) problem on the spectral representation not only for stationary processes, but for the wider class of processes with stationary increments.

A random process $\xi(t)$ is called a process with stationary increments (in the wide sense) if for any $t, \tau, s, \tau_1, \tau_2$ the expectations

$$\begin{aligned} &\mathbf{E}\{\xi(t+\tau) - \xi(t)\} = a(\tau), \\ &\mathbf{E}\{[\xi(t+s+\tau_1) - \xi(t+s)][\bar{\xi}(t+\tau_2) - \bar{\xi}(t)]\} = \mathbf{D}(s;\tau_1,\tau_2) \end{aligned} \tag{1}$$

exist and do not depend on t. (It is clear that any process stationary in the wide sense is always a process with stationary increments, but a process with

[1] Italicized figures in square brackets indicate the reference in Kolmogorov's paper. Italicized numbers in parentheses refer to the corresponding formula of No. 34.

stationary increments need not be stationary.) The results of the paper [*6*][2] imply that any process with stationary increments $\xi(t)$ allows a spectral representation of the form

$$\xi(t) = \int_{-\infty}^{\infty} (e^{i\lambda t} - 1) d\Phi(\lambda) + \xi_0 + \xi_1 t, \tag{2}$$

where ξ_0, ξ_1 are random variables and the random function $\Phi(\lambda)$ is such that its increments $\Phi(\Delta'_\lambda) = \Phi(\lambda'_2) - \Phi(\lambda'_1)$ and $\Phi(\Delta''_\lambda) = \Phi(\lambda''_2) - \Phi(\lambda''_1)$ on non-intersecting intervals $\Delta_\lambda = [\lambda'_2, \lambda'_1]$ and $\Delta''_\lambda = [\lambda''_2, \lambda''_1]$ are uncorrelated to each other (that is, $\mathbf{E}\{\Phi(\Delta'_\lambda)\overline{\Phi(\Delta''_\lambda)}\} = 0$), while the real monotone non-decreasing function $F(\lambda)$ of the argument λ given by $\mathbf{E}\{|\Phi(\Delta_\lambda)|^2\} = F(\Delta_\lambda)$ satisfies, for any $\epsilon > 0$, the condition

$$\int_{-\infty}^{-\epsilon} dF(\lambda) + \int_{-\epsilon}^{\epsilon} \lambda^2 dF(\lambda) + \int_{\epsilon}^{\infty} dF(\lambda) < \infty. \tag{3}$$

In the particular case of a stationary process $\xi(t)$ (considered for discrete time, in particular, by Kolmogorov's [*4*], [*5*]), $\xi_1 = 0$ and, furthermore, not only must (3) hold, but also the more restrictive condition

$$\int_{-\infty}^{\infty} dF(\lambda) < \infty. \tag{4}$$

In this case, clearly $\int_{-\infty}^{\infty} d\Phi(\lambda) = \xi'_0$ is a random variable of finite variance; therefore (2) may be rewritten as

$$\xi(t) = \int_{-\infty}^{\infty} e^{i\lambda t} d\Phi_1(\lambda)$$

(where $\Phi_1(\lambda) = \Phi(\lambda) + (\xi_0 - \xi'_0)K(\lambda)$ and $K(\lambda)$ is a "jump function" equal to 0 for $\lambda < 0$ and to 1 for $\lambda \geq 0$). Applied to stationary processes this brings us back to (*4*) and to the results (applied to the multidimensional process $\xi(t)$) of the paper No. 34 by Kolmogorov.

The spectral theory of random processes with stationary increments developed in [*6*]–[*7*] can be applied to the more general class of random processes $\xi(t)$ with stationary increments of order $n > 1$ (see [9], [10], where one can also

[2] The results of [*6*] dealing with geometry of Hilbert space were obtained a little later also by von Neumann and Schoenberg [8], who noted, however, that the same results can at the same time be interpreted as certain facts concerning random processes $\xi(t)$ with stationary increments.

find precise definitions of this class of random processes). For these processes the spectral representation is

$$\xi(t) = \int_{-\infty}^{\infty} \left[e^{i\lambda t} - 1 - i\lambda t - \ldots - \frac{(i\lambda t)^{n-1}}{(n-1)!}\right] d\Phi(\lambda) + \xi_0 + \xi_1 t + \ldots + \xi_n t^n, \quad (5)$$

where $\xi_0, \xi_1, \ldots, \xi_n$ are random variables, and the increments of the random function $\Phi(\lambda)$ on non-intersecting intervals Δ'_λ and Δ''_λ are again uncorrelated to each other, whereas the function $F(\lambda)$ determined by $\mathbf{E}\{|\Phi(\Delta_\lambda)|^2\} = F(\Delta_\lambda)$, satisfies, for any $\epsilon > 0$, the condition

$$\int_{-\infty}^{-\epsilon} dF(\lambda) + \int_{-\epsilon}^{\epsilon} \lambda^{2n} dF(\lambda) + \int_{\epsilon}^{\infty} dF(\lambda) < \infty. \quad (6)$$

Formulas describing spectral representations of processes with stationary increments of a certain order can be derived, for example, from results related to another generalization of the notion of a stationary random process, namely from the spectral theory of generalized stationary random processes. By analogy with the notion of a distribution or generalized function which has become widely used in modern analysis, a generalized random process is a random linear functional $\xi(\phi)$ (that is, a linear functional whose values are random variables) given on a certain linear space D of "good" (that is, sufficiently smooth) functions $\phi = \phi(t)$, for example, on the space D_∞ of infinitely differentiable functions (introduced by L. Schwartz), each of which vanishes identically outside some finite interval (only such a space $D = D_\infty$ will be considered in what follows). Clearly, an ordinary random process $\xi(t)$ can always be associated with a generalized process

$$\xi(\phi) = \int_{-\infty}^{\infty} \xi(t)\phi(t)dt. \quad (7)$$

From this viewpoint ordinary random processes may be considered as a particular case of generalized processes. In principle, however, there can also exist a generalized process $\xi(\phi)$ for which the "values at the point" $\xi(t)$ do not exist (that is, such that $\xi(\phi)$ cannot be described by (7)). A typical example of such a process $\xi(\phi)$ is the "white noise" process, very often encountered in various applications.

Generalized random processes were introduced in [11], [12] (see also [13], Chapter III). In [11]–[13] the notion of a stationary (in the wide sense) generalized random process $\xi(\phi)$ is defined also and a spectral theory for these $\xi(\phi)$ is developed. To define generalized stationary processes one only has to consider the operation T_τ in D given by $T_\tau\phi(t) = \phi(t+\tau)$ and then require $\xi(\phi)$ to satisfy

$$\begin{aligned} &\mathsf{E}\{\xi(T_\tau\phi)\} = \mathsf{E}\{\xi(\phi)\}, \\ &\mathsf{E}\{\xi(T_\tau\phi_1)\bar{\xi}(T_\tau\phi_2)\} = \mathsf{E}\{\xi(\phi_1)\bar{\xi}(\phi_2)\} \end{aligned} \tag{8}$$

for any τ. For simplicity we suppose that $\mathsf{E}\{\xi(\phi)\} = m(\phi) = 0$ for all ϕ. In this case a generalized stationary random process $\xi(\phi)$ has a spectral representation of the form

$$\xi(\phi) = \int_{-\infty}^{\infty} \tilde{\phi}(\lambda)d\Phi(\lambda), \tag{9}$$

where $\tilde{\phi}(\lambda) = \int_{-\infty}^{\infty} e^{i\lambda t}\phi(t)dt$ is the Fourier transform of $\phi(t)$, and the random function $\Phi(\lambda)$ has the same properties as $\Phi(\lambda)$ featuring in the spectral decomposition (*4*) of an ordinary random process $\xi(t)$, with the only difference that now $F(\lambda)$ is such that $\mathsf{E}\{|\Phi(\Delta_\lambda)|^2\} = F(\Delta_\lambda)$ satisfies the condition

$$\int_{-\infty}^{\infty} \frac{dF(\lambda)}{(1+\lambda^2)^m} < \infty \tag{10}$$

for some integer $m \geq 0$ (for ordinary stationary processes $\xi(t)$ we always have $\int_{-\infty}^{\infty} dF(\lambda) < \infty$, that is, $m = 0$). It can easily be seen that in the general case when the process $\xi(\phi)$ is ordinary, (that is, it is given by (7)), (9) immediately implies also the more usual spectral representation of the process $\xi(t)$ of the form (*4*).

Generalized random processes are in certain aspects simpler than ordinary processes; in particular, whereas an ordinary random process $\xi(t)$ has derivative $d\xi(t)/dt = \xi'(t)$ only under certain special conditions, a generalized process $\xi(\phi)$ is always differentiable, its derivative $\xi'(\phi)$ being $\xi'(\phi) = -\xi(\phi')$ (if a process $\xi(\phi)$ is given by (7) and $\xi'(t)$ exists, then clearly $\xi'(\phi) = \int_{-\infty}^{\infty} \xi'(t)\phi(t)dt$). Since generalized processes are differentiable, a generalized random process with stationary increments of given order n can be defined simply as a process $\xi(\phi)$ (non-stationary in general) whose nth derivative $\xi^n(\phi)$ is a generalized stationary process. Using this definition, it is easy to derive from (9) and (10) the general spectral representation of a generalized random process with stationary increments of order n, which includes (5) and (6) as particular cases (relating to processes $\xi(\phi)$ of the form (7)).

Another generalization of the spectral representation of a stationary random process $\xi(t)$ as a Fourier-Stieltjes integral (*4*) concerns homogeneous random fields in higher-dimensional spaces R^n, that is, random functions $\xi(\mathbf{x})$, $\mathbf{x} = (x_1, \ldots, x_n) \in R^n$, of many variables related to stationary processes. A random field $\xi(\mathbf{x})$, $\mathbf{x} \in R^n$, is called homogeneous (in the wide sense) if for any $\mathbf{r} \in R^n$,

$$\mathsf{E}\xi(\mathbf{x}+\mathbf{r}) = \mathsf{E}\xi(\mathbf{x}), \quad \mathsf{E}\xi(\mathbf{x}_1+\mathbf{r})\bar{\xi}(\mathbf{x}_2+\mathbf{r}) = \mathsf{E}\xi(\mathbf{x}_1)\bar{\xi}(\mathbf{x}_2) \tag{11}$$

(so that $\mathsf{E}\xi(\mathbf{x}) = \text{const}$, $\mathsf{E}\xi(\mathbf{x}_1)\bar{\xi}(\mathbf{x}_2) = B(\mathbf{x}_1 - \mathbf{x}_2)$). Such a field allows a spectral representation of the form

$$\xi(\mathbf{x}) = \int_{R^n} e^{i\mathbf{k}\mathbf{x}}\Phi(d\mathbf{k}), \tag{12}$$

where $\Phi(\Delta\mathbf{k})$ is a random function of an n-dimensional interval $\Delta\mathbf{k}$ in the space of vectors $\mathbf{k}$, whose values on non-intersecting intervals $\Delta\mathbf{k}_1$ and $\Delta\mathbf{k}_2$ are uncorrelated, and $F(\Delta\mathbf{k}) = \mathsf{E}\{|\Phi(\Delta\mathbf{k})|^2\}$ is integrable over the whole space (that is, $\int_{R^n} F(d\mathbf{k}) < \infty$). We note that as far back as 1941 Obukhov [14], [15] used the spectral decomposition (12) of a homogeneous random field $\xi(\mathbf{x})$ in his important work on statistical turbulence theory, referring to the paper [*6*] by Kolmogorov. A rigorous proof of this formula and some of its generalizations (concerning, in particular, fields with homogeneous increments and generalized homogeneous fields) can now be found, for example, in [16] (see also [13], §III.5). A number of further examples of "generalized spectral representations" of various classes of random functions related to the representation of stationary random processes $\xi(t)$ as Fourier-Stieltjes integrals (*4*) is given in particular in [17], [18].

A precise formulation of Zasukhin's results [*13*] mentioned by Kolmogorov and a rigorous proof of these results can be found in Rozanov's book [2], Chapter II. For a modern presentation and further development of Slutskii's results [*14*] see the papers by Moran [19], [20] and Kendall and Stuart [21], §47.15.

References

1. J.L. Doob, 'Time series and harmonic analysis', In: *Proc. First Berkeley Sympos. Math. Statist. Probab.*, 1949, pp.303–344.
2. Yu.A. Rosanov, *Stationary random processes*, Fizmatgiz, Moscow, 1963 (in Russian).

3. A.M. Yaglom, *Correlation theory of stationary random functions*, Gidrometeoizdat, Leningrad, 1981 (in Russian).
4. J.L. Doob, *Stochastic processes*, Chapman & Hall, 1953.
5. A.D. Ventsel, *Course on the theory of random processes*, Nauka, Moscow, 1975 (in Russian).
6. I.I. Gikhman and A.V. Skorokhod, *Introduction to the theory of random processes*, 2nd ed., Nauka, Moscow, 1977 (in Russian).
7. I.I. Gikhmann and A.V. Skorokhod, *The theory of random processes*, Vol. I, Nauka, Moscow, 1971 (in Russian).
8. J. von Neumann and I.J. Schoenberg, 'Fourier integrals and metric geometry', *Trans. Amer. Math. Soc.* **50**:2 (1941), 226–251.
9. A.M. Yaglom, 'Correlation theory of processes with random stationary nth increments', *Mat. Sb.* **37**:1 (1955), 141–196 (in Russian).
10. M. Pinsker, 'The theory of curves in Hilbert space with stationary nth increments', *Izv. Akad. Nauk SSSR Ser. Mat.* **19**:5 (1955), 319–345 (in Russian).
11. I.M. Gel'fand, 'Generalized random processes', *Dokl. Akad. Nauk SSSR* **100**:5 (1955), 853–856 (in Russian).
12. K. Itô, 'Stationary random distributions', *Mem. Coll. Sci. Univ. Kyoto* (A) **28**:3 (1954), 209–223.
13. I.M. Gel'fand and N.Ya. Vilenkin, *Generalized functions, Vol. 4, Applications of harmonic analysis. Rigged Hilbert spaces*, Academic Press, New York-London, 1964.
14. A.M. Obukhov, 'On the energy distribution in the spectrum of a turbulent flow', *Dokl. Akad. Nauk SSSR* **32**:1 (1941), 22–24 (in Russian).
15. A.M. Obukhov, 'On the energy distribution in the spectrum of a turbulent flow', *Izv. Akad. Nauk SSSR, Geography and Geophysics series* **5**:4/5 (1941), 453–466 (in Russian).
16. A.M. Yaglom, 'Some classes of random fields in n-dimensional space related to stationary random processes', *Teor. Veroyatnost. i Primenen.* **2**:3 (1957), 292–338 (in Russian). (Translated as *Theory Probab. Appl.*)
17. K. Karhunen, 'Über lineare Methoden in der Wahrscheinlichkeitsrechnung', *Ann. Acad. Sci. Fenn.* A **1**:37 (1947), 3–79.
18. A.M. Yaglom, 'Spectral representations for various classes of random functions', In: *Proc. Third All-Union Math. Congress, Moscow, Izd. Akad.*

Nauk SSSR, Vol. 1, 1963, pp. 250–273 (in Russian).

19. P.A.P. Moran, 'The spectral theory of discrete stochastic processes', *Biometrika* **36**:1 (1949), 63–70.
20. P.A.P. Moran, 'The oscillatory behaviour of moving averages', *Proc. Cambridge Phil. Soc.* **46**:2 (1950), 272–281.
21. M.J. Kendall and A. Stuart, *The advanced theory of statistics. Design and analysis and time series*, Griffin, 1983.

SPECTRAL REPRESENTATION OF RANDOM PROCESSES (Nos. 47, 49)

(Yu.G. Balasanov and I.G. Zhurbenko)

Let $T^{(k)}$ be the class of random processes $\xi(t)$ such that

$$\mathsf{E}|\xi(t)|^p \leq C_k < \infty,$$

and $S^{(k)}$ the class of processes $\xi(t) \in T^{(k)}$ such that for all $1 \leq l \leq k$, $-\infty < \tau < \infty$,

$$\mathsf{E}[\xi(t_1)\dots\xi(t_l)] = \mathsf{E}[\xi(t_1+\tau)\dots\xi(t_l+\tau)].$$

Then the example constructed in the commented papers shows that although $S^{(2)} \subset \Phi^{(2)}$, for $k > 2$ there exist processes of class $S^{(k)}$ that are not of class $\Phi^{(k)}$.

A suitable definition of the classes of random processes that have spectral representation of higher orders was discussed in a number of subsequent works of various authors. We first mention [1], where, at Kolmogorov's suggestion, the classes $\Delta^{(k)}$ are introduced and studied. They are defined as certain subclasses of the classes $S^{(k)} \cap \Phi^{(k)}$. For these classes it is more natural to consider the semi-invariant spectral measure $F_\xi^{(k)}(\lambda_1,\dots,\lambda_k)$ defined by the relation (see [1], [2], [9])

$$s_\xi^{(k)}(t_1,\dots,t_k) = \int_{R^k} \exp\Big\{i\sum_{j=1}^k t_j\lambda_j\Big\} F_\xi^{(k)}(d\lambda_1,\dots,d\lambda_k),$$

where

$$s_\xi^{(k)}(t_1,\dots,t_k) = \frac{i^{-k}\partial^k}{\partial u_1\dots\partial u_k} \ln \mathsf{E}\Big(\exp\Big\{i\sum_{j=1}^k u_j\xi(t_j)\Big\}\Big|_{u_1=\dots=u_k=0}\Big),$$

instead of the moment spectral measure $M_\xi^{(k)}(\lambda_1,\ldots,\lambda_k)$ (defined by the relation (see [1], [2], [9])

$$m_\xi^{(k)}(t_1,\ldots,t_k) = \mathbf{E}[\xi(t_1)\ldots\xi(t_k)] = \int_{R^k} \exp\Big\{i\sum_{j=1}^{k} t_j\lambda_j\Big\} M_\xi^{(k)}(d\lambda_1\ldots d\lambda_k)\Big).$$

In particular, the classes $\Delta^{(k)}$ are characterized by the fact that for $\xi(t)\in\Delta^{(k)}$ the semi-invariant measures $F_\xi^{(l)}$, $1\le l\le k$, are absolutely continuous with respect to Lebesgue measure on the sets $\lambda_1+\ldots+\lambda_l = 0 \pmod{2\pi}$. The study of the classes $\Delta^{(k)}$ was then continued in [2]. A generalization of $\Delta^{(k)}$ for vector random processes is obtained in [3]. Similar questions for random fields were considered in [4].

The work initiated by A.N. Kolmogorov in which the classes $\Delta^{(k)}$ were defined and studied, initiated a series of studies in a new branch, called the theory of higher spectra of stationary random processes and their statistical analysis. The results of this branch are now of fundamental character and are used extensively for solving various applied problems in astronomy, geophysics, studies of liquid and gas turbulence, etc. (see, for example, [5]–[8]).

A fundamental contribution to the further development of higher-order spectral analysis was made in [9] where the mathematical apparatus of higher moments and semi-invariants, necessary for further research, was developed (see [10], [11], etc.). Higher-order spectral theory for homogeneous random fields was considered by Yaglom [12]–[14]. Upper estimates for higher spectral densities and their derivatives under different mixing conditions can be found in [11] and are essential in statistical analysis.

Numerous works dealt with constructing and studying higher spectral densities (for example, [15]–[17], etc.). However, until recently, all the statistics had the same essential drawback: they did not allow one to construct an estimator of the higher semi-invariant spectral density for all arguments. Though the semi-invariant density of nth order is defined at all points $\lambda_1+\ldots+\lambda_n = 0 \pmod{2\pi}$, the statistics proposed gave no answer for the subsets $\lambda_{k_1}+\ldots+\lambda_{k_p} = 0 \pmod{2\pi}$, $1\le k_i\le n$, $i=1,\ldots,p$, for $p<n$. Naturally statistical estimators were unstable in a neighbourhood of these sets. This drawback has recently been eliminated by using statistics constructed by the time translation method [18].

References

1. A.N. Shiryaev, 'Some questions of the spectral theory of higher moments' I, *Teor. Veroyatnost. i Primenen.* **5**:3 (1960), 293–313 (in Russian). (Translated as *Theory Probab. Appl.*)
2. V.P. Leonov and A.N. Shiryaev, 'Some questions of the spectral theory of higher moments' II, *Teor. Veroyatnost. i Primenen.* **5**:4 (1960), 460–464 (in Russian). (Translated as *Theory Probab. Appl.*)
3. D.R. Brillinger, 'An introduction to polyspectra', *Ann. Math. Statist.* **36** (1965), 1351–1374.
4. N.N. Leonenko, 'Some questions of the spectral theory of higher moments of random fields', *Dokl. Akad. Nauk Ukrain. SSR* **9** (1975), 777–782 (in Russian).
5. P.I Kuznetsov, R.L. Stratonovich and V.I. Tikhonov, 'Some random functions and their transmission through non-linear systems', *Avtomatika i telemekhanika* **14**:4 (1963) (in Russian).
6. A.S. Monin and A.M. Yaglom, *Statistical hydromechanics* Vol. 1, Nauka, Moscow, 1965 (in Russian).
7. A.S. Monin and A.M. Yaglom, *Statistical hydromechanics* Vol. 2, Nauka, Moscow, 1967 (in Russian).
8. D.R. Brillinger, 'Identification of polynomial systems by polyspectra', *J. Sound and Vibr.* **2**:3 (1970), 301–313.
9. V.P. Leonov and A.N. Shiryaev, 'On a technique of computing semi-invariants', *Teor. Veroyatnost. i Primenen.* **4**:3 (1959), 342–355 (in Russian). (Translated as *Theory Probab. Appl.*)
10. V.P. Leonov, *Some applications of higher semi-invariants to the theory of stationary random processes*, Nauka, Moscow, 1964 (in Russian).
11. I.G. Zhurbenko, *Spectral analysis of time series*, Izdat. Moskov. Gos. Univ., 1982 (in Russian).
12. A.M. Yaglom, 'Some classes of random fields in n-dimensional space related to stationary random processes', *Teor. Veroyatnost. i Primenen.* **2**:3 (1957) 292–338 (in Russian). (Translated as *Theory Probab. Appl.*)
13. A.M. Yaglom, *An introduction to the theory of stationary random functions*, Prentice-Hall, Englewood Cliffs, 1962.
14. A.M. Yaglom, 'Second-order homogeneous random fields', In: *Proc. Fourth Berkeley Sympos. Math. Statist. Probab.*, 1961, pp. 593–622.

15. M.A. Rosenblatt and J.S. Van Ness, 'Estimation of the bispectrum', *Ann. Math. Statist.* **36** (1965), 1120–1136.
16. D.R. Brillinger and M.A. Rosenblatt, 'Asymptotic theory of estimates of *k*th order spectra', In: *Advanced Seminar on Spectral Analysis of Time Series*, Wiley, New York, 1967, pp. 189–232.
17. I.G. Zhurbenko and N.N. Trush, 'On estimators of spectral densities of stationary processes', *Litov. Mat. Sb.* **19**:1 (1979), 65–82 (in Russian).
18. I.G. Zhurbenko, 'Statistical estimators of higher spectra', *Teor. Veroyatnost. i Primenen.* **30**:1 (1985), 66–77 (in Russian). (Translated as *Theory Probab. Appl.*)

BROWNIAN MOTION (Nos. 14, 19, 24)
(A.M. Yaglom)

Papers Nos. 14, 19, 24 are closely related to the important papers Nos. 9 and 17 dealing mainly with the general theory of continuous Markov ("Stochastically determined" in Kolmogorov's terms) random processes. Starting from the well-known works by Einstein and Smoluchowski [1], [2], processes of this kind have been widely used in physics to describe Brownian motion of both individual particles and systems with many degrees of freedom; we recall in this connection that, as it turned out, the physicists Fokker and Planck, who studied Brownian motion, came to some of the conclusions given in No. 9 even earlier. It was therefore natural to expect that the results of No. 17 can also be directly applied to many problems concerning Brownian motion. Some concrete examples of these applications are given in Nos. 14, 19, 24.

Paper No. 14, written together with M.A. Leontovich and published in a physics magazine, is more related to physics than the other two. It solves the following problem in the theory of Brownian motion posed by S.I. Vavilov: to find the mean $\mathsf{E}(F)$ of the area F on the plane (x, y) covered by the projection of a spherical particle of fixed radius σ in the process of Brownian motion over time t. Based on partial differential equations for the transition probability densities of a multidimensional continuous Markov process found in No. 17, the authors obtain for $\mathsf{E}(F)$ a very precise asymptotic formula (applicable when $Dt/\sigma^2 \gg 1$, where D is the corresponding diffusion coefficient, which, due to Einstein's results [1], [2] for a spherical particle of radius σ immersed in a liquid or gaseous medium of viscosity μ is given for the absolute temperature T by the relation $D = k_0T/6\pi\sigma\mu$, where k_0 is the Boltzmann constant).

Note that the solution given in the paper is based on the following important result established in the same paper. Let $P_L(x, y; t)$ denote the probability of a particle located at (x, y) at time $t = 0$ and performing Brownian motion in the two-dimensional plane (the case of Brownian motion in n-dimensional space can be solved similarly for any n) crossing the boundary Γ of a given domain G (containing the point (x, y)) at least once within a given time t, so that the first crossing will be within a given part L of Γ (where L can coincide with Γ). In this case $P_L(x, y; t)$ satisfies a certain partial differential equation (the "first Kolmogorov equation", see (14)), and the following conditions: $P_L(x, y; 0) = 0$ for any interior point (x, y) of G and $P_L(x, y; t) \to 1$ for any $t > 0$, when (x, y) tends to a point of L, $L \subset \Gamma$, and $P_L(x, y; t) \to 0$ for $t > 0$ and (x, y) tending to a point of $\Gamma \backslash L$. These equations and conditions uniquely determine $P_L(x, y; t)$. Clearly, this general conclusion is of considerable interest even beyond the specific problem of Brownian motion solved in No. 14. It was obtained independently at the same time by Pontryagin, Andronov and Vitt [3]. This and some related results are now widely used in many applications and are included in a number of textbooks (see, for example, [4] §30, [5] §1.4, [6] §§26, 27).

A much more general question in the theory of Brownian motion is discussed by Kolmogorov in the short article No. 19. As is well known, in the classical theory of Brownian motion developed by Einstein and Smoluchowski [1], [2] the inertia of Brownian motion is neglected, that is, actually a particle is assumed to have zero mass. Therefore, in Einstein and Smoluchowski's theory a Brownian particle does not have finite velocity; thus, for example, for the particular case of Brownian motion of a free particle, Wiener had rigorously proved (see [7], Chapter IX) that with probability 1 the Brownian trajectory is a continuous but nowhere differentiable curve. We note also that in the Einstein-Smoluchowski approximation, Brownian motion of a physical system is modelled by a continuous Markov process in its coordinate space (for the motion of a free particle it is a random Wiener process).

It is clear that the non-differentiability of Brownian trajectories in the Einstein-Smoluchowski theory is closely related to the idealization introduced (neglecting inertia), which makes the theory inapplicable for an analysis of the motion over very small time intervals Δt. For the simplest case of Brownian motion of a free particle a refined theory taking into account the particle's inertia was developed in 1930 by Uhlenbeck and Ornstein [8] (see also Doob [9] and Chandrasekhar [10], Chapter II); in this refined theory trajectories of particles are differentiable (but do not have second derivative, so that a particle's acceleration is infinite). In fact, the same generalization of the classical theory of Brownian motion is discussed in No. 19 which, however, considers the general case of Brownian motion with n degrees of freedom, instead of the particular case of motion of a free particle. According to Kolmogorov, inertia is taken into account by considering the state of the system as given by the values of n positions $q_1, \ldots, q_n$ and their derivatives with respect to time (velocities) $\dot{q}_1, \ldots, \dot{q}_n$. Brownian motion is modelled here by a continuous Markov process in the $2n$-dimensional phase space of positions and velocities. The main equation for the transition probability density of this Markov process (equation (9) of No. 19) turns out in this case to be a degenerate parabolic-type equation (since its right-hand side contains only second derivatives with respect to velocities, but not with respect to two position coordinates or a position and a velocity coordinate). Soon after No. 19 was published, Kolmogorov's student, Piskunov, began a study of the mathematical theory of such equations (11). The Uhlenbeck-Ornstein theory of one-dimensional Brownian motion of a free particle is derived from Kolmogorov's general theory for the particular case $n = 1$ (so that the basic equation of the theory acquires the form (10), see No. 19), where $f = -\alpha q$ (and $\alpha = \beta/m$, where m is the particle's mass, and β the viscous friction coefficient equal to $6\pi\sigma\mu$ for a spherical particle of radius σ), and $k = \text{const} = k_0 T/m\beta$.

Finally, paper No. 24 discusses the interesting question of the meaning and conditions of statistical reversibility of Brownian motion. It is well known that in the thermodynamical sense processes of Brownian motion (or diffusion) are irreversible: in the presence of a large number of diffusing particles these processes always result in a levelling out of the distribution of the particles available as t increases, whereas the distribution becomes more and more inhomogeneous as t decreases. Apparently Schrödinger [12] was the first to draw attention to the fact that an essential role is played here by the presence of a well-defined initial condition at $t = 0$ and the assumption (always made) that the distribution of particles tends to a homogeneous one (in general, to a certain stationary ergodic distribution) as $t \to \infty$, whereas no constraints are usually imposed on the behaviour of the process as $t \to -\infty$. In this connection he considered a diffusion process during the time interval $t_0 < t < t_1$ under the condition that both the initial distribution at $t = t_0$ and the final distribution at $t = t_1$ are fixed and pointed out that in this setting, diffusion is to some extent reversible. In particular, assuming that diffusion takes place in a finite space interval (or in a finite volume), and the probability distribution tends to the same homogeneous (that is, stationary and ergodic) distribution both as $t \to \infty$ and $t \to -\infty$, then the changes in the probability distribution for $t_0 \leq t < \infty$ and $-\infty < t \leq t_0$ corresponding to a given value of the distribution at the time t_0 are described by absolutely identical relations, that is, the diffusion is completely reversible in time.

In his paper Schrödinger considered only the simplest one-dimensional diffusion described by the diffusion equation with constant coefficient D, whereas in No. 24 Kolmogorov studies the general case of an arbitrary n-dimensional Brownian motion described by a continuous Markov random process on a certain n-dimensional manifold R of positions $(x_1, \dots, x_n) = x$ of a physical system. If for this Markov process the "absolute (unconditional) probability density" $p(x)$ is given for all t, which is the strictly positive density of a stationary probability distribution, (when there exists an ergodic distribution, it is natural to take it as $p(x)$), then along with the usual probability density of the transition from a given state at $t = t_0$ to a certain state at $t = t_1 > t_0$ we can also determine the "inverse transition probability density", that is, the conditional probability density for the state at $t = t_0$, given the state at a later moment $t = t_1 > t_0$. After this it is natural to raise the problem of necessary and sufficient conditions for these two transition probability densities to be equal to each other. These necessary and sufficient conditions are found in No. 24. They are derived with the help of an invariant tensor form (derived in the same paper) of the general equations for the transition probability density of the n-dimensional Markov process given in No. 19. This probability density might be helpful for problems other than that on statistical reversibility considered here, and the desired conditions themselves appeared to be quite general and satisfied in the particular case considered by Schrödinger.

We note, however, that in No. 24 the problem of statistical reversibility of Brownian motion is still not treated in its most general form, since it is assumed there that the matrix $||B^{ij}(x)||$ of the coefficients of the second derivatives in the equations for the transition probability density is strictly positive definite for all x. At the end of §1 the author points out that the case of degenerate $||B^{ij}(x)||$ is in fact of considerable physical importance, since it appears, in particular, when considering Brownian motion in the phase space of positions and velocities of a physical system (see No. 19). But he immediately adds that the degenerate case is not considered in this paper. Later Kolmogorov asked a postgraduate of his, the author of the present comments, the question on the conditions of statistical reversibility of Brownian motion in the phase space of positions and velocities. In this case the very definition of statistical reversibility differs from that given in No. 24, since reversing the motion of a physical system entails changing the sign of all the velocities (see [13]). However, this difference is very simple and has little effect. To obtain necessary and sufficient conditions of statistical reversibility, we must now rewrite the main equations for the transition probability (equation (9) of No. 19 and the adjoint equation) in invariant tensor form. (This acquires an especially clear physical meaning when the coefficients of the metric quadratic form in the position space are taken to be the coefficients of the quadratic form of the velocities that gives the system's kinetic energy.) If we now confine ourselves to the most interesting case, when the coefficients of second derivatives with respect to the velocities in the equation for the transition probability densities depends only on the positions and the forces affecting the system in question depend linearly on the velocities, we can show (see [13]) that for a Brownian motion in the phase space to be statistically reversible it is only necessary that the corresponding stationary probability distribution have the same form as the canonical Gibbs distribution. Thus, in this case the statistical reversibility condition has clear physical meaning. If we then pass to the limit corresponding to letting the system's inertia tend to zero (that is, let all coefficients of the quadratic form of velocities tend to zero, which gives the kinetic energy), then we again come to the Einstein-Smoluchowski model of Brownian motion as a Markov process in the space of positions only; then the conditions for statistical reversibility of the Brownian motion in the phase space found in [13] by passing to the limit again revert to the conditions for statistical reversibility found in No. 24 by Kolmogorov. The latter circumstance sheds additional light on the physical meaning of the latter conditions.

References

1. A. Einstein and M. Smoluchowski, *Brownian motion*, Collected papers.
2. A. Einstein, 'Works on the theory of Brownian motion, In: A. Einstein, Collected scientific papers, Vol. 3, Nauka, Moscow, 1966, pp.75–91; 108–127; 149–151; 155–163.
3. L.S. Pontryagin, A.A. Andronov and A.A. Vitt, 'On statistical consideration of dynamic systems', *ZhETF* **3**:3 (1933), 165–180 (in Russian).
4. A.A. Svechnikov, *Applied methods of random function theory*, 2nd ed., Nauka, Moscow, 1968 (in Russian).
5. V.I. Tikhonov, *Blow-outs of random processes*, Nauka, Moscow, 1970 (in Russian).
6. V.I. Tikhonov and M.A. Mironov, *Markov processes*, Sovyetskoye Radio, (in Russian).
7. R. Paley and N. Wiener, 'Fourier transforms in the complex domain', *Amer. Math. Soc.* (1934).
8. G.E. Uhlenbeck and L.S. Ornstein, 'On the theory of the Brownian motion', *Phys. Rev.* **36** (1930), 823–841.
9. J.L. Doob, 'Brownian motion and stochastic equations', *Ann. of Math.* **43**:2 (1942), 351–369.
10. S. Chandrasekar, *Stochastic problems in physics and astronomy*, Inost. Lit., Moscow, 1947.
11. N.S. Piskunov, 'Boundary value problems for an equation of elliptic-parabolic type', *Mat. Sb.* **7** (1940), 385–424 (in Russian).
12. E. Schrödinger, 'Über die Umkehrung der Naturgesetze', *Sitzungsber. Preuss. Akad. Wiss., Phys.-Math. Kl.*, 12 März (1931), 144–153.
13. A.M. Yaglom, 'On statistical reversibility of Brownian motion', *Mat. Sb.* **24** (1949), 457–492 (in Russian).

MARKOV CHAINS WITH A COUNTABLE NUMBER OF STATES (No. 23)

(A.A. Yushkevich)

A.N. Kolmogorov's note [1] and its enlarged version [2] (No. 23 of the present edition) contain the foundations of the theory of countable homogeneous Markov chains; these works marked the beginning of the systematic study of Markov chains with a countable and, in general, an infinite set of states. To obtain a classification of the states and asymptotics of the transition probabilities $P_{ij}^{(n)}$ as $n \to \infty$ is an indispensable part of any detailed presentation of countable Markov chains [5], [8], [12], [31], [39], [40]. In an adapted form these results are included in textbooks for non-mathematicians.

In the papers cited A.N. Kolmogorov proved that 1) a countable space of states E is subdivided into a set of inessential states and classes of essential states (interacting with each other), while the classes, in their turn, are subdivided into periodic subclasses; 2) within one class the probability of an infinite number of returns is the same for all states and equals either 1 (recurrent class), or 0 (non-recurrent class); 3) within one class the mean recurrence time M_{jj} is either finite for all the states j (positive class) or infinite for all the states (zero class); 4) in a non-periodic class the limit

$$\lim_{n\to\infty} P_{ij}^{(n)} = M_{jj}^{-1} \tag{1}$$

exists and the sum of the limits in a positive class equals 1; 5) therefore, in general, the dependence of $P_{ij}^{(n)}$ on n is asymptotically periodic.

For a finite E all classes are recursive and positive: the division of E into inessential states, classes and subclasses and the asymptotic periodicity of the $P_{ij}^{(n)}$ for this case were independently obtained by Doeblin [3] (it was Markov who established the existence of the limits (1) independent of i for a finite chain with strictly positive P_{ij} and consisting, evidently, of one non-periodic class).

Kolmogorov's work was continued by many researchers. Erdös, Feller and Pollard [4], striving to "make Kolmogorov's results more accessible", proved (1) as a renewal theorem that later acquired a number of generalizations (see [5], Chap. XIII). Feller [5] was the first to present the foundations of the theory of countable homogeneous Markov chains for beginners; he pointed out that in a positive class the limits (1) form the only stationary distribution.[1] In [6] he found a number of refinements of (1), for example, an expression for $\sum_0^\infty (P_{jj}^{(n)} - M_{jj}^{-1})$ for a finite second moment $M_{jj}^{(2)}$ of the time of return to j, and later in [7] he gave a new proof to the renewal theorem.

Sarymsakov [8] reproduced Kolmogorov's classification and proposed a matrix method for calculating the limits (1); he also considered the central limit theorem for chains with a countable space E and with $E = [a, b]$.

[1] Feller first divided the states into recurrent and non-recurrent ones, and in his remarks on terminology, erroneously ascribes the same meaning to Kolmogorov's terms "essential and inessential states".

For the zero class, when the limits in (1) are 0, it is interesting to compare the orders of convergence to zero of the transition probabilities for different states, and also to compare future distributions for various initial states. Doeblin [9] established that within one class the finite non-zero limit

$$\lim_{n\to\infty} \frac{\sum_1^n P_{ij}^{(m)}}{\sum_1^n P_k^{(m)}} \tag{2}$$

always exists (Doeblin's result is actually important for the zero recurrent class, since for a positive class it follows from (1), while for a non-recurrent class it follows from the convergence of the series obtained by replacing n by ∞ in (2)). Chung [10] found that for a non-recurrent class the limit (2) is $\mu_j \mu_l^{-1}$ where μ_i is the average number of visits to i during one excursion from a fixed state h to h, and Derman [11] showed that under the same conditions $\mu = \{\mu_i\}$ is a σ-finite invariant measure which is unique to within a constant factor. These and related results are discussed in detail in the monograph by Chung [12].

Orey [13] showed that in a recurrent class under different initial states future distributions converge to each other in variation:

$$\lim_{n\to\infty} \sum_j |P_{ij}^{(n)} - P_{kj}^{(n)}| = 0. \tag{3}$$

Another proof was given by Blackwell and Freedman [14] (in a positive class (3) it follows directly from Kolmogorov's result (4)).

Kingman and Orey [15] considerably strengthened Doeblin's result on the limit (2) by proving an "individual" limit theorem for ratios: if in a non-periodic recurrent class

$$\sum_1^N P_{ii}^{(n)} > \epsilon \quad \forall i \tag{4}$$

for some N and $\epsilon > 0$, then

$$\lim_{n\to\infty} \frac{P_{ij}^{(n+m)}}{P_{kl}^{(n)}} = \frac{\mu_j}{\mu_l}. \tag{5}$$

As was pointed out by Molchanov [16], condition (4) is equivalent to the simpler condition $P_{ii}^{(M)} > \delta$ for any i. Molchanov generalized (5) to α-recurrent chains, that is, chains such that for some $\alpha \geq 1$,

$$\sum_{n=1}^{\infty} \alpha^n P_{ij}^{(n)} = \infty \quad \forall i, j,$$

but for each positive $\beta < \alpha$ the corresponding series converges; if in this chain $P_{ii}^{(N)} > \epsilon P_{ii}^{(M)}$ for some $N > M$, $\epsilon > 0$ and all i, then the limit in (5) exists and equals $\alpha^{-m}\mu_j\phi_i\mu_l^{-1}\phi_k^{-1}$, where $\mu = \alpha\mu P$ is (to within a constant factor) a unique α-harmonic measure, and $\phi = \alpha P\phi$ a unique positive function. The proof of Molchanov's theorem, which makes use of the Martin boundary, can be found in [17].

The asymptotics of the transition probabilities $P_{ij}(t)$ of a homogeneous Markov process with a countable state space and continuous time was studied by Lévy [18].

For an uncountable phase space E Kolmogorov's subdivision of E into the set of inessential states, classes and subclasses was made by Doeblin [19] and improved by Doob [20], Chung [21] and Orey [22]. The well-known condition of Doeblin assumed in these works guarantees the presence of a stationary distribution and the analogy with the case of positive classes. A definition of probability for a general space E was proposed by Harris [23]. In chains that are recurrent in Harris's sense there exists a unique σ-finite measure, which is invariant to within a constant factor. For these chains an analogue of the result on the existence (at $i = k$) of the limit (2), equal to $\mu_j\mu_l^{-1}$, holds (the Chacon-Ornstein theorem [24], later improved by a number of authors). These problems are discussed in detail in the monograph by Revuz [25].

Kolmogorov's papers [1], [2] have stimulated not only further studies on the classification of states and the asymptotics of transition probabilities, but to a significant extent, also the development of the whole theory of Markov chains with a countable or arbitrary phase space.

The first of these deals with the recurrent, primarily positive case and consists in extending the known results on sums of independent random variables or finite chains to additive functionals of the Markov chains considered. This includes the law of large numbers and related ergodic theorems, the central limit theorem and its refinements and corresponding asymptotic expansions, the multidimensional limit theorem for the number of visits to a given set of states, the law of the iterated logarithm, convergence of normalized increasing sums to a Wiener process (also called the invariance principle), convergence to non-Gaussian limit laws, etc. The ergodic theorem for the ratio of two functionals is proved for recurrent chains, while the other results require stronger assumptions of an ergodic character such as finiteness of the second moments

$M_{jj}^{(2)}$ or the above-mentioned condition of Doeblin. The first significant results in this field were obtained by Doeblin [19]. Far-reaching refinements of these results were obtained by Nagaev [26], [27]. Leading roles were played by Dobrushin [28] and Statulyavichus [29] in applying these results to inhomogeneous chains. More detailed references and information can be found in the book by Chung [12] (a lot of references to papers by Soviet authors were added in translation), and a recent monograph by Sirazhdinov and Formanov [30] that contains some old and many new results. The law of the iterated logarithm, the invariance principle and ergodic theorems for ratios are given in the book by Freedman [31], specifically for countable Markov chains.

Another trend that originated in the study of non-recurrent Markov chains (and processes) is Martin's theory of potentials and boundaries. The connection between Dirichlet's problem for the Laplace equation in a bounded domain G and Wiener's process in G considered before the time of reaching the boundary ∂G of the domain G has been known for a long time. This relationship allows us to introduce basic notions of classical potential theory in terms of Wiener's process or its transition functions. By analogy, the notions of harmonic function, potential Green's function and excessive (positive superharmonic) functions are transferred to other non-recurrent Markov processes. Basic works in developing the potential theory for Markov processes were done by Hunt [32] and Dynkin [33]. Potential theory for countable Markov chains is presented in the book by Henneken and Tortra [34].

The exit boundary of a non-recurrent Markov chain is constructed for describing the possible behaviour of trajectories as $n \to \infty$ (with probability 1 a trajectory is attracted to one of the boundary points). The entrance boundary plays a dual role. Feller was the first to construct a boundary for a countable Markov chain [35]. However, it was another, more meagre boundary that gained recognition. It was proposed independently by Doob [36] and Watanabe [37]. It is based on potential theory and is called Martin's boundary (in the case of Wiener's process in a plane domain G, Doob's construction coincides with that of Martin [38], which allows us to represent all functions harmonic in G by a formula similar to Poisson's integral for the disc). The probabilistic construction of the Martin boundary for a non-recurrent countable Markov chain was made by Hunt [39] and improved by Dynkin [40]. Potential theory and Martin's boundary for Markov chains with a common phase space is presented in the book by Revuz [25].

A new method for studying recurrent and non-recurrent countable Markov chains, based on the notion of the fundamental matrix, was proposed by Kemeny and Snell. These authors (and, independently, Orey) transferred the theory of potentials and boundaries to recurrent Markov chains. These results, as well as ergodic theorems and Martin boundaries are presented in detail in the monograph [41] which contains a comprehensive historical review and bibliography.

References

1. A.N. Kolmogoroff, 'Anfangsgründe der Theorie der Markoffschen Ketten mit unendlich vielen möglichen Zuständen', *Mat. Sb.* **1**:4 (1936), 607–610.
2. A.N. Kolmogorov, 'Markov chains with a countable number of possible states', *Byull. Moskov. Gos. Univ. Math. i Mekh.* **1**:3 (1937), 1–16 (in Russian).
3. W. Doeblin, 'Sur les chaînes discrètes de Markoff', *C. R. Acad. Sci. Paris* **203**:1 (1936), 24–26.
4. P. Erdös, W. Feller and H. Pollard, 'A theorem on power series', *Bull. Amer. Math. Soc.* **55** (1949), 201–204.
5. W. Feller, *An introduction to probability theory and its applications*, Vol. 1, Wiley, New York, 1950.
6. W. Feller, 'Fluctuation theory of recurrent events', *Trans. Amer. Math. Soc.* **67** (1949), 98–119.
7. W. Feller, 'A simple proof for renewal theorems', *Comm. Pure Appl. Math* **14** (1961), 285–293.
8. T.A. Sarymsakov, *Foundations of the theory of Markov processes*, Gostekhizdat, Moscow, 1954 (in Russian).
9. W. Doeblin, 'Sur deux problèmes de Kolmogoroff concernant les chaînes dénombrables', *Bull. Soc. Math. France* **66** (1938), 210–220.
10. K.L. Chung, 'An ergodic theorem for stationary Markov chains with a countable number of states', In: *Proc. Intern. Congr. Math. Cambridge (Mass)*, Vol. 1, 1950, p. 568.
11. C. Derman, 'A solution to a set of fundamental equations in Markov chains', *Proc. Amer. Math. Soc.* **5** (1954), 332–334.

12. K.L. Chung, *Markov chains with stationary transition probabilities*, Springer-Verlag, Berlin, 1960.
13. S. Orey, 'An ergodic theorem for Markov chains', *Z. Wahrscheinlichkeitstheorie* **1** (1962), 174–176.
14. D. Blackwell and D. Freedman, 'The tail σ-field of a Markov chain and a theorem of Orey', *Ann. Math. Statist.* **35** (1964), 1291–1295.
15. J.F.C. Kingman and S. Orey, 'Ratio limit theorems for Markov chains', *Proc. Amer. Math. Soc.* **15** (1964), 907–910.
16. S.A. Molchanov, 'Limit theorems for the ratio of transition probabilities of Markov chains', *Uspekhi Mat. Nauk* **22**:2 (1967), 124–125 (in Russian).
17. S.A. Molchanov, *Some problems in the theory of Martin boundaries for Markov processes*, Candidate of Sciences Dissertation, Moscow State Univ., 1967 (in Russian).
18. P. Lévy, 'Systèmes markoviens et stationnaires. Cas dénombrable', *Ann. Sci. École Norm. Supér. Ser.* 3 **68** (1951), 327–381.
19. W. Doeblin, 'Sur les propriétés asymptotiques de mouvement régis par certains types de chaînes simples', *Bull. Math. Soc. Roum.* **39**:1 (1937), 57–115; **39**:2, 3–61.
20. J.L. Doob, *Stochastic processes*, Wiley, New York, 1953.
21. K.L. Chung, 'The general theory of Markov processes according to Doeblin', *Z. Wahrscheinlichkeitstheorie* **2** (1964), 230–254.
22. S. Orey, *Limit theorems for Markov chain transition probabilities*, London, 1971.
23. T.E. Harris, 'The existence of stationary measures for certain Markov processes', In: *Proc. Third Berkeley Sympos. Math. Statist. Probab.*, Vol. 2, 1956, pp. 113–124.
24. R.V. Chacon and D.S. Ornstein, 'A general ergodic theorem', *Ill. J. Math.* **4** (1960), 153–160.
25. D. Revuz, *Markov chains*, Amsterdam, 1975.
26. S.V. Nagaev, 'Certain limit theorems for homogeneous Markov chains', *Teor. Veroyatnost. i Primenen.* **2**:4 (1957), 389–416 (in Russian). (Translated as *Theory Probab. Appl.*)
27. S.V. Nagaev, 'Refinement of limit theorems for homogeneous Markov chains', *Teor. Veroyatnost. i Primenen.* **6**:1 (1961), 67–86 (in Russian). (Translated as *Theory Probab. Appl.*)

28. R.L. Dobrushin, 'A central limit theorem for non-homogeneous Markov chains', *Teor. Veroyatnost. i Primenen.* **1**:1 (1956), 72–89 (in Russian). (Translated as *Theory Probab. Appl.*)
29. V.A. Statulyavichus, 'Local limit theorems and asymptotic expansions for non-homogeneous Markov chains', *Litov. Mat. Sb.* **1**:1 (1961), 231–314 (in Russian).
30. S.Kh. Sirazhdinov and Sh.K. Formanov, *Limit theorems for sums of random vectors connected into a Markov chain*, Fan, Tashkent, 1979 (in Russian).
31. D. Freedman, *Markov chains*, Holden-Day, San Francisco, 1971.
32. G.A. Hunt, 'Markoff processes and potentials', *Ill. J. Math.* **1** (1957), 44–93; 316–369; **2** (1958), 151–213.
33. E.B. Dynkin, *Markov processes*, Fizmatgiz, Moscow, 1963 (in Russian).
34. P.L. Henneken and A. Tortra, *Probability theory and some of its applications*, Mir, Moscow, 1974.
35. W. Feller, 'Boundaries induced by non-negative matrices', *Trans. Amer. Math. Soc.* **83** (1956), 19–54.
36. J.L. Doob, 'Discrete potential theory and boundaries', *J. Math. Mech.* **8** (1959), 433–458.
37. T. Watanabe, 'On the theory of Martin boundaries induced by countable Markov processes', *Mem. Colloq. Sci. Univ. Kyoto* (A) **33** (1960), 39–108.
38. R.S. Martin, 'Minimal positive harmonic functions', *Trans. Amer. Math. Soc.* **49** (1941), 137–172.
39. G. A. Hunt, 'Markoff chains and Martin boundaries', *Ill. J. Math.* **4** (1960), 313–340.
40. E.B. Dynkin, 'The boundary theory of Markov processes (Discrete case)', *Uspekhi Mat. Nauk* **24**:2 (1969), 3–42 (in Russian). (Translated as *Russian Math. Surveys*)
41. J.G. Kemeny, J.L. Snell and A.W. Knapp, *Denumerable Markov chains*, Princeton Univ. Press, 1966; New York, 1976.
42. A.N. Shiryaev, *Probability*, Nauka, Moscow, 1980 (in Russian).

WALD IDENTITIES (No. 35)

(A.A. Novikov)

The paper commented on here anticipates in many aspects numerous studies of generalized Wald identities performed later with the help of martingale theory.

In [1], [2] Wald obtained, among other things, the following relation for the first exit time ν of sums of independent identically distributed random variables ξ_k from a finite interval (a, b). *Let* $\xi_n = \sum_{k=1}^{n} \xi_k$ *and*

$$\nu = \inf\{n \geq 1;\ \xi_n \notin (a, b)\}.$$

Then $\mathsf{E}\nu < \infty$ *and if* $\mathsf{E}|\xi_k| < \infty$, *then*

$$\mathsf{E}(\xi_\nu - \nu(\mathsf{E}\xi_k)) = 0. \tag{1}$$

Moreover, if $\mathsf{E}\xi_k^2 < \infty$, *then*

$$\mathsf{E}[\xi_\nu - \nu(\mathsf{E}\xi_k)]^2 = (\mathsf{E}\xi_k^2)\mathsf{E}\nu. \tag{2}$$

These relations enabled him to obtain approximate formulas for the average observation time in the problem of sequential testing of two simple alternative hypotheses.

In paper No. 35 the author tried to get rid of the condition of identical distribution of the random variables ξ_k, which was quite important in Wald's method [1], [2]. But much more important was the generalization of (1) and (2) for arbitrary stopping times ν. Later, results of this type (with arbitrary stopping times) acquired a fundamental role in various problems of statistical sequential analysis [3], the theory of controlled random processes [4], boundary conditions for random processes [5], [6], etc.

Interestingly, the paper [35] did not use the notion of stopping times (or Markov times, see [3]) for a sequence of random variables, and Theorems 1 and 2 hold for some non-Markov times. However, in the proof of Theorems 3, 4 and 5 a condition close to the Markov property of the time ν was assumed implicitly. This was pointed out by Seits and Winkelbauer in [7] (in which also a refined formulation of Theorems 3, 4 and 5 is given).

When ν is a Markov stopping time with respect to the family of σ-algebras $F_n = \sigma(\xi_1, \ldots, \xi_n)$, $F_0 = \{\varnothing, \Omega\}$ (that is, the event $\{\nu = n\} \in F_n$ for any $n = 1, 2, \ldots$), the results of Kolmogorov's paper have the following generalizations.

Let ξ_k *be a sequence of random variables with finite expectation* $\xi_\nu = \sum_{k=1}^n \xi_k$ *and suppose that*

$$\mathbf{E}\Big[\sum_{k=1}^{\nu} \mathbf{E}(|\xi_k|^\alpha | F_{k-1})\Big]^{1/\alpha} < \infty. \tag{3}$$

for some $\alpha \in [1, 2]$. *Then*

$$\mathbf{E}\Big(\xi_\nu - \sum_{k=1}^{\nu} \mathbf{E}(\xi_k | F_{k-1})\Big) = 0. \tag{4}$$

Moreover, if $\mathbf{E}\xi_k^2 < \infty$ *and* $\mathbf{E}\Big[\sum_{k=1}^{\nu} \mathbf{E}(\xi_k^2 | F_{k-1})\Big] < \infty$, *then*

$$\mathbf{E}\Big[\xi_\nu - \sum_{k=1}^{\nu} \mathbf{E}(\xi_k | F_{k-1})\Big]^2 = \mathbf{E}\Big[\sum_{k=1}^{\nu} \mathbf{E}(\xi_k^2 | F_{k-1})\Big]. \tag{5}$$

Identity (4) was proved by Burkholder and Gundy [8] under condition (3) with $\alpha = 2$; its generalizations for the case $1 \le \alpha \le 2$ are given in the book by Chow and Teicher [9]. Similar results for the case of continuous time were obtained by Novikov [10], [11]. Identities with moments of ξ_n of all orders are considered in the work by Hall [12] and in [10], [11].

Wald identities (1) and (2) for the case of identically distributed summands are trivial consequences of (4) and (5). However, as follows from (3), the first of these relations holds only if

$$(\mathbf{E}\nu^{1/\alpha}) \cdot \mathbf{E}|\xi_k|^\alpha < \infty \textit{ for some } \alpha \in [1, 2].$$

References

1. A. Wald, 'On cumulative sums of random variables', *Ann. Math. Statist.* **15** (1944), 285–287.
2. A.Wald, 'Differentiation under the expectation sign in the fundamental identity of sequential analysis', *Ann. Math. Statist.* **17** (1946), 493–497.
3. A.N. Shiryaev, *Statistical sequential analysis*, Nauka, Moscow, 1976 (in Russian).
4. I.I. Gikhman and A.V. Skorokhod, *Controlled processes*, Naukova Dumka, Kiev, 1977 (in Russian).
5. A.A. Borovkov, *Probabilistic processes in queuing theory*, Nauka, Moscow, 1972 (in Russian).

6. A.A. Novikov, 'A martingale approach to problems of the first hitting time on non-linear boundaries', *Trudy. Math. Inst. Steklov* **158** (1981), 130–152 (in Russian). (Translated as *Proc. Steklov Inst. Math.*)
7. I. Seits and K. Winkelbauer, 'A note to the article by Kolmogorov and Prokhorov "On sums of a random number of random summands"', *Czechoslovak Math. J.* **3** (1953), 89–91.
8. D.L. Burkholder and R.F. Gundy, 'Extrapolation and interpolation of quasilinear operators on martingales', *Acta Math.* **124** (1970), 249–304.
9. Y.S. Chow and H. Teicher, *Probability theory*, Springer-Verlag, New York, 1978.
10. A.A. Novikov, 'On the stopping times of a Wiener process', *Teor. Veroyatnost. i Primenen.* **16**:3 (1971), 458–465 (in Russian). (Translated as *Theory Probab. Appl.*)
11. A.A. Novikov, 'On discontinuous martingales', *Teor. Veroyatnost. i Primenen.* **20**:1 (1976), 13–28 (in Russian). (Translated as *Theory Probab. Appl.*)
12. W.J. Hall, 'On Wald's equation in continuous time', *J. Appl. Probab.* **8**:1 (1971), 59–68.

S-CONVERGENCE (No. 42)

(A.V. Skorokhod)

In paper No. 42 [1] the following, more convenient metric, equivalent to the metric S introduced in §2, was proposed:

$$r_D(f,g) = \inf_{\lambda\in\Lambda} \sup_{0\le t\le 1} [\rho(f(t),g(t)) + \lambda(t) - t],$$

where $f,g \in D_X$; ρ is a metric in X; Λ is the set of all continuous invertible mappings λ of [0,1] onto itself such that $\lambda(0)=0$, $\lambda(1)=1$. In this metric D_X is separable if X is separable, but is not complete.

Billingsley [2] proposed a new kind of metric that, while inducing the same convergence, turns D_X into a complete space (naturally, if X is complete). It is defined in the following way:

$$d_0(f,g) = \inf_{\lambda\in\Lambda} \sup_{\substack{0\le t\le 1\\ 0\le s\le 1}} \left[\rho(f(t),g(t)) + \left|\ln\frac{\lambda(t)-\lambda(s)}{t-s}\right|\right]$$

(we give a simpler equivalent definition).

This convergence on the space of functions defined on $[0,\infty]$ without discontinuities of the second kind, similar to uniform convergence of continuous functions on all finite intervals, is generalized in [3].

References

1. I.I. Gikhman and A.V. Skorokhod, *Introduction to the theory of random processes*, Nauka, Moscow, 1965 (in Russian).
2. P. Billingsley, *Convergence of probability measures*, Wiley, New York, 1968.
3. C. Stone, 'Weak convergence of stochastic processes defined on semi-infinite time intervals', *Proc. Amer. Math. Soc.* **14** (1963), 694–696.

UNIFORM LIMIT THEOREMS (Nos. 43, 51)
(T.V. Arak)

The main theorems of paper No. 51 are now known as the Kolmogorov uniform limit theorems. As is mentioned in the introduction, Theorem 1 (or, equivalently, relation (0.9)) is the third refinement of the corresponding theorem from No. 43. Later, in 1965 Le Cam [1] showed that in Theorem 1 under suitable centralization we can take $D = \exp(n(F - E))$, $C_1 = 132$. In 1973 Ibragimov and Presman [2] gave a new proof with $C_1 = 8$. In parallel with estimates uniform over the set of all distributions, estimates for various narrower classes of distributions of summands were studied. In particular, Meshalkin proved that

$$\sup_{0\leq p\leq 1} \rho(B_p^n, \mathfrak{D}) = O(n^{-2/3}),$$

where B_p^n is the binomial distribution with parameters n and p.

The final solution was obtained by Arak [3], [4]: There exist positive (absolute) constants $\underline{C}$ and $\overline{C}$ such that

$$\underline{C}n^{-2/3} \leq \psi(n) \leq \overline{C}n^{-2/3}$$

(cf. (0.9) and (0.10)).

Using new methods for estimating uniform distances, Zaitsev and Arak [5] also found a definitive quantitative refinement of Theorem 2.

Higher-dimensional versions of Kolmogorov's uniform limit theorems were studied by Sazonov, Presman and Zaitsev.

References

1. 'On the distribution of sums of independent random variables', In: *Bernoulli, Bayes, Laplace (anniversary volume)*, L. le Cam, J. Neyman eds., Springer-Verlag, Berlin-Heidelberg-New York, 1965, pp. 179–202.
2. I.A. Ibragimov and E.L. Presman, 'On the rate of attraction of distributions of sums of independent random variables with accompanying laws', *Teor. Veroyatnost. i Primenen.* **18**:4 (1973), 753–765 (in Russian). (Translated as *Theory Probab. Appl.*)
3. T.V. Arak, 'On the rate of convergence in Kolmogorov's uniform limit theorem' I, II, *Teor. Veroyatnost. i Primenen.* **26**:2 (1981), 225–245; **26**:3, 449–463 (in Russian). (Translated as *Theory Probab. Appl.*)
4. T.V. Arak, 'A refinement of the lower estimate for the rate of convergence in Kolmogorov's uniform limit theorem', *Teor. Veroyatnost. i Primenen.* **27**:4 (1982), 767–772 (in Russian). (Translated as *Theory Probab. Appl.*)
5. T.V. Arak and A.Yu. Zaitsev, 'Estimates of the convergence rate in Kolmogorov's second uniform limit theorem', *Dokl. Akad. Nauk SSSR* **267**:1 (1982), 14–18 (in Russian). (Translated as *Soviet Math. Dokl.*)

CONCENTRATION FUNCTIONS (No. 45)

(V.M. Kruglov)

This remarkable work by A.N. Kolmogorov initiated numerous deep studies on concentration functions. There is now available a monograph [1] devoted specifically to concentration functions. The chapters on inequalities of Kolmogorov type constitute the core of this book. It also gives numerous generalizations and applications of the Lévy concentration function and a fairly detailed bibliography. In what follows we discuss in detail some results connected with Kolmogorov's inequality.

The notion of concentration function was introduced by P. Lévy in his celebrated monograph, published in 1937, "Théorie de l'addition des variables aléatoires", referred to in Kolmogorov's work. In the same work Lévy gave a number of very important properties of these functions. It took some time before the notion of concentration function became recognized and sufficiently known. Apparently, this is the reason why five years later it was introduced once more as a characteristic of probability distributions by Littlewood and

Offord.[1] Studies on refining and generalizing Kolmogorov's inequality were done in several directions. Two of them were pointed out by Kolmogorov himself: one is to obtain an inequality unifying Lévy's inequality with a refinement of it given by Kolmogorov; the other is to analyze concentration functions. As Kolmogorov says in his paper, his interest in concentration functions and their properties was aroused by the problem of approximating convolutions of distributions by infinitely divisible distributions.

The second line of approach relates to a result of Kolmogorov that he did not draw special attention to and is still little known. In any case, we will state it.

Let Q be a concentration function. We set

$$K_Q = \inf\Big\{K : K\frac{Q(x)}{Q(y)} \geq \frac{x}{y},\ y \geq x > 0\Big\}.$$

If $K_Q = 1$, then $Q = pE + (1-p)R$, where $0 \leq p \leq 1$, E is the unit distribution with 0 as support and R is an absolutely continuous distribution.

The first generalizations of Kolmogorov's inequality were obtained by Rogozin in 1961. In particular, in one of his papers he showed that the condition $L^2 \geq l^2 \log s$ in Kolmogorov's theorem can be omitted. Further generalizations and refinements of Kolmogorov's inequality were made by Siegel, Kesten, Le Cam, Miroshnikov, Petrov, Postnikova, Rogozin, Rosen, Sazonov, Sevast'yanov, Teodorescu, Hengartner, Enger, Esseen and others. The first generalization of Kolmogorov's inequality was obtained along the lines indicated by Kolmogorov by invoking combinatorial methods. It was followed by a tendency to use predominantly analytical methods. In the paper by Postnikova and Yudin published in 1978 one inequality of Kolmogorov type was proved by analytical methods making use of certain results of additive number theory; their inequality can be derived from Kesten's inequalities proved by him in 1969 using mainly combinatorial methods.

The latest result of this kind is the inequality by Miroshnikov and Rogozin [2], [3]. Denote by Q the concentration function of the sum $\xi = \xi_1 + \ldots + \xi_n$ of independent random variables, let Q_j be the concentration function of ξ_j and let $\xi_j^{(s)}$ be the symmetrization of the random variable ξ_j.

[1] Here we do not give a reference to their paper and to a number of other papers mentioned in what follows. The interested reader can find them in the monograph [1].

There exists an absolute constant C such that

$$Q(L) \le CL\Big(\sum_{j=1}^{n} \mathbf{E}(\min\{|\xi_j^{(s)}|, h_j\})^2 Q_j^{-2}(l_j)\Big)^{-\frac{1}{2}}$$

for any positive $l_1, \ldots, l_n, L, 2L \ge \max l_j$.

This inequality was proved by analytical methods. Moreover in the generality in which it is proved, and without comparing absolute constants, the Miroshnikov-Rogozin inequality is the strongest one obtained so far.

Of the numerous generalizations of Lévy's concentration function we will give only the one indicated by Enger. Let E be a convex set in Euclidean space R^d. Denote by $P_E(x)$ the Minkowski functional determined by E. We call

$$U(\xi, E) = \int_{R^d} \exp\{-P_E(x)\}\widetilde{P}(dx)$$

the integral concentration function of the random vector ξ, where $\widetilde{P}$ is the distribution of the symmetrization $\xi^{(s)}$.

Recently Miroshnikov [4] proved an inequality for complete concentration functions similar to the one given above.

There exists an absolute constant C such that

$$U(\xi, LE) \le CL\Big(\sum_{j=1}^{n} \mathbf{E}(\min\{P_E(\xi_j^{(s)}), l_j\})^2 U^{-2}(\xi_j, LE)\Big)^{-\frac{1}{2}}$$

for any positive $l_1, \ldots, l_n, L, 2L \ge \max l_j$. Here ξ and ξ_j denote random vectors with values in R^d; the convex set E belongs to a special class of sets which, in particular, includes the sets satisfying the condition

$$P_E(x) = \sum_{i=1}^{m} |A_i x|, \; x \in R^d,$$

where $A_1, \ldots, A_m$ are linear operators in R^d.

Note that the definition of an integral concentration function is of the same nature as that of the average concentration function of Kawata [5]. As a suitable generalization from the one-dimensional [1], p.155, to the multidimensional case, it shows a similarity in nature with the integral concentration function, regarded as integral transforms with certain kernels.

References

1. W. Hengartner and R. Teodorescu, *Concentration functions*, Academic Press, 1973.
2. A.L. Miroshnikov and B.A. Rogozin, 'Inequalities for concentration functions', *Teor. Veroyatnost. i Primenen.* **25**:1 (1980), 178–183 (in Russian). (Translated as *Theory Probab. Appl.*)
3. A.L. Miroshnikov and B.A. Rogozin, 'Remarks on an inequality for concentration functions of sums of independent random variables', *Teor. Veroyatnost. i Primenen.* **27**:4 (1982), 787–789 (in Russian). (Translated as *Theory Probab. Appl.*)
4. A.L. Miroshnikov, 'Inequalities for an integral concentration function', In: *Limit theorems for sums of random variables*, Trudy Inst. Math. **3** (1984), 159–170 (in Russian).
5. T. Kawata, 'The mean concentration function of a chance variable', *Duke Math. J.* **8** (1941), 666–677.

EMPIRICAL DISTRIBUTIONS (No. 15)

(E.V. Khmaladze)

1. This paper is the one in which Kolmogorov's classical statistic

$$D_n = \sqrt{n} \sup_t |F_n(t) - F(t)|$$

is introduced and in which its limit distribution, the Kolmogorov distribution, is established.

By 1933, the empirical distribution function F_n was an attractive object which had received little mathematical attention. There were no precise statements on the convergence of F_n to the distribution function F (see No. 3), and there were no correct results concerning the behaviour of the deviations of F_n from F (see No. 2), except for Pearson's theorem of 1900 on the limit distribution of χ^2 statistics (see [1]).

2. Let us briefly describe two well-known works by Cramér and von Mises illustrating the degree of formalization of the statements concerning deviations of F_n from F obtained by 1933.

In his book [2], published in 1931, von Mises ([10], pp. 316–327) introduced a statistic that he called the "ω^2-statistic":

$$\omega_n^2 = \int [F_n(t) - F(t)]^2 dt.$$

Discussing the "possible deviations" of ω_n^2 for large n, von Mises writes (see, for example, p.320 in [2]), that the theoretical interval of values for ω_n^2 (theoretische Wert) is the interval

$$C_n = (\mathbf{E}\omega_n^2 - 0.674\sqrt{\mathbf{D}\omega_n^2}, \quad \mathbf{E}\omega_n^2 + 0.674\sqrt{\mathbf{D}\omega_n^2}),$$

where 0.674 is the third quartile of the standard normal distribution Φ, that is, $\Phi(0.674) - \Phi(-0.674) = 0.5$. Apparently, this statement must be based on the idea of asymptotic normal distribution of the random variable $(\omega_n^2 - \mathbf{E}\omega_n^2)/\sqrt{\mathbf{D}\omega_n^2}$ which is clearly wrong.

It is interesting to note, however, that for the uniform distribution function on [0, 1],

$$\lim_{n\to\infty} \mathbf{P}\{\omega_n^2 \in C_n\} = 0.60,$$

which is not all that far from 0.5.

In Cramér's paper [3] of 1928, where he also proposed considering the statistic ω_n^2 for checking the hypothesis concerning F, there is no question raised on the limit distribution of the statistic ω_n^2. Instead, the following plan is carried out: assume that F is close to the normal distribution function $\Phi_{\mu,\sigma}$ with mean μ and variance σ^2, that is, F can be represented by a small but unknown number of terms of the Charlier series:

$$F = \Phi_{\mu,\sigma} + P_1\phi_{\mu,\sigma} + \ldots + P_k\phi_{\mu,\sigma}.$$

How many terms of this sum are essential ? To answer this question, Cramér suggested that one should consider the integrals

$$\Delta_{0n}^2 = \int [F_n(t) - \Phi_{\mu,\sigma}(t)]^2 dt,$$

$$\Delta_{1n}^2 = \int [F_n(t) - \Phi_{\mu,\sigma}(t) - P_1(t)\phi_{\mu,\sigma}(t)]^2 dt$$

etc. (up to the fourth order in Cramér's paper). If Δ_{in} decreases "strongly" as compared to $\Delta_{i-1,n}^2$, then the ith term in the decomposition of F is "essential". This technique is applied in the paper to many specific data, with parameters μ and σ^2 being estimated by the same data.

3. It follows from Theorem 1 of No. 15 that $\sup_t |F_n(t) - F(t)| \to 0$ in probability as $n \to \infty$. A stronger statement on the convergence of $\sup_t |F_n(t) -$

$F(t)|$ to zero with probability 1 as $n \to \infty$ (Glivenko's theorem) was published by Glivenko simultaneously with Kolmogorov's paper in the same issue of the journal (see [4]).

As regards the connection between Kolmogorov's theorem and Glivenko's theorem, it is interesting to note that 11 years later, in 1944, Smirnov proved in the fundamental work [6] that Kolmogorov's statistic satisfies the inequality for large deviations

$$\mathbf{P}\{D_n > \lambda\} < 2e^{-2\lambda^2}[1+o(1)], \ \lambda = o(n^{1/6}), \ n \to \infty,$$

which implies Glivenko's theorem.

4. Apparently, the first table of values of the Kolmogorov distribution function was published by Kolmogorov himself.

In 1939 Smirnov published six-digit tables of the Kolmogorov distribution function as an appendix to his paper [7] (see also [5], p. 117). Later these tables were published separately in 1948 in [8] and then reproduced in [9]. Very detailed seven-digit tables of the Kolmogorov distribution function were computed in 1965 and published on pages 267–277 in [5]. The step in these tables is fixed and equals 0.001 and the greatest argument value is 3.000. For this argument the Kolmogorov distribution function equals 0.999 999 969 5.

5. A remarkable lemma given in the paper, stating that the distribution of the statistic D_n does not depend on F_n, directed the further development of the theory of non-parametric goodness-of-fit tests in accordance with the following principle: only functionals of F_n and F whose distribution does not depend on F should be chosen as statistics of goodness-of-fit tests.

Here it is worth recalling the following: in 1900 in constructing the χ^2 statistic,

$$\chi_n^2(k) = n\sum_{i=1}^{k} \frac{[\Delta F_n(t_i) - \Delta F(t_i)]^2}{\Delta F(t_i)}, \quad t_0 < t_1 < \ldots < t_{k+n},$$

Pearson [1] had chosen the coefficients $1/\Delta F(t_i)$ so that the limit distribution of this quadratic form does not depend on the probabilities $\Delta F(t_i) = F(t_{i+1}) - F(t_i)$ and consequently, on the division points t_i and the distribution function F. Subsequently many papers appeared that established that these statistics were asymptotically normal. Clearly it suffices to normalize these statistics to obtain random variables with the standard limit distribution. In 1931, perhaps under the influence of this trend, von Mises [2] considered the statistics

$$\omega_n^2(\lambda) = \int [F_n(t) - F(t)]^2 \lambda(t) dt$$

and discussed only two weight functions $\lambda : \lambda_1(t) = n/f(t)$ where f is the density of a distribution function F and $\lambda_2(t) = 1/\mathbf{E}[\int [F_n(t) - F(t)]^2 dt]$. The choice of λ_1 was regarded as being analogous to the χ^2 statistic but unsuitable, since the integral in $\omega_n^2(\lambda_1)$ "usually" diverges, whereas λ_2 was considered suitable, since it standardized the means of the ω^2 statistics. The limit distribution of the $\omega_n^2(\lambda_2)$, however, clearly depends on F.

Then, in 1933, the above-mentioned lemma was formulated, and in 1937 Smirnov in [10], with reference to Glivenko, finally introduced the statistics $\omega_n^2(\lambda)$, where $\lambda(t) = \phi[F(t)]f(t)$, with distribution independent of F. Since then, or perhaps since 1952, after the papers by Anderson and Darling [11], this principle became standard in constructing non-parametric tests.

When a statistical hypothesis fixes a certain family $\mathbf{F}$ of distribution functions instead of a specific distribution function F, the realization of this principle becomes somewhat difficult. A brief description of papers dealing with this problem and a very incomplete bibliography, to which Gikhman's paper [12] should be added, can be found, for example, in [13], §2.1. In the same §2 it is described how to construct functionals of F_n and $\mathbf{F}$ whose limit distribution does not depend on $\mathbf{F}$.

When F is a continuous distribution in the multidimensional space R^m, $m > 1$, the lemma does not hold: it is false that the random variable $F(X_1, \ldots, X_m)$ has the same distribution for all F if the random vector $(X_1, \ldots, X_m)$ has continuous distribution function F. Therefore there seems to be no standard way for constructing non-parametric goodness-of-fit tests for distributions in multidimensional space, unlike the one-dimensional case. Still, we point out two modern [14], [15] and one old [16] paper on this subject.

6. The convolution (16) and the recurrence formula (9) of paper No. 15 were later extensively used for calculations for finite n. In particular, in 1950 Massey [17] was the first to use (16) to calculate small tables for the distribution function of the Kolmogorov statistic for $5 < n < 80$. Quite recently (see [18]) a recurrence formula similar to (9) was used for calculating the probability of F_n remaining within domains with various curvilinear boundaries.

Speaking of tabulating Φ_n, we note that modern tables usually contain the table of values of distribution functions Φ_n and the percentage points for $n \leq 100$. The known asymptotic formulas, however, fit even for $n \geq 30$. This side of the matter is discussed in the explanation part of [9], where appropriate references are also given.

7. We introduce the so-called homogeneous empirical process v_n, $v_n(t) = \sqrt{n}[F_n(t) - t]$ and Brownian bridge v, $v(t) = w(t) - tw(1)$, where w is a Wiener process, $t \in [0,1]$, and we consider the following functional of the empirical process:

$$K(v_n) = \sup_t \frac{|v_n(t)|}{h(t)}, \quad h \geq 0.$$

Let

$$P_n(\lambda, h) = \mathbf{P}\Big\{\forall t \in [0,1] : |F_n(t) - t| \leq \frac{\lambda}{\sqrt{n}} h(t)\Big\},$$

$$\Phi_{n,K}(\lambda, K) = \mathbf{P}\{K(v_n) \leq \lambda\}.$$

Clearly, $P_n(\lambda, h) = \Phi_{n,K}(\lambda, K)$, that is, the calculation of the probability that the empirical process remains within a certain boundary is equivalent to calculating the distribution functions for the functionals $K(v_n)$. These problems were solved in different ways, however.

The first problem, namely, that of studying the probability of remaining within a certain domain, is reviewed in the well-known papers by Gikhman and Gnedenko [19] and Borovkov and Korolyuk [20]. In particular, they describe Gikhman's work on limit theorems for the probability of remaining within the boundary for processes converging to diffusion processes (these works are very close in their ideas to Kolmogorov's paper) and the papers by Gnedenko and Korolyuk that make use of random walk methods.

We recall the development of the second problem. In No. 15 the limit for the distribution function of the Kolmogorov statistic was established: $\Phi_{nK} \Rightarrow \Phi$ for $h = 1$. However, the probabilistic "construction" of Kolmogorov distribution function Φ, namely the equality

$$\Phi(\lambda) = \mathbf{P}\{K(v) < \lambda\}, \tag{1}$$

remained unclear for a long time. In any case, it remained unclear of what use such a "construction" could be.

With time other statistics were discovered which, as was noticed later, can be conveniently represented as functionals of an empirical process. In particular, in 1939, Wald and Wolfowitz [22] considered the so-called weighted Kolmogorov statistic, that is, the functional $K(v_n)$ with weight function $h \neq 1$, while in 1937 Smirnov [10] considered the ω^2 statistic, that is, the functional

$$\omega^2(v_n) = \int_0^1 [v_n(t)]^2 dt. \tag{2}$$

It was natural to single out what the convergence of the distributions of various functionals of v_n had in common, namely, convergence of v_n to v in distribution.

The discovery of the fact that it is natural to represent various statistics as functionals of an empirical process was only one important feature of the matter. Another was to foresee that general efficient methods could be worked out for proving convergence in distribution of processes v_n to a process v and, in general, for proving weak convergence in function spaces.

Both methods were ripe in the late 1940's: on 30 November 1948 (as reported in 'Uspekhi Matematicheskikh Nauk" (Russian Mathematical Surveys), **4**:2 (1949), p. 173), A.N. Kolmogorov gave a lecture for the Moscow Mathematical Society called "Measures and distributions in function spaces", which discussed, among other matters, the problems of determining weak convergence of measures in function spaces. On 29 January 1949, Smirnov gave a talk on the "Cramér-von Mises test" [23] (see also [5], p.200), where again, after an interval of 12 years, he obtained the limit distribution of ω^2 statistics, this time using the representation (2) and the corresponding Parseval identity. Finally, in September 1949, Doob's well-known paper [24] appeared. Together with the description of the general viewpoint, to the effect that the convergence $\Phi_n \Rightarrow \Phi$ is a consequence of the convergence of v_n to v in distribution, he also proved (1).

This started the useful development of the theory of weak convergence in function spaces that presently constitutes the basis for the limit theorems in non-parametric statistics.

I recommend the small monograph by Durbin [25] and the review article by Gaensaler and Stute [26] for an acquaintance with the state of affairs in this field.

8. Let us make several general concluding remarks.

A. Kolmogorov's statistic, as well as the statistic $K(v_n)$ (see No. 7), stands apart from the other statistics of non-parametric goodness-of-fit tests in that it leads to confidence sets that are easy to visualize. Thus, for example, the confidence set $\{F' : n\int[F_n(t) - F'(t)]^2 dF'(t) < \lambda\}$ constructed by means of the statistic ω^2 is not graphic and therefore inconvenient. By contrast, the confidence set

$$\{F' : \sqrt{n}\sup|F_n(t) - F'(t)| < \lambda\} =$$
$$= \left\{F' : F_n(t) - \frac{\lambda}{\sqrt{n}} < F'(t) < F_n(t) + \frac{\lambda}{\sqrt{n}}\right\}$$

constructed with the help of Kolmogorov's statistic is easy to perceive and is widely used. This property of Kolmogorov's statistic was observed very early: back in 1939 Wald and Wolfowitz [22] introduced the confidence sets

$$\left\{F' : F_n(t) - \frac{1}{\sqrt{n}}\delta_n(F_n(t)) < F'(t) < F_n(t) + \frac{1}{\sqrt{n}}\delta_n(F_n(t))\right\}$$

relating to weighted Kolmogorov statistics.

B. In the same year 1939, two papers by Smirnov [27], [7] appeared. They gave distributions for the statistics $D_n^+ = \sqrt{n}\sup[F_n(t) - F(t)]$ and $D_{n_1n_2} = \sqrt{n}\sup|F_{1n_1}(t) - F_{2n_2}(t)|$, $n = n_1n_2/(n_1+n_2)$, namely,

$$\mathbf{P}\{D_n^+ < \lambda\} \to 1 - e^{-2\lambda^2}, \quad \mathbf{P}\{D_{n_1n_2} < \lambda\} \to \Phi(\lambda). \tag{3}$$

Here F_{1n_1} and F_{2n_2} are the empirical distribution functions constructed from two independent sequences of independent identically distributed random variables.

It is interesting that both statements (3), which are the best known statements in non-parametric statistics, were given as simple consequences of general theorems on the number of intersections of the curves F_n and $F + \lambda/\sqrt{n}$ and the curves F_{1n_1} and $F_{2n_2} \pm \lambda/\sqrt{n}$ respectively.

In 1944 Smirnov [6] gave the distribution of the statistic D_n^+ for finite n. In 1955 Korolyuk [28] obtained the distribution of the statistic $D_{n_1n_2}$ for finite n_1 and n_2 and in 1962 Borovkov [29] established the asymptotic expansions for the distributions and, in particular, for the probabilities of large deviations of $D_{n_1n_2}$. The specific use of certain results from [29] and [21] was demonstrated in [9], p. 84.

Apparently, Massey [30] was the first to give a table for the distribution of the two-sample Smirnov statistic for finite n_1 and n_2. Tables of percentage points of $D_{n_1n_2}$ for $n_1 < n_2 \leq 50$ are published by Borovkov, Markov and Sychev [31]. A remarkable example of the application of the Kolmogorov statistic to the verification of the hypothesis on the distribution function F was given in 1940 by Kolmogorov himself [32].

References

1. K. Pearson, 'On the criterion that a given system of deviations from the probable in the case of a correlated system of variables is such that it can be reasonably supposed to have arisen from random sampling', *Phil. Mag.* **5**:50 (1900), 157.
2. R. von Mises, *Warscheinlichkeitsrechnung und ihre Anwendungen in der Statistik und theoretischen Physik*, Franz Deuticke, Leipzig, Vienna, 1931.
3. H. Cramér, 'On the composition of elementary errors. Second paper: statistical applications', *Skand. Aktuarietidskr.* **1/2** (1928), 141–180.
4. V. Glivenko, 'Sulla determinazione empirica delle leggi di probabilita', *G. Ist. Ital. Attuar.* **4**:1 (1933), 92–99.
5. N.V. Smirnov, *Probability theory and mathematical statistics: Collected works*, Nauka, Moscow, 1970 (in Russian).
6. N.V. Smirnov, 'Approximation of distribution laws of random variables by empirical data', *Uspekhi Mat. Nauk* No.10 (1944), 179–206 (in Russian).
7. N.V. Smirnov, 'An estimate of the discrepancy between empirical distribution curves in two independent samples', *Byull. Moskov. Gos. Univ. Math. and Mech.* **2** (1939), 3–14 (in Russian).
8. N.V. Smirnov, 'Tables for estimating the goodness of fit of empirical distribution', *Ann. Math. Statist.* **19**:2 (1948), 279–281.
9. L.N. Bolshev and N.V. Smirnov, *Tables of mathematical statistics*, Nauka, Moscow, 1983 (in Russian).
10. N.V. Smirnov, 'On the distribution of von Mises' ω^2-test', *Mat. Sb.* **2**:1 (1937), 973–993 (in Russian).
11. T.W. Anderson and D.A. Darling, 'Asymptotic theory of certain "goodness of fit" criteria based on a stochastic process', *Ann. Math. Statist.* **23**:1 (1952), 193.
12. I.I. Gikhman, 'Remarks on Kolmogorov's goodness of fit test', *Dokl. Akad. Nauk SSSR* **91**:4 (1953), 715–718 (in Russian).
13. E.V. Khmaladze, 'Certain applications of martingale theory in statistics', *Uspekhi Mat. Nauk* **37**:6 (1983), 194–212 (in Russian). (Translated as *Russian Math. Surveys*)

14. P.J. Bickel and L. Breiman, 'Sums of functions of nearest neighbour distances, moments bounds, limit theorems and a goodness of fit test', *Ann. Probab.* **11** (1983), 185–214.

15. M.F. Schilling, 'Goodness of fit testing in R^m based on the weighted empirical distribution of certain nearest neighbour statistics', *Ann. Statist.* **11**:1 (1983), 1–12.

16. M. Rosenblatt, 'Remarks on multivariate transformations', *Ann. Math. Statist.* **23**:3 (1952), 470–472.

17. F.J. Massey, Jr., 'A note on the estimation of a distribution by confidence limits', *Ann. Math. Statist.* **21**:1 (1950), 116–118.

18. V.F. Kotel'nikova and E.V. Khmaladze, 'Calculating the probability of an empirical process remaining within a curvilinear boundary, *Teor. Veroyatnost. i Primenen.* **27**:3 (1982), 599–607 (in Russian). (Translated as *Theory Probab. Appl.*)

19. I.I. Gikhman and B.V. Gnedenko, 'Mathematical statistics', In: *Mathematics in the USSR during forty years:* 1917–1957, Vol. 1, Fizmatgiz, Moscow, 1959 (in Russian).

20. A.A. Borovkov and V.S. Korolyuk, 'On results of asymptotic analysis concerning boundary problems', *Teor. Veroyatnost. i Primenen.* **10**:2 (1965), 255–266 (in Russian). (Translated as *Theory Probab. Appl.*)

21. V.S. Korolyuk, 'Asymptotic analysis of the distribution of maximal deviations in a Bernoulli scheme', *Teor. Veroyatnost. i Primenen.* **4**:4 (1959), 369–397 (in Russian). (Translated as *Theory Probab. Appl.*)

22. A. Wald and J. Wolfowitz, 'Confidence limits for continuous distribution functions', *Ann. Math. Statist.* **10**:2 (1939), 105–118.

23. N.V. Smirnov, 'On the Cramér-von Mises test', *Uspekhi Mat. Nauk* **4**:4 (1949), 196–197 (in Russian).

24. J. Doob, 'Heuristic approach to the Kolmogorov-Smirnov theorems', *Ann. Math. Statist.* **20**:3 (1949), 393–403.

25. J. Durbin, 'Distribution theory for tests based on the sample distribution function', *Regional Conf. Ser. in Appl. Math., 9th issue*, SIAM, 1973.

26. P. Gaensaler and W. Stute, 'Empirical processes: A survey of results for independent and identically distributed random variables', *Ann. Probab.* **7**:2 (1979), 193–243.

27. N.V. Smirnov, 'On the deviations of the empirical distribution curve', *Mat. Sb.* **6** (48):1 (1939), 3–24 (in Russian).
28. V.S. Korolyuk, 'On the divergence of empirical distributions for the case of two independent samples', *Izv. Akad. Nauk SSSR Ser. Mat.* **19**:1 (1955), 81–96 (in Russian).
29. A.A. Borovkov, 'On the problem of two samples', *Izv. Akad. Nauk SSSR Ser. Mat.* **26**:4 (1962), 605–624 (in Russian).
30. F.J. Massey, Jr., 'Distribution tables for the deviation between two sample cumulatives', *Ann. Math. Statist.* **23**:3 (1952), 435–441.
31. A.A. Borovkov, N.P. Markov and N.M. Sycheva, *Tables for N.V. Smirnov's test of uniformity of two samples*, Izdat. Sibirsk. Otdel. Akad. Nauk SSSR, Novosibirsk, 1964 (in Russian).
32. A.N. Kolmogorov, 'On a new proof of Mendel's laws', *Dokl. Akad. Nauk SSSR* **27** (1940), 38–42 (in Russian).

THE METHOD OF LEAST SQUARES (Nos. 30, 31)
(M.B. Malyutov)

The method of least squares is one of the most popular statistical methods for parameter estimation in applications. Suffice it to say that specialists in geodesy, navigation, artillery, etc., even nowadays write their own textbooks on this method (which do not always meet the standards of modern statistics).

At the same time, before No. 30 was published, the mathematical presentations of this method were not based on modern geometrical ideas of higher-dimensional geometry and were essentially no different from the initial method of Gauss. It is in No. 30 that the connection of the method with orthogonal projection onto a subspace in R^N was explicitly used and it was discovered which properties of the method depend on the assumption of normality of the distribution of measurements and which are true for orthogonal observations. For applications it is important to present in detail the confidence intervals for the parameters and for the unknown variance of measurements and to supplement them with the tables of Student and χ^2-distributions.

Paper No. 30 played an important role in developing the mathematical theory of the method of least squares in the USSR and putting into some order the applications of this method. In particular, its influence can be seen in the books by Linnik [1] and Romanovskii [2]. The latter, for instance, reproduces with a detailed proof Kolmogorov's report "The real meaning of results of the analysis of variance" (Proceedings of the 2nd All-Union Meeting in Mathematical Statistics), which generalized the results of No. 30 to the theory of the analysis of variance and criticized certain not sufficiently rigorous formulations by Fisher.

Paper No. 30 remained unnoticed abroad and the modern presentation of the method of least squares was published only in 1959 by Scheffé [3].

Paper No. 31 clarifies a formula of Gauss for the method of least squares. Gauss found a formula for the variance Ds^2 of the standard estimator s^2 of the variance σ^2 for independent measurements y_i in the linear regression model. This formula includes the fourth moment $f\sigma^4$, and therefore Gauss obtained upper and lower bounds for Ds^2 that hold for any f. However, the lower bound was rough (sometimes even negative). In No. 31 a best possible lower bound for Ds^2 is found.

The presentation is essentially based on the geometric theory of the method of least squares, developed by A.N. Kolmogorov previously in No. 30 (for example, an important role is played by the orthogonality of certain matrices). With small modifications this work is reproduced in the book by Linnik mentioned above. In the subsequent literature main emphasis is laid on constructing asymptotic confidence intervals for σ^2, where instead of f an estimator of it with respect to the same sample is used [4].

References

1. Yu.V. Linnik, *The method of least squares and the foundations of the mathematical and statistical theory of observation processing*, Fizmatgiz, Moscow, 1952 (in Russian).
2. V.I. Romanovskii, *Mathematical statistics*, Vol. 2, Izdat. Akad. Nauk UzSSR, 1963.
3. H. Scheffé, *Analysis of variance*, Wiley, New York, 1959.
4. H. Drügas, 'Asymptotic confidence intervals for the variance in the linear regression model', In: *Statistics and probability: Essays in honour of C.R. Rao*, North-Holland, Amsterdam, 1982, pp.233–239.

UNBIASED ESTIMATORS (No. 38)

(Yu.K. Belyaev and Ya.P. Lumel'skii)

The notion of unbiased estimator is fundamental in mathematical statistics. Although this notion was known earlier, No. 38 gives the first systematic treatment of the properties of unbiased estimators and various methods for constructing them in terms of sufficient statistics.

Kolmogorov's paper and also [1] graphically demonstrate the applied value of unbiased estimators in problems of statistical control. Kolmogorov was the first to apply unbiased estimators to the determination of the effectiveness of practically used schemes of sample control with respect to an alternative feature. The construction of unbiased estimators for detecting any missed defective products, as well as for the *a priori* distribution of the number of defective items in the batches subject to control, is considered in [2]–[4]. Unbiased estimators were obtained for control with respect to a qualitative feature with known classification errors [5]. The construction of unbiased estimators for control with respect to alternative qualitative features, that is, for classifying into k quality groups, was considered in detail in [3], [4], where a vast bibliography is given. The unbiased estimators of the main control indices are included in GOST 24660-81 (State Standards).

The unbiased estimator of the normal distribution density obtained by A.N. Kolmogorov is widely used in the problem of control based on a quantitative feature [6]. Subsequently this result was generalized to the multidimensional normal distribution [7]–[9] and to problems of statistical classification [10].

A.N. Kolmogorov drew attention to the paper [11] in which unbiased estimators were constructed for binomial random walks. Later, problems of obtaining unbiased estimators and the completeness of the families of distributions generated by the first entrance schemes for various schemes of random walks were considered in [12]–[14], [5].

The upper and lower bounds introduced by Kolmogorov can be effectively used also in the cases when unbiased estimators do not exist. This is the case with estimating the number of missed defective items for the binomial distribution and a one-step control scheme. Upper and lower bounds for various functions of an unknown parameter, as well as estimators with minimal bias, were obtained in [15]–[18].

References

1. A.N. Kolmogorov, *Statistical inspection control for the admissible number of defective items equal to zero*, Leningrad, 1951, pp. 1–24 (in Russian).
2. S.Kh. Sirazhdinov, 'One-step statistical inspection control', *Proc. Inst. Math. Mekh. UzSSR Acad. Sci.* **15** (1955), 41–56 (in Russian).
3. Yu.K. Belyaev, *Probability methods of sample control*, Nauka, Moscow, 1975 (in Russian).
4. Ya.P. Lumel'skii, *Statistical estimators for results of quality control*, Izdat. Standart., Moscow, 1979 (in Russian).
5. Yu.K. Belyaev, 'Sequential estimators under sample control with respect to qualitative features with classification errors', *Dokl. Akad. Nauk SSSR* **231**:3 (1976), 521–524 (in Russian). (Translated as *Soviet Math. Dokl.*)
6. G.J. Liberman and G.J. Resnikoff, 'Sampling plans for inspection by variables', *J. Amer. Statist. Assoc.* **50**:270 (1955), 457–516.
7. Ya.P. Lumel'skii, 'Unbiased sufficient probability estimates for the multidimensional normal law', *Vestnik Moskov. Gos. Univ. Mat.* No.6 (1968), 14–17 (in Russian).
8. Ya.P. Lumelskii and P.N. Sapozhnikov, 'Unbiased estimators for probability densities', *Teor. Veroyatnost. i Primenen.* **14**:2 (1969), 372–380 (in Russian). (Translated as *Theory Probab. Appl.*)
9. S.G. Ghurye and J. Olkin, 'Unbiased estimation of some multivariate probability densities and related functions', *Ann. Math. Statist.* **40**:4 (1969), 1261–1271.
10. R.A. Abusev and Ya.P. Lumel'skii, 'Unbiased estimators and problems of classifying multidimensional normal sets', *Teor. Veroyatnost. i Primenen.* **25**:2 (1980), 381–389 (in Russian). (Translated as *Theory Probab. Appl.*)
11. M.A. Girshick, F. Mosteller and L.J. Savage, 'Unbiased estimates for certain binomial sampling problems with applications', *Ann. Math. Statist.* **17**:1 (1946), 13–23.
12. Yu.V. Linnik and I.V. Romanovskii, 'On the theory of sequential estimation', *Dokl. Akad. Nauk SSSR* **194**:2 (1979), 270–272 (in Russian). (Translated as *Soviet Math. Dokl.*)
13. Ya.P. Lumel'skii, 'Random walks satisfying generalized urn schemes', *Dokl. Akad. Nauk SSSR* **209**:6 (1973), 1281–1284 (in Russian). (Translated as *Soviet Math. Dokl.*)

14. Yu.K. Belyaev and A.N. Mal'kov, 'Completeness of families of probability measures in sampling by first entrance schemes', *Dokl. Akad. Nauk SSSR* **225**:1 (1975), 27–30 (in Russian). (Translated as *Soviet Math. Dokl.*)
15. S.Kh. Sirazhdinov, 'On estimators with minimum bias for the binomial distribution', *Teor. Veroyatnost. i Primenen.* **1**:1 (1956), 168–174 (in Russian). (Translated as *Theory Probab. Appl.*)
16. L.N. Bolshev, 'On probability estimators', *Teor. Veroyatnost. i Primenen.* **5**:4 (1960), 453–457 (in Russian). (Translated as *Theory Probab. Appl.*)
17. Yu.K. Belyaev and Ya.P. Lumel'skii, 'Estimators of polynomials based on limited plans in schemes of random walks', *Izv. Akad. Nauk SSSR Ser. Tekhn. Kibernet.* No.1 (1979), 61–67 (in Russian).
18. Yu.K. Belyaev and T.N. Dugina, 'On estimating the total number of defective items accepted under inspection with errors', *Elektron. Tekhn. Ser.* 8 No.5 (83) (1980), 24–29 (in Russian).

STATISTICAL PREDICTION (No. 18)

(A.M. Yaglom)

The first attempts at the statistical prediction of future values of certain meteorological parameters using linear regression equations to give a future value of the quantity Δy of interest in the form of a simple linear combination of the $\Delta x_1, \dots \Delta x_k$ known from observations made in the past or present were started in the 1920's. (In addition to Bauer's paper of 1925 referred to by A.N. Kolmogorov, we can also refer to later works [1]–[6] containing a number of additional references.) At first sight this prediction method seems to be quite simple: it requires only preliminary estimates of a certain number of correlation coefficients which determine the unknown coefficients $a_1, \dots, a_k$ in the regression equations and does not require cumbersome calculations such as those involved in "dynamic weather forecasting" based on numerical solution of partial differential equations approximately describing the dynamics of the atmosphere. Here, the only problem is to choose appropriate predictors $\Delta x_1, \dots, \Delta x_k$, that is, atmospheric characteristics in the past and present whose values are used for predicting Δy. It turns out, however, that to choose appropriate predictors is not that simple.

Indeed, it might seem that it is better to take as many predictors as possible, since in that case the forecast uses very broad initial information and will be accurate with high probability. Unfortunately, the empirical correlation coefficients r_i and r_{ij} between Δy and Δx_i, $i = 1, \ldots, k$, and between the pairs Δx_i and Δx_j, $j = 1, \ldots, k$, used for calculations are not exact and depend on the volume (and quality) of the available empirical data. Therefore, the coefficients $a_1, \ldots, a_k$ for which the linear combination $a_1 \Delta x_1 + \ldots + a_k \Delta x_k$ corresponding to these data gives the best approximation of Δy are not exact, that is, they are not the best for the whole sample of random variables $(\Delta x_1, \ldots, \Delta x_k, \Delta y)$ and applied to the subsequent independent sample, these values may be appreciably less appropriate. To avoid this, the choice of predictors should satisfy a number of special conditions. Here we will only consider the two most important of them.

First, in order that the accuracy of the determination of the coefficients $a_1, \ldots, a_k$ should not be too low, the total number k of predictors should be rather small as compared to the volume of the sample used for determining the coefficients r_i, r_{ij} and a_i. Thus, for example, if the order of the sample's volume is one or two hundred, then k should not exceed several units. This condition was intuitively clear even for the first researchers of statistical weather forecasting (see, for example, Kolmogorov's remark to the effect that when using a sample with 30–50 average annual values, k is usually chosen to vary from 3 to 7); later, the well-known American meteorologist Lorenz called it "the taboo of statistical forecasting". However, apart from this, there is another essential requirement, which is clarified in No. 18 and which is by no means always taken into account, even now. This second requirement is to forbid searching among a large number of systems of predictors (even if each of these systems contains only a small number of values) in order to choose the best of these systems. The point is that if we try a large number of various systems of predictors, then with high probability at least in one of them the empirical value of the cumulative correlation coefficient between these predictors and Δy is much larger than its true value. In this case it is quite possible that this system of predictors will be chosen for forecasting. However, as applied to a subsequent independent sample the empirical correlation coefficient of Δy and $\Delta u = a_1 \Delta x_1 + \ldots + a_k \Delta x_k$ will most probabably be essentially smaller than that of the initial sample. Kolmogorov had good reason to assert that this "blow-up of the maximal empirical correlation coefficient" will very often take place when searching for a large number of systems of predictors, which readily explains the failures of numerous attempts to use linear regression equations for long-term weather forecasting. As a "model example" that allows exact calculations, he analyzed the case when all the values under consideration have a multidimensional normal probability distribution, the initial sample has fixed size N and we check i independent systems of k predictors, for each of which the joint correlation coefficient with Δy takes the same value ρ. Here λ is such that the maximal value of the empirical composite correlation coefficient for our i groups of predictors to exceed λ with a certain given probability (say, $\frac{1}{2}$), increases rapidly as i increases, and at moderately high values of i exceeds ρ (see, for example, the table for the case $N = 42$, $k = 5$ and $i = 14$). We note also that the accurate derivation of Fisher's results used by Kolmogorov in his calculations can be found, for example, in the well-known monograph [7].

The considerations given in Kolmogorov's paper question the advantage of certain new methods of statistical weather forecasts proposed long after the paper was published. A typical example here is the "screening procedure" that was advocated, for example, by Miller [4], [8]. The essence of the method is to consider first a very large aggregate of predictors which is "screened" in the following way. First we take out the predictor (denote it by Δx_1) that corresponds to the largest empirical correlation coefficient with Δy to be forecast; then we take the one of those remaining (say Δx_2) that makes the largest contribution to the joint correlation coefficient of the pair $(\Delta x_1, \Delta x_2)$ with Δy; then we take the one of those remaining (say, Δx_3) that corresponds to the largest joint correlation coefficient of the three values $(\Delta x_1, \Delta x_2, \Delta x_3)$ with Δy, and so on. Usually, after several steps the addition to the joint correlation coefficient obtained at the next step appeared to be insignificant, so the procedure could well be terminated. Usually, the number of selected predictors is small enough not to contradict "the taboo" but, in view of the considerations published by Kolmogorov in 1933, in the case of a large initial set of possible predictors it is quite possible that for the selected k of them the empirical value of the joint correlation coefficient will only occasionally be large, so that on passing to independent samples, we obtain disappointing results.

The same is true for another widely used method of selecting only a small number of the most significant predictors, the method of empirical orthogonal functions, proposed in [9], [10] (see also [6], [11], [12]), which actually coincides with the method of principal components of multidimensional statistical analysis (cf. [7], Chap. 11). In this method, first a linear combination $u_1 = \Delta x_1$ of predictors with the largest variability (that is, variance) is chosen from the initial set of all possible "virtual predictors", then only linear combinations of predictors uncorrelated with u_1 are considered, and the combination $u_2 = \Delta x_2$ with greatest variance is chosen, etc. Usually after selecting a small number k of linear combinations $u_1, u_2, \ldots, u_k$ the variance of all the others is so small that they might as well not be considered. In cases when the original number of admissible predictors is comparatively small (as, for example, in Obukhov's work [10] where the number is 5), this method results in a notable simplification of the calculations (for example, allowing one to confine oneself to considering only the two most significant linear combinations of the five variables). If, however, we first consider a large number of predictors (for example, all the values of all or several meteorological fields for a large network of stations or at the nodes of a regular network), then we can again apply Kolmogorov's considerations, which suggest that the combinations are optimal only for the sample for which they were calculated, and on passing to an independent sample the results become much worse. It is quite possible that this accounts for some of the disappointing conclusions on statistical forecasting using empirical orthogonal functions contained in [4].

In conclusion, we should emphasize that Kolmogorov's work clearly demonstrates the need for a thorough study of the selective properties of systems of predictors used for developing methods of statistical weather forecasting. This study was started by A.N. Kolmogorov in 1933; unfortunately, many problems remain open even now.

References

1. F. Baur, 'Der gegenwärtige Stand der meteorologischen Korrelationsforschung', *Meteor. Z.* **47** (1930), 42–52.
2. F. Baur, *Einführing in die Grosswetterkunde*, Wiesbaden 1948.
3. G.P. Wadsworth, ' Application of statistical methods to weather forecasting', In: *Compendium of meteorology*, Boston, 1951, pp. 849–855.
4. E.N. Lorenz, 'Prospects for statistical weather forecasting', *Statist. Forecasting Proj. Final Rept.*, Cambridge (Mass.), 1959.
5. A.M. Yaglom, 'Statistical methods of extrapolating meteorological fields', In: *Proc. All-Union Meteor. Congr.*, Vol. 2, (Dynamic Meteorology),

Gidrometeoizdat, Leningrad, pp. 221–234 (in Russian).

6. P.N. Belov, 'Statistical methods of numerical weather forecasting', In: *Numerical methods of weather forecasting*, Gidrometeoizdat, Leningrad, 1975, pp. 259–281; 386–387 (in Russian).
7. T. Anderson, *Introduction to multivariate statistical analysis*, Wiley, 1958.
8. R.G. Miller, 'Statistical prediction by discriminant analysis', *Meteorol. Monogr.* 4:25 (1962).
9. E.N. Lorenz, 'Empirical orthogonal functions and statistical weather prediction', *Statist. Forecasting Proj. Sci. Rept.* No. 1, Cambridge (Mass.), 1976.
10. A.M. Obukhov, 'On statistically orthogonal expansions of empirical functions', *Izv. Akad. Nauk SSSR Ser. Geofiz.* No. 3 (1960), 432–439 (in Russian).
11. A.V. Mesherskaya, L.V. Rukhovets, M.I. Yudin and N.I. Yakovleva, *Natural components of meteorological fields*, Gidrometeoizdat, Leningrad, 1970 (in Russian).
12. M.I. Fortus, 'The method of empirical orthogonal functions and its application in meteorology', *Meteorology and Hydrology* No. 4 (1980), 113–119 (in Russian).

ON INTER-BED WASHOUT (No. 37)

(A.B. Vistelius)

The problem solved by A.N. Kolmogorov in this paper was posed while studying productive rock-masses (strata) of the Apsheronskii peninsula and red-rock oil-bearing sediments on Cheleken. These strata are composed exclusively of terrogeneous formations. They include, on the Apsheronskii peninsula, sparse conglomerate beds, while at Cheleken the strata consist completely of sands, aleurites and clays. The depressions in lower beds are filled by the shelves of the higher ones. Moreover, the bases of thick sand beds often contain fragments composed of clays that are identical in their appearance to the clays of the clay bed underlying the sand beds. These fragments are rounded, and they are considered by all researchers as the traces of washout of the lower bed when the upper one was formed. This pattern is especially typical of Cheleken.

These features of the strata leave no doubt that when they were formed, the lower beds were washed out during the formation of the upper ones ("Inter-bed washout"). So there were sound reasons for posing the problem solved by Kolmogorov and it has retained its importance to this day. At the same time one should bear in mind that when Kolmogorov's paper was published, a number of geologists believed that the thicknesses of beds in a profile do not depend on the material of the beds. In general, this opinion turned out to be false. Moreover, when the paper was published, geologists did not operate with random variables, probability distribution functions and sequences of values of a random variable. It was the period in which the foundation of a number of geological disciplines had just been laid, based on the notion of the stochastic character of the values studied. This fundamental restructuring later gave rise to mathematical geology and was largely as a result of Kolmogorov's paper, as well as to personal advice and remarks by him during 1945–1950.

For the application of No. 37 to the solution of the problem relating to the mechanism of bed formation, three requirements must be met:

a) numerical solutions of Kolmogorov's equation must be found;

b) the applicability of Kolmogorov's axioms (which, from the modern viewpoint, does not have a universal character at all) to specific profiles should be confirmed;

c) a model of bed-formation that gives $G(x)$ should be available. To obtain reasonable results, this model should be derived from bed-formation conditions and should allow testing based on data of observations.

Condition a) is in no way restrictive. Kolmogorov's equation coincides with the Wiener-Hopf equation, and there is extensive literature on numerical methods for solving this latter one. In any case, even in the late 1940's there were quite a number of methods for solving this problem. An appropriate algorithm was chosen immediately after Kolmogorov's paper was published, and it was used to compute a number of examples. These data were not published, since no function could be taken for $G(x)$.

The axioms adopted by Kolmogorov (condition b)) are as follows:

1) the random variables δ_n and ϕ_{n+1} are independent, and all the δ_n have the same distribution law $\mathbf{P}\{\delta_n < x\} = G(x)$;

2) the expectations $\mathbf{E}\delta_n = \int_{-\infty}^{\infty} x dG(x)$ are positive;

3) the distribution of δ_n is continuous, that is, it can be expressed in terms of a corresponding probability density $g(x)$ by the formula $G(x) = \int_{-\infty}^{x} g(x)dx$.

Recall that δ_n is the difference between the thickness of the sediment and the subsequent depth of the washout of this sediment directly after its accumulation is finished. The sequence of beds in the profile may have certain beds with thickness contained in the sequence δ_n. It is not clear how to identify the layers with thicknesses from the sequence δ_n during field work.

Items 2) and 3) of these axioms do not give rise to any doubts from a geological viewpoint. However, in analyzing item 1), the following facts discovered after publishing No. 37 have to be taken into account. Apparently, it is reasonable to assume that: 1) in specific profiles the beds that had undergone a washout and belong to the sequence δ_n can dominate; 2) when the layers of sand beds are deposited, the values η_n play a decisive role in the final determination of the thickness of a bed. For a sequence of clay beds, in many cases $\eta_n = 0$.

If this is true, then by studying the strata profiles during field work we can estimate to what extent item 1) of these axioms is true. The following relevant data were obtained by studying the profiles of red-rock and productive strata and the flysch of north-western Caucasus, Kakhetia and Southern Urals.

For beds of different compositions the average bed thicknesses are usually different. Thus for red-rock strata in Cheleken it was found out that the sand layers have average thickness of 84.11 cm. with standard error 188.34 and asymmetry +5.0 (for number of beds $n = 318$); for aleurites 15.22 with 16.93 and +2.7 ($n = 471$), and for clays 27.66 with 53.23 and +8.0 ($n = 792$) respectively.

Sample correlation functions for sequences of bed thicknesses are divided into two types, B and F (Vistelius [2]). For type B all autocorrelation coefficients are positive and monotone decreasing as the distance s between the beds increases (in the number of beds). For type F, the odd autocorrelation coefficients are negative and the even ones are positive. Both sequences of even and odd autocorrelation coefficients rapidly and monotonely decrease in absolute value as s increases. Hence, if the beds with thicknesses from the sequence δ_n really dominate in the profile, then the assumption of the independence of the δ_n can be questioned.

It is interesting to note the following. Let

$$\mathsf{E}(x_s, x_{s+r} | a_{s=i}, a_{s+r=j}) - \mathsf{E}(x_s | a_{s=i}, a_{s+r=j}) \mathsf{E}(x_{s+r} | a_{s=i}, a_{s+r=j}) = 0,$$

where a denotes bed composition, i, j denote elements of the set of bed compositions, x denotes the bed's thickness, s denotes the number of the bed starting from the foot of the profile, and r denotes the distance between beds. In this case, for simple Markov chains consisting of two states (for example, sand and clay), each of which corresponds to a random value (bed thickness) with distinct expectations for each state (that is, the average thickness of sand beds differs from that of clay beds), only three types of correlograms are possible: belonging either to type B, to type F, or identically zero (Vistelius [4]).

The latter statement shows that in the sequences of beds that were washed out only once after sedimentation, and without total destruction, the sequence of bed composition is a simple Markov chain, and the correlogram is of type F. Hence for these sequences, bed sections could be found in which successive beds have the same $g(x)$ and the δ_n are independent. Something similar was observed in flysch profile in the Southern Urals. Here a sequence of 1530 layers (described by Bezhaev) is close to a homogeneous Markov chain, and its correlogram is of type F with the following sample autocorrelation coefficients:

$$-0.23 \quad 0.17 \quad -0.19 \quad 0.07 \quad -0.09 \quad 0.05 \quad 0.02$$

(Vistelius [3]). The table below gives estimates of the correlation coefficients between bed thicknesses next to each other $(s+1)$ and every second one $(s+2)$ for fixed bed compositions.

Sample autocorrelation coefficients between the thicknesses of beds of fixed composition

Composition of the sth bed	Composition of the $(s+1)$th bed			Composition of the $(s+2)$th bed		
	π	α	γ	π	α	γ
π	–	$\frac{0.22}{43}$	$\frac{0.24}{153}$	$\frac{-0.04}{46}$	$\frac{0.09}{94}$	$\frac{0.28}{57}$
α	–	–	$\frac{0.17}{543}$	$\frac{0.07}{125}$	$\frac{0.18}{357}$	$\frac{0.27}{61}$
γ	$\frac{0.3}{195}$	$\frac{-0.11}{502}$	$\frac{0.35}{87}$	$\frac{0.06}{21}$	$\frac{-0.01}{94}$	$\frac{0.19}{667}$

where π are sand, α aleurite, and γ clay beds, whose compositions are given by the table's entries. The numerators of the entries give autocorrelation coefficients between the beds of the corresponding compositions and denominators the number of the bed pairs observed. Dashes mean that the number of the beds compared was less than 5.

It can be seen from the table that in a first approximation the beds composed of sands and aleurites make it possible to take a sample in such a way that the δ_n are independent, provided they dominate in the profile.

Summing up it can be said that, when embarking on the analysis of bed sequence we should very thoroughly check condition 1) of Kolmogorov's axioms. Here we may encounter both cases satisfying the axioms (especially if we use a special testing procedure and study not all the beds, but only some special types of them, for example, sands), and those contradicting it. One should also study the robustness of the model, that is, how violations of the axioms propagate on the final geological conclusions.

While the problem of verifying the axioms finally reduces to that of the robustness of the model and the necessity of developing special procedures for selecting beds that satisfy Kolmogorov's conditions, the question of choosing $g(x)$ in specific studies is much more difficult. Since there are non-observable values among the δ_n, $g(x)$ can only be found from a model of layer formation based on appropriate lithological (sedimentological) assumptions. Attempts to construct such a model have not resulted in anything really worth while so far.

Kolmogorov's paper was ahead of its time and only in 1962, when geologists became aware of the importance of mathematics for their science and mathematical geology took its proper place, did this work draw the attention of researchers. This paper was referred to in the first book on the application of mathematical methods in geology (Miller and Kahn [7]), and was mentioned later in all serious publications on this subject (Agterberg [1]). Unfortunately, these textbooks gave only general references to this paper, without analyzing it. In 1975 the monograph by Schwarzacher [8] was published. It was the first to analyze Kolmogorov's work. Schwarzacher considered the corresponding model for a discrete case using the method of random walks. He also imposed stronger independence conditions.

The literature on specific studies based on Kolmogorov's paper can be subdivided into two types. In one of them no analysis of the correspondence between Kolmogorov's axiomatics and specific conditions of profile formation is given. They make no attempt to justify the introduction of a density $g(x)$. In other words, they do not use the statement of the problem, but the algorithm that comes out of Kolmogorov's paper. In doing this, they forget that this algorithm should be used only after a thorough study of the conditions of profile formation. Among these works there is a publication that recommends a special "Kolmogorov index" for estimating washout (Mizutani and Hattori [6]).

In 1979 the first paper appeared that discussed Kolmogorov's axiomatics and gave data on the type of $G(x)$ found under special conditions (Dacey [5]). Dacey assumed that the ξ_n and η_n are independent identically distributed random non-negative variables which are also mutually independent. Under these conditions it appears that if ξ_n and η_n have exponential distributions in the continuous case (or geometric distribution in the discrete case), then the residues of beds after washout also obey this distribution. Dacey does not give any geological data to confirm the importance of his case for geology. Still, his paper must be singled out since it contains a competent analysis of Kolmogorov's work.

As we have already mentioned, Kolmogorov's paper was very much ahead of its time, and still some time was to pass before geologists were to realize the importance of this outstanding result and the approach as a whole. There are a number of branches of sedimentology (lithology) that cannot be developed without working out the whole range of problems posed by this work. Hopefully this will be understood, which will greatly contribute to progress in some important branches of geology.

References

1. F.P. Agterberg, *Geomathematics*, Elsevier, Amsterdam, 1974.
2. A.B. Vistelius, *Materials on lithostratography of productive strata of Azerbaidzhan*, Izdat. Akad. Nauk SSSR, Moscow, 1961 (in Russian).
3. A.B. Vistelius, *Red-rock deposits of the Cheleken peninsula* (*Lithology*), Nauka, Leningrad, 1966 (in Russian).
4. A.B. Vistelius, *Fundamentals of mathematical lithology*, Nauka, Leningrad, 1980 (in Russian).

5. M.F. Dacey, 'Models of bed formation', *Internat. Assoc. Math. Geol. J.* **11**:6 (1979), 655–668.

6. S. Mizutani and I. Hattori, 'Stochastic analysis of bed-thickness distribution of sediments', *Internat. Assoc. Math. Geol. J.* **4**:2, 123–146.

7. R.L. Miller and J.S. Kahn, *Statistical analysis in the geological sciences*, Wiley, New York, 1962.

8. W. Schwarzacher, *Sedimentation models and qualitative stratigraphy*, Elsevier, Amsterdam, 1975.

GPSR Compliance
The European Union's (EU) General Product Safety Regulation (GPSR) is a set of rules that requires consumer products to be safe and our obligations to ensure this.

If you have any concerns about our products, you can contact us on

ProductSafety@springernature.com

In case Publisher is established outside the EU, the EU authorized representative is:

Springer Nature Customer Service Center GmbH
Europaplatz 3
69115 Heidelberg, Germany

www.ingramcontent.com/pod-product-compliance
Ingram Content Group UK Ltd.
Pitfield, Milton Keynes, MK11 3LW, UK
UKHW020129290926
15471UKWH00003B/379

9789401050036